Principles of
Power Electronics

This book is in the **Addison-Wesley Series in Electrical Engineering**.

Many of the designations used by manufacturers and sellers to distinguish their products are claimed as trademarks. Where those designations appear in this book, and Addison-Wesley was aware of a trademark claim, the designations have been printed in initial caps or all caps.

Library of Congress Cataloging-in-Publication Data

Kassakian, John G.
    Principles of power electronics / John G. Kassakian, Martin F. Schlecht, George C. Verghese.
        p.   cm.
    Includes index.
    ISBN 0-201-09689-7
    1. Power electronics.   I. Schlecht, Martin F.   II. Verghese, George C.   III. Title.
TK7881.15.K37   1991
621.31'7–dc20   90-44906
                                                                                  CIP

2 3 4 5 6 7 8 9 10-HA-95949392

Principles of Power Electronics

John G. Kassakian

Martin F. Schlecht

George C. Verghese
Massachusetts Institute of Technology

ADDISON-WESLEY PUBLISHING COMPANY
Reading, Massachusetts • Menlo Park, California • New York
Don Mills, Ontario • Wokingham, England • Amsterdam • Bonn
Sydney • Singapore • Tokyo • Madrid • San Juan • Milan • Paris

To our families

Preface

WE designed this text specifically to *teach* the subject of power electronics. Although the coverage is broad, we develop topics in sufficient depth to expose the *fundamental* principles, concepts, techniques, methods, and circuits necessary for you to understand and design power electronic systems for applications as diverse as a 5-W switching converter and a 500-MW high-voltage dc transmission terminal.

Traditionally, power electronics has been considered and taught as an aggregate of separate disciplines; for example, motor drives, dc/dc converters, or static rectifier systems. Almost invariably students are exposed to power electronics in only one context, depending on the perspective of the book or the instructor. Moreover, those who acquire an understanding of power electronics "on the job" usually have limited exposure to the subject. However, all applications of power electronics share a common base, and we have tried to make this fact clear in this book.

The book is divided into four parts. Each part begins with an overview chapter that establishes a context for the remaining chapters of the part. These overviews are substantial enough to stand independently and are intended to do so for certain teaching purposes.

Part I, "Form and Function," is the backbone of the book. In it we present the relationship between the form, or topology, of power circuits and the function these circuits perform. The common features of circuits that perform the basic electrical energy conversion functions—ac/dc, dc/dc, dc/ac, and ac/ac—are introduced in this part. But the real purpose of Part I is to present a way of thinking about power electronic circuits and visualizing their behavior that can be extended to new situations and can serve as the basis for synthesis as well as analysis.

In Part II, "Dynamics and Control," we consider the unique problems of modeling and controlling power electronic systems. We present analytical approaches to modeling their dynamic behavior and show how to use these approaches in designing and evaluating practical feedback control systems. Because of its role in stability evaluation and its potential importance in the design of fully digital control systems, we also present the advanced topic of sampled-data modeling and control in Part II.

In Part III, "Components," we discuss the behavior and characterization of the elements from which power electronic circuits are constructed. A substantial portion of Part III is devoted to semiconductor devices, going beyond the ideal switch models for these devices that sufficed for Parts I and II. We believe this emphasis to be correct, not only because the major technological advances in power electronics have been due to the availability of new semiconductor devices or the significant improvement in the performance of conventional devices, but also because the future expansion of applications of power electronics will rely heavily on continued semiconductor device innovations. A thorough, but admittedly rapid, review of magnetics is also presented in Part III. Because magnetic components are almost always unique to the application, the goal of the magnetics chapter is to provide a practical foundation for the design of magnetic components for power electronic circuits.

In Part IV, "Ancillary Issues," we address a variety of important, additional topics that must be considered in the design of any practical system. We examine gate and base drives, snubbers, forced commutation circuits, and thermal modeling and heat sinking.

A course in power electronics might use this book in one of several ways. Part I in its entirety, and the Overview chapters in Parts II through IV would serve well as the basis for an advanced undergraduate or first graduate subject. Chapter 22 ("Gate and Base Drives for Power Semiconductor Devices") might also be included. A more advanced graduate course might skim Part I and address Part II in detail. Other courses may be tailored to need by selecting various chapters from Parts II through IV. Each chapter in Parts I, III, and IV is relatively self-contained. Selections from Part II can be made in at least two ways. Chapter 11 ("Dynamics and Control: An Overview") and the first few sections of Chapter 14 ("Feedback Control Design") may be used together in a course that emphasizes control of power electronics systems. An advanced graduate course could well include Chapters 12 ("State-Space Models") and 13 ("Linear and Piecewise Linear Models"), and the later sections of Chapter 14, which consider advanced topics in modeling, stability evaluation, and control.

We use examples extensively in this book to illustrate concepts or techniques introduced in the text and also to introduce ways of thinking about problems, methods of analysis, and the use of approximations. The examples also form the basis for many of the end-of-chapter problems, and the creative instructor can use them to generate additional exercises, problems, or examples.

We designed the end-of-chapter problems to stimulate thinking about the material presented in the chapter. That is, they are not intended as exercises to drill students in the use of particular equations in the text. Often, we introduce new circuits, concepts, or ways of approaching problems by using previous discussions in the text as the basis for considering the new material. We also present practical variations of circuits discussed in the text.

The notes and bibliography at the end of each chapter point you to selected papers in the research literature and to books that underlie, complement, or extend the material in the chapter. These bibliographies however, are not exhaustive.

For a number of years, much of this text has been used at MIT in note form for graduate subjects in power electronics and machine control. It has therefore benefited significantly from suggestions, critiques, and reviews provided by many of our students and colleagues. Though they are too numerous to list by name, we are very grateful to each of them. We are particularly indebted to Prof. Malik Elbuluk of the University of Akron and Prof. David Torrey of Worcester Polytechnic Institute, each of whom carefully appraised this work from the perspectives of both student and teacher. In addition, we wish to acknowledge the contributions and valuable advice provided by our other manuscript reviewers: Donald J. Bosack (Northern Illinois University), W. Gerard Hurley (University of Limerick), J. Ben Klaasens (Delft University of Technology), Philip J. Krein (University of Illinois at Urbana), David Luchaco (Lutron Electronics Company), Daniel M. Mitchell (Collins Defense Communications, Rockwell International Corporation), F. Luis Pagola (Universidad Pontificia Comillas), and Rudy Severns (Springtime Enterprises). The photograph from which the cover is rendered was taken by Larry Silva of MIT, and we thank him for permission to use it. We are grateful to the corporate members of the MIT/Industry Power Electronics Collegium for their support and encouragement. Lastly, we thank the Addison-Wesley team. Writing this book would have been a much more difficult task without their patience, guidance, and flexibility.

Cambridge, MA
                                                        John G. Kassakian
                                                        Martin F. Schlecht
                                                        George C. Verghese

Contents

Chapter 1

Introduction

IN this chapter we describe power electronics and present a brief introduction to semiconductor switching devices and magnetic components. An introduction to these circuit elements is necessary because we utilize them in Part I, although we do not discuss them in detail until Part III. We also introduce nomenclature that we use throughout the book.

1.1 POWER ELECTRONIC CIRCUITS

The dominant application of electronics today is to process information. The computer industry is the biggest user of semiconductor devices, and consumer electronics, including cameras, is second. While all these applications require power (from a wall plug or a battery), their primary function is to process information; to take the digital optical signal produced by a compact disk and transform it into an analog audio signal, for instance. Power electronic circuits are principally concerned with processing energy. They convert electrical energy from the form supplied by a source to the form required by a load. For example, the part of a computer that takes the ac mains voltage and changes it to the 5-V dc required by the logic chips is a power electronic circuit (often abbreviated as *power circuit*). In many applications the conversion process concludes with mechanical motion. In these cases the power circuit converts electric energy to the form required by the electromechanical transducer, such as a dc motor.

Efficiency is an important concern in any energy processing system, for the difference between the energy into the system and the energy out is usually converted to heat. Although the cost of energy is sometimes a consideration, the most unpleasant consequence of generating heat is that it must be removed from the system. This consideration alone dictates the size of power electronic apparatus. Therefore a power circuit must be designed to operate as efficiently as possible. The efficiency of very large systems exceeds 99%. High efficiency is achieved by using

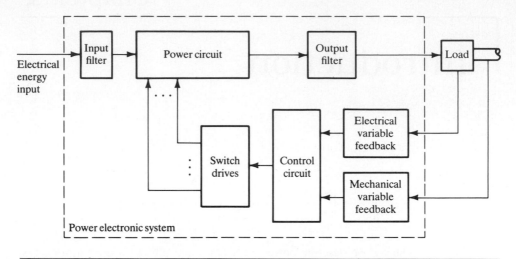

Figure 1.1 A block diagram of a typical power electronic system.

the power semiconductors as switches (where their voltage is nearly zero when they are on, and their current is nearly zero when they are off) to minimize their dissipation.[†] The only other components in the basic power circuit are inductors and capacitors, so the ideal power circuit is lossless.

A power electronic system consists of much more than a power circuit. The block diagram of a typical system is shown in Fig. 1.1. Switching creates waveforms with harmonics that may be undesirable because they interfere with proper operation of the load or other equipment, so filters are often employed at the inputs and outputs of the power circuit. The system load, which may be electrical or electromechanical, is controlled via the feedback of electrical and/or electromechanical variables to a control circuit. This control circuit processes the feedback signals and drives the switches in the power circuit according to the demands of these signals. The system also includes mechanical elements, such as heat sinks and structures to support the physically large components of the power circuit.

1.2 POWER SEMICONDUCTOR SWITCHES

The basic semiconductor devices used as switches in power electronic circuits are the bipolar and Schottky diodes, the bipolar junction transistor (BJT), the metal-oxide-semiconductor field-effect transistor (MOSFET), and a class of latching bipolar devices known as thyristors, the most common of which is the silicon controlled rectifier (SCR). Their circuit symbols and operating regions in the v–i plane are shown in Fig. 1.2. We discuss these and other hybrid devices in detail in Part III.

[†]Exceptions, such as linear voltage regulators, are so few that we do not consider them explicitly in this book.

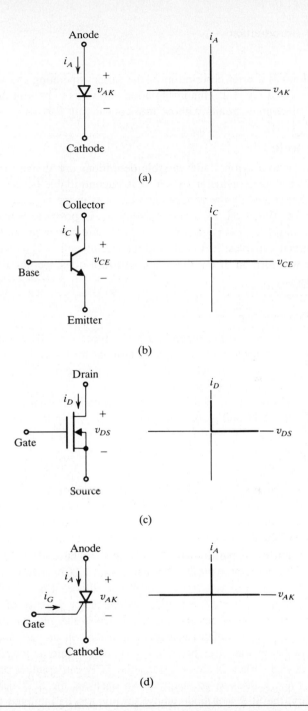

Figure 1.2 Circuit symbols and operating regions for semiconductor devices used as switches in power electronic circuits: (a) the diode; (b) the (npn) bipolar junction transistor (BJT); (c) the (n-channel) power metal-oxide-semiconductor field-effect transistor (MOSFET); (d) the silicon controlled rectifier (SCR).

What follows is a brief description of the salient operating characteristics of each device shown in Fig. 1.2. This information allows us to present the basic operation of power electronic circuits without first considering Part III.

1.2.1 The Diode

The diode, whose symbol and variable definitions are shown in Fig. 1.2(a), is an uncontrollable semiconductor switch. It is uncontrollable because whether it is on or off is determined by the voltages and currents in the network, not by any action we can take. When on, its anode current, i_A, is positive. When off, its anode–cathode voltage, v_{AK}, is negative.[†] The diode switches in response to the behavior of its terminal variables. If it is off and the circuit causes v_{AK} to try to go positive, the diode will turn on. If on, the diode will turn off if the circuit tries to force i_A to go negative.

1.2.2 The Transistor

Transistors, whether of the bipolar or MOS type, are fully controllable switches. They possess a third terminal (the *base* terminal for the BJT and the *gate* terminal for the MOSFET) from which we can turn the device on and off. The symbols and terminal variables for the npn BJT and n-channel power MOSFET are shown in Fig. 1.2(b) and (c). Both of these devices can carry current in only one direction, and for the npn BJT and n-channel MOSFET shown in the figure, these directions are $i_C > 0$ and $i_D > 0$, respectively. When off, they can support only one polarity of voltage, which, for the transistors shown, are $v_{CE} > 0$ and $v_{DS} > 0$. These voltage and current polarities are reversed for the pnp BJT and the p-channel MOSFET. But, for reasons discussed in Part III, npn and n-channel devices are the most commonly used types of power transistors.

1.2.3 The Thyristor

The only member of the thyristor family that we describe in this introduction is the SCR, whose circuit symbol is shown in Fig. 1.2(d). It is a switch that in some ways can be thought of as a "semicontrollable" diode. If no signal is applied to the gate, the device will remain off, independent of the polarity of v_{AK}. To turn the SCR on, a brief pulse of current, i_G, is applied to the gate terminal during a time when $v_{AK} > 0$. This initiates a regenerative turn-on process that quickly latches the SCR in the on state, in which $v_{AK} \approx 0$ and the gate no longer has any control over the device. When in this on-state, the SCR can conduct only positive i_A. It turns off when i_A tries to go negative. So once on, the SCR behaves as a diode. In summary, the SCR is a diode whose turn-on can be inhibited by not applying a gate pulse.

[†]The use of "K" instead of "C" reflects the Greek origin of the word cathode, or *kathodos*, meaning "way down," that is, the negative terminal.

1.3 TRANSFORMERS

Transformers are a prominent feature of power electronic circuits. We treat them extensively in Part III (Chapter 20), but the following introduction to their behavior permits us to use them as circuit elements in Parts I and II.

Transformers are employed to provide electric isolation and the step-up or step-down of ac voltages and currents. The *ideal transformer* shown in Fig. 1.3(a) has two-windings of N_1 and N_2 turns. Dots indicate the direction of the windings. If a voltage is applied to one winding so that the dot is positive, the dotted ends of all the other windings (only one in this case) are also positive. If its terminal variables are defined relative to the dots as shown in Fig. 1.3(a), the ideal transformer has the following terminal relationships:

$$\frac{v_1}{v_2} = \frac{N_1}{N_2} \qquad (1.1)$$

$$\frac{i_1}{i_2} = -\frac{N_2}{N_1} \qquad (1.2)$$

A straightforward application of these relations shows that if an impedance of value Z_1 is connected to terminals 1–1′, an impedance of value $Z_2 = (N_2/N_1)^2 Z_1$ is measured at terminals 2–2′. Using *(1.1)* and *(1.2)*, we can also show that $v_1 i_1 = -v_2 i_2$; that is, the instantaneous power into one port is equal to the instantaneous power out of the other. The ideal transformer neither dissipates nor stores energy.

A transformer is ideal if it obeys *(1.1)* and *(1.2)*, but no practical transformer is ideal. In most transformers, the principle departures from ideal result in some voltage and current being "lost" in the transformation, so terminal variables are not precisely related by *(1.1)* and *(1.2)*. A model that represents these effects is shown in Fig. 1.3(b). Some of the terminal current i_1' is shunted through the *magnetizing*

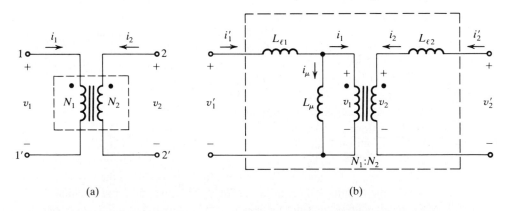

(a)　　　　　　　　　　　　　　(b)

Figure 1.3　(a) The ideal transformer model.　(b) A more practical model, in which the effects of magnetizing inductance (L_μ) and leakage $(L_{\ell 1}$ and $L_{\ell 2})$ are included.

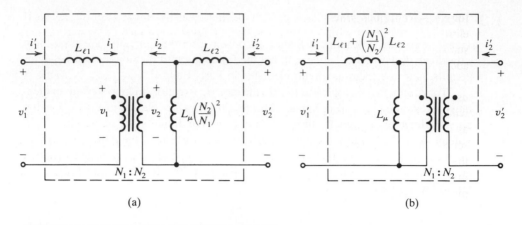

Figure 1.4   (a) The model of Fig. 1.3(b), with the magnetizing inductance placed on the N_2 side of the ideal transformer.   (b) The model of Fig. 1.3(b), simplified by reflecting $L_{\ell 2}$ through the ideal transformer and combining it with $L_{\ell 1}$.

inductance L_μ and is called the *magnetizing current*. So whereas i_1 and i_2 are still related by *(1.1)* and *(1.2)*, the real terminal currents i_1' and i_2' are not. Similarly, the real terminal voltages v_1' and v_2' differ from v_1 and v_2 by the drops across $L_{\ell 1}$ and $L_{\ell 2}$, which are called *leakage inductances*. In Chapter 20 we describe the physical origins of these effects.

Figure 1.3(b) shows L_μ across the winding N_1. We can, however, *reflect* it through the ideal transformer so that it appears across the N_2 winding, as shown in Fig. 1.4(a). Sometimes we do this because the result is analytically more convenient to use. Although two leakage inductances, one for each winding, are shown in Fig. 1.3(b), they are often combined by reflecting one through the ideal transformer. If the voltage drop across this inductor is small relative to the voltage across L_μ, then L_μ can be moved inside this reflected inductance without introducing much of an error, and the two leakage inductances can be combined. The resulting approximate model is shown in Fig. 1.4(b).

Another useful model transformation is to reflect the entire circuit on one side of the ideal transformer to the other side. A transformation of this kind is shown in Fig. 1.5. There, not only has the magnetizing inductance been reflected to the N_1 side, but the rest of the N_2 side circuit, C_o and R_o, has also been "brought through" the ideal transformer. Of course, the isolation function is lost in the transformation, which makes the technique inappropriate for the analysis of some circuits.

We can calculate or measure the leakage and magnetizing inductances of transformers, and we sometimes construct transformers to have specific values for these parameters. And, even though we have been discussing only two winding transformers, similar but somewhat more complicated considerations apply to the modeling of transformers with more than two windings. Other practical considerations, such as the resistance of the windings or losses in the core, are represented by the addition of appropriate elements to the model of Fig. 1.3(b).

Figures 1.3(b) and 1.4 show the schematic transformer representation that we use throughout this book. The circuit model being used to describe a transformer will be enclosed in a dashed box. The model frequently has an ideal transformer as one of its elements, represented by windings with adjacent double bars. Some schematic conventions utilize the double bars to represent an iron core, but we use the bars to indicate the coupled windings of an ideal transformer when it appears inside a dashed box. This convention avoids ambiguity and schematic clutter when more than two windings are involved.

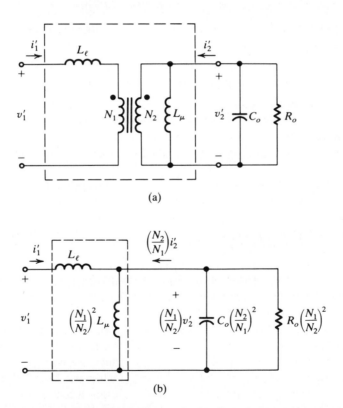

(a)

(b)

Figure 1.5   (a) A transformer with an RC load on the N_2 side.   (b) The circuit of (a) with all the N_2 side components reflected to the N_1 side so that the ideal transformer can be eliminated from the transformer model.

1.4 NOMENCLATURE

Because we discuss several different kinds of variables, we need to establish their definitions now to avoid confusion later.

1. Variables that may be time dependent are represented by lowercase names, such as v_1. When necessary for clarification, the time dependence is explicitly indicated, as for example, $v_1(t)$.

2. Variables that are constant are represented by uppercase names, such as V_1, $I_{1\text{rms}}$, or V_{dc}.

3. The average value or dc component of a periodic variable is denoted by angle brackets around the variable, for example, $\langle v_o \rangle = V_o$. Note that the average value is a constant and so is represented by an uppercase name.

4. A *local average* is defined in Chapter 11 and is indicated by an overbar, $\bar{x}$ or $\bar{x}(t)$. Note that the local average is a function of time.

5. Perturbations around a constant value are indicated by a tilde; for instance, $i_L = I_L + \tilde{i}_L$.

6. Harmonic components of a nonsinusoidal periodic waveform are indicated by an additional subscript representing the harmonic number. For example, $v_a = v_{a_1} + v_{a_3} + v_{a_5} + \cdots$.

7. Complex amplitudes of sinusoidal functions are represented by hatted uppercase names:

$$v(t) = V\cos(\omega t + \phi) = \text{Re}\left(Ve^{j\phi}e^{j\omega t}\right) = \text{Re}\left(\widehat{V}e^{j\omega t}\right)$$

The complex amplitude of $v(t)$ is $\widehat{V} = Ve^{j\phi}$. The prefix "Re" means "real part of."

Notes and Bibliography

We have included an annotated bibliography at the end of most chapters. It provides sources of additional information on topics that you might want to pursue further.

PART I

Form and Function

Chapter 2

Form and Function: An Overview

POWER electronic circuits change the character of electrical energy: from dc to ac, from one voltage level to another, or in some other way. We refer to such circuits generically as *converters*, *static converters* (because they contain no moving parts), *power processors*, or *power conditioners*. The part of the system that actually manipulates the flow of energy is the *power circuit*. It is the frame for the rest of the system's components, such as the control circuit or the thermal management parts.

The power circuit has a basic topology to which we add other circuit elements that perform ancillary functions, such as protection against transient overvoltages and filtering to eliminate electromagnetic interference. Although these other elements are important, they do not affect the function of the power circuit. Their purpose is to modify certain aspects of the power circuit's behavior, such as the rates of rise of currents or voltages. The study of a power circuit with all these additional elements can quickly become dominated by particulars rather than fundamentals. Therefore we concern ourselves only with the basic forms of power circuits in Part I of this book. We describe methods of building on these basic structures to transform them into practical power circuits, and eventually into systems, in Parts II, III, and IV.

In Part I we show how a desired conversion function influences the form, or topology, of the power circuit. We also use these forms to illustrate the analytic tools and ways of thinking that you should apply when studying a power circuit. In most cases we keep both the function and the form simple. Where we present a more advanced topology, our goal is to show the connection between it and its simpler form—and the benefits gained from the added complexity.

2.1 THE FUNCTIONS OF A POWER CIRCUIT

Before we can specify the form of a power circuit, we must define its function. In general, its function is to alter the characteristics of electrical energy provided by one external system to those required by another. For instance, the power supply for a computer must convert the sinusoidal mains energy (60 Hz, 110 V rms in the United States) to a 5 V dc waveform. Another example is a power circuit for driving a variable speed ac motor, which might draw power from a battery and deliver a sinusoidal current waveform to the motor.

The types of functions that a power circuit can perform are limited only by the characteristics of electric power that are to be altered. As already mentioned, the transformations from ac to dc and from dc to ac are two possible functions. Interfacing systems with waveforms that are similarly shaped, but have different amplitudes, is another. If both waveforms are constant in time, we call the power circuit a *dc/dc converter*, and if both are alternating, we call the circuit an *ac/ac converter*. In the latter case, we might want to change the frequency or phase as well as the amplitude.

It is important that you not think of a power circuit's function as fixed for all time. The value of such a circuit is not just its ability to alter the form of electric energy, but also its ability to do so in response to a control signal. For instance, we can make the output of a computer power supply remain at 5 V even though the amplitude of the utility waveform changes by more than ±20%. In some applications, such as light dimmers, the entire function of the power circuit is to provide this controllability.

A power circuit provides an interface between two other systems external to it and therefore imposes relationships between the voltage and current waveforms at one port and those at the other. Exactly what waveforms exist at these ports depends not only on these relationships, but also on how the two external systems respond to being related in this manner. Thus it is important always to describe the operation of a power circuit in the context of the external systems to which it is connected.

In certain situations one external system, such as a voltage source, will dictate a waveform at one port independent of the power circuit or the other external system. In these cases, we treat the waveform as an input to be processed by the power circuit to create an output. The external system at the output then defines the covariable (the current in this case) at its port, and we work backward through the power circuit to determine the shape of the covariable's waveform at the input. Because of their simplicity, we often use these cases when presenting the initial topologies in the following chapters.

The direction of energy flow usually determines whether ports serve as input or output. However, many power circuits are capable of processing bidirectional power, and identifying their ports as input or output leads to ambiguity. This ambiguity is aggravated by the fact that the types of semiconductor devices used to construct the circuit also constrain the direction of power. Thus we could construct two identical topologies to process power in opposite directions. This is an important issue, which we address frequently, for it emphasizes that two visually distinct circuits can behave similarly.

The basic form of a power circuit stems primarily from the need to provide efficient energy conversion. This need precludes the use of a transistor operating as a linear amplifier, regardless of its designed power level. For almost all energy conversion applications, such a circuit technique simply dissipates too much energy relative to the amount it processes.[†] Similarly, using a resistor in conjunction with an energy storage element is not a practical way to make a low- or high-pass filter in a power circuit. All elements in the basic power circuit, at least in their ideal form, must be lossless. This requirement leaves us with two kinds of components with which to build power circuits: switches (semiconductor devices that are either fully on or off) and energy storage elements (inductors and capacitors).

2.2 AC/DC CONVERTERS

Without regard for the direction of energy flow, ac/dc converters comprise the broadest class of power electronic circuits. They are present in every piece of line-operated electronic equipment—from table radios to large computers. They also are used extensively in industrial controls and processes, such as variable speed motor drives, induction heating, plating, and the electrolytic production of chemicals.

Because of its symmetry, the basic converter topology is capable of bidirectional power; that is, the same topology can convert ac to dc, or dc to ac. For this reason we do not attach any significance to the order of the "ac" and "dc" in the name "ac/dc converter."

2.2.1 Basic Topology and Energy Flow

Our first example of a power circuit is one that creates a dc voltage from an ac voltage source. In this case, the power circuit produces a waveform that has an average value (the dc voltage) from one that does not (the ac source). Using switches configured in the topology of Fig. 2.1(a) produces the desired waveform. Because one of the two external networks connected to the converter is a resistor, there is no ambiguity about input or output ports. Energy must flow from the source to the resistor. When the ac voltage is positive, closing the two switches marked P and leaving the two switches marked N open connects the input voltage to the output in the positive sense. When the ac voltage is negative, reversing the states of the switches reverses the connection of the ac voltage source to the output terminals, resulting in an output voltage that is again positive.

The waveform v_d in Fig. 2.1(a) has a dc component (equal to $2V_s/\pi$), but it also contains unwanted ac components that we can remove by the addition of energy storage elements. For instance, if we use a low-pass LC filter as shown

---

[†]We might choose to use a linear amplifier for a few applications. Although we specifically restrict our definition of power circuit to exclude these circuits, many of the issues we address in this book relate directly to their design.

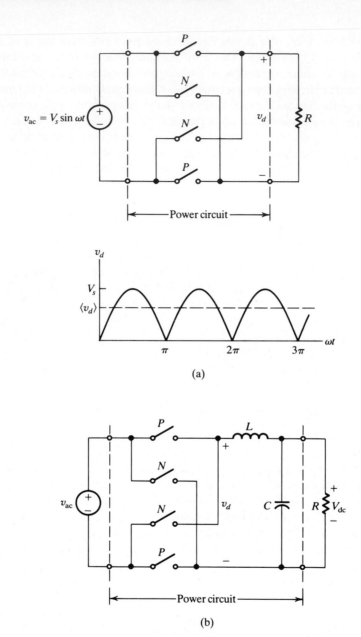

Figure 2.1 (a) A power circuit, consisting only of switches, that converts an ac voltage, v_{ac}, to one containing a dc component, v_d. (b) The ac/dc converter of (a) with the addition of filter elements L and C to remove the unwanted ac components from v_d.

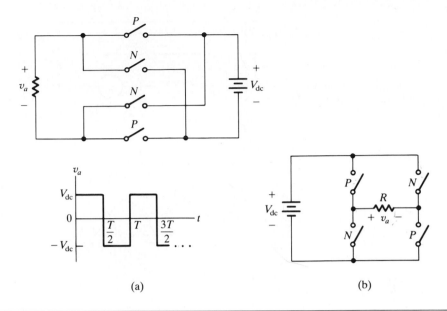

(a)   (b)

Figure 2.2   (a) The converter topology of Fig. 2.1(a) connected and controlled to provide inversion.   (b) The conventional way of drawing the bridge converter of (a).

in Fig. 2.1(b), most of the ac components in the voltage waveform created by the switches will appear across the inductor instead of at the output. As part of the design process, we must choose element values large enough to achieve the level of attenuation desired.

An ac/dc converter in which energy flows from the ac network to the dc network is called a *rectifier*. However, using an energy source such as a battery for the dc external system, as Fig. 2.2(a) shows, allows energy to flow in the other direction. The circuit is then called an *inverter*. Note that the same power circuit provides both functions. In practice the external networks and switch implementation and control determine the function. The topology of this connection of four switches is called a *bridge*. It is used extensively in power electronics and is usually drawn as shown in Fig. 2.2(b).

EXAMPLE 2.1

A Converter Linking Two Sources

If the inductor and/or the capacitor in the low-pass filter in the topology of Fig. 2.1(b) is large enough, the output voltage will be constant at $V_{dc} = 2V_s/\pi$. We may now replace the capacitor and resistor with a voltage source of this value, as shown in Fig. 2.3(a), without

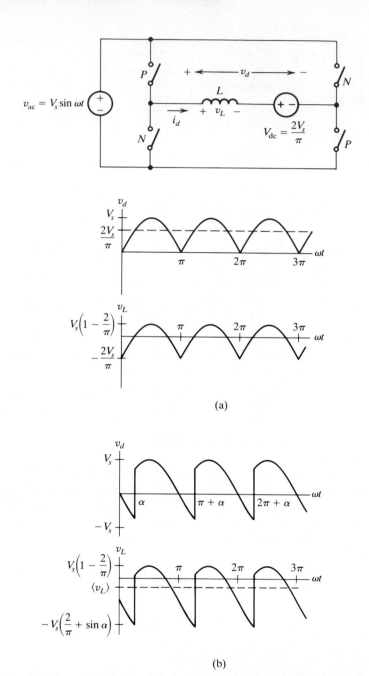

(a)

(b)

Figure 2.3 (a) The converter topology of Fig. 2.1(b) with the load resistor and filter capacitor replaced by a voltage source equal to the average value of the rectified voltage v_d. (b) Waveforms resulting when the switches are controlled to give an average (negative) value to v_L.

changing the operation of the circuit. We now have a converter whose ports are connected to sources capable of supplying energy. What is the direction of power?

The answer is not at all obvious, because we cannot determine i_d without knowledge of the operating history of the switches, and the direction of i_d determines the direction of power. What we can do is write an expression for i_d in terms of v_d (which is explicitly determined by the switches) and V_{dc}:

$$i_d = \frac{1}{L} \int_{-\infty}^{t} (v_d - V_{\text{dc}})dt \qquad (2.1)$$

Figure 2.3(a) shows the voltage $v_d - V_{\text{dc}} = v_L$ for the case where the switches are controlled so that conduction alternates between the P and N switches at zero-crossings of v_{ac}. The average value of this voltage calculated over an interval π is

$$\langle v_L \rangle = \frac{1}{\pi} \int_{0}^{\pi} \left(V_s \sin \omega t - \frac{2V_s}{\pi} \right) d(\omega t) = 0 \qquad (2.2)$$

making no net change in i_d. This condition is known as operation in the *periodic steady state*, because the circuit is in the same state at the beginning and end of each switching period. If L were very large ($L \approx \infty$), i_d would not vary, even during the interval π. If the switches were controlled in this manner for all time, there would never be an average value of voltage across L and i_d forever would be zero.

However, we control the opening and closing of switches. Shifting the switching instants by an angle α from the zero-crossings of v_{ac}, as shown in Fig. 2.3(b), creates a nonzero average voltage across L. This voltage causes the current to change, decreasing (going negative) in this case. When the current reaches the value corresponding to the desired power, the switching times are changed to the zero-crossings of v_{ac} (Fig. 2.3a). The average value of v_L is now once again zero and i_d remains constant.

Note that because we have chosen a dc source voltage equal to the maximum possible value of the dc component of v_d, we cannot control the switches to give a positive average value to v_L. Thus we restrict the circuit to energy flow from the dc network to the ac network. A smaller value of V_{dc} would permit flow in either direction. This circuit, controlled as described, is one example of a class of circuits called *phase controlled converters*. We consider them in detail in Chapter 5.

2.2.2 Filtering

The use of basic power electronic converter topologies, such as the basic ac/dc converter topology of Fig. 2.1(a), frequently results in deviation from the desired waveform by one or more of the port variables. In these cases we must modify the topology by adding filters to remove the unwanted components from the port variables. Figure 2.1(b) shows one way of doing this for the dc side of the ac/dc converter. Let's now consider this issue more generally.

We can obtain a simpler alternative to the filter of Fig. 2.1(b) by removing the capacitor and making the inductor very large. The resulting filter has a single pole at $\omega = 1/\tau = R/L$. Placing this pole at a frequency that is very low compared to the switching frequency yields an inductor (and resistor) current that is nearly constant

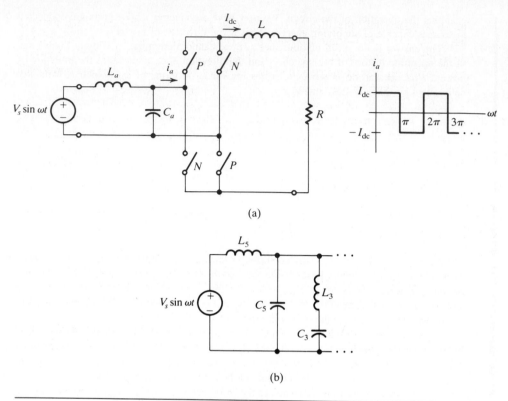

(a)

(b)

Figure 2.4 (a) The ac/dc converter topology of Fig. 2.1(a), with a first-order low-
pass filter (RL) on the dc side and a second-order low-pass filter
(L_aC_a) on the ac side. (b) An alternative and more effective ac side
filter.

at some value I_{dc}. But the current in the ac source, i_a, is now a square wave instead
of a sinusoid, which is undesirable for reasons we discuss in Chapter 3. We must
employ another filter to eliminate all but the fundamental component from i_a.

Figure 2.4(a) shows the ac/dc converter topology of Fig. 2.1(b) with the filter
on the dc side modified as in the preceding discussion and a second order filter
consisting of L_a and C_a on the ac side. The alternate action of the P and N
switches creates the square-wave current i_a by alternately reversing the direction
of I_{dc} as it is reflected through the switches to the dc side. The characteristics of the
ac network connected to the converter strongly influence the form of the ac filter.
In this case the network is simply the source, v_{ac}, which ideally has an incremental
impedance of zero at any frequency. Therefore a shunt filter alone will not work,
and the filter topology must present an impedance (ωL_a in this case) in series with
the source at all but (ideally) the fundamental frequency.

In practice the low-pass ac filter does not work very well. The reason is that
the first, and largest, undesirable harmonic of i_a is the third. Because it is so close

to the fundamental, the filter pole cannot be placed at a frequency that strongly attenuates the third without also influencing the fundamental. Figure 2.4(b) shows an alternative filter circuit. It uses a series trap, L_3 and C_3, to shunt the third harmonic from the output, and the low-pass filter, L_5 and C_5, to remove harmonics from the fifth and above.

From these examples you can see that the introduction of filters complicates the basic power circuit topology required to perform the conversion function. It is equally important that you recognize the influence of the external networks on the form and effectiveness of the filter circuits. When these filters are part of the power circuit, you can determine the performance of the converter only in the context of an application that specifies the characteristics of the external networks.

EXAMPLE 2.2

A Resonant Converter

In Example 2.1 we discussed a way of controlling power by varying the phase angle between the zero-crossings of the ac waveform and the switching times. When the ac port of the converter incorporates a resonant filter, we sometimes can use an alternative control technique based on the strong variation with frequency of the transfer function of the filter.

Figure 2.5(a) is an ac/dc converter with a series resonant filter on the ac side. This topology, consisting of a split source and only two switches, is known as a *half-bridge*.

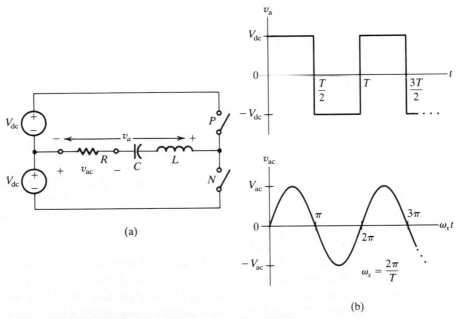

(a)

(b)

Figure 2.5   (a) A half-bridge ac/dc converter topology employing a series tuned filter on the ac side.   (b) The relationship between the unfiltered voltage, v_a, and the output voltage, v_{ac}, when the filter is tuned to the switching frequency ω_s.

Figure 2.5(b) shows the waveforms v_a and v_{ac} when the filter is tuned to the fundamental of v_a. That is, the switching frequency ω_s is equal to the resonant frequency $\omega_o = 1/\sqrt{LC}$. If the Q of the series RLC circuit is high, the filter provides good selectivity and v_{ac} is nearly sinusoidal. At switching frequencies other than $\omega_s = \omega_o$, we can determine the amplitude of the output from the magnitude of the admittance $Y(j\omega_s)$:

$$|V_{ac}| = RY(j\omega_s)V_{a_1}$$

$$= R\left|\frac{-j\omega_s C}{\omega_s^2 LC - j\omega_s RC - 1}\right|\left(\frac{4}{\pi}V_{dc}\right) \qquad (2.3)$$

where V_{a_1} denotes the amplitude of the fundamental (ω_s) component of v_a.

The magnitude of $Y(j\omega)$ is shown in Fig. 2.6. Switching at a frequency higher than ω_o allows the filter to still do a good job of removing harmonics from v_a. The rapid attenuation provided by the filter reduces the almost sinusoidal output voltage. Thus by varying the switching frequency, we can control the power delivered to the load resistor.

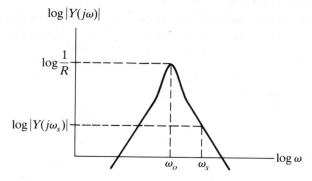

Figure 2.6   Admittance as a function of frequency for the RLC circuit on the ac side of Fig. 2.5(a).

The topology of Fig. 2.5(a) is known as a *series resonant converter*. It is one member of a family of ac/dc converters called *resonant converters*. Designers use them principally to obtain clean, high-frequency ac waveforms, often for induction heating applications. Note that the technique of power control that we just discussed results in a varying output frequency. Therefore any application of this type of control must not involve a precise output frequency.

2.3 DC/DC CONVERTERS

Used extensively in power supplies for electronic equipment, dc/dc converters control the flow of energy between two dc systems. The dc/dc converter takes the dc output of the ac/dc converter and transforms it to the different dc voltages required by the electronics—5 V and ±15 V, for example. These converters are also used in battery powered equipment and to control the speed of dc motors in many traction applications, such as battery powered forklifts or trains operating from a dc third

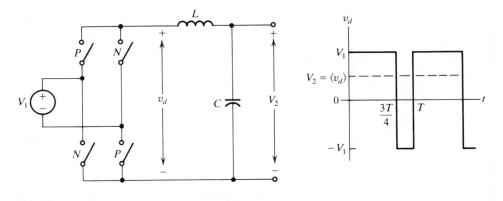

Figure 2.7 Basic dc/dc converter topology. The conversion ratio V_2/V_1 can vary between ± 1.

rail or catenary. In relatively high-power applications, such as traction, the dc/dc converter is known as a *chopper*.

2.3.1 Basic Topology

The basic dc/dc converter topology is shown in Fig. 2.7. The striking feature of this circuit is its similarity to the ac/dc converter topology shown in Fig. 2.1(b). In fact, the only difference is the control of switches. In the dc/dc converter, the switches are controlled to produce a voltage v_d that contains a nonzero dc component. This component is then extracted by the low-pass LC filter to produce the output dc voltage V_2.

As before, we alternately operate the P and N switches. Instead of leaving each set closed for exactly half the cycle as we did in the ac/dc converter, here we control them to have asymmetric on-times so that v_d contains a dc component. For example, if we leave the P switches on for three quarters of the switching period T, the voltage v_d shown in Fig. 2.7 results. It has an average value of $V_1/2$. In this circuit we can obtain an average voltage between V_1 and $-V_1$ by adjusting the relative conduction periods of the two sets of switches. The fraction of its switching period during which a switch is on is known as the *duty ratio* D of the switch. Here the duty ratio of the P switches is 0.75, and that of the N switches is 0.25.

EXAMPLE 2.3

A Simplified dc/dc Converter Topology

We can simplify the topology of Fig. 2.7 by requiring an output voltage of only one polarity; for example, $0 < V_2 < V_1$. The circuit of Fig. 2.8, which contains only two switches, is sufficient to do the job. As we turn these switches on and off sequentially, the voltage waveform they create steps between V_1 and 0.

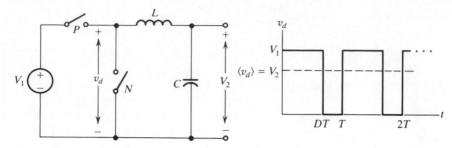

Figure 2.8   A simplified dc/dc converter topology limited to producing an output voltage $0 < V_2 < V_1$.

Assume that we leave switch P on for time DT, where T is the switching period and D is the duty ratio of the P switch. We then turn switch P off and switch N on for the rest of the period, that is, for a time $(1 - D)T$. The output voltage V_2, the average value of v_d, is then DV_1. We can alter this average value by changing the duty ratio D, but the lower limit of the output voltage is zero.

What did we gain from this simplification? One benefit is the reduction in the number of switches required. Another, which can be shown through a harmonic analysis, is that the ac components of v_d in this circuit are smaller than those in v_d for the bridge circuit of Fig. 2.7. We can therefore use smaller filter elements to achieve the same level of ripple in the waveforms presented to the external system.

Because the frequency of the desired output (dc) of the converter of Fig. 2.7 is much less than the switching frequency $(1/T)$, it is a member of a class of converters known as *high-frequency switching converters*. We discuss high-frequency dc/dc converters in Chapter 6. As you will see in Chapter 8, another member of this class is the high-frequency ac/dc converter.

We can also achieve the transformation from dc to dc in a way different from that which we have been discussing. Instead we can use a dc/ac converter to create an ac waveform from the dc at one port and then an ac/dc converter to transform this ac waveform back to dc at the other port. At first glance this approach appears to be wasteful because it requires so many switches and filter elements. It does, however, give us a point in the circuit where the waveforms are ac. As shown in Fig. 2.9, we can install a transformer at this point to give electrical isolation and to

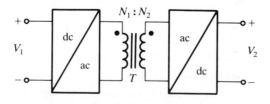

Figure 2.9   A dc/dc converter consisting of two cascaded ac/dc converters. This configuration permits the use of an isolating transformer T, as shown.

make use of its turns ratio when the difference between the two external voltages is large. We discuss this type of converter, called an *isolated high-frequency converter*, in Chapter 7.

2.4 AC/AC CONVERTERS

An ac/ac converter converts an ac waveform of one amplitude, phase, and/or frequency to another ac waveform with different parameters. As with the dc/dc converter, several different topologies can be used, depending on system requirements.

2.4.1 Topologies

We cannot identify a basic topology for the ac/ac converter, as we did for the ac/dc and dc/dc conversion functions. One of three distinct topological approaches can be used, depending on system requirements. The first and simplest can be used to change the amplitude parameters of an ac waveform (the rms value, for instance). It is known as an ac *controller* and functions by simply taking symmetric "bites" out of the input waveform. The second can be utilized if the output frequency is much lower than the input source frequency. This topology is called a *cycloconverter*, and it approximates the desired output waveform by synthesizing it from pieces of the input waveform. The third approach consists of two ac/dc converters with their dc ports connected. The result is known as a *dc link converter*.

The ac Controller Figure 2.10 illustrates the operation of the ac controller as a light dimmer. The switch is opened at every zero-crossing and reclosed at some later point in the half-cycle. This circuit does not control the fundamental frequency of the output waveform. However, it can control the rms value of the output or the amplitude of its fundamental component by varying the turn-on point, α, of the switch. To keep the harmonic components of the waveform created by the switch

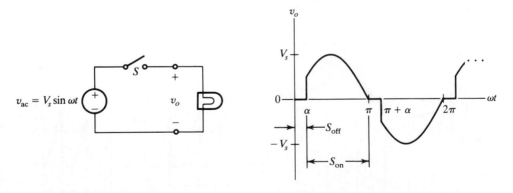

Figure 2.10 The ac controller utilized as an incandescent light dimmer.

from reaching the external systems, we can again add filters to both the input and output ports of the power circuit. Although the waveform of Fig. 2.10 shows the "bite," or *notch*, starting at the zero-crossing, we can place the notch anywhere in the waveform so long as the switch implementation permits it.

The Cycloconverter   Figure 2.11 illustrates the basic operation of the cycloconverter. The topology of this circuit is identical to that of the ac/dc converter of Fig. 2.4 without the ac side filter. The difference between the two circuits lies in switch implementation and control. For v_2 to be positive, the P switches are closed when v_1 is positive, and the N switches conduct when the source is negative. But if the P switches are closed when v_1 is negative—and the N switches when v_1 is positive—the filtered output voltage v_2 will be negative. Therefore, by phase controlling the switches as described in Example 2.1, we can obtain any value for v_2 between $2V_1/\pi$ and $-2V_1/\pi$. If we vary the controlling phase angle α sinusoidally at a frequency w_2 that is very low compared to the source frequency w_1, and choose the inductor so that:

$$\frac{1}{w_1} \ll \frac{L}{R} \ll \frac{1}{w_2} \tag{2.4}$$

the low-pass filter does a good job of filtering out the source frequency ripple. However, the filter provides little attenuation at the modulating frequency w_2, and v_2 is then an ac voltage at a frequency w_2.

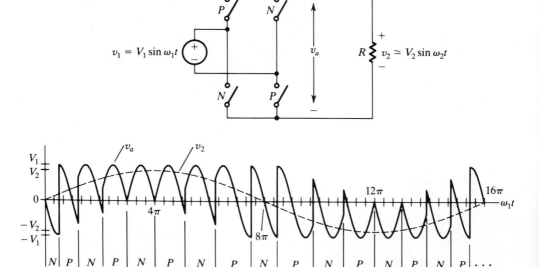

Figure 2.11   An illustration of a bridge circuit operated as a cycloconverter.

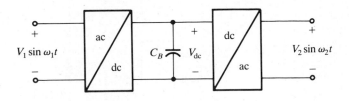

Figure 2.12   An ac/ac converter topology utilizing a dc link.

The dc Link Converter   Figure 2.12 illustrates the dc link converter topology. The first ac/dc converter creates a dc waveform from the input ac. The second converts the dc energy back to ac with the desired parameters at the output. The part of the circuit where the energy exists in dc form is the dc link, or *dc bus*. The energy storage elements located at this bus can provide more than filtering. They can also store any momentary mismatch in energy between the input and output power. This function is called *load balancing energy storage*. In Fig. 2.12 this function is provided by the dc bus capacitor C_B. For some applications the energy stored in the dc link is made large enough to support the continued operation of the system in the event of a power failure. A converter configured in this way is known as an *uninterruptible power supply* (UPS).

2.5 INFLUENCE OF SWITCH IMPLEMENTATION

So far in this chapter we have discussed the operation of power circuit topologies with ideal switches. We can open or close these switches at will. When closed, they can carry current in either direction; when open they can support a voltage of either polarity. But there is no ideal switch in practice. Semiconductor switches share only some of the characteristics of ideal switches. For example, the bipolar transistor can carry current in only one direction when it is on; when it is off, it can support only one voltage polarity. And we cannot exert any control over the turn on or turn off of a diode. The limitations of semiconductor switches have an enormous influence on the performance of the topologies we have presented in this chapter. Let's now consider the implications of various switch implementations, using some examples.

EXAMPLE 2.4

The ac/dc Converter Using Diodes

Figure 2.13 shows the ac/dc converter of Fig. 2.1(a) with diodes serving as the switches. A diode can carry only positive current (anode to cathode), will turn on if its anode–cathode voltage attempts to go positive, and will turn off if the current tries to go negative. Applying these contraints to Fig. 2.13(a) you can see that both i_d and v_d must be positive. Similarly, Fig. 2.13(b), in which the diodes are reversed, produces negative values of i_d and v_d.

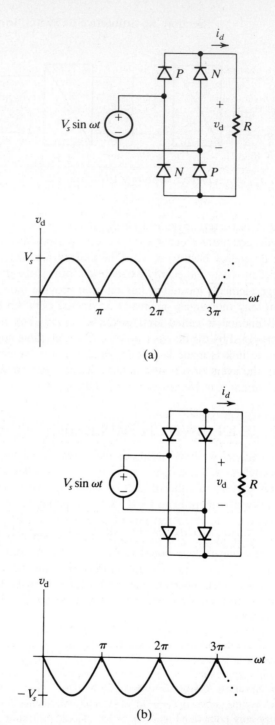

Figure 2.13 The ac/dc converter of Fig. 2.1(a) with diodes used as the switches: (a) diode directions giving a positive v_d; (b) diode directions giving a negative v_d.

The current constraint imposed by the diode directions is easy to see, but the restricted polarity of v_d is a bit less obvious. A straightforward way of convincing yourself of this constraint is to assume that the P switches are on in Fig. 2.13(a), and the source $V_s \sin \omega t$ is crossing zero from positive to negative. What is happening to the voltage across the N diodes? Writing KVL around the loop consisting of a P diode (which is on and has no voltage across it), an N diode (which is off), and the source shows that the anode–cathode voltage of the N diode is $-v_{ac}$. This anode–cathode voltage is negative (and consistent with the N diode's being off) while $V_s \sin \omega t$ is positive but becomes positive when $V_s \sin \omega t$ goes negative. Therefore the N diodes will turn on when $V_s \sin \omega t$ passes through zero from positive to negative. Then because the current will reverse at this time also, the P diodes will turn off. The use of diodes forces switching at the zero-crossings of the source $V_s \sin \omega t$ and does not allow any control of the output voltage.

EXAMPLE 2.5

Switch Implementation in a dc/dc Converter

A dc/dc converter topology similar to that of Fig. 2.8 is shown in Fig. 2.14. The P switch has been created with an npn transistor, and both ports are connected to dc sources through filters. In what direction is energy flowing? What semiconductor device can we use for the N switch?

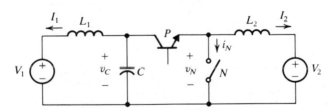

Figure 2.14   A dc/dc converter linking two dc sources. The filters are designed to make the source currents dc.

The npn transistor will permit current to flow only from left to right. Therefore the current I_2 must be positive, and energy flows into the source V_2. The circuit can operate only to transfer energy from V_1 to V_2.

We can determine the requirements of the N switch by looking at its voltage and current. When P is open, N must be closed, so $i_N = -I_2 < 0$ during conduction. When P is on, N must be open and $v_N = v_C$. If the filter $L_1 C$ is designed properly, the capacitor voltage v_C will be approximately constant and equal to V_1. (L_1 cannot support an average voltage.) Therefore $v_N > 0$ when N is open. This v_N–i_N characteristic is that of a diode. But we cannot control when a diode turns on and off. Can a diode be used here? The answer is yes, because the diode will be forced off (reverse biased) when the controlled switch P turns on. The current I_2, which must be continuous, will force the diode on when P turns off.

Problems

2.1   Determine the current in the voltage source, v_{ac}, of Fig. 2.1(a). What is the average power delivered by this source?

2.2   Determine the current in the battery, V_{dc}, of Fig. 2.2. What is the power delivered to the resistor in this system?

2.3   The bridge converter of Fig. 2.3(a) has an ac source of value $v_{ac} = 170 \sin \omega t$ V and a dc source of value $V_{dc} = 75$ V. What value of α results in periodic steady state operation?

2.4   If the inductor in the converter of Fig. 2.3(a) is very large, i_d can be considered constant at a value I_{dc} determined by the history of the switches. The inductor and dc voltage source can be replaced by a current source, as shown in Fig. 2.15. Sketch the ac source current i_a. Superimpose this sketch on the waveform of the source voltage $V_s \sin \omega t$ to show their relationship in time. Determine the average power delivered to the current source in terms of I_{dc} and V_s.

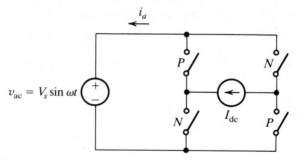

Figure 2.15   The converter circuit of Fig. 2.3(a) with its dc voltage source and inductor (assumed large) replaced by an equivalent current source. This circuit model is the subject of Problem 2.4.

2.5   A dc/ac converter designed to connect a battery to a resistive load is shown in Fig. 2.16. The load requires a sinusoidal voltage at a frequency of 50 Hz. The switches

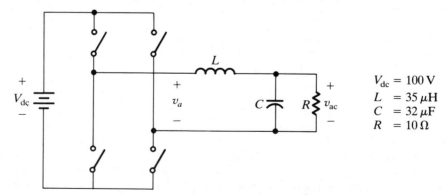

$V_{dc} = 100$ V
$L = 35\ \mu H$
$C = 32\ \mu F$
$R = 10\ \Omega$

Figure 2.16   A dc/ac converter containing a second-order low-pass filter on the ac side. This circuit is the subject of Problem 2.5.

are operated to make v_a a square-wave voltage. Determine the amplitudes of the
fundamental and third harmonic components of the load voltage v_{ac}.

2.6   A resistor replaces the dc source in the phase-controlled converter of Fig. 2.3(a).
Determine and plot the average value of the voltage across this resistor as α is varied
between 0 and π.

2.7   A series resonant converter using the topology of Fig. 2.5(a) is constructed with the
following element values:

$$L = 159 \ \mu H, \qquad C = 0.25 \ \mu F, \qquad R = 5 \ \Omega, \quad \text{and} \quad V_{dc} = 100 \ V$$

What is the power delivered to the 5 Ω load resistor if $w_s = w_o$? Make an intelligent approximation about the effect of damping. What is the amplitude of the third
harmonic in v_{ac}? Express your answer as a percentage of the fundamental of v_{ac}.
    The switching frequency is now adjusted so that $w_s = 3w_o$. Sketch v_a and v_{ac}
on the same axes. What is the power delivered to the load? What is the amplitude of
the third harmonic?

2.8   Determine the output voltage V_2 of the dc/dc bridge converter of Fig. 2.7 in terms
of V_1 and the duty ratio D of the P switches.

2.9   The filter and load of a dc/dc converter are as shown in Fig. 2.17. The input dc voltage
source is 100 V, and the required dc load voltage (the average value of v_o) is 50 V.
The size of L must be chosen such that the peak–peak ripple on the output current
i_o does not exceed 0.5 A. These requirements can be met with the switch topology
of either Fig. 2.7 or Fig. 2.8. Sketch the voltage v_d produced in this application by
each topology. For each topology calculate the value of L required to meet the ripple
specification. (*Hint:* Because the ripple is small, a close approximation to the exact
answer can be obtained by assuming v_o to be constant.) Assume $T = /ms.$

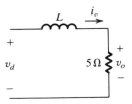

Figure 2.17   The filter (L) and load (R) of a dc/dc converter. The required value of L
is determined in Problem 2.9.

2.10   Two dc sources are connected through a dc/dc converter, as shown in Fig. 2.18. In
its periodic steady state the converter delivers 100 W from the 10 V source, V_2, to

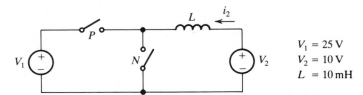

$V_1 = 25 \ V$
$V_2 = 10 \ V$
$L \ = 10 \ mH$

Figure 2.18   A dc/dc converter designed to transfer energy from V_2 to V_1. This circuit
is the subject of Problem 2.10.

the 25 V source, V_1. What is the lowest frequency at which the switches can operate to produce a peak–peak ripple of no more than 60 mA in the current i_2?

When this circuit is started, the switches must be controlled so that i_2 builds up to its periodic steady state dc value (10 A). Determine a control strategy for the switches (their on-state durations as a function of time) that will minimize the duration of this start-up transient.

2.11   Determine and plot the rms output voltage as a function of α for the ac controller of Fig. 2.10.

2.12   Can the bridge inverter of Fig. 2.2 be constructed using diodes for the switches?

2.13   How should the diode used to implement the switch N in Fig. 2.14 be connected in the circuit? What is the conversion ratio V_2/V_1 in terms of the duty ratio D of the P switch?

2.14   Determine a switch implementation (both N and P) for the topology of Fig. 2.14 that would permit the transfer of energy from V_2 to V_1. What is the conversion ratio V_1/V_2 in terms of the duty ratio D of the N switch?

Chapter 3

Introduction to Rectifier Circuits

A rectifier is a circuit for changing ac to dc. A basic rectifier circuit produces dc in the electrical engineering sense, that is, unipolar current flow. However, it does not produce dc in the mathematical sense, that is, a waveform whose spectrum consists of a single component having a frequency of zero. Instead, a rectifier circuit produces an output that contains considerable ac components. These ac components result in fluctuations, called *ripple*, about the dc output level. Eliminating this ripple and obtaining an approximation to "pure" dc requires insertion of a filtering process after the basic rectification function.

Because George Westinghouse emerged the victor over Thomas Edison in the great ac/dc battle of the twentieth century, rectifiers are the backbone of power electronic circuits. Although they come in a large number of different configurations (for example, single phase, three phase, half wave, and full wave), rectifier circuits possess fundamentally similar principles of operation. Our purpose in this chapter is to introduce you to a number of different concepts, using two simple single-phase rectifier circuits as vehicles. These circuits behave in a qualitatively transparent manner and thus are well suited to this purpose. We reserve the discussion of more complex rectifier circuits for Chapters 4 and 5.

Because power electronic circuits include switches, their topologies almost always vary with time. Thus we can describe their behaviors most easily in the time domain. Another purpose of this chapter, then, is to reintroduce you to the techniques of time domain analysis. Most electrical engineers either forget or never fully understand these techniques in their preoccupation with frequency domain methods. We also introduce the analytic method of *assumed states*, which is particularly appropriate to the analysis of power electronic circuits. Our discussion is based on the assumption that the diodes in the rectifier circuits are ideal; that is, they have no forward drop when they are on or reverse leakage current when they are off.

Throughout this book, we refer to the *average value of variables*. We use angle brackets to denote the average value of a quantity, say, $f(t)$ as "$\langle f(t) \rangle$." We use this notation to specify average values of periodic functions. We later extend the notion of "average" to nonperiodic functions but denote differently the average of such functions.

3.1 POWER FLOW IN ELECTRICAL NETWORKS

The ideal power electronic circuit provides a lossless transformation of the form of electric power. The reason is that the circuit is made up of (ideally) lossless components—inductors, capacitors, transformers, and switches. Average power in and average power out of lossless transformations are equal. We state this condition in terms of *average* power rather than instantaneous power in order to accommodate the possibility of energy storage, as in inductors and capacitors, within the transformation. Keeping your eye on the power is often insightful in the analysis of multiport power electronic networks (such as rectifiers), which because they are nonlinear and vary with time, do not yield easily to conventional analysis.

Consider the network N (which perhaps varies with time and/or is nonlinear) shown in Fig. 3.1. Assume that the terminal variables $i(t)$ and $v(t)$ are periodic with period $T = 2\pi/\omega$, but not necessarily sinusoidal, in which case they may be expressed in terms of their Fourier series as:

$$v = \sum_{n=-\infty}^{\infty} V_n e^{jn\omega t} \quad \text{and} \quad i = \sum_{m=-\infty}^{\infty} I_m e^{jm\omega t}$$

The time-average power $\langle p(t) \rangle$ at the terminals of N is

$$\langle p(t) \rangle = \frac{1}{T} \int_0^T vi \, dt \tag{3.1}$$

Because the product of sinusoidal variables of different frequencies integrated over a common period is zero (the components are said to be *orthogonal*), only components of v and i that are of the same frequency contribute to the average power at the terminal pair. If either v or i consists of a single frequency component (including dc), only the corresponding component in the covariable contributes to the integral in *(3.1)*.

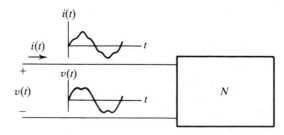

Figure 3.1   General time varying nonlinear network, N, having periodic terminal variables v and i.

For a lossless network containing no energy storage (one containing only switches, for instance), the power spectra of its input and output ports must be identical; that is, the instantaneous input and output powers must be equal. This fact helps explain the separate roles of the switches and the energy storage elements in a power circuit. For example, it shows that a rectifier system that converts 60 Hz ac to pure dc cannot do so with switches only. Because the input voltage is a 60 Hz sinusoid, there must be a 60 Hz component of current to produce a nonzero average input power (to match the average output power). But this 60 Hz component of input current, when multiplied by the 60 Hz input voltage, produces a 120 Hz component of instantaneous input power. As there is no 120 Hz component of output power, the circuit must contain energy storage in addition to switches. Stated simply, there is no magic you can do with switches to eliminate the need for energy storage at this low frequency (120 Hz)—a need that translates into physically large capacitors or inductors.

3.2 HALF-WAVE RECTIFIERS

The half-wave rectifer is the simplest type of rectifier circuit, consisting of only a single diode. Unfortunately, its output voltage contains ripple at the ac input frequency, which makes filtering more difficult than for other circuits having ripple frequencies that are multiples of the input frequency. Moreover, its ac source current contains a dc component that makes the circuit impractical to use with an input transformer, as you will see in Chapter 20. When vacuum tubes with their large forward drops were used as rectifying diodes, the half-wave circuit was frequently employed instead of alternative circuits in which two or more conducting diodes appeared in series. Today, however, the low forward drop of semiconductor diodes makes it unnecessary to accept the disadvantages of this circuit in most cases. The exceptions are in low-voltage applications, such as power supplies for computers and electrochemical processes (plating, for instance), or in very cost-sensitive applications, such as battery chargers for portable power tools. But its simplicity makes the half-wave circuit a good way to introduce basic rectifier concepts.

3.2.1 Half-Wave Rectifier with Resistive Load

The purely resistive load of the half-wave rectifier circuit shown in Fig. 3.2 constrains the rectified voltage v_d and the current i_d to have the same polarity at all times. The diode cannot conduct current in the reverse direction, so it must be off when the source voltage is negative. If we instead assume that it is on under this condition, we find an inconsistency between the current direction and the diode state; that is, we find the diode current to be negative. Achieving consistency between a diode's state and its voltage or current is the essence of the *method of assumed states*. Although this example is simple, more complicated arrangements of switches and passive components present a greater challenge when you use the method.

To apply the method of assumed states to an arbitrary network, we assume that

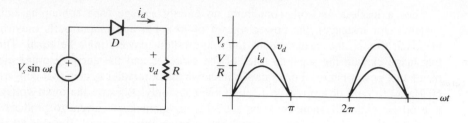

Figure 3.2   Half-wave rectifier with resistive load.

each diode in the network is in either an off or on state. We then solve the network for the currents in the diodes that are on and the voltages across those that are off. If an on diode's current is negative or an off diode's voltage is positive, there is a basic inconsistency between our assumptions about the diode states and the terminal characteristics of the diodes, because a diode cannot conduct negative current or block positive voltage. In this case the assumed states are not a possible set, and we try another combination of on and off diodes. Once we obtain a consistent set of states, we determine the diode voltages and currents as functions of time and test them to see which diode first presents an inconsistency. We then assume that this diode (or diodes) changes state at this point and solve the circuit for the new set of states. By progressing in this way we eventually return to our original consistent set of states, in the process having analyzed the circuit for a complete period.

Because we generally want the dc component of the rectifier output, let's calculate this component for the simple rectifier of Fig. 3.2. We do so by calculating the average of v_d over one cycle:

$$\langle v_d \rangle = \frac{1}{2\pi} \int_0^{2\pi} v_d(\omega t)\, d(\omega t) = \frac{V_s}{\pi} \tag{3.2}$$

The function of most rectifiers is to produce an output voltage with a low ripple content.[†] When low ripple is important, a low-pass filter is placed between the rectifier and the load. However, the presence of a filter often changes the operating characteristics of the rectifier, and therefore the output of the filter is not always the average value of v_d determined before the filter is added. For example, putting a low-pass filter in series with the load resistor R in Fig. 3.2 changes the circuit's operation so that $\langle v_d \rangle$ is no longer equal to V_s/π, as we show next.

3.2.2 Half-Wave Rectifier with Inductive Load

Figure 3.3 shows a half wave rectifier containing a simple low-pass filter consisting of an inductor L in series with the load resistor R. The influence of L on v_d is clearly evident when we compare the waveforms of Fig. 3.2 with those of Fig. 3.3.

[†]The major exception is the rectifiers used in the electrochemical industry for plating and electrolysis, where ripple causes no harm.

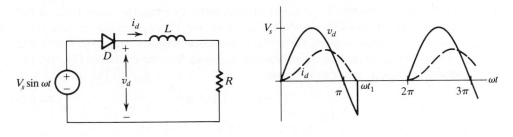

Figure 3.3 Half-wave rectifier with inductive load.

First, v_d goes negative, and, second, the ac line current (which is the same as i_d) is out of phase with the line voltage. Although "phase" is an inexact concept when used to refer to distorted waveforms, what we generally mean is the phase of the fundamental component of the waveform, which we discuss in Section 3.4.

We can again utilize the method of assumed states to derive the waveforms of Fig. 3.3. If we assume that the diode is on, i_d must be positive. We will now show that with the circuit operating in the cyclic steady state the diode must turn off for part of the cycle. First, let's assume that it does not. Then $v_d = V_s \sin \omega t$, which results in $\langle v_d \rangle = 0$. Because the average voltage across L must be zero and the average voltage across the load resistor equals $\langle v_d \rangle$, which is also zero, the average voltage across R must also be zero. Thus the current through the resistor either will be zero or have both positive and negative excursions. Zero current is not possible for finite values of L, and the diode precludes negative current. Therefore 360° conduction of the diode is inconsistent with the constraints imposed by both the circuit and the diode, and the load current must be zero for part of each cycle. Hence this operating condition is known as *discontinuous conduction*.

During periods when the diode is off, the source voltage must be negative. The time at which the diode turns on must then be the time when the source voltage crosses zero from negative to positive. If the diode did not turn on at this time, it would be supporting a forward voltage. The problem now is to determine when the diode turns off.

Figure 3.4 shows an equivalent circuit for the calculation of i_d, where the switch is closed when the source voltage crosses zero from negative to positive.

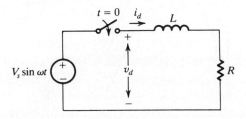

Figure 3.4 Equivalent circuit for the calculation of i_d.

The resulting response, i_d, consists of two parts: one has the same form as the excitation and is variously known as the *driven, forced,* or *particular* response; the other is characterized by the eigenvalues, or natural frequencies, of the circuit and is known as the *natural* or *homogeneous* response. For the circuit of Fig. 3.4, the network equations in terms of i_d are

$$i_d = 0 \qquad \omega t < 0 \tag{3.3}$$

$$\frac{di_d}{dt} + \frac{R}{L} i_d = \frac{V_s}{L} \sin \omega t; \qquad \omega t > 0 \tag{3.4}$$

Solving *(3.4),* we obtain i_d for $0 < \omega t < \omega t_1$:

$$i_d = \frac{V_s}{Z} \sin(\omega t - \phi) + A e^{-Rt/L} \tag{3.5}$$

where

$$Z = \sqrt{R^2 + (\omega L)^2} \tag{3.6}$$

and

$$\phi = \tan^{-1}\left(\frac{\omega L}{R}\right) \tag{3.7}$$

The time t_1 in Fig. 3.3 is the first zero crossing of i_d, or the time at which the diode turns off. We determine the value of A by applying the boundary condition $i_d(0+) = i_d(0-) = 0$:

$$A = \frac{V_s}{Z} \sin \phi = \frac{V_s \omega L}{Z^2} \tag{3.8}$$

Figure 3.5 is a graph of i_d and its particular and homogeneous components. The continuity of state variables across boundaries (and elsewhere) reflects the

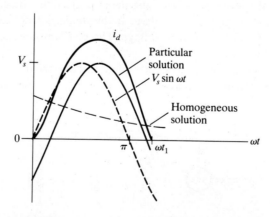

Figure 3.5   The load current, i_d, and its constituents for the circuit of Fig. 3.4.

fact that the energy of a physical system cannot change instantaneously unless an impulse of power is present. Our circuit contains no stored energy for $t < 0$, and thus it cannot contain any energy at $t = 0+$. The only energy storage element present is L, so this constraint means that $i_d(0+) = 0$. We can find the value of t_1 from (3.5) by iteratively solving the transcendental equation $i_d(t_1) = 0$.

An interesting but unfortunate characteristic of this simple rectifier is that $\langle v_d \rangle$, the voltage in which we are interested, is a function of the load R. Stated in proper terms, this rectifier exhibits *load regulation*. That is, the load (through its current) affects, or regulates, the output voltage. This may be confusing at first, because when we say that something is to be regulated, we usually imply that it is held constant. When we use the term *regulation* in the context of rectifiers, however, the implicit meaning is load regulation. The source of regulation in this rectifier circuit is the negative excursion of v_d between π and wt_1. Looking at Fig. 3.5, you can see that this interval is a function of $\tau = L/R$, a function of R (the load) for constant L.

The behavior of this circuit is insensitive to the location of L relative to the diode; that is, it can be on either the ac or dc side. Thus, as a practical ac source invariably contains inductance, especially if the source includes a transformer, some degree of regulation is difficult to avoid.

3.2.3 Half-Wave Rectifier with Freewheeling Diode

The half-wave rectifier with low pass filter shown in Fig. 3.6 is a more practical circuit than the one shown in Fig. 3.3. The addition of D_2 now permits the load current i_d to be continuous and prevents v_d from going negative. Using the method of assumed states, we can prove that a further constraint on circuit operation is that D_1 and D_2 cannot be on simultaneously. When D_1 is off, D_2 allows the energy in the circuit to maintain continuity by providing a path through which the inductor current can "free wheel." For this reason D_2 is known as a *freewheel* (or *freewheeling*) diode. Diodes perform identical functions in other power circuits, where they are called *bypass, flyback,* or *catch* diodes.

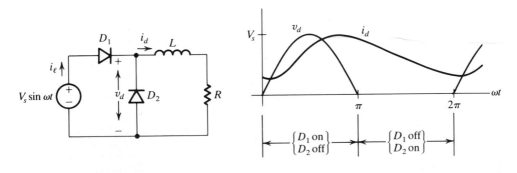

Figure 3.6   Half-wave rectifier with a freewheeling diode.

If the inductor is large enough, i_d never decays to zero (an operating condition known as *continuous conduction*, where $L/R \gg \pi/\omega$). We can then further show that D_1 and D_2 cannot be off simultaneously because there must always be a path for i_d. With this condition, the circuit has two topological states: in one, D_1 is on and D_2 is off; in the other, D_1 is off and D_2 is on.

The behavior of i_d shown in Fig. 3.6 is the result of solving the circuit equations for each of the two topological states of the network. For the state D_1 on and D_2 off, the circuit equation in terms of i_d is

$$\frac{di_d}{dt} + \frac{R}{L}i_d = \frac{V_s}{L}\sin\omega t; \qquad 0 < \omega t < \pi \tag{3.9}$$

For the freewheeling state in which D_1 is off and D_2 is on, the equation is

$$\frac{di_d}{dt} + \frac{R}{L}i_d = 0; \qquad \pi < \omega t < 2\pi \tag{3.10}$$

From the waveform of v_d in Fig. 3.6, we can calculate $\langle v_d \rangle$:

$$\langle v_d \rangle = \frac{1}{2\pi} \int_0^\pi V_s \sin\omega t\, d(\omega t) = \frac{V_s}{\pi} \tag{3.11}$$

Note that *(3.11)* gives the maximum possible average value of v_d independent of either L or R. This rectifier circuit, then, exhibits *no regulation* (unless L contains significant resistance). The result is that we can make L as large as necessary to achieve the desired degree of filtering. As L goes to infinity, i_d becomes constant, and the voltage across R is only the dc component of v_d.

EXAMPLE 3.1

A Half-Wave Rectifier with a Capacitive Filter

In many applications, especially those in which cost is a major consideration, a capacitive filter, rather than the inductive filter just discussed, is used. Figure 3.7 shows a half-wave rectifier circuit containing such a filter. We will determine both v_d and $\langle v_d \rangle$ for this circuit.

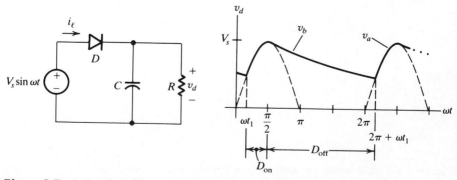

Figure 3.7   A simple half-wave rectifier with a capacitive filter.

For $RC \gg \pi/\omega$, the circuit of Fig. 3.7 functions much like a peak detector, so the form of the voltage v_d can be drawn qualitatively, as shown. The only parameter necessary to specify v_d is t_1, the time at which D turns on. The voltage v_d during the diode on and off states is the source voltage and a decaying exponential, respectively. Thus:

$$v_a(\omega t) \equiv v_d(\omega t) = V_s \sin \omega t; \qquad \omega t_1 < \omega t < \frac{\pi}{2} \qquad (3.12)$$

$$v_b(\omega t) \equiv v_d(\omega t) = V_s e^{-(\omega t - \pi/2)/\omega RC}; \qquad \frac{\pi}{2} < \omega t < (2\pi + \omega t_1) \qquad (3.13)$$

We can determine the time t_1 by establishing the condition that the two solutions are equal at $\omega t = 2\pi + \omega t_1$ and solving for ωt_1:

$$v_a(2\pi + \omega t_1) = v_b(2\pi + \omega t_1) \qquad (3.14)$$

and

$$V_s \sin(\omega t_1) = V_s e^{-(2\pi + \omega t_1 - \pi/2)/\omega RC} \qquad (3.15)$$

We solve for ωt_1 implicitly (solving for ωt_1 in terms of ωt_1) and then use iteration to determine a value for ωt_1 as follows. Solving the left side of (3.15) for ωt_1, we obtain

$$\omega t_1 = \sin^{-1}\left(e^{-(\omega t_1 + 3\pi/2)/\omega RC}\right) \qquad (3.16)$$

An iterative solution to (3.16) requires numerical parameters in the equation. For this example let's assume that $V_s = 24$ V, $RC = 20$ ms, and $\omega = 377$ rad/s, and make $\omega t_1 = 0$ our initial guess. Using these parameters, and plugging $\omega t_1 = 0$ into the right-hand side of (3.16) gives a new value of ωt_1:

$$\omega t_1 = \sin^{-1} 0.54 = 0.56$$

Using this new value on the right-hand side of (3.16) gives:

$$\omega t_1 = \sin^{-1}\left(0.54 e^{-0.56/7.54}\right) = 0.52$$

One more iteration gives a result consistent to two places, which is close enough for our purposes:

$$\omega t_1 = \sin^{-1}\left(0.54 e^{-0.52/7.54}\right) = 0.52$$

$$t_1 = \frac{0.52}{377} = 1.38 \text{ ms}$$

We can now calculate the average value of v_d:

$$\langle v_d \rangle = \frac{V_s}{2\pi}\left\{\int_{0.52=\omega t_1}^{\pi/2} \sin \omega t \, d(\omega t) + \int_{\pi/2}^{(2\pi + 0.52)} e^{-(\omega t - \pi/2)/7.54} d(\omega t)\right\}$$

$$= 0.74 V_s = 17.7 \text{ V}$$

Note that unlike the half-wave circuit with freewheeling diode of Fig. 3.6, for which $\langle v_d \rangle = V_s/\pi$ independent of the value of L, the average value of the voltage at the output of a capacitive filter is a strong function of the load and filter time constant RC and is always larger than V_s/π.

The peak-to-peak amplitude of the ac ripple on v_d, v_{pp}, expressed as a percent of $\langle v_d \rangle$, is often of interest. In this case it is

$$v_{pp} = \frac{V_s(1 - \sin \omega t_1)}{0.74V_s} = 68\%$$

3.2.4 Circuit Replacement by Equivalent Source

The circuit of Fig. 3.6 presents an opportunity to demonstrate another very useful technique of circuit analysis called *circuit replacement by equivalent source*. Under the condition that i_d is never zero, the voltage v_d is defined irrespective of the details of the load circuit. We can therefore determine the detailed behavior of the dc, or load, side of the circuit by replacing the ac source and diodes with a voltage source having a waveform v_d, as shown in Fig. 3.8. Such circuit replacement by an equivalent source is an especially useful technique for numerical simulations.

3.2.5 The Periodic Steady State

The waveform of i_d in Fig. 3.6 illustrates the *periodic steady state*. It contains both the particular and homogeneous solutions to the differential equations that describe the network: *(3.9)* and *(3.10)*. In this sense it is different from the "steady state" solution that we associate with the driven response of a linear, time-invariant (LTI) network to a sinusoid. Furthermore, it is composed of two distinct analytic expressions, one valid for $0 < \omega t < \pi$ and the other for $\pi < \omega t < 2\pi$. The first is *(3.5)*, repeated here for the appropriate interval:

$$i_d = \frac{V_s}{Z} \sin(\omega t - \phi) + Ae^{-Rt/L}; \quad 0 < \omega t < \pi \qquad (3.17)$$

The second is simply the natural response of the LR load circuit, or:

$$i_d = Be^{-Rt/L}; \quad \pi < \omega t < 2\pi \qquad (3.18)$$

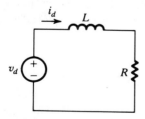 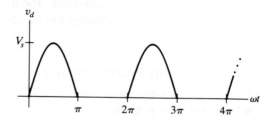

Figure 3.8   The technique of circuit replacement by equivalent source illustrated by replacing the sinusoidal voltage source and diodes of Fig. 3.6 with the equivalent source v_d.

Two boundary conditions are required in order to determine the coefficients A and B. The two boundaries are $\omega t = 0$ and $\omega t = \pi$, the points at which the diodes switch. However, we must exploit periodicity in setting up these boundary conditions. The first of them is

$$i_d(0+) = i_d(2\pi-) \qquad (3.19)$$

Again, this condition simply represents the constraint imposed by the required continuity of state variables or energy. The second boundary condition is

$$i_d(\pi-) = i_d(\pi+) \qquad (3.20)$$

We do not solve these equations for A and B here because the solutions are just straightforward algebra. Of use for sketching the waveform of i_d is the fact that its derivative is continuous at the boundaries. Convince yourself that a discontinuity in the derivative would require a step in v_d (which does not exist). The ability to quickly determine such details of waveforms is of considerable use in many circumstances where a qualitative description of a circuit's behavior is desired.

3.3 AC SIDE REACTANCE AND CURRENT COMMUTATION

We mentioned in the previous section that the behavior of the half-wave rectifier of Fig. 3.3 is insensitive to the location of L. The same is not true of the circuit of Fig. 3.6. In particular, the presence of inductance on the ac side as well as on the dc side, as shown in Fig. 3.9, creates a third topological state of the network: both diodes are on simultaneously. This state is known as the *commutation* state because, as we will show, the load current is being transferred, or *commutated*, from one diode to the other during this state. The inductance responsible for the existence of this state is known as the *commutating inductance*, L_c, or *commutating reactance*, $X_c = \omega L_c$.

Throughout the following discussion we assume that $L_d/R \gg \pi/\omega$, so that, for all practical purposes, i_d is constant. Once you understand the behavior of the circuit under this condition, you will see clearly how to include the effects of ripple or discontinuous conduction. The assumption of constant i_d during the commutation state is often employed in the analysis of rectifier circuits, as in most cases $L_d \gg L_c$.

3.3.1 Commutation Processes and Equivalent Circuits

If we assume that the circuit of Fig. 3.9 is in the state with D_1 on and D_2 off, and note that there is no drop across L_c because the current through it is constant, we see that v_d is forced to equal the source voltage, $V_s \sin \omega t$. This state becomes inconsistent with the diode voltages when the source voltage goes negative. Two states are now possible, only one of which can be consistent with the circuit constraints. Either D_2 turns on and D_1 turns off, as in the circuit of Fig. 3.6, or D_2

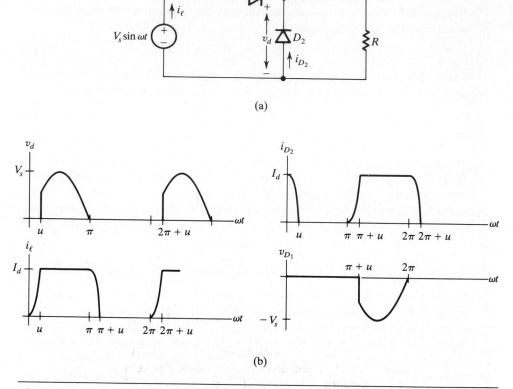

Figure 3.9   (a) A half-wave rectifier with commutating inductance L_c. (b) The behavior of branch variables.

turns on and D_1 remains on. The latter state must occur next, because the former would require a step change in the energy stored in L_c. The time during which both D_1 and D_2 are on is known as the *commutation period* and has a duration u in electrical degrees.

Commutation is the process of moving a current from one branch of a circuit to another. In the case of Fig. 3.9, the current I_d is being moved from the D_1 branch to the D_2 branch. The process is identical to that performed by the commutator and brushes in a dc motor: The commutation period is the time during which two bars of the commutator are shorted by the brush, and the armature current is being transferred from one bar to the other.

The equivalent circuit of the rectifier during the commutation period is shown in Fig. 3.10, from which we can calculate i_ℓ (or $i_{D_2} = I_d - i_\ell$), whose waveform

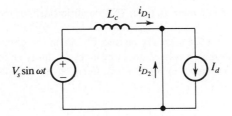

Figure 3.10   The equivalent circuit for Fig. 3.9(a) during the commutation interval, $\pi < \omega t < (\pi + u)$.

is shown in Fig. 3.9(b):

$$i_\ell(\omega t) = i_\ell(\pi) + \int_\pi^{\omega t} \frac{V_s}{\omega L_c} \sin \omega t \, d(\omega t)$$

$$= I_d - \frac{V_s}{\omega L_c}(1 + \cos \omega t); \qquad \pi < \omega t < \pi + u \tag{3.21}$$

We determine the duration of the commutation period, u, from the condition $i_\ell(\pi + u) = 0$; that is,

$$u = \cos^{-1}\left(1 - \frac{\omega L_c I_d}{V_s}\right) = \cos^{-1}\left(1 - \frac{X_c I_d}{V_s}\right) \tag{3.22}$$

The commutating inductance L_c is the inductance across which the voltage forcing the commutating branch current up or down appears. The voltage doing the forcing is called the *commutating voltage*, which in this case equals $V_s \sin(\omega t)$.

The diode states for which Fig. 3.10 is drawn are consistent with their branch variables until $i_{D_1} = 0$ at $\omega t = \pi + u$. The next consistent state is D_1 off, D_2 on. The voltage across D_1 is now the source voltage, so this state persists until $\omega t = 2\pi$, when the source changes polarity. The resulting process of the current commutating from the D_2 branch back to the D_1 branch is identical to that just described, except that the commutating voltage is reversed. But the same electrical angle u is required for the current in L_c to build up from zero to I_d. Note that during this commutation period, $v_d = 0$ instead of $V_s \sin \omega t$. Figure 3.9(b) shows the diode branch variables during normal operation.

EXAMPLE 3.2

An Assumed-State Calculation

The consequences of step changes in energy are often important in determining the consistency of a switch state, as we have just shown. A technique that avoids the indeterminate voltages or currents caused by assuming step changes in stored energy is to assume that the switch turns on or off in a nonzero time. This condition defines the derivative of the switch current or voltage.

We again start with the state D_1 on and D_2 off in the circuit of Fig. 3.9(a) but now assume that when the source voltage goes negative, D_2 turns on instantly and D_1 turns off linearly in time Δt, as shown in Fig. 3.11(a). If $\Delta t \ll \pi/\omega$, then $V_s \sin \omega t \approx 0$ during Δt and the voltage across D_1 is

$$v_{D_1} = -L_c \frac{di_\ell}{dt} = L_c \frac{I_d}{\Delta t} \qquad (3.23)$$

as shown in Fig. 3.11(b). This is a large forward voltage, which is contrary to our assumption that D_1 is on and therefore has zero volts across it. So our assumption is inconsistent with the resulting diode variables. The state D_1 on and D_2 on is the consistent state, as we have already argued.

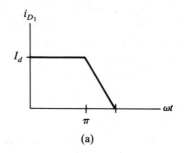

(a)

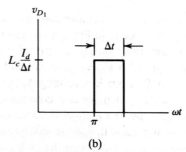

(b)

Figure 3.11   (a) The current i_{D_1} in Fig. 3.9(a) shown turning off in time Δt.   (b) The voltage across D_1 during the time the diode is turning off.

3.3.2 Effects of Commutation

The presence of commutating inductance has two effects on the terminal characteristics of the rectifier of Fig. 3.9(a). First, it causes load regulation of the output voltage; second, it changes the waveform of the ac source current. We can determine the regulation characteristic by calculating $\langle v_d \rangle$ as a function of u and substituting (3.22), which contains the load current explicitly:

$$\langle v_d \rangle = \frac{V_s}{2\pi} \int_u^\pi \sin \theta \, d\theta = \frac{V_s}{2\pi}(1 + \cos u) = \frac{V_s}{\pi}\left(1 - \frac{X_c I_d}{2V_s}\right) \qquad (3.24)$$

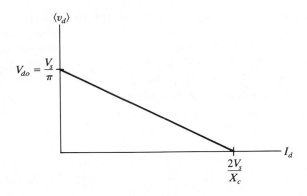

Figure 3.12 Load regulation curve for the half-wave rectifier with commutating reactance of Fig. 3.9.

We plotted this result in Fig. 3.12. You can see that the commutating reactance $X_c = \omega L_c$ has the same effect on the average output voltage as a resistor of value $X_c/2\pi$ in series with the load. (Note that in other respects there is a considerable difference between X_c and a resistor. For example, X_c is a lossless element.) The term $X_c I_d/V_s$ will show up again later when we discuss more complicated rectifier circuits. It is a normalized parameter known as the *reactance factor*, and as *(3.24)* shows, it is directly related to the degree of regulation exhibited by a rectifier circuit.

You can also see the effect of commutating inductance on the source current by comparing the waveform of i_ℓ in Fig. 3.6, which is equal to i_d for $0 < \omega t < \pi$ and 0 for $\pi < \omega t < 2\pi$, to i_ℓ in Fig. 3.9(b). If we assume L/R in the circuit of Fig. 3.6 to be very large so $i_d \approx I_d$, as we did in the circuit of Fig. 3.9(a), then i_ℓ for the circuit of Fig. 3.6 is a rectangular pulse with a duration of half a cycle and an amplitude I_d. When we add commutating inductance to obtain the circuit of Fig. 3.9(a), the pulse duration is extended by u and the pulse shape becomes approximately trapezoidal. For the same peak value, the source current of the rectifier with commutating inductance has a higher rms value than the source current in the rectifier circuit of Fig. 3.6. This result means that the *power factor*, a measure we discuss next, is reduced when commutating inductance is added to the circuit.

3.4 MEASURES AND EFFECTS OF DISTORTION

All the rectifier circuits discussed so far have in common a distorted line (source) current waveform. One practical effect is that the source must have a volt-ampere (VA) rating that is higher than it would need to be if it were supplying the same average power to a linear resistor. The VA *rating* of a source is the product of its maximum deliverable rms voltage and current. For example, a conventional

110 V_{rms} outlet in an American home generally has a maximum current rating of 15 A_{rms}, as determined by wire size, fixture capability, and fuse or circuit breaker rating. Therefore this outlet (source) has a rating of 1650 VA.[†] A sinusoidal, 60 Hz voltage source (the outlet) delivers average power only through the 60 Hz component of its current, as shown by *(3.1)*. Thus if the rms value of its total current exceeds the rms value of its fundamental, the outlet cannot deliver its rated power, although it may be delivering its rated rms current. We also speak of the VA *product* at a terminal pair, which means the product $V_{rms}I_{rms}$ at the terminals. This product is called the *apparent power* and is given the symbol S.

The *power factor*, k_p, of a terminal pair is the ratio of the average power to the apparent power S at the terminals. The factor embodies the effects of both distortion and phase shift between voltage and current. A measure of the distortion—caused by the undesirable frequency components in a waveform—is given by the *total harmonic distortion* (THD).

3.4.1 Power Factor

We define the power factor k_p of a two terminal network as the ratio of the average power measured at the terminals to the product of the rms values of the terminal voltage and current; that is,

$$k_p = \frac{\langle p(t) \rangle}{V_{rms}I_{rms}} = \frac{\langle p(t) \rangle}{S} \tag{3.25}$$

The importance of this measure is that it reflects how effectively available power is being used. Because sources of power generally have thermal limits, and resistors can model many of their loss mechanisms (dissipation proportional to rms voltage or current squared), we can express the source capability in terms of rms voltage and current. Thus a source supplying an average power less than the apparent power $V_{rms}I_{rms}$ at its terminals is not operating at its full capability at this voltage and current.

EXAMPLE 3.3

Power Factor of an ac Controller

Light dimming is one application of a class of circuits known as *ac controllers*. These circuits are also used to control the speed of small appliances and ac powered hand tools. You will learn how they work in Chapter 5. We assume that the ac controller of Fig. 3.13 is feeding a resistive load R that results in the current waveform shown. What is the power factor k_p at the ac source?

[†]In practice, a wall receptacle is loaded to only 80% of its maximum VA rating, or 1320 VA in this case.

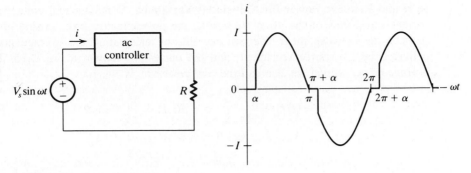

Figure 3.13   An ac controller.

We calculate k_p as follows:

$$\langle p(t) \rangle = \frac{V_s I}{\pi} \int_\alpha^\pi \sin^2 \omega t \, d(\omega t) = \frac{V_s I}{2} \left[\left(1 - \frac{\alpha}{\pi}\right) + \frac{1}{2\pi} \sin 2\alpha \right] \qquad (3.26)$$

$$V_{rms} = \frac{V_s}{\sqrt{2}} \qquad (3.27)$$

$$I_{rms} = \sqrt{\frac{1}{\pi} \int_\alpha^\pi I^2 \sin^2 \omega t \, d(\omega t)} = \frac{I}{\sqrt{2}} \sqrt{\left(1 - \frac{\alpha}{\pi}\right) + \frac{1}{2\pi} \sin 2\alpha} \qquad (3.28)$$

$$S = \frac{V_s I}{2} \sqrt{\left(1 - \frac{\alpha}{\pi}\right) + \frac{1}{2\pi} \sin 2\alpha} \qquad (3.29)$$

$$k_p = \frac{\langle p(t) \rangle}{S} = \sqrt{\left(1 - \frac{\alpha}{\pi}\right) + \frac{1}{2\pi} \sin 2\alpha} \qquad (3.30)$$

This power factor as a function of α is plotted in Fig. 3.14. From this plot you can see that if $\alpha = \pi/2$, the source can supply only 71% of the power it could supply at $\alpha = 0$, for the same value of S.

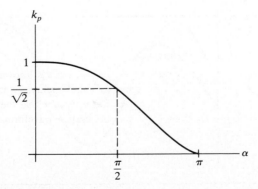

Figure 3.14   Power factor k_p as a function of α for the ac controller of Fig. 3.13.

Real and Reactive Power for Sinusiodal Variables   For sinusoidal voltage and current waveforms of the same frequency, the power factor is the cosine of the phase angle between them. We can see this by calculating the average power delivered by the source voltage v_s and the current i_s shown in Fig. 3.15. This average power, which we also call the *real power P*, is

$$P = \langle p(t) \rangle = \frac{1}{2\pi} \int_0^{2\pi} v_s i_s d(\omega t) = \frac{V_s I_s}{2} \cos \theta$$

$$= V_{srms} I_{srms} \cos \theta = S \cos \theta \qquad (3.31)$$

We call the angle θ the *power factor angle*. Electricians and utility engineers commonly use this parameter because their concern is primarily with sinusoidal voltages and currents. If $\theta > 0$ (an inductive load), the current lags the voltage in time, and the result is a *lagging power factor*. A *leading power factor* results for $\theta < 0$ (a capacitive load).

In addition to the real power P, we *define* a quantity called *reactive power Q* as:

$$Q \equiv V_{srms} I_{srms} \sin \theta = S \sin \theta \qquad (3.32)$$

Reactive power is defined mathematically to be in quadrature with real power. Therefore:

$$|P + jQ| = V_{srms} I_{srms} = S \qquad (3.33)$$

The usefulness of Q is that it tells us how to *compensate* a load using, for instance, reactive elements (inductors and capacitors) to make the resulting power factor unity. For example, the reactive power being delivered by the source in Fig. 3.15 is positive and given by *(3.32)*. A capacitor placed across the source would draw negative reactive power Q_C because $\theta_C = -\pi/2$. If we choose the value of this capacitor so the net reactive power Q' delivered by the source is 0, that is:

$$Q' = Q_C + Q = -V_{srms}^2 C\omega + V_{srms} I_{srms} \sin \theta = 0 \qquad (3.34)$$

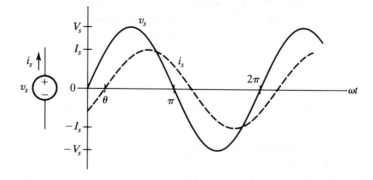

Figure 3.15   Voltage and current waveforms at the terminals of a source. The angle θ is called the power factor angle.

then only real power would be delivered by the source v_s. We can solve *(3.34)* to determine the necessary value of C:

$$C = \frac{I_{srms} \sin \theta}{\omega V_{srms}} \qquad (3.35)$$

Now *(3.33)* will tell us what the new source current I_s' is if P remains unchanged from its value given by *(3.31)*:

$$|P + jQ'| = P = V_{srms} I_{srms}' = V_{srms} I_{srms} \cos \theta \qquad (3.36)$$

$$I_{srms}' = I_{srms} \cos \theta < I_{srms} \qquad (3.37)$$

The compensated source current I_{srms}' is reduced from its uncompensated value I_{srms}, permitting the source to deliver more real power at a given current.

Power Factor of Distorted Waveforms   Because of the switching occurring in power electronic circuits, the voltage and current waveforms at a port are seldom both sinusoidal. For example, none of the rectifier circuits we have presented so far produces a sinusoidal line current, although their line voltages are sinusoidal. Therefore the concept of power factor angle is not particularly useful when we are dealing with power electronic circuits.

The following derivation of k_p explicitly accounts for both waveform distortion and the phase displacement of similar frequency components of the covariables. Thus we can express k_p as the product of two terms, one representing the effect of distortion and the other the effect of displacement:

$$k_p = \frac{\langle p \rangle}{S} = k_d k_\theta \qquad (3.38)$$

In this expression, k_θ is the *displacement factor* and k_d is the *distortion factor*. The case we consider here is that of distortion in only one of the variables. We assume that the other is sinusoidal, a situation which is more often true than not.

To derive an expression for the power factor, we assume a port, such as that of network N in Fig. 3.1, to have periodic voltage and current waveforms that we can describe analytically as:

$$v(t) = V_s \sin \omega t \qquad (3.39)$$

$$i(t) = \sum_{n=0}^{\infty} I_n \sin(n\omega t + \theta_n) = I_1 \sin(\omega t + \theta_1) + \sum_{|n| \neq 1} I_n \sin(n\omega t + \theta_n) \qquad (3.40)$$

In terms of these variables, the average power is

$$\langle p \rangle = \frac{1}{T} \int_0^T vi \, dt = \frac{V_s I_1}{2} \cos \theta_1 = V_{srms} I_{1rms} \cos \theta_1 \qquad (3.41)$$

where the term $\cos \theta_1$ is the displacement factor k_θ, and I_{1rms} is the rms amplitude of the fundamental component of i. By factoring out I_{rms}, we can find the distortion factor k_d:

$$\langle p \rangle = V_{srms} I_{rms} \frac{I_{1rms}}{I_{rms}} \cos \theta = S k_d k_\theta \qquad (3.42)$$

$$k_d = \frac{I_{1rms}}{I_{rms}} \qquad (3.43)$$

3.4.2 Total Harmonic Distortion

One measure of distortion in waveforms is a figure known as the *total harmonic distortion*, or THD, which we define as:

$$\text{THD} = \sqrt{\frac{I_{\ell rms}^2 - I_{1rms}^2}{I_{1rms}^2}} \qquad (3.44)$$

where I_{1rms} is the rms amplitude of the fundamental component of i_ℓ. We now compute I_1 as the fundamental term of the Fourier series for i_ℓ:
dissipated in a resistor because of the distortion components of a waveform to the power that would be dissipated because of the fundamental component alone.

Let's calculate the THD of the line current i_ℓ in the half-wave rectifier with freewheeling diode of Fig. 3.6. We assume that $L/R = \infty$, so the line current is periodic rectangular pulses of amplitude I_d, as shown in Fig. 3.16.

In this case we may express the THD in terms of the rms amplitudes of the components as:

$$\text{THD} = \sqrt{\frac{I_{\ell rms}^2 - I_{1rms}^2}{I_{1rms}^2}} \qquad (3.45)$$

where I_{1rms} is the rms amplitude of the fundamental component of i_ℓ. We now

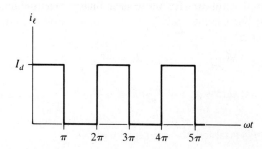

Figure 3.16 Waveform of the line current i_ℓ for the rectifier of Fig. 3.6, assuming that $L/R = \infty$.

compute I_1 as the fundamental term of the Fourier series for i_ℓ:

$$I_{1\text{rms}} = \frac{1}{\sqrt{2\pi}} \int_0^{2\pi} i_\ell \sin \omega t \, d(\omega t)$$

$$= \frac{1}{\sqrt{2\pi}} \int_0^{\pi} I_d \sin \omega t \, d(\omega t) = \frac{\sqrt{2} I_d}{\pi} \tag{3.46}$$

The square of the rms value of the total line current $I_{\ell\text{rms}}^2$ is

$$I_{\ell\text{rms}}^2 = \frac{1}{2\pi} \int_0^{2\pi} i_\ell^2 \, d(\omega t) = \frac{I_d^2}{2} \tag{3.47}$$

We can now calculate the THD for this waveform:

$$\text{THD} = \sqrt{\frac{I_d^2/2 - 2I_d^2/\pi^2}{2I_d^2/\pi^2}} = 121\% \tag{3.48}$$

You will encounter this particular waveform often, so remembering its THD will be useful. In addition, this THD provides a handy reference for comparison of the THDs of other waveforms.

In terms of the THD of the distorted waveform as defined in (3.44), we can express k_d as:

$$k_d = \sqrt{\frac{1}{1 + (\text{THD})^2}} \tag{3.49}$$

EXAMPLE 3.4

Using Power Relationships to Compute THD

By exploiting the power relationships discussed in Section 3.1, we can calculate the THD of the line current for the rectifier of Fig. 3.2 (which is equal to i_d in this case) without explicitly calculating its Fourier components. We do so by first equating the average powers on the ac and dc sides of the rectifier to determine I_1, the amplitude of the fundamental component of the line current:

$$\langle p_{dc} \rangle = \frac{1}{2\pi} \int_0^{2\pi/\omega} v_d i_d \, dt = \frac{V_s^2}{2\pi R} \int_0^{\pi} \sin^2 \omega t \, d(\omega t) = \frac{V_s^2}{4R} \tag{3.50}$$

and

$$\langle p_{ac} \rangle = \frac{V_s I_1}{2} = \langle p_{dc} \rangle = \frac{V_s^2}{4R} \tag{3.51}$$

Hence:

$$I_1 = \frac{V_s}{2R} \tag{3.52}$$

and

$$I_{1\mathrm{rms}} = \frac{V_s}{2\sqrt{2}R}$$

We may now calculate the THD in terms of the rms amplitudes:

$$I_{1\mathrm{rms}}^2 + \sum_{n\neq1} I_{nr\mathrm{ms}}^2 = I_{\mathrm{rms}}^2 = \frac{1}{2\pi} \int_0^{2\pi} i_d^2\, d(\omega t) = \frac{V_s^2}{4R^2} \qquad (3.53)$$

giving

$$\sum_{n\neq1} I_{nr\mathrm{ms}}^2 = \left[\frac{V_s}{2R}\right]^2 - \left[\frac{V_s}{2\sqrt{2}R}\right]^2 = \left[\frac{V_s}{2\sqrt{2}R}\right]^2 \qquad (3.54)$$

Thus:

$$\mathrm{THD} = \sqrt{\frac{\sum_{n\neq1} I_{nr\mathrm{ms}}^2}{I_{1\mathrm{rms}}^2}} = 100\% \qquad (3.55)$$

Notes and Bibliography

A good discussion of the clash of personalities and professional opinions between Edison and Westinghouse can be found in [1]. Although it was these two men who debated the ac/dc issue, if they had not others would have. The battle also raged in other countries with similar results. As recently as 1950 a section of New York City was supplied with dc instead of ac, and as late as 1969 dc was in some of the student dormitories of Boston University.

Single-phase rectifier circuits are treated extensively in [2]. Included are a variety of loads and filters, including capacitive filters and parallel LR loads. This book is extraordinarily comprehensive, but will prove an inconvenient source of information if you do not read German.

A very interesting historical perspective on devices used as rectifier switches (vacuum tube, gas filled, mercury pool, and solid state through germanium technology) is presented in Chapters 2–5 of [3]. The text contains a number of photographs and sketches of these early devices.

Emanuel's paper [4] is an excellent and comprehensive treatment of power factor in systems producing nonsinusoidal variables. The six detailed discussions at the end of the paper and Emanuel's response add considerably to the value of this paper. They also illustrate the degree of confusion and controversy that still exists about the concept of power factor for nonsinusoidal variables.

1. T. S. Reynolds and T. Bernstein, "The Damnable Alternating Current," *Proc. IEEE* 64 (9): 1339–1343 (September 1976).
2. Th. Wasserab, *Schaltaungslehre Der Stromrichtertechnik* (Berlin: Springer-Verlag, 1962).
3. F. G. Spreadbury, *Electronic Rectification* (New Jersey: D. Van Nostrand Company, 1962).
4. A. E. Emanuel, "Powers in Nonsinusoidal Situations—A Review of Definitions and Physical Meaning," *IEEE Trans. on Power Delivery* 5 (3): 1377–1389 (July 1990).

Problems

3.1   By using the method of assumed states, show that in the circuit of Fig. 3.3 the diode off state is consistent with its branch variables after t_1.

3.2   Use the method of assumed states to show that D_1 and D_2 cannot be on simultaneously in the rectifier circuit of Fig. 3.6.

3.3   The circuit of Fig. 3.17 is representative of a class of ancillary circuits called *snubbers*, which we discuss in Chapter 24. Its purpose is to limit the rate of rise of the voltage v_Q. Determine, sketch, and dimension i_L, v_C, and v_Q. What purpose does the diode and resistor serve?

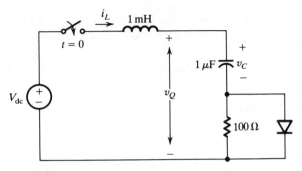

Figure 3.17   The snubber circuit analyzed in Problem 3.3.

3.4   The circuit in Fig. 3.18 varies with time because L_2 is switched in and out of the circuit by the diode D. The circuit is initially at rest, and at $t = 0$ the switch S is closed. Calculate and plot i_ℓ, i_1, i_2, and v_C for $0 < t < 50 \ \mu$s.

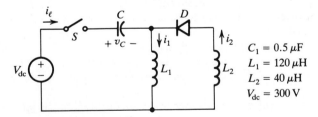

$C_1 = 0.5 \ \mu\text{F}$
$L_1 = 120 \ \mu\text{H}$
$L_2 = 40 \ \mu\text{H}$
$V_{\text{dc}} = 300 \ \text{V}$

Figure 3.18   The second-order switched circuit analyzed in Problem 3.4.

3.5   For most engineering purposes it is not necessary to calculate exactly ωt_1 in Example 3.1. Determine the error that would result in the calculation of $\langle v_d \rangle$ if we approximated ωt_1 as zero, that is,

$$e^{-(3\pi/2+\omega t_1)/\omega RC} \approx e^{-3\pi/2\omega RC} \qquad (3.56)$$

Would this be an appropriate approximation to make in calculating the ripple in v_d?

3.6   Construct a regulation curve for the circuit of Fig. 3.9 if we replace L_c with a resistor of value $R_c = X_c$.

3.7 Determine the power factor of the half-wave rectifier with freewheeling diode of Fig. 3.6. Assume that $L/R \gg 2\pi/\omega$, so $i_d \approx I_d$.

3.8 Determine the THD of the line current for the rectifier with commutating inductance shown in Fig. 3.9(a). What is the power factor of this circuit? How does it compare to the power factor if $L_c = 0$? Approximate the line current as a trapezoid whose rising and falling edges have duration u.

3.9 Sketch the ac source current i_ℓ in the half-wave rectifier with capacitive filter of Example 3.1. What is the power factor of this circuit?

3.10 Figure 3.19 is a schematic drawing of a *full-wave, centertapped rectifier*. The two ac sources are usually created by a transformer with a centertapped secondary winding.

(a) Assume that $L_c = 0$ and sketch $v_d(t)$.
(b) Determine and sketch i_{ℓ_1} and i_{ℓ_2} for $0 < \omega t < 2\pi$.
(c) Determine and sketch the load regulation curve for this circuit and compare it quantitatively to that in Fig. 3.12 for the half-wave rectifier of Fig. 3.9.

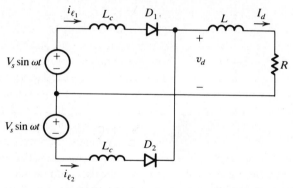

Figure 3.19 The full-wave, centertapped rectifier of Problem 3.10.

3.11 The circuit of Fig. 3.20 is often used to create a dual voltage supply, ±15 V, for instance. Sketch and dimension v_{d1} and v_{d2}. (*Hint:* try to recognize the independence of the two outputs.)

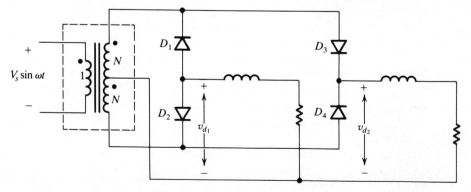

Figure 3.20 The dual voltage rectifier of Problem 3.11. The transformer is ideal.

Chapter 4

Bridge and Polyphase Rectifier Circuits

IN Chapter 3 we described the basic operation of rectifiers, using the single-phase, half-wave rectifier circuit as a vehicle. Although easily understood and effective for learning purposes, this circuit is seldom used today. The reasons are that the half-wave circuit has a high ratio of ac source VA rating to dc power (that is, the ac source operates at a low power factor), produces a dc component in the ac source current (which causes problems for transformers in the source network), and requires a large amount of dc filtering to achieve a specified ripple. As you will see when we present alternative circuits the singular advantage of the half-wave circuit is that only a single diode drop is present between its ac and dc sides. When there was no practical alternative to vacuum diodes, this advantage made the half-wave rectifier a popular circuit, because vacuum diode drops range from tens to hundreds of volts and represent appreciable loss and thermal problems. Semiconductor diodes have effectively neutralized this advantage of the half-wave circuit, except in low voltage, high-power applications, the most notable being in the electrochemical industry (for example, low potential electrolytic cells for the production of chlorine), or in applications where ripple is not a problem, such as battery charging.

The disadvantages of the half-wave circuit are largely overcome by circuits that function as *full-wave* rectifiers. In these circuits the diodes connect the dc load to the ac source during both the positive and negative half cycles of the source. The result is a bilateral source current having no dc component and an increase in the fundamental dc-side ripple frequency that reduces the dc filtering requirements. The most common of these circuits is the *bridge rectifier*. An alternative, the *full-wave centertapped* or *half-bridge* rectifier analyzed in Problem 3.10, is also used extensively.

A *polyphase rectifier* is a circuit containing several ac sources whose rectified voltages are combined at the output. The sources have equal amplitudes and frequencies but differ in phase from one another. Both half-wave and full-wave rectifier circuits can be designed for polyphase as well as single phase ac systems.

Polyphase rectifiers produce less distortion of the ac source current and higher dc-side ripple frequencies than single-phase circuits. For these reasons, systems operating at power levels in excess of approximately 10 kW utilize polyphase rectifiers almost exclusively.

In this chapter we introduce the concept of full-wave rectification by an appropriate interconnection of two half-wave circuits. We then use the resulting bridge circuit as a building block to create polyphase, full-wave rectifiers. We also discuss the operating characteristics of these circuits from both the ac and dc sides.

4.1 THE SINGLE-PHASE FULL-WAVE BRIDGE RECTIFIER

The full-wave bridge rectifier is the workhorse of single-phase rectifier circuits. Its ac source current contains no dc component, and the fundamental frequency of the ripple on the dc voltage is equal to twice that of the half-wave circuits of Chapter 3. For the same ac source voltage, the full-wave rectifier produces an average output voltage twice that of the half-wave circuit with freewheeling diode, even though the voltage and current ratings of the diodes in the two circuits are the same. A disadvantage of all bridge circuits is that the input and output ports have no common terminal. In a rectifier this means that the ac source and the dc load cannot share the same ground.

The half-wave rectifier with freewheeling diode of Fig. 4.1 serves as the basic building block for the bridge circuit, and in this context the circuit is sometimes called a *half-bridge*. So long as i_d is continuous (nonzero), v_d has the waveform shown.

A circuit similar to that of Fig. 4.1 in every respect except the directions of the diodes, which are reversed, produces a load current i_d and voltage v_d of the same form as those shown but of opposite polarity. If we now connect the load between these two half-wave rectifiers, its voltage will be the difference between the output voltages of the two circuits, or $v_d = v_{d_1} - v_{d_2}$, as shown in Fig. 4.2(a). The waveform v_d in Fig. 4.2(b) is known as a full-wave rectified voltage because both the positive and negative half cycles of the ac source waveform are present on the dc side of the circuit.

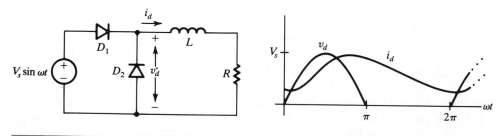

Figure 4.1 The half-wave rectifier with freewheeling diode, sometimes called a half-bridge circuit.

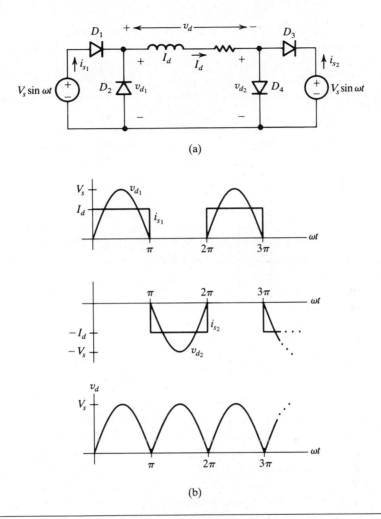

Figure 4.2 (a) The single-phase full-wave bridge circuit as a connection of two half-wave rectifiers with freewheeling diodes (half-bridges). (b) Waveforms of variables in the circuit of (a).

Because the two ac sources in Fig. 4.2(a) are identical, the anode of D_1 and cathode of D_3 are at the same potential and may be connected. The circuit thus requires only a single ac source. Two different ways of drawing the result are shown in Fig. 4.3(a). We call this connection of diodes a *bridge* because the circuits connected to its input and output (the ac source and the RL load) *bridge* one another topologically. Although the source currents i_{s_1} and i_{s_2} of Fig. 4.2 contain dc components, the current i_s in the equivalent source in Fig. 4.3 does not. The reason is that the source current in the bridge is the sum of the two source currents in Fig. 4.2, which have dc components of equal magnitude but opposite polarity.

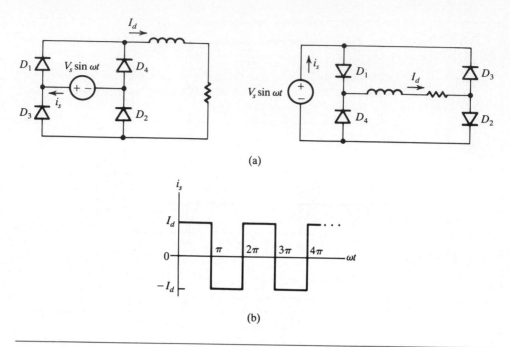

(a)

(b)

Figure 4.3   (a) Equivalent ways of drawing the circuit of Fig. 4.2 as a bridge. (b) The resulting source current, $i_s = i_{s_1} + i_{s_2}$.

4.1.1 Output Voltage and Power Factor of the Single-Phase Bridge

All the operating characteristics of the single-phase bridge rectifier may be inferred from those of the half-wave rectifier with freewheeling diode. First, since the two half-bridges in Fig. 4.2(a) can be viewed as independent voltage sources, v_{d_1} and v_{d_2}, the average output voltage of the bridge is twice that of the half-wave circuit with freewheeling diode given by *(3.9)*. That is

$$\langle v_d \rangle = \langle v_{d_1} \rangle - \langle v_{d_2} \rangle = \frac{2V_s}{\pi} = 0.64V_s \qquad (4.1)$$

For the same dc currents, the rms value of the ac line current for the bridge of Fig. 4.3 is equal to only $\sqrt{2}$ times the rms value of the line current for one of the half-bridges of Fig. 4.2(a). The power factor k_p of the ac source in the bridge circuit is

$$k_p = \frac{\langle p \rangle}{S} = \frac{(0.64V_s)(I_d)}{(V_s/\sqrt{2})(I_d)} = 0.91 \qquad (4.2)$$

For comparative purposes, the power factor of the half-wave circuit with freewheeling diode is 0.64.

EXAMPLE 4.1

The Half-Bridge Rectifier

A half-bridge circuit can be used to create a full-wave rectified voltage if it is supplied by a centertapped voltage source, as shown in the *centertapped rectifier* of Fig. 4.4(a). What are $\langle v_d \rangle$ and k_p at the sources?

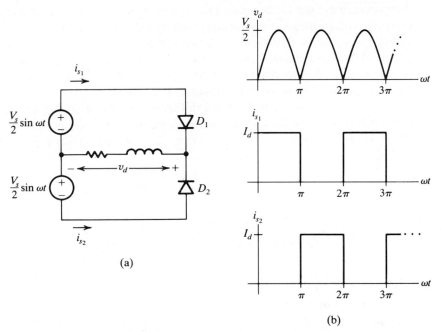

(a)

(b)

Figure 4.4 (a) Half-bridge rectifier circuit with centertapped source. (b) Waveforms of variables in the circuit of (a).

An assumed-state analysis shows that D_1 and D_2 cannot be on or off at the same time. Therefore the top and bottom halves of the circuit behave as two independent half-wave rectifiers. The result is the waveforms shown in Fig. 4.4(b). Although v_d is a full-wave rectified voltage, each source carries a dc component of current resulting in a source power factor of:

$$k_p = \frac{\langle v_d \rangle \langle i_{s_1} \rangle}{(V_s/2\sqrt{2})(I_d/\sqrt{2})} = \frac{(V_s/\pi)(I_d/2)}{V_s I_d/4} = \frac{2}{\pi} = 0.64 \qquad (4.3)$$

If the two sources in the half-bridge are created by center tapping the secondary of a transformer, the power factor at the primary terminals will be $k_p = 0.9$ because the primary current will have no dc component. However, the size of wire chosen for the secondary winding will have to be based on the poorer power factor of 0.64.

4.1.2 Commutation and Regulation in the Single-Phase Bridge

We can derive the commutation process in a bridge rectifier with ac-side reactance from the circuit of Fig. 4.5(a), which shows the two half-bridges, each with commutating inductance $2L_c$. The current source I_d represents the load. Commutation of the two half-bridges is simultaneous, but oppositely directed. That is, while the load current is commutating from D_1 to D_2 it is also commutating from D_4 to D_3. Thus all four diodes are on during the commutation process, $v_{x_1} = v_{x_2}$, and the derivatives of i_{s_1} and i_{s_2} are equal. During periods when no commutation is taking place, the voltage drops across the commutating inductances are zero because of the constant load-current constraint. At all times, then, the anode of D_1 and cathode of D_3 are at the same potential and may be connected. The resulting circuit is shown in Fig. 4.5(b), which is simply the bridge circuit with a commutating inductance equal to one half that in the half-bridges of Fig. 4.5(a).

The operations of the two circuits in Fig. 4.5 are the same. Thus the commutation period u for the bridge with commutating inductance L_c is equal to that for a half-bridge having commutating inductance $2L_c$. We obtain that expression from (3.22):

$$u = \cos^{-1}\left(1 - \frac{2X_c I_d}{V_s}\right) \tag{4.4}$$

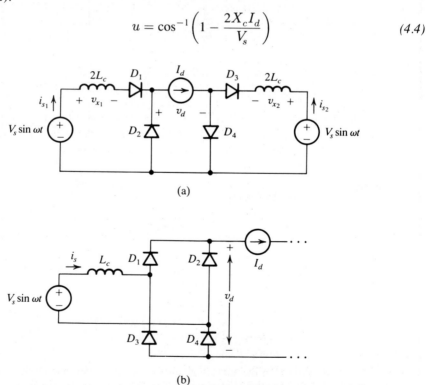

(a)

(b)

Figure 4.5   (a) Full-wave rectifier consisting of two half-bridges with commutating inductance.   (b) Full-wave bridge equivalent to (a).

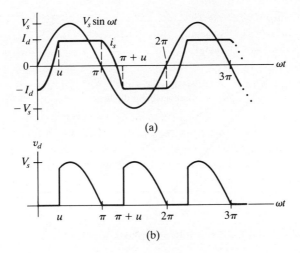

Figure 4.6   Waveforms for the bridge rectifier of Fig. 4.5(b):   (a) source voltage $V_s \sin \omega t$ and source current i_s;   (b) rectified output voltage v_d.

The commutation periods are equal because the current in the inductance in the full-bridge circuit swings between I_d and $-I_d$ during commutation ($\Delta i_s = 2I_d$), whereas it varies only between I_d and 0 in the half-wave circuit.

The average dc voltage as a function of I_d (the load regulation characteristic) is twice that for a half-wave rectifier with free-wheeling diode having a commutating reactance of $2X_c$. Applying this factor of 2 to *(3.24)*, we obtain

$$\langle v_d \rangle = \frac{V_s}{\pi}(1 + \cos u) = \frac{2V_s}{\pi}\left(1 - \frac{X_c I_d}{V_s}\right)$$

$$= V_{do}\left(1 - \frac{X_c I_d}{V_s}\right)$$

(4.5)

where V_{do} is the name we use to designate the maximum possible value of the output of a rectifier circuit.

The waveforms of the source current i_s and the output voltage v_d in the circuit of Fig. 4.5(b) are shown in Fig. 4.6. The regulation characteristics of the half-wave and full-wave single-phase rectifiers are compared in Fig. 4.7. Normalized to V_{do}, the bridge circuit produces twice the regulation of the half-wave circuit. As Δi_s and X_c always appear as a product in the expression for $\langle v_d \rangle$, the effect of $\Delta i_s = 2I_d$ in the bridge is the same as that of doubling the commutating reactance in the half-bridge. But only infrequently is the regulation advantage of the half-wave circuit important enough to overcome its ac-side disadvantages of a poor power factor and the presence of a dc component in the current.

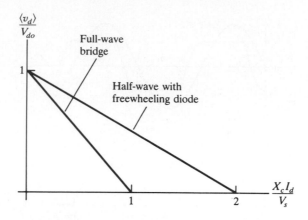

Figure 4.7 Normalized load regulation curves for the single-phase half-wave and bridge rectifiers.

4.2 AN INTRODUCTION TO POLYPHASE RECTIFIER CIRCUITS

We can most easily describe polyphase rectifier circuits as diode OR gates, in which the output assumes the value of the highest input. Figure 4.8 shows a four input OR gate that is also a 4-phase half-wave rectifier circuit. So long as one of the inputs is greater than zero, a diode will be on. The source voltages shown have arbitrary phase relationships with one another. In practical polyphase rectifier circuits, the ac sources are generally disposed symmetrically in phase, but the basic operation of these circuits does not depend on this condition.

We can now obtain full-wave polyphase rectification by connecting two of the half-wave polyphase circuits in series with the load, as we did in evolving the single-phase full-wave bridge circuit. Such a connection is shown in Fig. 4.9(a), where a pair of 3-phase half-wave circuits are used. The sources are shown as two

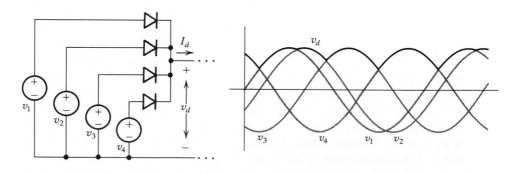

Figure 4.8 A 4-phase half-wave rectifier viewed as a 4-input OR gate.

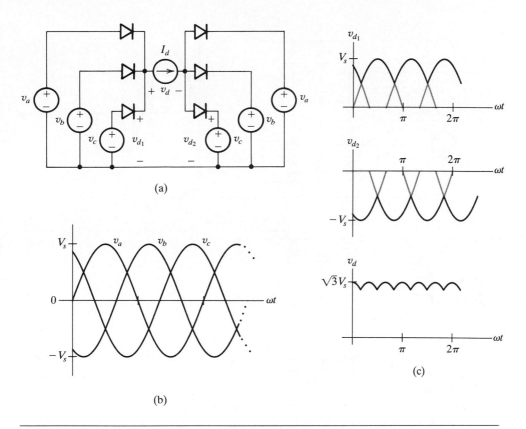

Figure 4.9 A 3-phase, full-wave rectifier constructed from a pair of 3-phase, half-wave (3-pulse) circuits: (a) interconnection of the 3-pulse circuits; (b) source voltages; (c) the individual half-wave voltages and their difference, v_d.

conventional 3-phase sets. The three voltages in each set are equal in amplitude and frequency and are phase displaced from each other by 120°. We designate them as the a, b, or c phase voltage, as shown in Fig. 4.9(b).

The individual half-wave voltages v_{d_1} and v_{d_2} and their difference, v_d, are shown in Fig. 4.9(c). The six pulses in v_d for every cycle of line voltage is the reason we call the 3-phase full-wave circuit a *6-pulse rectifier*. We call the 3-phase half-wave circuit a *3-pulse rectifier* for similar reasons.

We can redraw the circuit of Fig. 4.9 in a more conventional way to illustrate its bridging topology if we recognize that nodes at the same voltage can be connected, thereby eliminating three sources. Figure 4.10(a) shows the result: a 3-phase bridge rectifier circuit, along with one of the phase voltages, v_a, and its relationship to its line current i_a. The relationships between the other phase voltages and their currents are identical, but they are shifted by $\pm 2\pi/3$ from the waveforms shown.

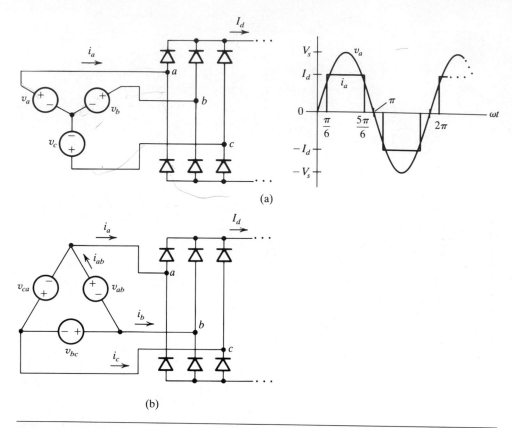

(a)

(b)

Figure 4.10   (a) The circuit of Fig. 4.9 redrawn as a conventional 3-phase bridge (6-pulse) circuit. The 3-phase source is Y-connected.   (b) The same circuit as (a) but with the sources Δ-connected.

The 6-pulse circuit is the basic building block for constructing higher pulse-number rectifiers. For instance, we may construct a 12-pulse circuit by connecting two 6-pulse circuits in series as we show in Example 4.2, or in parallel, as you will do in Problem 4.11.

We can represent a 3-phase set of voltages in two ways. One way is shown in Fig. 4.10(a), where each source in the set is referenced to a common point. This point is called *ground, neutral,* or the *wye* (Y) *point,* the latter because if the connection is drawn symmetrically, the three sources form a "Y." The voltages in this representation are called *line-to-ground* or *line-to-neutral* voltages. We identify such voltages by a single subscript indicating its phase. For instance, v_a is the line-to-neutral voltage of phase a.

An alternative way is shown in Fig. 4.10(b). The sources in this circuit do not have a common connection but are arranged in a *delta* (Δ) connection. We call them *line-to-line* voltages and designate them by a double subscript— for example, v_{ab}. The two sets of voltages are equivalent, so $v_{ab} = v_a - v_b$.

If $v_a = V_s \sin \omega t$, and $v_b = V_s \sin(\omega t - 2\pi/3)$, then:

$$v_{ab} = V_s \sin \omega t - V_s \sin\left(\omega t - \frac{2\pi}{3}\right) = \sqrt{3} V_s \sin\left(\omega t + \frac{\pi}{6}\right) \qquad (4.6)$$

Although the Y- and Δ-connected sources produce the same line-to-line voltages, they are not interchangeable in a circuit if current is flowing in the neutral line of a Y-connected source, as in the 3-pulse circuit. A Δ-connected source could not be used in this case because it has no neutral. In the circuit of Fig. 4.10(a), however, there is no neutral current, so the circuit of Fig. 4.10(b) is a functional equivalent. We can also connect 3-phase loads in Y or Δ.

EXAMPLE 4.2

A 12-Pulse Rectifier Circuit

A series connection of two 6-pulse bridges results in a 12-pulse rectifier only if the ripples of the two bridges are phase-shifted relative to each other. Otherwise the ripple waveforms of the two bridges are simply congruent, resulting in 6-pulse performance. However, 12 symmetric pulses result from shifting the 3-phase ac sources supplying the two bridges by $\pi/6$ with respect to one another. This shift usually involves special 3-phase transformer connections called *wye/delta* (Y/Δ) and *wye/wye* (Y/Y) connections.

Two 6-pulse bridges connected in series and supplied by phase-shifted ac sources are shown in Fig. 4.11(a). The Y/Y and Δ/Y boxes represent the necessary transformer connections. The resulting line-to-line voltages v_{ab} and $v_{a'b'}$ illustrate the 30° phase shift created by the transformer connections. Because each of the 6-pulse bridges operates independently, the output voltage v_d is the sum of v_{d_1} and v_{d_2}. The waveforms of v_{d_1}, v_{d_2}, and v_d in Fig. 4.11(b) show how the 12-pulse output occurs.

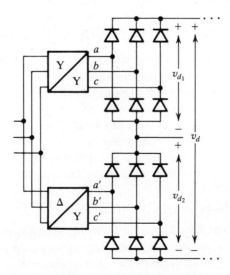

Figure 4.11   (a) Two 6-pulse circuits connected in series to make a 12-pulse rectifier.

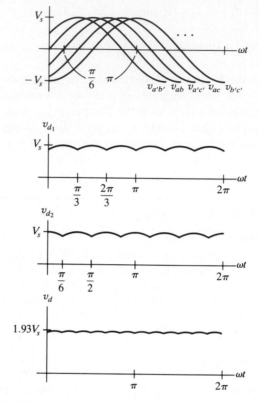

Figure 4.11 (cont.)   (b) Waveforms of variables in the circuit of (a).

4.3 COMMUTATION IN POLYPHASE RECTIFIERS

We first consider commutation in the simple 3-pulse half-wave circuit of Fig. 4.12(a) and then address the commutation behavior of the 6-pulse circuit. However, we cannot simply derive the behavior of the 6-pulse circuit as the sum of the behaviors of two 3-pulse circuits, as we did in the case of the single-phase bridge. The reason is that all the diodes in the 6-pulse circuit are not on simultaneously during u. This condition prevents us from connecting the anodes and cathodes of complementary diodes, as we did in deriving the single-phase bridge circuit of Fig. 4.5(b). We must derive the commutation behavior of the 6-pulse circuit explicitly.

4.3.1 Commutation in the 3-Pulse Rectifier

If we assume a constant dc current I_d, commutation between diodes in the 3-pulse rectifier begins at the same time the current would instantaneously commutate if $L_c = 0$. Except for the value of the commutating voltage, the process is identical to that for the single-phase half-bridge circuit.

Figure 4.12(b) illustrates the equivalent circuit of the 3-pulse rectifier during commutation from phase c to phase a. Both diodes D_1 and D_3 are conducting during the commutation period u. For this time we can express the voltage v_d in terms of either v_a or v_c:

$$v_d = v_a - L_c \frac{di_a}{dt} \quad \text{or} \quad v_d = v_c - L_c \frac{di_c}{dt}$$

As di_a/dt must equal $-di_c/dt$ because of the constraint $i_a + i_c = I_d$, we can solve these two equations for v_d during u:

$$v_d = \frac{v_a + v_c}{2} \tag{4.7}$$

Therefore v_d is simply the average of v_a and v_c during u. We can again determine the commutation period u by equating the volt–time integral for the sum of the

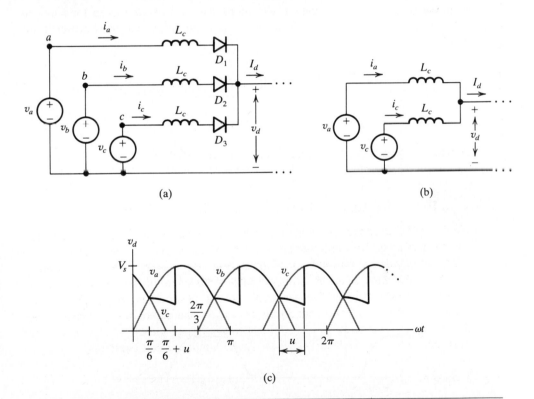

(a)

(b)

(c)

Figure 4.12   Commutation between phases in the 3-phase half-wave (3-pulse) rectifier circuit: (a) the 3-pulse circuit; (b) the equivalent circuit during commutation from phase c to phase a; (c) the voltage v_d of (a), showing the effect of the commutation period u.

two commutating inductances to the dc current I_d:

$$I_d = \frac{1}{2\omega L_c} \int_{\pi/6}^{(\pi/6)+u} V_s \left[\sin \omega t - \sin \left(\omega t + \frac{2\pi}{3} \right) \right] d(\omega t)$$

$$= \frac{1}{2X_c} \int_0^u \sqrt{3} V_s \sin \omega t \, d(\omega t) = \frac{\sqrt{3} V_s}{2X_c} (1 - \cos u)$$

(4.8)

Solving for u yields

$$u = \cos^{-1} \left(1 - \frac{2X_c I_d}{\sqrt{3} V_s} \right)$$

(4.9)

The effect of u on v_d is shown in Fig. 4.12(c). The resulting line currents, i_a, i_b, and i_c, are shown in Fig. 4.13. Note that like the single-phase half-wave rectifier with freewheeling diode, the line currents contain a dc component. Also, because the line current pulse width is shorter for the 3-pulse circuit ($2\pi/3$ compared to π, in the absence of commutating reactance), the ratio of the rms to fundamental values of the line current is higher for the 3-pulse circuit, and a lower power factor (0.48 versus 0.64) results.

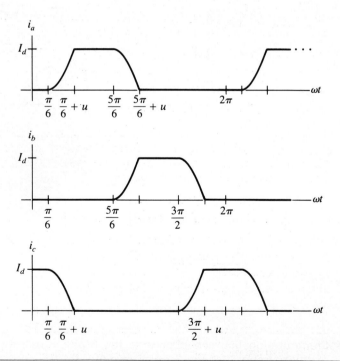

Figure 4.13   Line currents in the 3-pulse rectifier of Fig. 4.12(a).

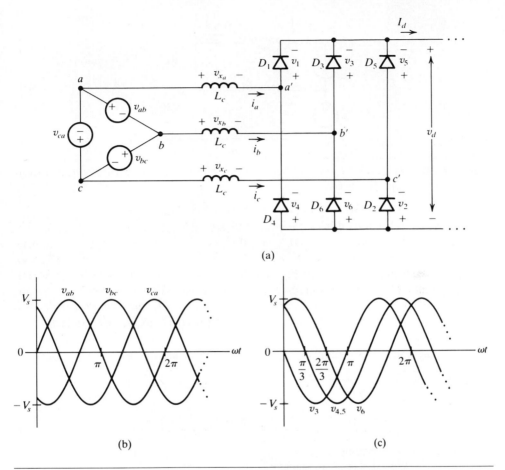

(a)

(b)

(c)

Figure 4.14   (a) A 6-pulse bridge with commutating reactance and a Δ-connected 3-phase source.   (b) The 3-phase source voltages.   (c) Diode voltages, assuming that D_1 and D_2 are on.

4.3.2 Commutation in the 6-Pulse Bridge Rectifier

A 6-pulse bridge with commutating reactance and a Δ-connected source is shown in Fig. 4.14(a). The rather odd diode numbering is the result of numbering them in their conducting sequence.

We begin our analysis by assuming that only D_1 and D_2 are conducting. The line currents are

$$i_a = I_d, \qquad i_b = 0, \quad \text{and} \quad i_c = -I_d$$

During this time there are no voltage drops across the commutating reactances, and the behavior of the circuit is identical to that of Fig. 4.9 or Fig. 4.10. The voltages

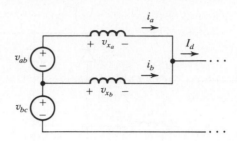

Figure 4.15 Equivalent circuit of the rectifier of Fig. 4.14(a) during commutation from D_1 to D_3.

across the other (off) diodes are

$$v_3 = v_{b'a'} = v_{ba}, \qquad v_4 = v_5 = v_{c'a'} = v_{ca}, \quad \text{and} \quad v_6 = v_{c'b'} = v_{cb}$$

These voltages are shown in Fig. 4.14(c), from which we can conclude that D_3, D_4, D_5, and D_6 can be simultaneously reverse biased only for $2\pi/3 < \omega t < \pi$. Because v_3 is the first diode voltage to go positive, D_3 turns on at $\omega t = \pi$, and $v_{a'b'}$ goes to zero. Commutation of the current from D_1 to D_3 now commences at $\omega t = \pi$.

The equivalent circuit during commutation is shown in Fig. 4.15. Because $i_a + i_b = I_d$, $di_b/dt = -di_a/dt$. Therefore $v_{x_b} = -v_{x_a}$ and:

$$v_{ab} + 2v_{x_b} = 0$$

$$v_{x_b} = -\frac{v_{ab}}{2} \tag{4.10}$$

We can now determine u:

$$I_d = \frac{1}{X_c} \int_\pi^{\pi+u} v_{x_b} \, d(\omega t) = \frac{1}{X_c} \int_\pi^{\pi+u} -\frac{v_{ab}}{2} \, d(\omega t)$$

$$= \frac{1}{X_c} \int_\pi^{\pi+u} -\frac{V_s}{2} \sin \omega t \, d(\omega t) = \frac{V_s}{2X_c}(1 - \cos u) \tag{4.11}$$

$$u = \cos^{-1}\left(1 - \frac{2X_c I_d}{V_s}\right) \tag{4.12}$$

During the commutation interval u, the voltage v_d is equal to the average of v_{bc} and v_{ac}. We can show this condition by expressing v_d as $-v_{ca} - v_{x_a}$ and as $v_{bc} - v_{x_b}$. We then use $v_{ab} + v_{bc} + v_{ca} = 0$ and (4.10) to eliminate v_{x_a} and v_{x_b} and obtain

$$v_d = \frac{v_{bc} - v_{ca}}{2} = \frac{v_{bc} + v_{ac}}{2} \tag{4.13}$$

Once commutation is complete, only two diodes are conducting (D_3 and D_2), there are no drops across the commutating reactances, and v_d assumes the value of the line-to-line source connected to the output by the conducting diodes. If we continue this analysis for a complete cycle, the waveform for v_d shown in Fig. 4.16(a) results. From this waveform we can calculate the regulating characteristic $\langle v_d \rangle = f(X_c I_d / V_s)$:

$$\langle v_d \rangle = \frac{3V_s}{2\pi}(1 + \cos u) = \frac{3V_s}{\pi}\left(1 - \frac{X_c I_d}{V_s}\right) = V_{do}\left(1 - \frac{X_c I_d}{V_s}\right) \qquad (4.14)$$

The analysis leading to (4.14) is valid only if $u < \pi/3$, because it is based on the assumption that commutation of the current between two diodes commences at the angles $\omega t = n\pi/3$ (Fig. 4.16a). We call operation in this region *mode I*. However, increasing the reactance factor $X_c I_d / V_s$ beyond the point where $u = \pi/3$ delays commutation beyond the angles $\omega t = n\pi/3$, preventing the next pair of diodes from starting to commutate at $n\pi/3$. This result becomes evident if you consider the voltages across the off-state diodes. While three diodes are conducting during commutation, for instance D_1, D_3, and D_2, the voltage across all other diodes is the same and equal to $-v_d$. Therefore, so long as three diodes are conducting and $v_d > 0$, a fourth diode cannot turn on. When $u = \pi/3$, Fig. 4.16(b) shows that v_d is still greater than zero. An increase in the reactance factor $X_c I_d / V_s$ now results in the commutation interval "sliding" to the right, as shown in Fig. 4.16(c). Because we are assuming that a periodic steady state exists, requiring six identical pulses per cycle, u remains constant at $\pi/3$ for this mode, which we call *mode II*.

If we assume that commutation in mode II lags by an angle ϕ, meaning that commutation starts at angles $n\pi/3 + \phi$ with reference to Fig. 4.16(c), $\langle v_d \rangle$ is

$$\langle v_d \rangle = \left(\frac{3V_s}{\pi}\right)\frac{1}{2}\left[\cos \phi + \cos(\phi + \pi/3)\right] \qquad (4.15)$$

We can determine the *commutation delay angle* ϕ from (4.11) with the appropriate limits:

$$
\begin{aligned}
I_d &= \frac{-V_s}{2X_c} \int_{\pi+\phi}^{\pi+\phi+\pi/3} \sin \omega t \, d(\omega t) \\
&= \frac{V_s}{2X_c}\left[\cos \phi - \cos(\phi + \pi/3)\right] = \frac{V_s}{2X_c} \cos\left(\phi - \frac{\pi}{3}\right)
\end{aligned}
\qquad (4.16)
$$

As $(\phi - \pi/3) < 0$ for $\phi < \pi/3$, when solving (4.16) for ϕ, we must be careful to select the range of the arccosine to be $-\pi < \cos^{-1} x < 0$. We do so by negating the function and evaluating it for its principal value; that is,

$$\phi = \frac{\pi}{3} - \cos^{-1}\frac{2X_c I_d}{V_s} \qquad (4.17)$$

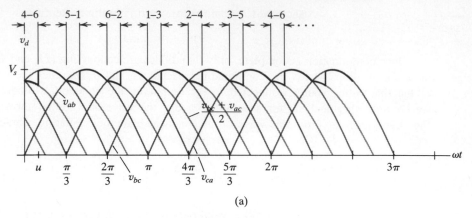

(a)

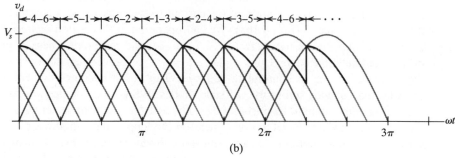

(b)

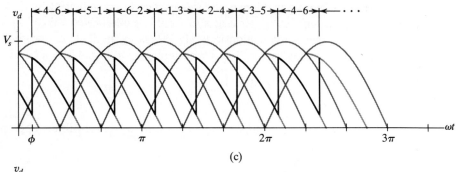

(c)

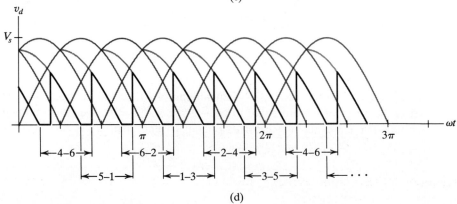

(d)

◀ **Figure 4.16** Output voltage v_d for the rectifier of Fig. 4.14(a) (The diodes undergoing commutation are identified by their numbers.): (a) mode I $(u < \pi/3)$; (b) boundary between modes I and II $(u = \pi/3)$; (c) mode II $(u = \pi/3)$; (d) mode III $(u > \pi/3)$.

Substituting *(4.17)* into *(4.15)*, we obtain the regulating characteristic for mode II:

$$\langle v_d \rangle = \frac{3V_s}{2\pi}\left\{\cos\left[\frac{\pi}{3} - \cos^{-1}\frac{2X_cI_d}{V_s}\right] + \cos\left[\frac{2\pi}{3} - \cos^{-1}\frac{2X_cI_d}{V_s}\right]\right\}$$

$$= \frac{3V_s}{2\pi}\left\{\cos\left[\cos^{-1}\frac{2X_cI_d}{V_s} - \frac{\pi}{3}\right] + \cos\left[\cos^{-1}\frac{2X_cI_d}{V_s} - \frac{2\pi}{3}\right]\right\}$$

$$= \frac{3V_s}{2\pi}\left\{\sqrt{3}\cos\left[\cos^{-1}\frac{2X_cI_d}{V_s} - \frac{\pi}{2}\right]\right\} = \frac{3\sqrt{3}V_s}{2\pi}\sin\left[\cos^{-1}\frac{2X_cI_d}{V_s}\right] \qquad (4.18)$$

$$= \left(\frac{3V_s}{\pi}\right)\frac{\sqrt{3}}{2}\frac{\sqrt{V_s^2 - (2X_cI_d)^2}}{V_s}$$

$$= \frac{V_{do}\sqrt{3}}{2}\sqrt{1 - \left(\frac{2X_cI_d}{V_s}\right)^2}$$

Mode II lasts until $\phi = \pi/6$, at which point $v_d = 0$ at the end of the commutation interval. A fourth diode now turns on if we further increase X_cI_d/V_s, initiating *mode III*. Analysis of the rectifier operating in this mode is straightforward, but tedious, so we do not present it here. However, Fig. 4.16(d) shows v_d in this mode.

The complete regulating characteristic for the rectifier, including mode III, is shown in Fig. 4.17. Note that the regulation becomes more severe as X_cI_d/V_s is increased and a new mode is entered.

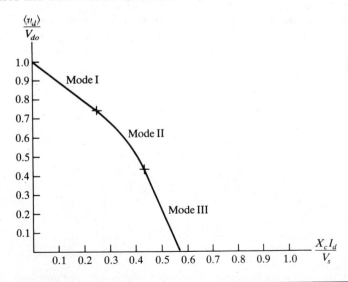

Figure 4.17 Regulation curve for the rectifier of Fig. 4.14(a) operating in modes I, II, and III.

Notes and Bibliography

Schaefer's book [1] is the English language "bible" of rectifier circuits. It contains tables, diagrams, and graphs that characterize a wide variety of rectifier circuits, both single phase and three phase. Although out of print, the book should be available in any good engineering library. Also of unusual value if you read German is [2]. Both of these texts focus most of their attention on three-phase rectifier circuits. The classic papers on multi-pulse rectifiers containing commutating reactance are [3] and [4].

The case of finite dc inductance, and therefore a dc current that is not constant, is treated in [5] for both single-phase and three-phase rectifiers. Attention is placed on the consequence for power factor. The use of ac-side capacitance to compensate for the displacement component of ac current is described.

1. J. Schaefer, *Rectifier Circuits, Theory and Design*, (New York: John Wiley & Sons, 1965).
2. Th. Wasserab, *Schaltungslehre der Stromrichtertechnik*, (Berlin: Springer-Verlag, 1962).
3. R. L. Witzke, J. V. Kresser, and J. K. Dillard, "Influence of ac Reactance on Voltage Regulation of 6-Phase Rectifiers," *AIEE Transactions* 72, pt. 1 (July 1953), 244–253.
4. R. L. Witzke, J. V. Kresser, and J. K. Dillard, "Voltage Regulation of 12-Phase Double-Way Rectifiers," *AIEE Transactions* 72, pt. 1 (November 1953), 689–697.
5. A. W. Kelley and W. F. Yadusky, "Rectifier Design for Minimum Line Current Harmonics and Maximum Power Factor," in *IEEE Applied Power Electronics Conference (APEC) Proceedings* (1989), 13–22.

Problems

4.1   In Section 4.1.1 we calculated the power factor of the single-phase bridge by relating its characteristics to those of the half-bridge, which we calculated in Chapter 3. Calculate the power factor of the full-bridge by using the ac input variables of Fig. 4.3.

4.2   Figure 4.18 is a rectifier circuit known as a *voltage doubler*. It is often used to provide for dual voltage operation—for instance, from both the 110 V residential service in

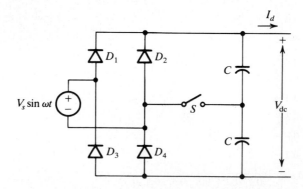

Figure 4.18   The voltage doubler circuit of Problem 4.2. The inductor is sometimes eliminated to reduce the cost of the circuit.

the United States and the 220 V service in Europe—thereby avoiding the expense of manufacturing separate products for the two markets.

(a) If the switch S is open, draw the equivalent circuit and calculate V_{dc}.

(b) Repeat (a) if S is closed.

(c) For what voltage V_{dc} must the equipment powered by this supply be designed?

4.3 If we wanted to operate the two series-connected 6-pulse bridges in Example 4.2 as a 6-pulse rectifier, that is, without the phase shift between the ac sources created by the Y/Y and Δ/Y transformer connections, could we operate the circuit without transformers?

4.4 What is $\langle v_d \rangle$ for the 12-pulse circuit of Fig. 4.11?

4.5 The two 6-pulse waveforms in Fig. 4.11, v_{d_1} and v_{d_2}, have the sixth harmonic of the ac frequency as their first nonzero component, whereas the 12-pulse waveform, v_d, has the twelfth harmonic as its first nonzero component. Show why the sixth harmonic is missing from v_d.

4.6 Calculate the power factor of the 3-phase bridge of Fig. 4.10(a) and compare it with that of a single-phase bridge.

4.7 An important advantage of high-pulse–number rectifiers is that as the pulse number goes up so does the order of the first nonzero harmonic in the line current. The higher frequencies are easier to filter from the line current. What are the first two nonzero harmonics in the line current of the 6-pulse rectifier of Fig. 4.10(a)?

4.8 One result of the phase-shifting transformer connections used in the 12-pulse rectifier of Fig. 4.11(a) is that the net primary line current has a lower THD than that of a 6-pulse circuit. Figure 4.19 shows the transformer connections necessary to generate the 6-phase source for the 12-pulse rectifier of Fig. 4.11(a).

(a) Sketch the line current $i_{a'}$ on the primary (Δ) side of the Δ/Y transformer.

(b) Sketch the primary-side line current i_A. What are the first two nonzero harmonics of this current?

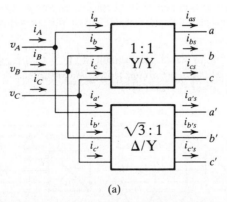

(a)

Figure 4.19 Transformer connections for the 12-pulse rectifier of Fig. 4.11(a). The line current harmonics for this connection are analyzed in Problem 4.8. (a) The interconnection of a Y/Y and Δ/Y 3-phase transformer to generate a 6-phase source.

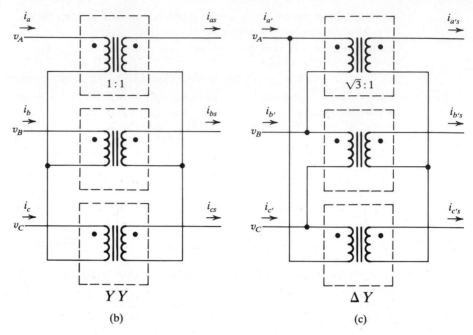

Figure 4.19 (cont.) (b) Three single-phase transformers connected to create a 3-phase Y/Y transformer. (c) Three single-phase transformers connected to create a 3-phase Δ/Y transformer.

(c) Compare the first two nonzero line current harmonics of this 12-pulse circuit to those of the 6-pulse circuit, as determined in Problem 4.7.

(d) Calculate the THD of the 12-pulse line current and compare it to the THD of the 6-pulse rectifier.

4.9 For a balanced 3-phase set of Y-connected voltage sources (equal frequencies, amplitudes, and phase displacements), derive the equivalent Δ-connected set.

4.10 Derive *(4.14)*.

4.11 An alternative to the 12-pulse circuit of Fig. 4.11(a) is shown in Fig. 4.20, where the two 6-pulse bridges are connected in parallel instead of series.

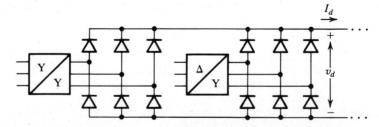

Figure 4.20 A 12-pulse rectifier consisting of two 6-pulse circuits connected in parallel. This circuit is the subject of Problem 4.11.

(a) Sketch v_d for this circuit and calculate $\langle v_d \rangle$.

(b) Compare $\langle v_d \rangle$ for this circuit with $\langle v_d \rangle$ for the 12-pulse circuit discussed in Example 4.2.

(c) What is the rms current in the diodes in this circuit, and how does it compare to that for the diodes in Fig. 4.11(a)? (The use of an *interphase transformer* to reduce the current ratings of the diodes in Fig. 4.11(a) is discussed in Problem 20.6.)

4.12 In some applications a rectifier is supplied from an ac current source instead of a voltage source. If the frequency is high enough, the junction capacitance of the rectifiers has an appreciable effect on the dc output current. Consider the circuit of Fig. 4.21, in which the diode junction capacitance C_j has been modeled as constant and in parallel with an ideal diode. Determine the load regulation characteristic for this circuit,

$$\langle i_d \rangle = f\left(\frac{V_d}{X_j I_s}\right) \qquad (4.19)$$

where X_j is the reactance of the junction capacitance $X_j = 1/\omega C_j$. The junction capacitance of a 20 A, 100 V Schottky diode is approximately 200 pF. At what frequency would you expect the effects of C_j to become important?

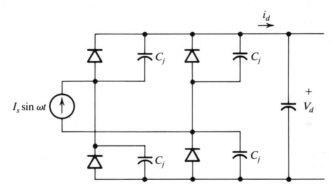

Figure 4.21   The current-fed bridge rectifier of Problem 4.12, including the diode junction capacitances, C_j, which have an important effect at high frequencies.

4.13 Plot v_{ab} and i_a for the 6-pulse rectifier with a Δ-connected source shown in Fig. 4.10(b), and compare it to the plot in Fig. 4.10(a). What are the power factors of these two circuits? Remember that the circuits are functionally indistinguishable.

4.14 Determine the source current i_{ab} in Fig. 4.10(b) and plot it relative to v_{ab}. What is the power factor at this source? (Assume there are no circulating current components in the Δ. That is, there are no current components present in the legs of the Δ that are not also present in the line currents i_a, i_b, and i_c.)

4.15 How is the operation of the 6-pulse circuit of Fig. 4.10(a) affected if two large capacitors of equal value are connected as shown in Fig. 4.22, with their centerpoint connected to the neutral of the Y-connected source? What is V_d?

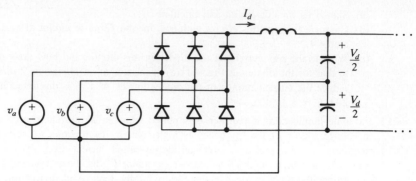

Figure 4.22   A 6-pulse rectifier with dc centerpoint connection. This circuit is the subject of Problem 4.15.

4.16   The mode III region of the regulation curve for a 6-pulse rectifier is presented in Fig. 4.17 without a supporting derivation. Show that the termination of this region, the point where $\langle v_d \rangle = 0$, occurs at $X_c I_d / V_s = 1/\sqrt{3}$. (The solution to this problem is not complicated. Assume four diodes are on all the time, so the line currents are sinusoidal and I_d circulates through a pair of series diodes. Then determine the condition under which the current in one of the diodes goes to zero. That is, determine the high current boundary of mode III.)

Chapter 5

Phase-Controlled Converters

THE rectifier circuits discussed so far are uncontrollable; that is, their output voltage is a function of system parameters and cannot be adjusted in response to parametric changes, such as variations in load (I_d) or ac voltage (V_s). You saw this characteristic in Fig. 3.12 for the single-phase circuit with freewheeling diode. However, we can make rectifier circuits controllable if we replace the diodes with devices known as *silicon controlled rectifiers* (SCR). This device is one member of a family of controllable switches known collectively as *thyristors*. Think of the SCR as a diode that will not conduct when forward biased until an appropriate control signal is applied to a third terminal, called the *gate*.

A rectifier circuit containing one or more SCR devices for controlling the output voltage is known as a *phase-controlled rectifier* or simply a *controlled rectifier*. In addition to providing voltage control, phase-controlled rectifiers can be designed to permit power to flow from the dc side to the ac side (as long as there is a source of energy on the dc side), opposite to the direction of power flow in a diode rectifier circuit. This process is called *inversion*, and a circuit operating in this way is called a *phase-controlled inverter*. In this chapter we discuss the behavior of both the phase-controlled rectifier and the phase-controlled inverter. These circuits differ primarily in the way they are controlled, so we refer to them as *phase-controlled converters* when no operating mode is specified.

We discuss the physics and detailed behavior of SCRs in Chapter 18. In this chapter you need to become familiar with only three characteristics of the SCR:

1. Turning on an SCR requires applying a positive signal to its gate while its anode–cathode voltage is positive.

2. Once on, an SCR remains on independent of the presence or absence of a signal at its gate, until its anode current goes to zero. That is, once on, an SCR behaves like a diode.

3. When the anode current goes to zero, a small amount of time, known as the *turn-off time t_q*, must elapse before a positive anode–cathode voltage can be reapplied to the SCR without its turning on.

79

Phase-controlled rectifiers have a broad range of applications. They are frequently used as preregulators to create a dc supply from which one or more dc/dc converters operate. In the electrochemical industry they are used to control the power to electrolytic processes, such as electroplating and chlorine production. Very high power SCRs are used to build the converters for high voltage dc transmission systems. The speed of dc motors and certain types of induction motors can be controlled using a simple phase-controlled rectifier. And the *ac controller*, a derivative of the phase-controlled rectifier circuit, is the basis for consumer products such as light dimmers and variable-speed appliances and tools. The ac controller generally contains a member of the thyristor family called a *bidirectional triode thyristor* (TRIAC), which is functionally equivalent to a pair of SCRs connected in antiparallel; that is, the TRIAC is a thyristor capable of controlled conduction in either direction.

Because the SCR is a regenerative device (once on, it keeps itself on) with a uniform current density in its on state, it can be manufactured with very high current and voltage ratings. The ratings of commercially available devices range from about 250 mA and 50 V to 4 kA and 6 kV.

5.1 SINGLE-PHASE CONFIGURATIONS

Single-phase controlled converters are generally used at power levels below about 10 kW. Some applications are dc motor drives, lighting controls, battery chargers, and as preregulators for ac motor drives and switching power supplies. Besides the importance of their applications, these single-phase circuits demonstrate all the basic phase-controlled converter concepts—such as commutation, regulation, inversion, and power factor—and therefore serve as good vehicles for introducing phase control.

5.1.1 Half-Wave Controlled Rectifier with Resistive Load

If a phase-controlled converter is supplying a resistive load, we know that it is operating as a rectifier, because there is no source of energy on the dc side to permit inversion. We first consider the circuit of Fig. 3.2 with the diode replaced by an SCR, as shown in Fig. 5.1. Note the circuit symbol for an SCR, which is a diode symbol with a third terminal, called the gate, connected to its cathode. We do not show the circuit that provides the gate signal, because in this chapter we are concerned only with the operation of the power circuit. (We generally consider the gate circuit to be part of the control circuitry and discuss it in Chapter 22.) Here, only the gate circuit's ultimate function at the gate terminal is important, and it is represented by either the presence or the absence of a signal.

If no gate signal is ever applied to the SCR of Fig. 5.1, the output voltage of the rectifier is zero because the SCR never turns on. Proper operation of this

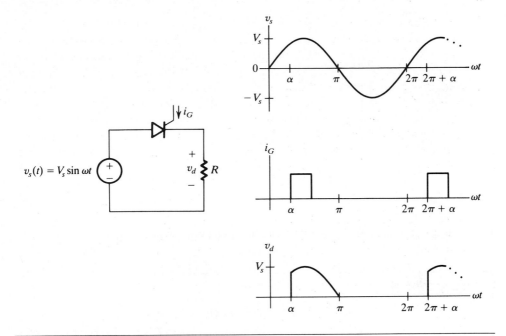

Figure 5.1 Half-wave phase-controlled rectifier with resistive load and its associated waveforms.

circuit requires application of the gate signal at some time during the period of normal diode conduction, that is, when the anode–cathode voltage of the SCR is positive. Figure 5.1 shows the gate signal being applied at an electrical angle α between 0 and π, which is during the interval in which a diode would conduct if it replaced the SCR. This signal is a current pulse (whose width is not critical) applied once every cycle of the line voltage waveform. Once the SCR is on, it remains on (even after the gate pulse ends) until its anode current goes to zero, which in this circuit occurs at the zero-crossing of the ac input voltage. The angle α is variously called the *firing angle*, the *angle of retard*, or the *delay angle*. Retard or delay is measured relative to the angle at which the device would have turned on if it were a diode, or multiples of 2π in this case. The resulting output voltage v_d is also shown in Fig. 5.1, and from it we can infer the effect of gate control. The important characteristic of this voltage is its dc component:

$$\langle v_d \rangle = \frac{V_s}{2\pi}(1 + \cos\alpha) = \frac{V_{do}}{2}(1 + \cos\alpha) \qquad (5.1)$$

where V_{do}, the maximum possible value of $\langle v_d \rangle$, is also the output of an equivalent diode rectifier. The voltage $\langle v_d \rangle$ as a function of α is known as the *control characteristic* of the rectifier and is shown in Fig. 5.2.

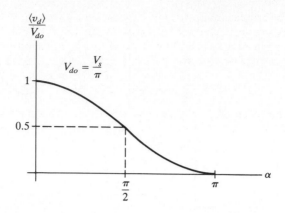

Figure 5.2 Control characteristic, *(5.1)*, for the rectifier of Fig. 5.1.

EXAMPLE 5.1

Linearizing the Phase-Control Characteristic

The problem with the control characteristic shown in Fig. 5.2 is that its incremental gain, $d\langle v_d\rangle/d\alpha$, approaches zero near full voltage or zero voltage. This is not a desirable feature in a system's transfer function, if we want to close a feedback loop around it.

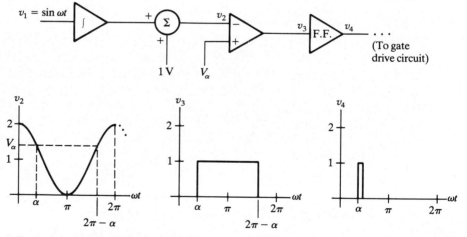

Figure 5.3 A functional block diagram of a circuit for linearizing the control characteristic of a phase-controlled rectifier.

Figure 5.3 illustrates a common technique for linearizing the control characteristic, that is, making its gain constant over the entire operating range. A scaled version of the ac input voltage, v_1, is first integrated, then given a dc offset to produce v_2. This voltage is then compared to the new controlling variable V_α to produce a pulse v_3, commencing at $\omega t = \alpha$. In Fig. 5.3, v_1 has an amplitude of 1 V, the dc offset is 1 V, and $0 < V_\alpha < 1$ V. The output of the comparator v_3 is high when $V_\alpha > v_2$, producing a pulse starting at $\omega t = \alpha$. In terms of V_α, the pulse starts when $V_\alpha = 1 + \cos \alpha$, or:

$$\alpha = \cos^{-1}(V_\alpha - 1) \tag{5.2}$$

Substituting this value of α into (5.1), we obtain the new control characteristic having constant gain:

$$\langle v_d \rangle = \frac{V_s}{2\pi}(V_\alpha) \tag{5.3}$$

$$\frac{dv_d}{dV_\alpha} = \frac{V_s}{2\pi} \tag{5.4}$$

The duration of the pulse v_3 is $2\pi - 2\alpha$, which can be too long, depending on the rectifier circuit. We can use a one-shot flip-flop, triggered on the rising edge of v_3, to create a shorter pulse v_4 starting at α, which is also shown in Fig. 5.3.

5.1.2 Full-Wave Phase-Controlled Bridge Rectifier

Like the simple single-diode rectifier of Fig. 3.2, the phase-controlled rectifier circuit of Fig. 5.1 is not very interesting in terms of practical applications. A much more useful circuit is the phase-controlled bridge shown in Fig. 5.4. In this circuit, one diagonal pair of thyristors conducts for some part of each half-cycle. We assume in the following discussion that we can model the load as a current source of value I_d.

We can explain the behavior of the output voltage waveform v_d by referring to the diode bridge rectifier, in which one pair of diodes is turned off by the action of the second pair turning on. In the phase-controlled rectifier, if the second pair does not turn on, the load current must continue to flow in the first pair, keeping it on. This means that Q_1 and Q_2 continue to conduct after the source voltage changes sign and becomes negative. They turn off when Q_3 and Q_4 receive a gate signal and turn on. Then the load current commutates instantaneously (because there is no commutating reactance in the circuit) to Q_3 and Q_4.

The average value of v_d, calculated from the waveform of v_d in Fig. 5.4, is

$$\langle v_d \rangle = \frac{1}{\pi} \int_\alpha^{\pi+\alpha} V_s \sin \omega t \, d(\omega t) = \frac{2V_s}{\pi} \cos \alpha = V_{do} \cos \alpha \tag{5.5}$$

This voltage as a function of α is plotted in Fig. 5.5(a). The most interesting characteristic of the relationship is that, for $\pi/2 < \alpha < \pi$, the average value of the output voltage is negative. Therefore, over this range of α, power is flowing from the dc side of the circuit to the ac side. As described in the introduction to this chapter, operation with this direction of power flow is called *inversion*.

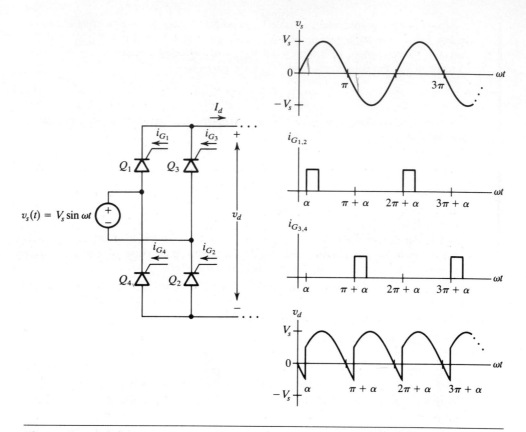

Figure 5.4   Phase-controlled bridge rectifier with no ac-side (commutating) reactance.

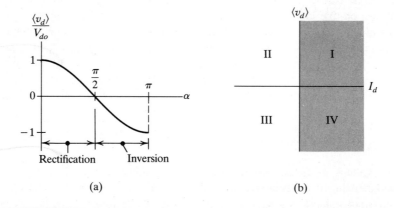

(a)

(b)

Figure 5.5   (a) Control characteristic and (b) quadrants of operation for the bridge rectifier of Fig. 5.4.

The circuit of Fig. 5.4 can operate only if the load current I_d is positive. Otherwise the thyristors could not conduct. The region of possible operation for this circuit, represented in the *output plane* (the plane defined by $\langle v_d \rangle$ and I_d), is shown as the shaded area in Fig. 5.5(b). Operation is possible only in quadrants I and IV of this plane; hence the circuit is called a *two-quadrant* converter. The region of the plane in which power is flowing from the dc side to the ac side (quadrant IV) is called the *inversion region*, and when operating here, the circuit is called an *inverter*.

EXAMPLE 5.2

A Magnet Discharge Application

Both magnetic resonance imaging (MRI) systems and magnetically confined fusion reactors require precise control of high-magnetic fields. In this example we use a phase controlled converter to control the magnetic field of a large electromagnet, such as might be used in these applications. Figure 5.6 shows the magnet modeled as the series combination of an inductance of value 0.5 H and a resistance of 2.5 Ω. The 60-Hz line voltage has a peak value of 2000 V. Assuming that the required steady-state magnet current is 400 A, what is the value of α that results in this value of current? How quickly can the magnet current be brought to zero from this value?

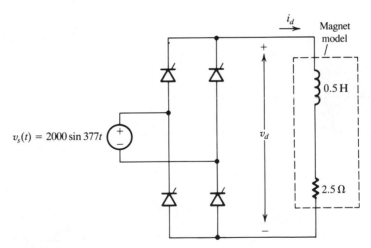

Figure 5.6 A phase-controlled converter being used to control the magnetic field of a high-field magnet.

The time constant of the load is $\tau = L/R = 0.2$ s $\gg 1/60$ Hz, so i_d has very little ripple. The inductor cannot support an average voltage in the steady-state, so $\langle v_d \rangle$ appears across R, and $\langle i_d \rangle$ is

$$\langle i_d \rangle = \frac{\langle v_d \rangle}{R} = \frac{2(2000)}{\pi R} \cos \alpha \qquad (5.6)$$

Setting $i_d = 400$ A and solving for α, we get

$$\alpha = 38.2°$$

This value of α results in $\langle v_d \rangle = 1000$ V.

The fastest way to bring the magnet current to zero is to remove the magnetic stored energy by operating the converter in the inversion region at the value of α that produces the maximum negative voltage. In this example the appropriate value of α is $\alpha = \pi$, which results in $\langle v_d \rangle = -1273$ V. (In Section 5.3.2 we show that an angle slightly less than $\alpha = \pi$ must be used.) We can now replace that part of the circuit consisting of the ac source and SCRs with an equivalent source whose value steps from $\langle v_d \rangle = 1000$ V to $\langle v_d \rangle = -1273$ V when we begin the magnet discharge. This equivalent circuit is shown in Fig. 5.7(a), and the resulting behavior of i_d is shown in Fig. 5.7(b). The current as a function of time is

$$i_d = -509 + 909e^{-t/\tau}$$

Setting $\tau = 0.2$ s and $i_d = 0$, we get

$$t_o = \tau \ln \frac{909}{509} = 0.12 \text{ s}$$

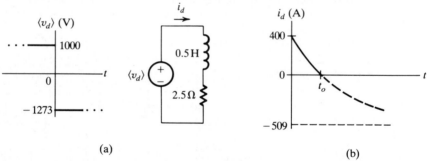

(a) (b)

Figure 5.7 (a) Equivalent circuit for the calculation of i_d. The line and SCRs have been replaced by the source $\langle v_d \rangle$. (b) The behavior of i_d in the equivalent circuit as a function of time.

5.1.3 Converter Power Factor

An unfortunate consequence of phase control is that it degrades the power factor from what it would be for a diode rectifier. Consider the bridge of Fig. 5.8, which also shows the line voltage and current. We assume that the load is a constant current source of value I_d and, for simplicity, no commutating reactance. We determine the power factor as a function of α.

In Chapter 3 we discussed the power factor for distorted waveforms and showed how to write the power factor as the product of two factors: a displacement factor k_θ and a distortion factor k_d. Figure 5.8 represents the case in which the line voltage is sinusoidal, but the current contains distortion (it is a square wave).

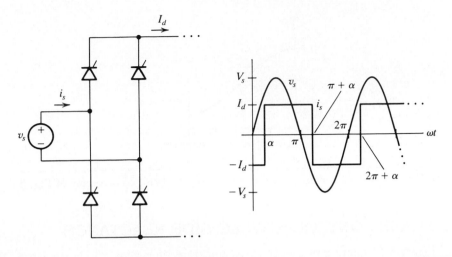

Figure 5.8 Phase-controlled bridge converter with a constant current load and no commutating reactance.

At this point we intuitively know what to expect. The square wave will retain its shape independent of α (although its amplitude might change). Therefore k_d will be constant. As α increases, the phase shift of the fundamental of the current waveform with respect to the voltage will increase, thus decreasing k_θ. Therefore we expect the power factor to decrease with increasing α.

We can write the power factor k_p as:

$$k_p = k_d k_\theta = \frac{I_{1\mathrm{rms}}}{I_{\mathrm{rms}}} \cos \theta_1 \qquad (5.7)$$

The fundamental component of a square wave has an amplitude equal to $4/\pi$ times the amplitude of the square wave. The fundamental is in phase with the square wave, which means that the fundamental component of the current lags the voltage by the firing angle α. For this problem, then, k_p becomes

$$k_p = \frac{(4I_d/\sqrt{2}\pi)\cos \alpha}{I_d} = 0.9 \cos \alpha$$

We plotted this function in Fig. 5.9 for $0 < \alpha < \pi$, which includes inversion operation in quadrant IV.

The power factor of this circuit varies from 0.9 to -0.9, depending on the value of α. If the circuit contained commutating reactance, the edges of the current waveform would not be so abrupt, and k_d would be slightly larger. However, k_θ would be smaller, because the fundamental component of the current would be shifted to the right of α (see Problem 5.6).

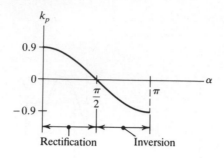

Figure 5.9   Power factor as a function of α for the bridge converter of Fig. 5.8.

5.2 PHASE CONTROL WITH AC-SIDE REACTANCE

Figure 5.10 shows a phase-controlled bridge rectifier with ac-side reactance. The effect of this reactance is identical to its effect in the diode rectifier, except that the commutation process is delayed by the electrical angle α. During the commutation interval u all four thyristors are on and $v_d = 0$. The source appears directly across L_c and is the commutating voltage. We can determine the commutating angle u as a function of α and I_d by relating the current change in L_c ($2I_d$) to the commutating voltage ($V_s \sin \omega t$) and u. That is,

$$2I_d = \frac{1}{\omega L_c} \int_{\alpha}^{\alpha+u} V_s \sin \omega t \, d(\omega t) \tag{5.8}$$

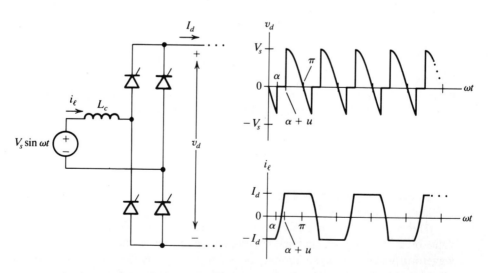

Figure 5.10   Phase-controlled bridge rectifier with ac-side (commutating) induc-tance L_c.

Solving this equation for u (and using $X_c = \omega L_c$) yields

$$u = \cos^{-1}\left(\cos\alpha - \frac{2X_c I_d}{V_s}\right) - \alpha \qquad (5.9)$$

We seldom need to determine the value of u explicitly, but (5.9) is useful in eliminating u from the expression for $\langle v_d \rangle$. We do so as follows:

$$\langle v_d \rangle = \frac{V_s}{\pi}\left\{\int_0^\alpha (-\sin\omega t)\, d(\omega t) + \int_{\alpha+u}^{\pi} \sin\omega t\, d(\omega t)\right\}$$

$$= \frac{V_s}{\pi}\{\cos\alpha + \cos(\alpha + u)\} \qquad (5.10)$$

But from (5.9), we know that:

$$\cos(\alpha + u) = \cos\alpha - \frac{2X_c I_d}{V_s}$$

which, when substituted into (5.10), gives

$$\langle v_d \rangle = \frac{2V_s}{\pi}\left\{\cos\alpha - \frac{X_c I_d}{V_s}\right\} = V_{do}\left\{\cos\alpha - \frac{X_c I_d}{V_s}\right\} \qquad (5.11)$$

Note the similarity between this expression and that for the diode bridge, (4.5), and also that the term $X_c I_d/V_s$, which is responsible for load regulation, is independent of the phase angle α. Figure 5.11 shows the family of regulation curves described by (5.11). The curves extend into quadrant IV, which represents inversion with $I_d > 0$ and $\langle v_d \rangle < 0$. We explain the diagonal boundary of the curves in this quadrant in Section 5.3.

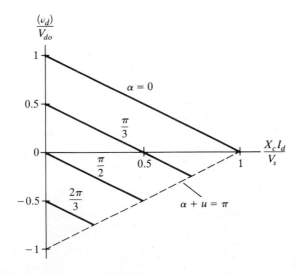

Figure 5.11 Family of regulation curves for the phase-controlled converter of Fig. 5.10.

EXAMPLE 5.3

Phase Control of a Battery Charger

Figure 5.12(a) shows a phase-controlled bridge rectifier being used as a battery charger. A large inductor L smooths the charging current I_B. The battery is modeled as a voltage source, V_B (= 72 V). Note that the polarity of V_B does not permit steady-state inversion, although a transient excursion into inversion is possible until the inductor L is discharged. We are interested in determining the charging current I_B as a function of α.

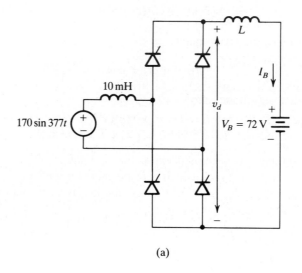

(a)

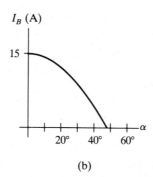

(b)

Figure 5.12   (a) Battery charger using a phase-controlled rectifier containing commutating reactance.   (b) Relationship between charging current and firing angle.

The average value of v_d in the periodic steady state must equal the battery voltage, that is, $\langle v_d \rangle = 72$ V. The reason is that there can be no average voltage across L in the periodic steady state. Furthermore, this condition is independent of α. Using (5.11) to state this condition, $\langle v_d \rangle = 72$ V, in terms of I_B and α, we obtain

$$I_B = \frac{V_s}{X_c} \cos \alpha - \frac{\pi V_B}{2X_c} = (45 \cos \alpha - 30) \text{ A} \qquad (5.12)$$

We plotted this function in Fig. 5.12(b). We can control the charging current between 0 and 15 A by varying α between approximately $48°$ and $0°$. Note that the only reason that we can control this circuit is the presence of commutating reactance. As we change α, the current changes just enough to cause u given by (5.9) to assume the value necessary to ensure the condition $\langle v_d \rangle = 72$ V. If X_c equaled 0, we would be forced to set α so that $\alpha = \cos^{-1}(\pi V_B/2V_s) = 48.3°$, and the charging current would be indeterminate. In practice, however, there is always some resistance in both L and the battery, providing a degree of controllability. (See Problem 5.12.)

5.3 INVERSION LIMITS

Inversion requires a source of energy on the dc side of the converter, as well as an ac voltage source on the ac side to commutate the thyristors. For the circuits we have been discussing, the dc source must be in quadrant IV of the output ($\langle v_d \rangle$ versus I_d) plane; that is, its current must be positive and its voltage negative, as these variables are defined in Fig. 5.4. Such a source might be a battery, a dc generator, or a solar photovoltaic array. A phase-controlled converter may also operate transiently in the inversion region to remove stored energy from the load, even though the load cannot provide steady-state power. Example 5.2 illustrated this mode of operation. In this section we explore more completely the behavior of phase-controlled converters operating in the inversion mode.

5.3.1 Commutation Failure

When a phase-controlled rectifier such as the bridge circuit shown in Fig. 5.13(a) operates in the inversion region, we must ensure that commutation is complete before the next zero-crossing of line voltage, at which point the commutating voltage changes polarity. If commutation were not complete by this time, the current in the SCR pair that is turning off would begin to increase, thus keeping the pair on. Figure 5.13(b) shows the commutation process during normal operation, when the load current properly commutates from one device pair to the other. Figure 5.13(c) shows commutation failure caused by changing the firing angle from α_1 in the first half-cycle to α_2 in the second. For the selected value of α_2, the load current in the SCR pair turning off at the end of the second half-cycle is not given sufficient time to reach zero before the commutating voltage changes sign and the current begins to increase. Nothing can be done to recover from this failure until the fourth half-cycle, when the control circuit returns the firing angle to α_1. We assume that the load current I_d is constant during this entire transient as it frequently is in practice.

Commutation failure is not necessarily catastrophic and may even be used as part of the gate control system. That is, we can set α for maximum inversion by sensing the onset of commutation failure and then backing α off slightly.

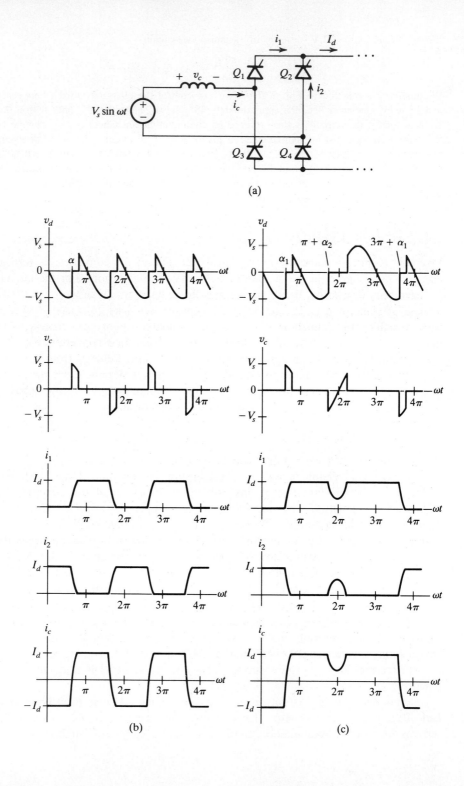

(a)

(b)

(c)

◄ **Figure 5.13** (a) A phase-controlled bridge converter with commutating inductance. (b) Voltages and currents during normal commutation for
constant firing angle α. (c) The same variables during commutation failure induced by moving the firing angle from α_1 to α_2 in the
second half-cycle. Recovery is achieved by moving the firing angle
back to α_1 during the fourth half-cycle.

EXAMPLE 5.4

Commutation Failure

In this example we assume that the magnet power supply of Example 5.2 (shown in Fig.
5.6) experiences commutation failure while discharging the magnet. The cause might be
control system noise that pushes α just beyond $180°$. The current at the time of failure
was 200 A. By how much does the magnet current increase if the converter recovers at the
earliest possible time?

The voltage v_d is shown in Fig. 5.14(a). Commutation failure occurs at $\omega t = 0$, and
recovery is not possible until $\omega t = 2\pi$. That is, the same pair of SCRs remain on for three
half-cycles. The magnet current, which had been decreasing, increases during the positive
half-cycle and reaches its maximum incremental excursion at $\omega t = \pi$.

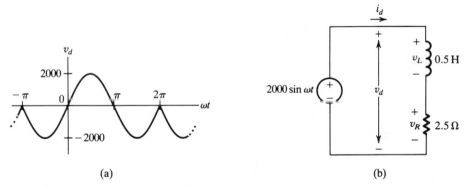

(a)   (b)

Figure 5.14 (a) The voltage v_d during commutation failure for the magnet power
supply of Example 5.2. (b) Equivalent circuit of the converter for
$-\pi < \omega t < 2\pi$.

The equivalent circuit for the converter during the three half-cycles in which the same
pair of SCRs conduct is shown in Fig. 5.14(b). At $\omega t = 0$, $i_d = 200$ A, and we can calculate
$i_d(\pi)$ by integrating the inductor voltage from 0 to π. Because the magnet time constant is
quite long, we assume that the resistive drop v_R will not change much between $\omega t = 0$ and
π; that is, it is constant at 500 V. Based on this assumption, $i_d(\pi)$ is

$$i_d(\pi) = 200 + \frac{1}{\omega L} \int_0^\pi (v_d - v_R)\, d(\omega t)$$

(5.13)

$$\approx 200 + \frac{2000}{(377)(0.5)} \int_0^\pi \sin \omega t\, d(\omega t) - \frac{500\pi}{(377)(0.5)} = 212.9 \text{ A}$$

Therefore the magnet current has increased by 12.9 A before proper inverter operation is restored. The result is a maximum change in v_R of 32 V, which, compared to 500 V, is probably small enough to ignore—as we did.

5.3.2 Margin Angle

Commutation failure is a hazard when we are trying to obtain the largest possible negative voltage from a phase-controlled rectifier operating in the inversion mode. From Fig. 5.13 you can see that for a given load current and commutating voltage, a certain minimum area is necessary under the commutation voltage waveform to ensure complete commutation of the load current. Thus we cannot exceed a certain value of α, say α_{max}, without causing commutation failure. The angle $\pi - \alpha_{max}$ is called the *margin angle*, γ.[†] It differs from u_{max} by the time required for the thyristors that are turning off to recover their ability to block (remain off) when a positive anode–cathode voltage is reapplied. This recovery time is given by the device manufacturer as the parameter t_q. The value of t_q can range from a few microseconds to several hundred microseconds, depending on thyristor size and the application for which it was designed. (We discuss this important parameter in more detail in Chapter 19.)

The margin angle for the single-phase bridge is equal to $u_{max} + \omega t_q$ and $\alpha_{max} = \pi - (u_{max} + \omega t_q)$. The boundary $\alpha + u = \pi$ in Fig. 5.11 represents the margin angle constraint, assuming that $t_q = 0$. For the bridge converter, we can determine the margin angle as a function of the reactance factor $(X_c I_d / V_s)$ by substituting $u_{max} = \gamma = \pi - \alpha_{max}$ into (5.9):

$$\gamma = \cos^{-1}\left\{ \cos(\pi - \gamma) - \frac{2X_c I_d}{V_s} \right\} + \gamma - \pi \qquad (5.14)$$

which implies that:

$$\cos(\pi - \gamma) - \frac{2X_c I_d}{V_s} = -1$$

and

$$\gamma = \pi - \cos^{-1}\left\{ \frac{2X_c I_d}{V_s} - 1 \right\} \qquad (5.15)$$

We plotted the margin angle γ as a function of the reactance factor and for $t_q = 0$ in Fig. 5.15.

[†] The term "margin angle" sometimes refers to the quantity $\pi - \alpha_{design}$, where α_{design} includes a safety factor to ensure against failure under any design condition, such as low line voltage or overloads. We do not use the term in this ambiguous sense.

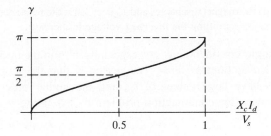

Figure 5.15   Margin angle γ as a function of reactance factor $X_c I_d / V_s$ for the bridge converter of Fig. 5.10.

EXAMPLE 5.5

Inversion with Commutating Reactance

Let's reconsider the magnet discharge calculation of Example 5.2 and introduce 1 mH of commutating reactance. The result is the circuit of Fig. 5.16. What is the maximum negative magnet voltage at the onset of inversion?

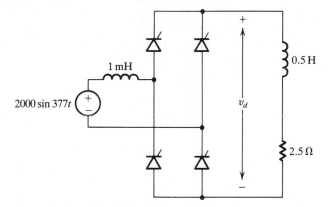

Figure 5.16   Phase-controlled magnet power supply containing commutating reactance.

   We know the reactance factor of this converter at the onset of inversion, so we can obtain the margin angle from *(5.15)*. First, we calculate the reactance factor:

$$\frac{X_c I_d}{V_s} = \frac{(377)(10^{-3})(400)}{2000} = 0.075 \qquad (5.16)$$

which, from *(5.15)*, gives $\gamma = 31.9°$. Therefore $\alpha_{max} = 180° - \gamma = 148.1°$, and the inverting voltage is

$$\langle v_d \rangle = \frac{V}{\pi} [\cos \alpha_{max} + \cos(\alpha_{max} + \gamma)] = -1177 \text{ V} \qquad (5.17)$$

Note that, as the magnet discharges and I_d decreases, the reactance factor also decreases, and α_{max} increases, permitting the inverting voltage to increase.

Margin-angle considerations are especially important in 3-phase, high-power phase-controlled rectifier systems. The reason is that the large thyristors used in these systems have large values of t_q, and the higher order modes of 3-phase rectifiers aggravate the commutation problem by extending u.

5.4 PHASE-CONTROLLED 3-PHASE CONVERTERS

A phase-controlled 3-phase bridge converter is shown in Fig. 5.17(a). Its operation is similar to the 3-phase rectifier of Fig. 4.10(b), except that commutation is delayed by α from when commutation occurs between diodes in Fig. 4.10(b). We can calculate $\langle v_d \rangle$ from the waveform of v_d in Fig. 5.17(b). The result is

$$\langle v_d \rangle = \frac{3V_s}{\pi} \cos \alpha \qquad (5.18)$$

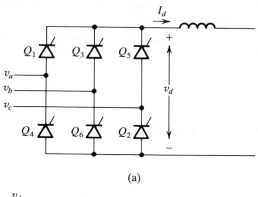

(a)

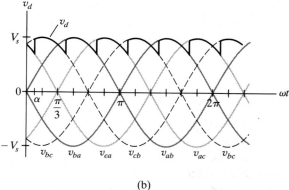

(b)

Figure 5.17   (a) A phase-controlled 3-phase bridge converter.   (b) The line-to-line voltages and the voltage v_d resulting from a firing angle α.

Notes and Bibliography

Written by some of the pioneers in the development of the SCR, [1] contains a wealth of information about the device and its applications. Chapter 8 of [1] discusses phase control circuits, and contains a very extensive chart summarizing possible circuits and their characteristics. The authors also present alternatives to the single-phase bridge circuit in addition to the hybrid bridge you are asked to analyze in Problem 5.8. Various gate trigger circuits and their characteristics also are presented.

A good chapter (3) on phase-controlled rectifiers also is included in [2]. The authors do an especially good job of analyzing 3-phase, phase-controlled circuits containing commutating reactance. Waveform diagrams are clear and extensively employed. The most extensive coverage of phase-controlled rectifiers is given in Chapter 2 of [3]. Such issues as regulation through all three modes, power factor, and transient behavior at turn-on are included. The discussion of power factor of nonsinusoidal waveforms is particularly comprehensive. Rissik's work [4] is one of the earliest and most comprehensive on the operation of rectifiers. Although the rectifying switch is an arc tube, the analyses are still relevant to modern rectifier circuits. The mathematical developments are detailed and cover not only single- and polyphase-controlled and uncontrolled circuits, but also the cycloconverter.

A method of improving the power factor of controlled rectifiers with pulse numbers greater than 6 is described in [5]. Six-pulse groups are fired unsymmetrically—one of the groups is fully advanced or fully retarded to minimize its reactive power requirement and the other group is controlled. A circuit to control the neutral current in a Y-connected source to improve power factor is the subject of [6].

1. F. E. Gentry, F. W. Gutzwiler, N. Holonyak, and E. E. Von Zastrow, *Semiconductor Controlled Rectifiers: Principles and Applications of p-n-p-n Devices* (New Jersey: Prentice-Hall, Inc., 1964).
2. B. D. Bedford and R. G. Hoft, *Principles of Inverter Circuits* (New York: John Wiley & Sons, Inc., 1964).
3. G. Möltgen, *Converter Engineering* (Chichester: John Wiley & Sons Ltd., 1984).
4. H. Rissik, *The Fundamental Theory of Arc Convertors* (London: Chapman & Hall, 1939).
5. W. McMurray, "A Study of Unsymmetrical Firing for Phase Controlled Converters," *IEEE Trans. Industry Applications* (May/June 1972): 289–295.
6. V. R. Stefanovic, "Power Factor Improvement with a Modified Phase-Controlled Converter," *IEEE Trans. Industry Applications* (March/April 1979): 193–201.

Problems

5.1 Derive a phase-controlled converter circuit capable of four-quadrant operation and specify the necessary gate drives. (Such a circuit is known as a *dual converter*.)

5.2 What fraction of the energy removed from the magnet in Example 5.2 returns to the source? Does the ratio of energy recovered (returned to the source) to energy dissipated in R depend on the rate at which energy is removed from the 0.5 H inductor?

5.3 How would the circuit of Fig. 5.6 behave if i_d were not zero and the gate signals to all four SCRs were removed? Sketch $i_d(t)$ and $v_d(t)$ under this condition.

5.4   Derive and plot $\langle v_d \rangle = f(\alpha)$ for the phase-controlled rectifier with freewheeling diode shown in Fig. 5.18. Is this circuit capable of inversion?

5.5   An SCR replaces the diode in the rectifier of Fig. 5.18 as shown in Fig. 5.19. Determine and plot the control characteristic for this converter.

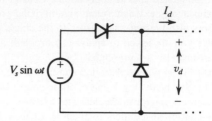

Figure 5.18   Half-wave phase-controlled rectifier with freewheeling diode analyzed in Problem 5.4.

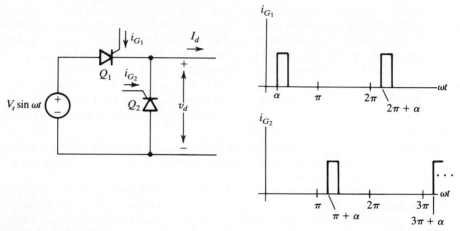

Figure 5.19   Phase-controlled converter whose control characteristic is determined in Problem 5.5

5.6   Repeat the power-factor calculation of Section 5.1.3 when ac-side reactance is present. A computer will help you get an exact solution because of the sinusoidal rising and falling edges of the line current waveform. But you can obtain a reasonably close approximation by assuming that the waveform is trapezoidal.

5.7   An SCR with a specified t_q of 150 μs is used in the 400-Hz converter of Fig. 5.20. What is the maximum magnitude of the voltage, $|\langle v_d \rangle|$, at which this converter can invert?

5.8   In certain applications a controlled bridge rectifier consisting of two SCRs and two diodes is adequate. Such a *hybrid bridge* circuit is shown in Fig. 5.21. Determine and plot the dc voltage $\langle v_d \rangle$ as a function of α for $0 < \alpha < \pi$. Why is this circuit incapable of inversion? Compare the power factor of this hybrid bridge to that of the four-SCR bridge of Fig. 5.4.

5.9   Figure 5.22 shows a phase-controlled bridge converter with commutating resistance R_c. Determine the regulation curves for this converter and compare them with the curves in Fig. 5.11 for the converter with commutating inductance.

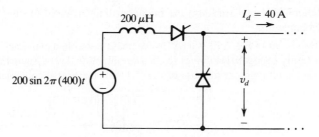

Figure 5.20 A 400-Hz converter analyzed in Problem 5.7. The SCRs have a specified t_q of 150 μs.

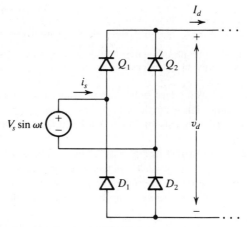

Figure 5.21 A hybrid phase-controlled bridge rectifier circuit, discussed in Problem 5.8.

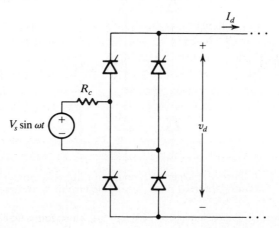

Figure 5.22 Phase-controlled bridge converter with commutating resistance R_c for Problem 5.9.

5.10 Determine (approximately) the minimum time required to discharge the magnet of Example 5.5.

5.11 The circuit of Fig. 5.23 is a half-wave phase-controlled converter. Determine and plot a family of regulation curves for this circuit. How do the regulation characteristics of this circuit compare to those of Fig. 5.11 for the bridge converter?

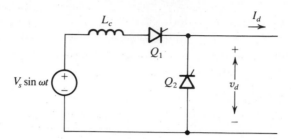

Figure 5.23 Half-wave phase-controlled converter, the subject of Problem 5.11. The sequence of gate drives is as shown in Fig. 5.19.

5.12 Batteries, such as the lead–acid storage battery used in automobiles, always contain some resistance. For example, a 12-V battery with a short-circuit (approximately the "cold-cranking") capacity of 240 A can be modeled as having an internal resistance of 50 mΩ.

  The battery charger of Fig. 5.24 is like the one discussed in Example 5.3, except that the battery is now modeled as the series connection of a 0.240-Ω resistance and a 72-V voltage source. Determine and plot the charging current I_B as a function of α. How does this curve differ from that in Fig. 5.12(b)?

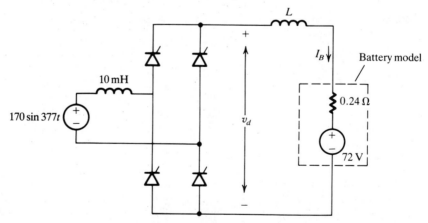

Figure 5.24 A battery charger with commutating reactance charging a battery containing internal resistance. This circuit is analyzed in Problem 5.12.

5.13 Derive the relationship $\langle v_d \rangle = f(X_c I_d / V_s)$, which forms the inversion limit $\alpha + u = \pi$ in Fig. 5.11.

5.14 Reconcile a power factor of zero ($\alpha = \pi/2$) with what is happening on the dc side of the converter of Fig. 5.8.

5.15 The ac controller of Fig. 5.25(a) is supplying a resistive load, perhaps a heater.

(a) Sketch $i_o(t)$ for this circuit, assuming that $\alpha \neq 0$.

(b) Determine and plot v_{orms} as a function of α for this circuit.

In many applications (incandescent light dimming, for example) the large values of di_o/dt present in the circuit of Fig. 5.25(a) cause both radio frequency interference (RFI) and acoustic noise (lamp "buzz"). To reduce these effects, the manufacturer often places an inductor in series with the resistive load, as shown in Fig. 5.25(b).

(c) For what range of α is the conduction of the controller discontinuous? Determine and carefully sketch i_o for some value of α within this range. Why are the RFI and acoustic noise both reduced?

(d) Determine and sketch i_o for α outside the range determined in (c).

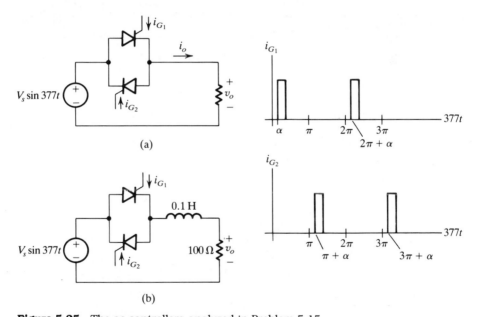

(a)

(b)

Figure 5.25 The ac controllers analyzed in Problem 5.15.

5.16 Commercial buildings are often supplied with 3-phase 208-V_{rms} line-to-line electrical service. Lighting circuits are then connected between line and neutral, giving $V_{\ell n} = V_{\ell\ell}/\sqrt{3} = 120\ V_{rms}$. If light dimmers, which can be thought of as ac controllers, are connected to the lights, the result is the equivalent circuit of Fig. 5.26.

Assume that the lamps are incandescent and can be modeled as resistors, that each phase has the same number of lamps, and that the dimmers are adjusted to give the same light levels on all circuits (α is the same for all dimmers). Sketch the neutral current i_n and determine the fundamental frequency of this current.

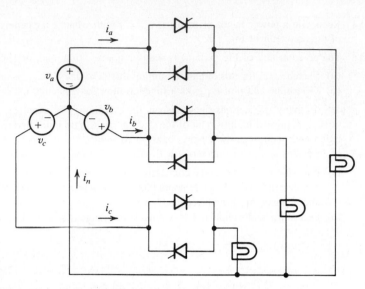

Figure 5.26   The 3-phase lighting control circuit for Problem 5.16.

5.17   Commutation inductance is added to the phases of the 3-phase converter of Fig. 5.17.

(a)   Sketch v_d for $\alpha = 30°$ and $u = 15°$. What is $\langle v_d \rangle$?

(b)   Repeat (a) for $\alpha = 100°$ and $u = 20°$. In this case (inversion) is $|\langle v_d \rangle|$ less than or greater than $|\langle v_d \rangle|$ for $u = 0$, that is, for a converter with no commutating reactance?

Chapter 6

High-Frequency Switching dc/dc Converters

HIGH-FREQUENCY switching converters are power circuits in which the semiconductor devices switch at a rate that is *fast compared to the variation of the input and output waveforms*. Certain characteristics of high frequency converters differentiate them from the converters described in other chapters of Part I. First, unlike the rectifiers and inverters of Chapters 3–5, the semiconductor devices in a high-frequency converter cannot use the reversal of the external waveforms to turn themselves off. Second, unlike the resonant converters of Chapter 9, the difference between the switching frequency and the frequency of the external waveforms is large enough to permit the use of low-pass filters to remove the unwanted switching frequency components.

High-frequency switching converters are used most often as interfaces between dc systems of different voltage levels. Our discussion in this chapter and Chapter 7 therefore is in the context of this application. These converters are known as *high-frequency dc/dc converters*, and examples of their use are the power supplies in computers and other electronic equipment.

High-frequency switching converters can also be used as an interface between dc and ac systems. Although we look at the unique aspects of this application in Chapter 8, much of what we say about dc/dc converters in this chapter applies to dc/ac converters as well. The reason is that, in the dc/ac application, the switching frequency is usually so much higher than the ac-output (or -input) frequency that we can consider the variables at the ac port to be constant over many switching periods. In other words, for times on the order of the switching period, we can view the dc/ac converter as a dc/dc converter.

In this chapter we present the structure and operation of high-frequency switching converters. Rather than just categorizing a large number of circuits, we develop the fundamental concepts and topological relationships on which all such circuits are based. Again, you will see that the specification of a topology does not simultaneously specify which port is the input and which is the output. Only when we specify the implementation and control of the switches do the ports obtain identities as input or output.

6.1 THE DC/DC CONVERTER TOPOLOGY

The simplest form of a switching dc/dc converter is that of Fig. 6.1. The switch opens and closes at a frequency $1/T$, with the ratio of the on-time to the period defined as D, the *duty ratio*. The resulting load voltage v_2 is a *chopped* version of the input—a series of pulses having an amplitude of V_1 and an average, or dc, value of DV_1. But this dc value comes with a substantial amount of ripple, which is present not only in the load voltage v_2 but also in the source current i_1. Few applications can make use of this dc component in the presence of so much distortion. The high frequencies contained in the ripple can cause both conducted and radiated interference with other apparatus, such as computers or communications equipment. Moreover, some loads, such as integrated digital circuits, function properly only if operated from a dc power supply with little ripple. Therefore a dc/dc converter generally has terminal voltages and currents that deviate only minutely and instantaneously from their dc values. This restriction means that we have to modify the elementary topology of Fig. 6.1. We develop these modifications by first exploring the topological constraint imposed by the requirement of nearly constant terminal variables.

Figure 6.2(a) shows a converter connecting two systems whose terminal voltages and currents are dc with the values shown. Whatever is in the box has to produce terminal variables that are free from pulsations or ripple. The difference between the input and output voltages, $V_i - V_o = 50$ V, must drop across an element connected between the ports, and the box must contain a shunt element to provide a path for the difference between the input and output currents, or $I_o - I_i = 5$ A. This minimal connection of elements is shown in Fig. 6.2(b). Note that the power absorbed by the series element is equal to the power supplied by the shunt element, so the box is lossless. But the energy must be transferred from the shunt to the series element. It is difficult to imagine a way to implement this simple two-element topology.

A circuit element that can support both a nonzero average voltage and a nonzero average current without dissipating energy is a switch. Furthermore, we can control the average value of its variables by varying the ratio of its on to off times. The problem with using switches as the shunt and series elements, however,

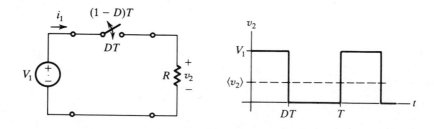

Figure 6.1   The simplest form of a switching dc/dc converter.

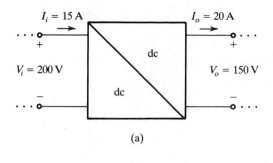

(a)

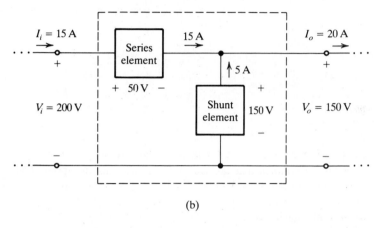

(b)

Figure 6.2   (a) A converter connecting two dc systems.   (b) The minimal necessary topology for performing the conversion function of (a).

is that instantaneous values of the input current and output voltage will differ from
their average values. But we can extract their average values by using low-pass
filters. By defining the converter to include these filters, we create an interface that
produces the desired terminal variables.

We now determine the necessary filter elements. First, we assume that the input
voltage and output current are constant at 200 V and 20 A. We then add the elements
necessary to make their covariables, the input current and output voltage, constant
at 15 A and 150 V. The result is terminal variables that are then consistent with
one another. For example, if the external output network were a resistor, whatever
we did to produce a dc output voltage would also produce a dc output current,
which is consistent with our initial assumption. (Establishing assumptions and then
showing consistency is a circuit-analysis, as well as circuit-synthesis, technique.)

Figure 6.3 shows switches as the shunt and series elements in a converter
connecting a dc voltage at the input, $V_i = 200$ V, to a dc current at the output,
$I_o = 20$ A. The switches operate at a constant switching frequency $1/T$, and are
controlled so as to be complementary. That is, when one is closed, the other is

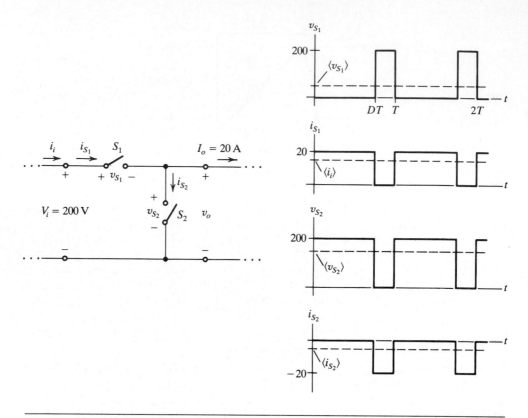

Figure 6.3 Two dc systems connected by a lossless interface, utilizing switches as the shunt and series elements.

open (so that a path is always provided for the load current and the input voltage is never shorted). The waveforms of v_{S_1} and v_{S_2} show that switches S_1 and S_2 are on for times DT and $(1 - D)T$, respectively, giving average values of:

$$\langle v_{S_1} \rangle = (1 - D)V_i \qquad (6.1)$$

and

$$\langle v_{S_2} \rangle = DV_i \qquad (6.2)$$

Because $\langle v_o \rangle = \langle v_{S_2} \rangle = 150$ V, we must control the switches so that $D = 0.75$. The average voltage across the series switch is 50 V, or the difference between the average input and output voltages that we said was necessary when discussing Fig. 6.2. Similarly, we can show that the average current in the shunt switch is the difference between the average input and output currents, as required, or:

$$\langle i_{S_2} \rangle = \langle i_i \rangle - I_o = -(1 - D)I_o = -5 \text{ A} \qquad (6.3)$$

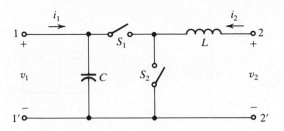

Figure 6.4 Simplest topology for a high-frequency dc/dc converter with low-pass filtering at its ports.

We assume that the input voltage contains no ripple. However, the input current contains substantial ripple caused by switching, as shown in Fig. 6.3. Although the output current is ripple-free, the output voltage is not. To obtain the desired ripple-free input current and output voltage, we must insert low-pass filters at the input and output. In their simplest form, these filters consist of a shunt capacitor at the input and a series inductor at the output. By making these elements very large, small values of external input network impedance and output network conductance will result in acceptably small ripples in all the terminal variables. The resulting high-frequency dc/dc converter topology is shown in Fig. 6.4. We have not identified input and output ports, because energy can flow in either direction, depending on how we control the switches.

EXAMPLE 6.1

Input Filter Effectiveness

Figure 6.5 shows a converter whose external input circuit is modeled by a voltage source V_i behind a resistor R. The resistor might represent wiring resistance or the internal resistance of a battery. We assume that the converter inductor L is large enough to make the output current constant at I_o. What is the peak–peak amplitude of the ripple on the input current i_i?

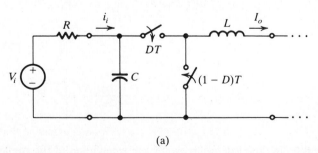

(a)

Figure 6.5 (a) Switching converter with a low-pass filter to eliminate ripple from the current flowing through the input circuit.

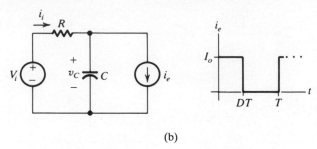

(b)

Figure 6.5 (cont.)   (b) The switches, inductor, and external output network replaced by an equivalent source, i_e.

If the switches are operating at a fixed duty ratio, D, we can replace the circuit to the right of the capacitor by the equivalent source i_e, as shown in Fig.6.5(b). Furthermore, being concerned only with the ripple component of i_i, we can replace i_e with its ac component i'_e. The circuit that we analyze is shown in Fig. 6.6(a), where i'_i is the ac component of the input current and the voltage source has been modeled as an incremental short.

If the low-pass filter is effective, that is, $\tau = RC \gg T$, then almost all the ripple current will pass through C. Therefore the capacitor voltage ripple is triangular with a peak–peak amplitude of:

$$\Delta v_C = \frac{(1-D)I_o}{C} DT \tag{6.4}$$

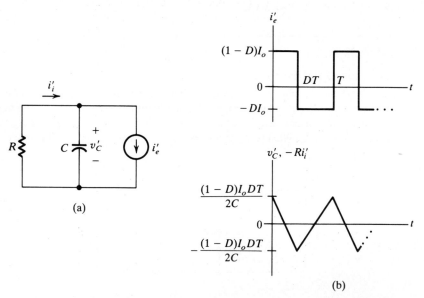

(a)

(b)

Figure 6.6   (a) Equivalent circuit for the calculation of the ripple component of the input current in the circuit of Fig. 6.5.   (b) Branch variables for the circuit of (a), assuming that $RC \gg T$.

as shown in Fig. 6.6(b). The peak–peak amplitude of the input ripple current thus will be

$$\Delta i_i = \frac{\Delta v_C}{R} = \frac{(1-D)I_oDT}{RC} \qquad (6.5)$$

From (6.5) we see that if $C \to \infty$ and R is arbitrarily small (but nonzero), then $\Delta i_i \to 0$.

In many cases, we add additional energy storage elements to the external systems to enhance filtering. For instance, if the external network connected to port 1–1' in Fig. 6.4 has a very low ac impedance, C will have to be very large. However, if we increase the impedance of the external network by placing an inductor in series with it, C can be much smaller. Similarly, if the external impedance at port 2–2' is very high, L will have to be large unless we place a capacitor in parallel with the port to reduce its impedance. These additional elements on each side of the converter make the filters second order. Third, fourth, and even higher order filters are used if the need to attenuate the ac components is great enough. These elements do not change the basic function of the original capacitor and inductor, so we do not include them in the topologies of this chapter.

6.2 THE CANONICAL SWITCHING CELL

We refer to the minimal practical dc/dc converter topology developed in Section 6.1 and shown in Fig. 6.4 as the *canonical switching cell*.[†] We redrew it symmetrically in Fig. 6.7, using a single-pole double-throw switch, which satisfies the condition that the switches be neither on nor off simultaneously. The canonical cell is the

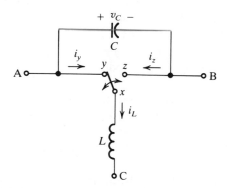

Figure 6.7 The canonical switching cell, the basic building block of all high-frequency dc/dc converters.

[†]E. Landsman, "A Unifying Derivation of Switching dc–dc Converter Topologies," *PESC Record*, IEEE, 1979, pp. 239–243.

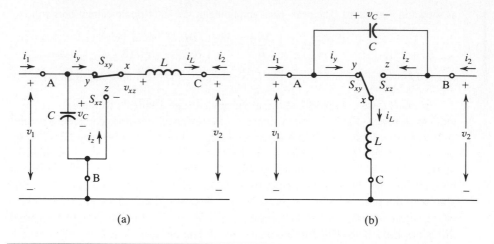

Figure 6.8   The two unique connections of the canonical cell between two dc systems: (a) the direct converter;   (b) the indirect converter.

basic building block for all high-frequency switching converters. The distinctions among converters arises mainly from the way in which the external systems are connected to the cell. These connections determine both the input/output conversion ratios and the levels of current and voltage stress imposed on the cell's components.

We may connect the three-terminal canonical cell of Fig. 6.7 between two dc systems in one of three ways. The connections differ with respect to the cell node that we make common to both ports. Because of the symmetry of the cell, however, the use of node A or B as a common node results in indistinguishable topologies. Thus there are only two unique connections to consider. If node A or B is common, the resulting topology is that of Fig. 6.8(a), which is identical to Fig. 6.4. Figure 6.8(b) shows the topology that results from the use of node C as the common terminal.

In the connection of Fig. 6.8(a), there is a direct dc path between the input and output ports when switch S_{xy} is conducting. For this reason we call this the *direct converter* connection. In the connection of Fig. 6.8(b), there is no dc path between the ports in any switch state. Therefore we call this the *indirect converter* connection. The use of the names *direct* and *indirect* highlights the essential difference among various specific high-frequency dc/dc converter circuits. The presence of transformers or additional filter elements does not negate this distinction.

6.3 THE DIRECT CONVERTER

We now consider the direct converter of Fig. 6.8(a) in detail. We are particularly interested in the dc conversion ratio and the implementation of the switches. Note that we cannot define the ports as input or output until we specify the semiconductor devices to be used as the switches.

6.3.1 The dc Conversion Ratio of the Direct Converter

In the discussion that follows we assume that L and C are large enough to eliminate switching frequency components from the terminal variables v_1, i_1, and v_2, i_2. Therefore we specify the terminal variables by using uppercase letters to indicate dc quantities. For example, $i_1(t) = \langle i_1(t) \rangle = I_1$.

If we turn series switch S_{xy} on and wait for a long time, voltages V_1 and V_2 are equal, and currents I_1 and I_2 are equal in magnitude but opposite in sign. But if we turn shunt switch S_{xz} on for a long time, V_2 and I_1 both are zero even though V_1 and I_2 are finite. The ratios V_2/V_1 and $-I_1/I_2$ are unity in the first case and zero in the second. When the switches are turned on and off at a high frequency, the ratios will have a value between these two extremes. As a result, we can make our first statement about the voltage conversion ratio of a direct converter: The two external voltages are of the same sign. As we show in Section 6.4.1, these voltages are of opposite polarities in the indirect converter. Consideration of power balance now indicates that the current conversion ratio for a direct converter is the negative inverse of its voltage conversion ratio.

The values of the conversion ratios, V_2/V_1 and $-I_1/I_2$, depend on the duty ratio of the switches, and we can determine them from the waveforms of v_{xz} and i_y shown in Fig. 6.9. For the direct converter, we generally express the conversion ratios in terms of the duty ratio D of the series switch. As the inductor cannot support an average voltage, $V_2 = \langle v_{xz} \rangle = DV_1$. And as the capacitor cannot carry an average current, $I_1 = \langle i_y \rangle = D(-I_2)$. Solving these constraints for the conversion ratios gives

$$\frac{V_2}{V_1} = D \qquad\qquad (6.6)$$

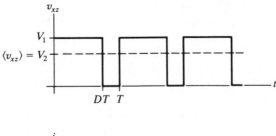

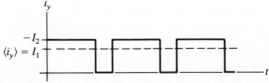

Figure 6.9   Waveforms of the switch variables v_{xz} and i_y for the direct converter of Fig. 6.8(a).

and

$$\frac{I_2}{I_1} = -\frac{1}{D} \qquad (6.7)$$

From *(6.6)* and *(6.7)* we can show that the average terminal powers are equal and opposite, that is, $V_1 I_1 = -V_2 I_2$. We expected this result because the ideal canonical cell is lossless. Note that we still have not specified the direction of power flow; the product $V_1 I_1$ is not necessarily positive, and $V_2 I_2$ is not necessarily negative.

So far, we have assumed that the converter is switching at a constant frequency, $1/T$, as we vary D to control the conversion ratio. Known as *constant frequency control*, it is only one of the ways by which we can operate the high-frequency converter. We can also control the conversion ratio by turning the series switch on for a fixed time and varying its off time, called *constant on-time control*. Alternatively, we could hold the off time of the series switch constant and vary its on time, called *constant off-time control*. Constant on-time and off-time control both result in changes in the switching frequency as we vary the conversion ratio. In all three cases, the conversion ratio depends only on the percentage of the period the switches are in one state or the other.

Because of the relationship between the two external voltages given by *(6.6)*, the direct converter has come to be known as either a *down* (or *buck*) *converter* or an *up* (or *boost*) *converter*, depending on the direction of power flow. If power flows toward the lower voltage system (V_2), it is a down converter. If power flows the other way, it is an up converter. These are the names commonly used when we discuss these circuits, but they do not describe two separate converters. As we show in Section 6.3.2, the desired direction of power flow only affects how we implement the switches.

Note that these names are based on how the *voltages* of the two external systems are related. Although in many applications the voltage variable is of prime concern, in some situations the current variable is more important. In these cases, the circuit we called a voltage down converter is actually a current up converter, and the voltage up converter is a current down converter. Thus the popular names of these circuits are misleading unless you understand the commonality of topology and function implied by the single (direct converter) connection of the canonical cell.

6.3.2 Implementation of Switches

When we replace the generalized switches we have used so far with semiconductor devices, we also specify the direction of power flow in the dc/dc converter. The reason is that semiconductor switches, specifically diodes and transistors, are unilateral devices; that is, they conduct current in only one direction and block voltage of only one polarity. The exception is the SCR, whose appropriateness to this application we consider separately. Therefore, when we specify particular semiconductor devices for the switches, we fix the polarities of the terminal vari-

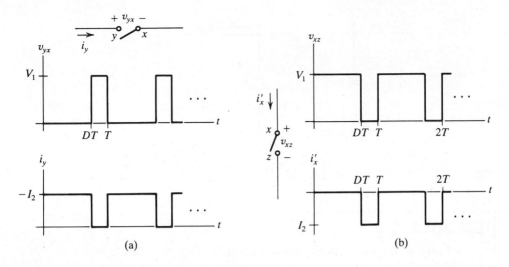

Figure 6.10   Voltage and current waveforms for the series and shunt switches in the direct converter of Fig. 6.8(a) when it is operating in the down mode: (a) series switch;   (b) shunt switch.

ables, defining the direction of power flow. Alternatively, when we specify the direction of power flow, we define the semiconductor devices.

The Down Converter   To illustrate this relationship between switch type and direction of power flow, let's consider the direct converter of Fig. 6.8(a) operating in the down mode. That is, power is flowing from the higher voltage side (V_1) to the lower voltage side (V_2). In this case, $V_1 > 0$, $I_1 > 0$, $V_2 > 0$, $I_2 < 0$; switch voltages and currents are shown in Fig. 6.10, where the shunt switch component of i_x is designated as i'_x $(i_x = -i_z$ in Fig. 6.8a).

The current and voltage of the series switch, i_y and v_{yx}, are both positive. The most appropriate device with this characteristic is the transistor. If the voltage or current is higher than a transistor can handle, we can use a thyristor. But the use of a thyristor introduces additional complications, such as the need for a commutation circuit to turn off the SCR (see Chapter 22). For historical reasons, a dc/dc converter utilizing thyristors is known as a *chopper*.

The current and voltage of the shunt switch, i_x and v_{xz}, are of opposite polarities, as are those of a diode. It may not be immediately apparent that an uncontrollable device is suitable for the shunt switch. However, the diode is forced on and off by the switching of the series switch. When the series switch closes, V_1 reverse biases the diode, and when the series switch opens, the continuity of current in L forces the diode to conduct. The function and operation of the shunt switch is identical to that of the freewheeling diode in the half-wave rectifier of Fig. 3.6. The down converter—with the series and shunt switches replaced by a transistor and diode, respectively—is shown in Fig. 6.11.

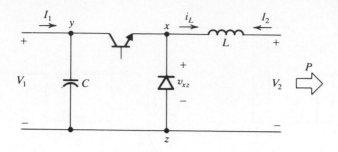

Figure 6.11   Switch implementation for the direct down converter.

EXAMPLE 6.2

A Down Converter with Common Positives

At times we need to connect two dc systems so that their positive terminals are common. This is the case in some automotive applications—for instance, connecting the positive terminal of the battery to ground (the car body). In this example, we design a down converter for such an application.

Again, we consider the general direct converter of Fig. 6.8(a). The "down" constraint requires that $V_1 I_1 > 0$ and $V_2 I_2 < 0$. The common positive requirement implies $V_1 < 0$ and $V_2 < 0$. Therefore $I_1 < 0$ and $I_2 > 0$. Referring to Fig. 6.10 and noting the polarity for the terminal variables in this application, you can see that the series switch is still a transistor and that the shunt switch is still a diode. However, since their directions of conduction are the reverse of those for the devices in the converter of Fig. 6.11, they are connected in the opposite sense. The resulting converter with common positive is shown in Fig. 6.12.

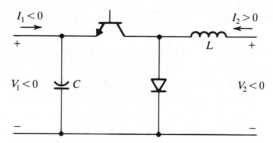

Figure 6.12   A down converter configured so that the positive terminals of the input and output ports are common.

We can also derive the converter of Fig. 6.12 by simply transforming the circuit of Fig. 6.11. We do so by noting that we can change the position of the inductor and the series switch to the lower branches in their respective loops without changing the operation of the circuit. The resulting circuit is shown in Fig. 6.13. Now, by simply flipping the circuit over, we have the result shown in Fig. 6.12.

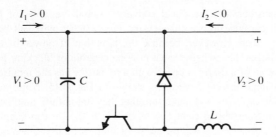

Figure 6.13   A down converter with common positives derived by changing the positions of the inductor and series switch in the circuit of Fig. 6.11.

The Up Converter   We now consider the switch implementation necessary to create an up converter. We choose the switches so that the power in the direct converter of Fig. 6.8(a) flows from the lower voltage side (V_2) to the higher voltage side (V_1). We do so by simply reversing either the terminal currents or the terminal voltages, relative to the down converter we just discussed.

   If we choose to reverse the currents, the switch currents of Fig. 6.10 are inverted. In this case, the shunt switch becomes a transistor and the series switch becomes a diode. The result is the direct up converter shown in Fig. 6.14. Note that the controllable switch is now the shunt switch instead of the series switch.

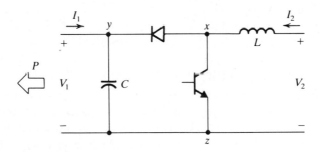

Figure 6.14   Switch implementation for the direct up converter.

EXAMPLE 6.3

Switch Implementation for a Bilateral Direct Converter

Some applications of dc/dc converters require energy to flow in both directions. Two examples are a dc motor that is actively braked and a magnetic resonance imaging (MRI) machine magnet, which must create a precisely controlled time-varying magnetic field. In both cases the terminal voltages change sign in order to remove or introduce energy into a magnetic field system. That is, with reference to Fig. 5.5(b), the converters operate in quadrants I and IV. How do we implement the switches in a direct converter designed for such applications?

We again consider the direct converter switch variables of Fig. 6.10, noting that if the terminal currents do not change sign the switches will still be required to conduct in only one direction. However, because the direction of power is reversed by inverting the terminal voltages, the switches must now be capable of blocking both voltage polarities. A switch made up of a diode in series with a transistor has the required characteristic, and the resulting bilateral direct converter is shown in Fig. 6.15. Note that we must now control both switches explicitly and make sure that they are not off simultaneously. However, when energy flows from left to right ($V_1 > 0$, $V_2 > 0$), we can supply Q_2 with base current continuously, because D_2 will do the switching. Similarly, when power flows from right to left ($V_1 < 0$, $V_2 < 0$) we can supply Q_1 with continuous base current.

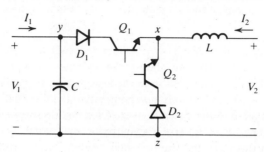

Figure 6.15   Switch implementation in the direct converter to provide bidirectional power flow with voltage reversal (operation in quadrants I and IV).

Note that the diodes in the circuit of Fig. 6.15 are on the collector sides of Q_1 and Q_2. Therefore we can connect the emitters of Q_1 and Q_2. This approach simplifies the base drive circuit by referencing the drives for both transistors to the same point.

6.3.3 About the Duty Ratio

In Section 6.3.1 we defined the duty ratio for the direct converter as the fraction of the switching period during which the series switch is on. Actually, common practice is to define the duty ratio as the fraction of the period during which the *controllable* switch (the transistor) conducts, regardless of whether it is in the series or shunt position. When we use this definition for D, (6.6) and (6.7)—written for the up converter of Fig. 6.14—become

$$\frac{V_2}{V_1} = 1 - D \tag{6.8}$$

$$\frac{I_2}{I_1} = -\frac{1}{1 - D} \tag{6.9}$$

We sometimes refer to the quantity $1 - D$ as D'. We also commonly express the conversion ratio for a dc/dc converter in the form of the output over the input. Therefore we can write the voltage conversion ratio for the up converter as:

$$\frac{V_{\text{out}}}{V_{\text{in}}} = \frac{1}{1 - D} = \frac{1}{D'} \tag{6.10}$$

Finally, the conventional conversion ratio expressions are ambiguous when power can flow in both directions, because we cannot label either of the external systems as the input or the output, and both switches are controllable. In this case we normally use the terms and expressions that are appropriate for the direction power is flowing at the time.

6.4 THE INDIRECT SWITCHING CONVERTER

The indirect converter connection of the canonical cell is shown in Fig. 6.8(b). As you will see in this section, the indirect converter exhibits operating characteristics that are very different from those of the direct converter just discussed.

6.4.1 The dc Conversion Ratio of the Indirect Converter

An important distinction between direct and indirect converters is the relative polarity of their terminal voltages and currents, as defined in Fig. 6.8. The voltages are of the same polarity in the direct converter, but they are of opposite polarities in the indirect converter. In contrast, the terminal currents are of opposite signs in the direct converter, but are of the same sign in the indirect converter.

We can derive the polarity relationships between the terminal variables of the indirect converter of Fig. 6.8(b) by considering the inductor voltage and the capacitor current. The voltage across the inductor is V_1 when switch S_{xy} is closed and V_2 when S_{xz} is closed, so one of these two voltages must be positive and the other negative to satisfy the condition that the average inductor voltage be zero. Similarly, the capacitor current is $-I_2$ when S_{xy} is closed and I_1 when S_{xz} is closed, so both I_1 and I_2 must be of the same polarity to satisfy the condition that the average capacitor current be zero.

If we leave switch S_{xz} on for a long time, both V_2 and I_1 become zero, regardless of the values of V_1 and I_2. Therefore both the ratios V_2/V_1 and I_1/I_2 become zero. However, if we leave switch S_{xy} on for a long time, these ratios become minus and plus infinity, respectively. As we inferred for the direct converter, when the two switches are alternating, these ratios will be somewhere between their two extremes. For the indirect converter, the magnitude of the conversion ratio ranges from zero to *infinity*, compared with zero to 1 for the direct converter.

The simplest way to determine the exact dependence of the conversion ratios on the duty ratio is to set the average inductor voltage to zero by balancing its positive and negative volt–time integrals. If we define D to be the fraction of the switching period T during which S_{xy} is closed, the average inductor voltage is zero when:

$$V_1 DT = -V_2(1 - D)T$$

which gives a voltage conversion ratio of:

$$\frac{V_2}{V_1} = -\frac{D}{1 - D} \qquad (6.11)$$

Similarly, we can determine the current conversion ratio by setting the average capacitor current to zero, that is

$$I_2 DT = I_1(1 - D)T$$

which gives a current conversion ratio of:

$$\frac{I_2}{I_1} = \frac{1 - D}{D} \qquad (6.12)$$

We can also obtain *(6.12)* from *(6.11)* by setting the net power into the converter to zero.

We commonly refer to the indirect converter as an *up/down* or *buck/boost* converter, because the output can be either higher or lower than the input. This condition is true for both the voltage and the current and for either direction of power flow.

6.4.2 Implementation of Switches

If we consider the switch voltages and currents as we did for the direct converter, we can determine the appropriate semiconductor devices to use for the indirect converter. Again, we must specify the direction of power flow and the polarity of one of the terminal voltages or currents. For power flowing left to right in the circuit of Fig. 6.8(b) (there is no higher or lower voltage side in the indirect converter) and $V_1 > 0$, switch S_{yx} conducts positive current and blocks positive voltage (note that $V_2 < 0$ for the assumed conditions on power and V_1). Therefore this switch can be a transistor. Switch S_{zx}, on the other hand, conducts positive current but blocks negative voltage, which are conditions met by a diode. The resulting indirect converter circuit is shown in Fig. 6.16.

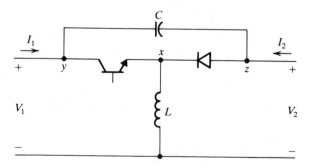

Figure 6.16   Switch implementation for the indirect converter with power flowing from left to right and $V_1 > 0$.

EXAMPLE 6.4

Indirect Converter with Power Flowing Right to Left

One way to make an indirect converter with power flowing from right to left in the circuit of Fig. 6.8(b) is to take the converter of Fig. 6.16 and flip it about the inductor. In this case, the voltage at the left terminal pair becomes negative. If the requirement were that this voltage be positive, we would invert the voltage and current for both switches and turn the diode and transistor around. The resulting circuit is shown in Fig. 6.17.

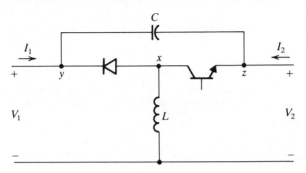

Figure 6.17   Indirect converter switch implementation for power flowing from right to left and $V_1 > 0$.

6.4.3 Variations on the Basic Indirect Converter Topology

There are two important variations of the basic indirect converter topology. For both variations, the conversion ratios given by *(6.11)* and *(6.12)* remain unchanged, and the voltage and current waveforms at the semiconductor devices are unaffected. The only thing that we change is deployment of the L and C to filter the high-frequency switch currents and voltages.

The capacitor and inductor in the indirect converter of Figs. 6.16 and 6.17 perform their high frequency filtering functions in the following ways. The ac components of the switch currents must circulate through C to prevent their appearing at the terminals. But note that because we can model L as an open circuit at the switching frequency, the series connection of the external networks is in parallel with C. Therefore the high-frequency impedance of C must be much smaller than the sum of the high-frequency impedances of the external networks. Similarly, the ac components of the switch voltages must appear across L if they are not to appear at the terminals. In this case, we model the capacitor as a short circuit at the switching frequency, which results in the parallel connection of the two external networks appearing in series with L. Thus the high-frequency impedance of L must be much larger than the parallel combination of the high-frequency impedances of the external networks. (We derive these conditions in more detail in Section 6.5.)

The implications of these conditions are: If the capacitor is to be of reasonable size, at least one of the external networks must have a high ac impedance; if the

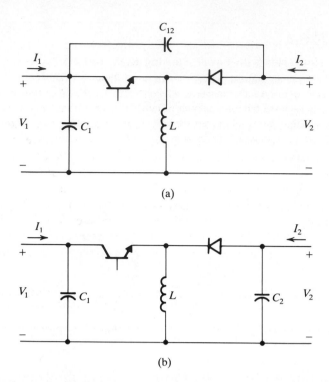

(a)

(b)

Figure 6.18   Variations of the basic indirect converter topology to provide ac terminal voltage filtering for two external networks having high ac
impedances: (a) addition of capacitor C_1 to the basic indirect topology;   (b) alternative placement for the capacitors in (a). This circuit
is commonly referred to as the *buck/boost* converter.

inductor is to be small, at least one of the external networks needs to have a low
ac impedance. In the ideal case, one of the external systems has a high and the
other a low ac impedance.

Unfortunately, both external systems often have a large impedance, particularly
when the switching frequency is high and long (inductive) connection wires are
used. Under these conditions the inductor must be unreasonably large. However, if
we modify the topology by placing an additional capacitor across one of the two
ports, as shown in Fig. 6.18(a), we make one of the external systems appear to have
a low ac impedance. Note that the purpose of this capacitor C_1 is to allow most
of the ac switch voltage to drop across a reasonably sized inductor. This capacitor
does not help the original capacitor C_{12} filter the ac switch current. This switch
current continues to flow through C_{12}, because one of the two external systems
still has a large ac impedance.

At the switching frequency, the three nodes of the canonical cell are shorted
together by the two capacitors of Fig. 6.18(a). We can achieve the same result

with the arrangement of capacitors shown in Fig. 6.18(b). If we treat the inductor as an open circuit at the switching frequency, we can see that both capacitors now help filter the ac switch currents. Each carries the full ac switch current if its ac impedance is small compared to the impedance of its corresponding external system. But across each appears a dc voltage equal only to that at its port, rather than the sum of the terminal voltages, which is what the capacitor in the indirect topology of Fig. 6.16 must withstand. Each of the two capacitors therefore is smaller than the original. In fact, under certain assumptions, we can show the sum of these two capacitors' peak energies to be equal to that of the original capacitor (see Section 6.5.5). Using the names *up/down* or *buck/boost* to identify a converter circuit usually implies the circuit variation of Fig. 6.18(b).

When both external systems have a low ac impedance, an inconveniently large capacitor must be used to remove ac ripple from the terminal currents. We can, however, increase the impedance of one of the external networks by placing an inductance in series with it, such as L_1 in Fig. 6.19(a). This approach results in

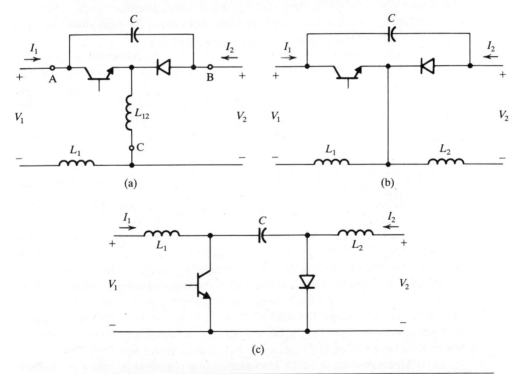

(a)

(b)

(c)

Figure 6.19 Indirect converter topology variations designed to provide ac terminal current filtering for two external networks having low ac impedances: (a) addition of inductor L_1 to the basic indirect converter topology; (b) alternative placement for the inductor in (a); (c) the circuit of (b) redrawn in the form commonly known as the Ćuk converter.

a return to the ideal situation in which an indirect converter interfaces one high-
and one low-impedance dc network. Although we can now reduce the size of the
cell capacitor, this additional inductor has no influence on the size of the original
cell inductor L_{12}. The parallel impedance of the two external systems is still low
because at least one of them has a low ac impedance, so most of the ac switch
voltage will still appear across L_{12}. So far we have simply traded off capacitor size
for an additional inductor.

Of the three branches stemming from node C in Fig. 6.19(a), two have inductors that, if they have been properly sized to eliminate ac voltage ripple from
the converter terminals, we can model as open circuits at the switching frequency.
Therefore the ac current flowing through the third branch is also zero. We can
achieve the same result if we arrange the two inductors as L_1 and L_2, shown in
Fig. 6.19(b). Now both inductors contribute to filtering the ac switch voltage. Each
must have an ac impedance that is large compared to the impedance of its corresponding external system. But each carries only the terminal current rather than
the sum of the terminal currents that L_{12} must carry. Therefore each inductor is
smaller than L_{12}. For the same ripple currents, the sum of the peak stored energies
in these two inductors is equal to the peak energy in L_{12}.

Although we have shown the inductors of Fig. 6.19(b) connected in series with
the negative terminals of the converter, positioning them in series with the positive
terminals, as shown in Fig. 6.19(c), is more practical. This arrangement changes
nothing functionally, but it results in a circuit node common to the input, output,
and emitter of the transistor, which is both convenient and often required. This
variant of the basic indirect converter is commonly known as the Ćuk converter.[†]

6.5 THE CHOICE OF CAPACITOR AND INDUCTOR VALUES

So far in our discussion of dc/dc converters, we have assumed that the high frequency filter elements L and C of the canonical cell are "large enough" to reduce
the switching frequency ripple in the terminal variables to an acceptable level. We
now consider the specification of these components in more detail. This subject is
important because the physical size of a converter is strongly influenced by the size
of its energy storage elements. Furthermore, comparisons among different converter
circuits are often based on ripple amplitude for a given amount of energy storage.
In this section we discuss the source of ripple in terminal currents and voltages,
develop models for analyzing ripple for various converter topologies, and specify
the relative sizes of L and C for direct and indirect converters.

In a typical power circuit the filter elements are made large enough to limit ac
ripple in the terminal variables to small values. As a result, the capacitor is nearly
an ac short circuit and the inductor nearly an ac open circuit. In what follows,
therefore, we make the simplifying assumptions that the inductor is an ac open

---

[†]Slobodan Ćuk and R. D. Middlebrook, *A New Optimum Topology Switching dc-to-dc Converter*,
IEEE Power Electronics Specialists Conference Record, 1977, pp. 160–179.

circuit when discussing ripple *current*, and the capacitor is an ac short circuit when discussing ripple *voltage*. These assumptions permit us to make straightforward first-order estimates of the ripple in the terminal variables.

6.5.1 A Ripple-Frequency Model for the Direct Converter

We first consider the source and distribution of ripple current in the direct converter of Fig. 6.8(a), repeated here as Fig. 6.20(a) for convenience. With L assumed to be an open circuit at the ripple frequencies, there is no ripple in i_2, and the ac components of i_y and i_z must be equal and opposite. We designate the ac component of i_y as i_y'. For the purpose of calculating the ripple component of i_1, i_1', we now replace the circuit to the right of C in Fig. 6.20(a) with an equivalent current source of value i_y'. The resulting ripple-current model is shown in Fig. 6.20(b). Note

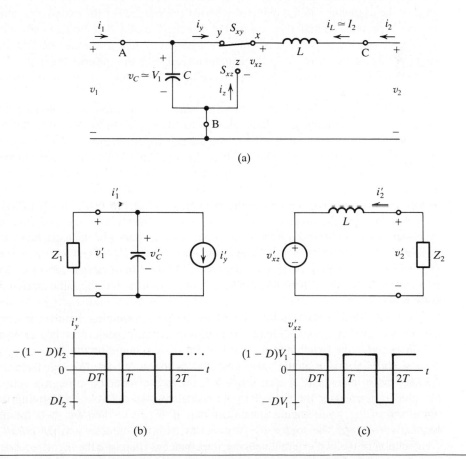

(a)

(b)

(c)

Figure 6.20   (a) The direct converter connection of the canonical cell.   (b) First-order ripple-current model for the direct converter of Fig. 6.8(a). (c) First-order ripple-voltage model for the same converter.

that the ripple current i_1' depends on how i_y' divides between C and the impedance of the external network, Z_1.

We now develop a model that we can use to calculate the ripple voltage at port 2 of the direct converter. Based on the assumption that the capacitor is an ac short circuit at the switching frequency, the ac component of the voltage v_{xz}, which we call v_{xz}', is as shown in Fig. 6.20(c). We now replace the circuit to the left of L in Fig. 6.20(a) by an equivalent voltage source of value v_{xz}'. The resulting ripple-voltage model is shown in Fig. 6.20(c). We can now calculate the first-order voltage ripple at port 2 by dividing v_{xz}' between the impedance of L and that of the external network Z_2.

6.5.2 A Ripple-Frequency Model for the Indirect Converter

For the indirect converter of Fig. 6.8(b), repeated here as Fig. 6.21(a), the interactions of the capacitor and the inductor with the impedances of the external systems are more complicated than they are for the direct converter. In the direct converter, the high-frequency impedance of only one external system influences the filtering effectiveness of the L or C in the canonical cell. For the indirect converter, we must simultaneously consider the impedance of both external systems to determine the ripple in the terminal variables.

We first consider the effect of capacitor value on the terminal ripple currents. We assume that the inductor is large enough for its ripple current to be negligible, permitting us to treat it as an infinite impedance (open circuit) at the switching frequency. Again, $i_y' = -i_z'$. The result is the switching frequency ripple-current model shown in Fig. 6.21(b). We dervive the waveform of i_y' by recognizing that the inductor current is $I_1 + I_2$ and is switched back and forth between nodes y and z. The ac current i_y' has two paths through which it can flow. One is through the capacitor, and the other is through the *series* connection of the two external system impedances. Therefore the high-frequency impedance of the capacitor must be small compared to the sum of the external impedances, $Z_1 + Z_2$, if we are to keep the switching frequency current from flowing through the external network. The ripple current that does show up at the converter terminals has the same amplitude at both ports.

Now we consider the requirements on the inductor, assuming that the capacitor is large enough that its voltage ripple is negligible and the capacitor can be treated as a short circuit in the switching frequency model. This assumption forces $v_{yx}' = v_{zx}'$, putting nodes y and z at the same potential in the ac model. We can therefore connect these nodes and in turn connect them to the inductor through a source of value v_{yx}', creating the voltage-ripple model of Fig. 6.21(c). (In deriving the waveform of v_{yx}', you should remember that if $V_1 > 0$, then $V_2 < 0$ for the indirect converter.) The source v_{yx}' is in series with the inductor and the *parallel* combination of the two external network impedances. Therefore the high-frequency impedance of the inductor must be large compared to that of Z_1 and Z_2 in parallel if the switching frequency voltage is not to appear at the converter terminals. To the extent that it does appear, the ripple voltage is the same at both ports.

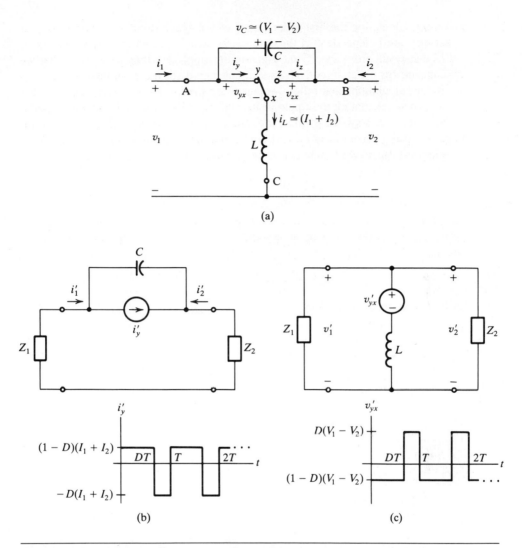

Figure 6.21 (a) The indirect converter.   (b) First-order ripple-current model for the converter.   (c) First-order ripple-voltage model for the converter.

6.5.3 Minimum L and C for the Direct Converter

Up to this point we have assumed that L was large enough that its current could be considered constant and that C was large enough that its voltage could be considered constant. We now address the problem of determining how large L and C have to be in order to justify these approximations. Any ripple in i_L or v_C shows up directly as first-order ripple in i_2 or v_1, respectively, in Fig. 6.20(a). It would have a second-order effect on i_1' or v_2' by adding ripple to the flat portions of the waveforms of the ripple sources i_y' and v_{xz}' in Fig. 6.20(b) and (c). Another reason for calculating the first-order capacitor voltage and inductor current ripples is that they are independent of the external system impedances. The values for L and C that result from specifying a maximum ripple are thus parameters that we can compare for different converters without regard for the external systems.

In the calculations that follow we again use the first-order approximations that $i_1 \approx I_1$ when calculating the ripple in v_1, and $v_2 \approx V_2$ for calculating the ripple in i_2. When the series switch is off in the direct converter of Fig. 6.20(a), the current flowing into the capacitor is I_1, and during this interval (of length $(1 - D)T$) the capacitor voltage v_1 will increase, giving a peak–peak ripple amplitude of:

$$\Delta v_1 = \frac{1}{C} \int_0^{(1-D)T} I_1 dt = \frac{I_1(1 - D)T}{C} \tag{6.13}$$

During this same interval, the voltage across the inductor equals V_2, and the inductor current $i_L = -i_2$ decreases by an amount equal to its peak–peak ripple amplitude. That is,

$$-\Delta i_L = \Delta i_2 = \frac{1}{L} \int_0^{(1-D)T} V_2 dt = \frac{V_2(1 - D)T}{L} \tag{6.14}$$

With specified limits on the terminal current and voltage ripple amplitudes Δi_2 and Δv_1, we can use (6.13) and (6.14) to determine minimum values of L and C:

$$C \geq \frac{I_1(1 - D)T}{\Delta v_1} \tag{6.15}$$

$$L \geq \frac{V_2(1 - D)T}{\Delta i_2} \tag{6.16}$$

After we have determined values for L and C from these expressions, we can determine the maximum energy stored in each element, which is useful because the stored energy is an indication of the element's size and cost. To express the stored energy in a useful way, we first define the *ripple ratios* $\mathcal{R}_C$ and $\mathcal{R}_L$ for the capacitor voltage and the inductor current, respectively. These are the ratios of the peak ripple amplitude to the dc (average) value of the capacitor voltage and inductor current, or:

$$\mathcal{R}_C = \frac{\Delta v_1/2}{V_1} \tag{6.17}$$

$$\mathcal{R}_L = \frac{\Delta i_2/2}{I_2} \tag{6.18}$$

By combining (6.15) with (6.17) and (6.16) with (6.18), we can find the minimum peak stored energy in the capacitor and inductor:

$$E_C = \frac{1}{2}CV_{1p}^2 = \left(\frac{1-D}{4}\right)\left(\frac{P_o}{f_s}\right)\frac{(1 + \mathcal{R}_C)^2}{\mathcal{R}_c} \tag{6.19}$$

$$E_L = \frac{1}{2}LI_{2p}^2 = \left(\frac{1-D}{4}\right)\left(\frac{P_o}{f_s}\right)\frac{(1 + \mathcal{R}_L)^2}{\mathcal{R}_L} \tag{6.20}$$

In these expressions, f_s is the switching frequency $1/T$; V_{1p} and I_{2p} are the peak values of the capacitor voltage and inductor current, respectively; and P_o is the average power flowing through the converter, that is, $P_o = I_1 V_1 = -I_2 V_2$. The value P_o/f_s is the amount of energy transferred from input to output during one cycle of the switching frequency. As we decrease the power or increase the switching frequency, the energy storage requirement goes down. By differentiating (6.19) and (6.20), we can show that the stored energy is minimized by making the ripple ratios equal to unity. This solution corresponds to a design in which the capacitor voltage or the inductor current just reaches zero each cycle. These equations also show that, for a fixed ripple ratio, as the value of $1 - D$ gets larger, the peak energy storage gets proportionally larger. Therefore as the difference between the input and output voltage or current becomes greater, the energy storage requirement becomes proportionally greater.

EXAMPLE 6.5

Specifying L and C Values for a Direct Converter

The circuit of Fig. 6.22(a) shows a direct (down) converter between a source having an internal resistance of 0.1 Ω and a resistive load of 0.1 Ω. What values of L and C will ensure that the input and output currents have ripples with peak-to-peak amplitudes of no more than 5% of their dc values if $f_s = 50$ kHz? If these minimum values of L and C are used in the circuit, what are the peak-to-peak terminal ripple-voltage amplitudes?

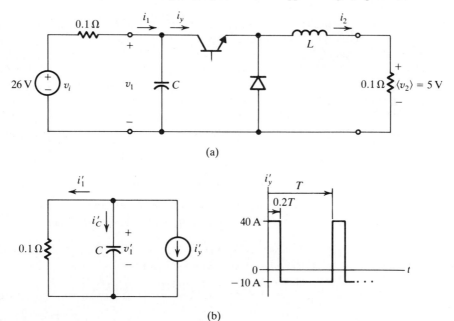

(a)

(b)

Figure 6.22   (a) Down converter circuit.   (b) Input ripple-current model for the circuit in (a).

With the specified load on the circuit, the average values of the converter terminal variables are

$$I_1 = 10 \text{ A} \qquad V_1 = 25 \text{ V} \qquad I_2 = 50 \text{ A} \quad \text{and} \quad V_2 = 5 \text{ V}$$

The duty ratio D is 0.2, and $T = 20$ μs.

We can easily calculate the value of L from *(6.16)*:

$$L \geq \frac{5(0.8)(2 \times 10^{-5})}{0.05(50)} = 32 \ \mu\text{H}$$

To calculate the value of C, we must first use the input ripple-current model of Fig. 6.22(b) to calculate i'_1. Because $i'_1 \ll i'_y$ with the proper values of L and C, we can make the first-order approximation that $i_C \approx -i'_y$. Therefore the peak-to-peak ripple voltage on C, $\Delta v'_1$, is

$$\Delta v'_1 \approx \frac{(1 - D)I_2 DT}{C} = \frac{(0.2)\,40\,(2 \times 10^{-5})}{C} \tag{6.21}$$

We can now calculate the peak-to-peak input current ripple $\Delta i'_1$ by recognizing that v'_1 appears across the input resistance. Thus:

$$\Delta i'_1 \approx \frac{\Delta v'_1}{0.1} \leq 0.05 I_1 \tag{6.22}$$

Solving *(6.21)* and *(6.22)* for C yields

$$C \geq 3200 \ \mu\text{F}$$

We can now calculate the first-order ripple in the terminal voltages. Because the load is resistive, the output voltage ripple as a percent of V_2 is the same as the ripple-current specification. That is,

$$\frac{\Delta v'_2}{V_2} = \frac{\Delta i'_2}{I_2} = 5\%$$

The input voltage ripple given by *(6.21)* is

$$\frac{\Delta v'_1}{V_1} = \frac{0.05}{25} = 0.2\%$$

6.5.4 Minimum L and C for the Indirect Converter

We use the same approach to determine the capacitor and inductor values for the indirect converter of Fig. 6.8(b). The resulting values of L and C are

$$L \geq \frac{V_1 DT}{\Delta i_L} \tag{6.23}$$

$$C \geq \frac{I_1(1 - D)T}{\Delta v_C} \tag{6.24}$$

In the indirect converter the average capacitor voltage is the sum of $|V_1|$ and $|V_2|$, and the average inductor current is the sum of $|I_1|$ and $|I_2|$. Expressed in terms of these terminal variables, the ripple ratios for the capacitor and inductor are

$$\mathcal{R}_C = \frac{\Delta v_C/2}{|V_1| + |V_2|} \tag{6.25}$$

$$\mathcal{R}_L = \frac{\Delta i_L/2}{|I_1| + |I_2|} \tag{6.26}$$

We can again express the peak stored energy in the capacitor and inductor in terms of the ripple ratios:

$$E_C = \left(\frac{1}{4}\right)\left(\frac{P_o}{f_s}\right)\frac{(1 + \mathcal{R}_C)^2}{\mathcal{R}_C} \tag{6.27}$$

$$E_L = \left(\frac{1}{4}\right)\left(\frac{P_o}{f_s}\right)\frac{(1 + \mathcal{R}_L)^2}{\mathcal{R}_L} \tag{6.28}$$

Note that, unlike for the direct converter, these storage requirements do not depend on the duty ratio. We can again minimize them by choosing a ripple ratio equal to unity.

The peak energy storage requirements of the indirect converter are typically larger than they are for the direct converter. Only when the direct converter operates at a duty ratio near zero are they equal. For example, at an operating point of $D = 0.5$ for the direct converter, the indirect converter requires twice as much storage.

6.5.5 Calculations for the Up/Down and Ćuk Converters

The buck/boost and Ćuk variations of the indirect converter use two capacitors and two inductors, respectively. For these converters we compare the total capacitive or total inductive peak energy storage requirements to the corresponding values calculated for the indirect converter. The requirement placed on the inductor in the buck/boost variant is the same as that for the inductor in the basic indirect converter. The requirement on the capacitor in the Ćuk variant is the same as that for the capacitor in the basic circuit. Let's consider the buck/boost circuit of Fig. 6.18(b).

The terminal voltage ripples in the indirect converter are related by the relative sizes of the external impedances connected to the two ports of the converter, as discussed in Section 6.5.2. In the analysis that follows we assume that the ripple amplitudes at the two ports are proportional to their dc voltages. That is, the ripple ratio at each port is the same as the ripple ratio of the capacitor in the basic indirect converter against which we are comparing the stored energy requirements of the buck/boost variant. As a result, the two capacitors in the buck/boost converter each have ripple ratios equal to that of the single capacitor in the basic indirect converter, or $\mathcal{R}_C$.

In the buck/boost circuit of Fig. 6.18(b), the capacitor at the left-hand port, C_1, is charged in the same way as the capacitor in the direct converter. Therefore it stores the same peak energy, which is given by *(6.19)*:

$$E_{C_1} = \left(\frac{1-D}{4}\right)\left(\frac{P_o}{f_s}\right)\frac{(1+\mathcal{R}_C)^2}{\mathcal{R}_C} \tag{6.29}$$

The ripple voltage on C_2 is

$$\Delta v_{C_2} = \frac{DTI_2}{C_2} \tag{6.30}$$

giving a minimum capacitance of:

$$C_2 \geq \frac{DTI_2}{2V_2\mathcal{R}_C} \tag{6.31}$$

The resulting peak energy stored in C_2 is

$$E_{C_2} = \left(\frac{D}{4}\right)\left(\frac{P_o}{f_s}\right)\frac{(1+\mathcal{R}_C)^2}{\mathcal{R}_C} \tag{6.32}$$

The sum of E_{C_1} and E_{C_2} is exactly equal to the peak energy stored in the single capacitor of the basic indirect converter and given by *(6.27)*.

A similar analysis done on the Ćuk variant of the indirect converter gives the same results. That is, the total energy storage requirement is identical to that for the basic indirect converter if we postulate the same ripple requirements at the ports, current ripple in this case.

6.6 SEMICONDUCTOR DEVICE STRESSES

Semiconductor switches must be rated to carry the peak current I_p and withstand the peak voltage V_p presented to them by the power circuit. These ratings not only affect the cost of a switch, but they also affect various device performance parameters, such as storage time, current gain, and switching speed. Therefore they are meaningful parameters for comparing the attributes of different circuits. The product of the two peak stresses V_pI_p is frequently useful for making comparisons among the various topologies. We call this product the *switch stress parameter*.

For both the direct and indirect converters of Fig. 6.8, the peak voltage at the switches is equal to the maximum capacitor voltage. Similarly, the peak current carried by either switch is equal to the maximum inductor current. In the case of the up/down variant of the indirect converter of Fig. 6.18(b), the peak switch voltage is the sum of the two peak capacitor voltages. For the Ćuk variant of Fig. 6.19(c), the peak current is the sum of the two peak inductor currents.

For the direct converter circuit of Fig. 6.8(a), the stress parameter for both switches is

$$V_pI_p = V_1(1 + \mathcal{R}_C)I_2(1 + \mathcal{R}_L) = P_o\frac{(1+\mathcal{R}_C)(1+\mathcal{R}_L)}{D} \tag{6.33}$$

For the indirect converter of Fig. 6.8(b), the stress parameter is

$$V_p I_p = (|V_1| + |V_2|)(1 + \mathcal{R}_C)(|I_1| + |I_2|)(1 + \mathcal{R}_L)$$

$$= P_o \frac{(1 + \mathcal{R}_C)(1 + \mathcal{R}_L)}{D(1 - D)} \qquad (6.34)$$

Note that for both converters the stress parameter is smallest when the conversion ratio is unity, that is, when $D = 1$ in the direct converter and when $D = 0.5$ in the indirect converter.

We call the switch stress parameters normalized to the power P_o the *switch stress factors*. Figure 6.23 shows them for both direct and indirect converters under the condition of zero ripple ratios. We need to plot only the stresses for output-to-input voltage ratios between zero and 1 because of circuit symmetry. We obtain the switch-stress factors for ratios between 1 and ∞ by exchanging the input and output roles of the ports. For example, the factors for a conversion ratio of 2.5 are the same as those for a conversion ratio of $0.4(= 1/2.5)$. When the conversion ratio is unity, the stresses on the switches in the indirect converter are four times greater than they are for the direct converter. At a 50% conversion ratio, the difference is more than 2. Even at a 25% ratio the difference is greater than 1.5. Clearly we would not choose to use the indirect converter unless we needed either a negative output voltage or *both* up and down conversion ratios from the same converter. The relative energy storage requirements for the two converters, as derived in Sections 6.5.3 and 6.5.4, also support this decision.

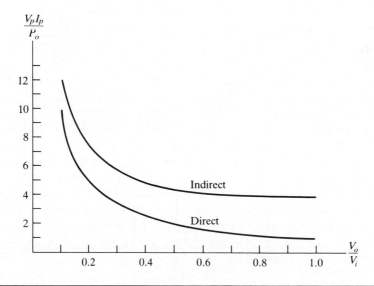

Figure 6.23   Switch stress factors for direct and indirect converters.

6.7 CONVERTER OPERATION WITH DISCONTINUOUS CONDUCTION

So far we have implicitly assumed that the peak inductor current ripple was smaller than the dc component of inductor current. The total current therefore was always positive, and the diode was forced to be on when the transistor was off. However, if the dc component of the current is smaller than the peak ripple, the total current will fall to zero during the time when the diode is on, that is, during the period $1 - D$. The diode will then turn off, and the inductor current will remain zero until the transistor is turned on again. When this sequence occurs, we say that the converter is operating in the *discontinuous conduction* mode.

Figure 6.24 shows the current and voltage waveforms of the direct converter of Fig. 6.11 operating in the discontinuous conduction mode. In drawing these waveforms, we assumed an additional capacitor across the output terminal pair so that v_2 is constant at V_2. Note that when the inductor current is zero, the voltage across the inductor is also zero, and the diode is reverse biased by the voltage $v_{xz} = V_2$. The average output voltage (which in this case is V_2) is still equal to $\langle v_{xz} \rangle$, but its value in this mode of operation is higher than *(6.6)* predicts for the continuous conduction mode.

For the direct converter operating in continuous conduction, the voltage across the inductor is set at all times by the state of the switches. The average of this voltage must be zero, so the duty ratio specifies, by *(6.6)* and *(6.7)*, the input/output conversion ratios. In discontinous conduction, however, the inductor voltage averaged over a switching cycle is zero independent of D. The reason is that discon-

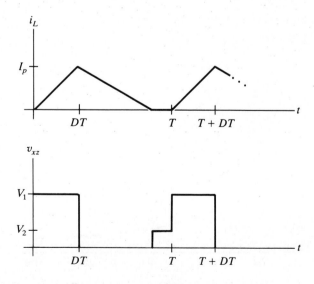

Figure 6.24   Waveforms of i_L and v_{xz} for the down converter of Fig. 6.11 operating in the discontinuous conduction mode.

tinuous conduction, by definition, ensures this condition. Therefore the duty ratio does not directly control the voltage conversion ratio. Instead, given V_1 and V_2, the duty ratio controls the average current in the inductor, and hence the dc components of i_1 and i_2, or I_1 and I_2. That is, in discontinuous conduction the current and voltage conversion ratios are functions of the duty ratio, but the functional dependencies are *not* those of *(6.6)* and *(6.7)*. We must specify the external systems in order to determine these conversion ratios.

EXAMPLE 6.6

The Direct Converter in Discontinuous Conduction

The converter of Fig. 6.25 is operating in the discontinuous conduction mode and supplying a constant voltage V_2 to a resistive load in parallel with a large capacitor. What is the value of V_2?

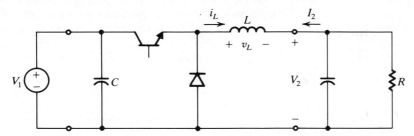

Figure 6.25   A down converter analyzed in Example 6.6 under the condition of discontinuous conduction.

When we turn the transistor on for a time DT, the inductor current increases linearly from zero to a peak value of:

$$i_{Lp} = \frac{(V_1 - V_2)DT}{L} \qquad (6.35)$$

When we then turn the transistor off, the voltage across the inductor is $-V_2$, and the time required to reduce its current to zero is

$$\Delta t = \frac{Li_{Lp}}{V_2} = \left(\frac{V_1}{V_2} - 1\right)DT \qquad (6.36)$$

The average value of i_L therefore is

$$\langle i_L \rangle = \frac{i_{Lp}(DT + \Delta t)}{2T} = -\langle i_2 \rangle = \frac{V_2}{R} \qquad (6.37)$$

When we combine *(6.35)*, *(6.36)*, and *(6.37)*, we obtain

$$V_2^2 + V_2\left(\frac{V_1 RD^2 T}{2L}\right) - \left(\frac{V_1^2 RD^2 T}{2L}\right) = 0 \qquad (6.38)$$

which we can solve for V_2.

We generally do not design high-frequency converters to be operated in the discontinuous conduction mode. The reason is that the peak inductor energy and the peak current stresses on the semiconductor devices are very high compared to their values in a converter operating at the same power level in the continuous conduction mode. If the load varies over a wide range, however, the converter may enter the discontinuous mode at the low-power end. When this happens, the response of the converter to changes in the duty ratio is altered. In this case, we must make sure that the control circuit continues to work properly. We address this issue in Part II.

Notes and Bibliography

The evolution of the canonical cell is described in [1]. Landsman's approach is insightful and clear, and the paper represents the first time the fundamental similarity among the various dc/dc converter topologies is identified. An excellent coverage of topologies and all sorts of variations on them can be found in [2], which reflects Severns' extensive practical experience with dc/dc converters. Mitchell's book [3] does a more concise job of treating circuits and is more mathematical than [2]. It also presents a good discussion of control of dc/dc converters. The Ćuk converter is first presented in [4].

1. E. Landsman, "A Unifying Derivation of Switching dc-dc Converter Topologies," in *IEEE Power Electronics Specialists Conference Record* (1979), 239–243.
2. R. P. Severns and G. Bloom, *Modern DC-to-DC Switchmode Power Converter Circuits* (New York: Van Nostrand Reinhold Co., 1985).
3. D. M. Mitchell, *DC-DC Switching Regulator Analysis* (New York: McGraw-Hill, Inc., 1988).
4. R. D. Middlebrook and S. Ćuk, "A New Optimum Topology Switching dc-to-dc Converter," in *IEEE Power Electronics Specialists Conference (PESC) Record* (1977), 160–179.

Problems

6.1   Proceeding along the lines of Example 6.1, determine the amplitude of the ripple on the voltage v_o in Fig. 6.26. Assume that the input capacitor C is large enough that the input voltage is constant at value V_i and that the Norton equivalent resistance R_N of the external load network is small.

6.2   Figure 6.27 shows a down converter supplying 5 V to a load of 0.1 Ω from a 25-V source having an internal resistance of 0.2 Ω. Determine the duty ratio D at which the converter is operating.

6.3   Derive a switch implementation for the direct up converter so that the input and output circuits share a common positive terminal.

6.4   Reconsider Example 6.3 and develop a bilateral direct converter in which the terminal currents reverse direction but the terminal voltage polarities remain unchanged.

6.5   Determine a switch implementation for a bilateral indirect converter in which the terminal currents change sign.

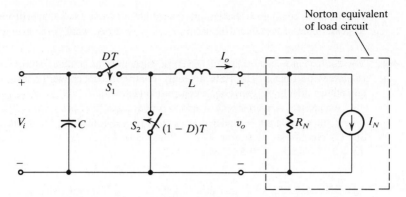

Figure 6.26   The dc/dc converter whose output voltage ripple is calculated in Problem 6.1.

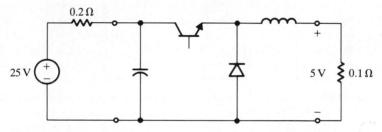

Figure 6.27   A down converter whose operating duty ratio is calculated in Problem 6.2.

6.6   Derive *(6.12)* from *(6.11)* by setting the net power into the converter to zero.

6.7   A *four-quadrant* converter can operate with its terminal variables I_t and V_t in any of the four quadrants of the I_t–V_t plane. Note that this implies bilateral power flow and that it makes no difference which terminal pair you consider in a two-pair network. Derive the switch implementation for a four-quadrant indirect converter.

6.8   The direct up converter shown in Fig. 6.28 serves as a battery charger. The control circuit provides constant current charging at a switching frequency of 20 kHz. The

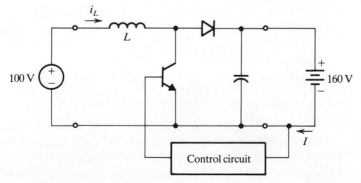

Figure 6.28   A direct up converter serving as a battery charger. This circuit is the subject of Problem 6.8.

current i_L is continuous. Determine the minimum value of L that will give a peak-to-peak ripple of less than 100 mA in i_L. If $I = 20$ A, what is the average value of i_L?

6.9 The *equivalent series resistance* (ESR) of capacitors in dc/dc converters is often an important parameter. The ESR not only contributes to converter loss, but it also sometimes affects the converter's control characteristics.

An indirect converter with a second-order output filter is shown in Fig. 6.29. The output filter capacitor is modeled with an ESR value of R_C. Assuming that all Ls and Cs are infinite, determine V_2 as a function of V_1, I_1, D, and R_C. If $I_1 = 10$ A, $D = 0.25$, and $R_L = 0.5$ Ω, what is V_2?

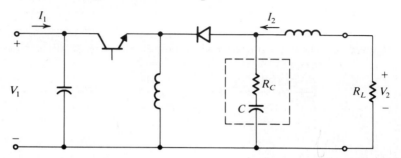

Figure 6.29 The indirect converter of Problem 6.9. The output capacitor C has an ESR value of R_C.

6.10 A direct up converter connects two external systems, as shown in Fig. 6.30. Assume that L and C are large enough that ripple is not a concern. As a function of R_o/R_i and the duty ratio D, find expressions for:

(a) V_o/V_i.
(b) the efficiency of the system, $\eta = P_o/P_i$, where P_i is the power from the source and P_o is the power delivered to R_o.
(c) the duty ratio at which the output voltage is maximized.

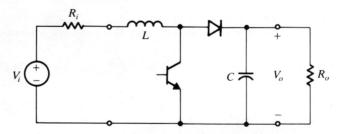

Figure 6.30 The direct up converter analyzed in Problem 6.10.

6.11 Compare the energy storage requirements of a direct converter operating at $D = 0.5$ to those of an indirect converter if their ripple ratios are the same. Check your answer with the statement at the end of Section 6.5.4.

6.12 Show that the total stored-energy requirement of the Ćuk variant of the indirect converter of Fig. 6.19(c), is identical to that of the basic indirect converter of Fig. 6.16 if their terminal ripple current ratios are the same.

6.13   A 100-V regulated dc supply must be designed using an unregulated 50–200-V source. Two possible ways of designing this supply are shown in Fig. 6.31. Figure 6.31(a) is the up/down, or buck/boost, variant of the indirect converter. Figure 6.31(b) is a cascade connection of a direct-down and a direct-up converter. Compare these alternative designs in the following manner.

(a) Ignoring the ripple in all capacitor voltages and inductor currents, express the transistor switch stress parameter in terms of the output power P_o and the voltage conversion ratio V_o/V_i for the up/down converter of Fig. 6.31(a).

(b) Repeat (a) for the circuit of Fig. 6.31(b), but express the transistor stresses in terms of P_o, V_i/V_m, and V_o/V_m.

(c) Find the value of V_m that minimizes the sum of the switch stress parameters for the circuit of Fig. 6.31(b). Interpret this result in terms of the duty ratios at which the two parts of the cascade operate as V_i varies from 50 V to 200 V.

(d) Compare the switch stress parameters for the circuit of Fig. 6.31(a) with the optimized sum of the parameters found in (c). Which is lower?

(e) Discuss other issues that would affect your choice of topology for this application.

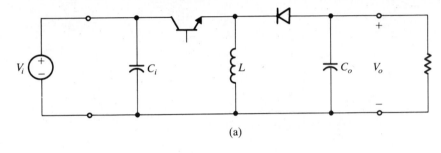

(a)

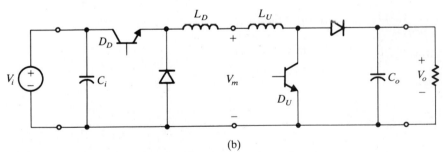

(b)

Figure 6.31   Power supply alternatives compared in Problem 6.13: (a) a single up/down converter; (b) a cascade of two direct converters, a down on the left and an up on the right.

6.14   The converter of Fig. 6.25 has the following parameters:

$$V_1 = 25\ \text{V} \qquad L = 300\ \mu\text{H} \qquad R = 2\ \Omega \qquad f = 1\ \text{kHz} \quad \text{and} \quad D = 0.2$$

(a) Determine and sketch the inductor current i_L.

(b) What is V_2?

(c) At what value of D does the transition from discontinuous to continuous conduction modes take place?

Chapter 7

Isolated High-Frequency dc/dc Converters

WE add transformers to the topology of a high-frequency converter for two reasons: (1) to provide electrical isolation between the two external systems; and (2) to reduce the component stresses that result when the input/output conversion ratio is far from unity. (We showed the relationship between switch stress factor and the conversion ratio in Fig. 6.23.) There are many ways in which we can include the transformer in the topology of a dc/dc converter. We present and discuss some of them in this chapter. We believe that the overview of transformers presented in Chapter 1 is sufficient background for understanding the material in this chapter. However, if you are still uncomfortable when confronted with a transformer, you should read Section 20.4 before continuing with this chapter.

Recall that a transformer winding cannot have a dc voltage across it, because the magnetizing inductor is a short circuit at zero frequency. Therefore we need to create, from the dc voltages of the external systems, an ac voltage with no average value. Thus we usually—but not necessarily—have the switches that produce this ac voltage also control the input/output conversion ratio.

There are essentially two high-frequency transformer topologies. The first, referred to as a *forward converter*, is based on the direct converter. The second, referred to as a *flyback converter*, is based on the indirect converter. In the forward converter, the algebraic sum of the instantaneous power over all windings is zero. That is, the transformer is not required to store significant energy. Although some energy is stored in the transformer's magnetizing inductance, we minimize this energy by making the inductance large. In the flyback converter, however, the transformer *is* required to store energy. During one part of the switching cycle, the primary winding takes energy from the input system and stores it in the magnetizing inductance. During the second part of the cycle, a second winding removes this energy and delivers it to the load. These two basic isolated converter topologies are the subjects of this chapter.

Even though it may not be immediately apparent, every topology presented in this chapter is based on the canonical cell of Section 6.2. The transformer, of course, adds additional elements to the circuit, but we can identify basic isolated converter operation during periods DT and $(1 - D)T$ with one of the canonical cell connections.

7.1 THE SINGLE-ENDED ISOLATED FORWARD CONVERTER

Figure 7.1(a) shows one version of an isolated down converter. The transformer is modeled as an ideal transformer with a shunt inductor L_μ representing its magnetizing inductance. (For now we are going to ignore the transformer's leakage inductance.) We call this converter a *single-ended* isolated forward converter— "single-ended" because power flows through the transformer for only one polarity of the primary voltage. The diode D_3 and the voltage source V_c provide a path for the magnetizing current i_μ when the transistor turns off.

When the transistor turns on, the transformer primary voltage v_P is equal to the input voltage V_1. The secondary voltage v_S is related to v_P by the turns ratio, and its polarity is positive. Therefore the output rectifier diode D_1 is on, the freewheeling diode D_2 is reverse biased, and $v_d = V_1/N$. The output current I_2 is reflected to the primary circuit as $i_1 = I_2/N$. Note that if we ignore the magnetizing current, and if $N = 1$, operation of the single-ended isolated forward converter is indistinguishable from the nonisolated down converter at this point. Furthermore, a nonunity turns ratio simply provides a step up or step down of the voltage and current.

If we ignore the magnetizing inductance, the transformer's primary and secondary currents i_P and i_S are zero when the transistor is off. The freewheeling diode D_2 will therefore turn on to carry the output current and $v_d = 0$. This condition corresponds to that part of the cycle when the shunt switch S_{xz} is conducting in the nonisolated direct converter of Fig. 6.8(a).

A practical isolated single-ended forward converter has two features that complicate both its performance and its design. The first is the need to provide for the consequences of nonzero stored energy in the transformer magnetizing branch L_μ. The second is the fact that processing this magnetizing energy results in switch stresses that exceed those of the nonisolated forward converter. We now consider these two features in detail.

7.1.1 Magnetizing Current and Clamping

If the transformer magnetizing current i_μ is nonzero, we can no longer ignore it when the transistor Q turns off. We must provide the current with a path in order to prevent it from being discontinuous. Furthermore, while i_μ is flowing during the time Q is off, v_P must be negative so that $di_\mu/dt < 0$ and $i_\mu(T) = i_\mu(0)$. This is the periodic steady state condition, and it guarantees that $i_\mu(T)$ is not increasing

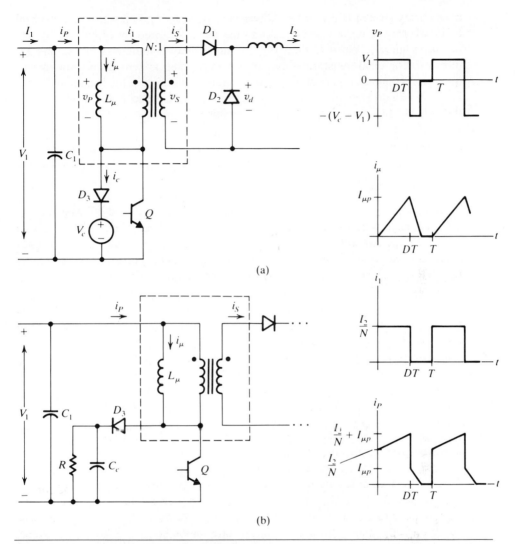

(a)

(b)

Figure 7.1   Single-ended isolated down converter with clamp: (a) clamp voltage represented by a voltage source V_c;   (b) clamp voltage implemented with a capacitor C_c and a discharge resistor R.

every cycle. Stated another way, the magnetic flux B in the core must return to its starting value at the end of every cycle. (This condition is called *resetting* the core.) Therefore we must provide a means for reversing the polarity of v_P, when Q is off. We do so with the diode D_3 and the clamp voltage V_c.

You can best understand the operation of the clamp by first assuming that the magnetizing current is zero when Q turns on. Because v_P is positive and constant at V_1 when Q is on, i_μ, which adds to the current carried by Q but not by D_1, increases linearly. When Q turns off, I_2 immediately commutates to D_2, and D_1 turns off. The ideal transformer currents i_S and i_1 step to zero at this time, but i_μ must be continuous. It cannot flow into the primary winding of the ideal transformer because the corresponding secondary current would have to flow the wrong way through D_1. Instead, i_μ forces D_3 on, clamping the transistor collector at V_c and v_P at $V_1 - V_c$.

The magnetizing current i_μ must be zero by the start of the next cycle. If i_μ were not zero, it would continue to increase each cycle until the transformer saturated. If i_μ is to return to zero by the start of the next cycle, v_P must be negative when Q is off. The voltage V_c must therefore be greater than V_1. How much greater depends on the relative lengths of time Q is on and off. Specifically, the time integral of v_P (generally referred to as the *primary volt-seconds*) across the transformer when Q is off must equal the negative of the primary volt-seconds when Q is on. That is, the net volt-seconds must be zero. For example, if Q is on for 50% of the cycle, V_c must equal $2V_1$. Note that this condition is the same as $\langle v_P \rangle = 0$.

If V_c were precisely equal to the value necessary to balance the positive and negative primary volt-seconds, the transformer flux at the end of each cycle would return to its value at the beginning of the cycle. Because of a transient or an imperfection in control, this starting value may not be zero. In such a case, the peak flux in the transformer will be higher than necessary, and we may still have to worry about saturation. Therefore, to guarantee that $i_\mu(0) = 0$, we usually make the clamp voltage slightly greater than the critical value necessary for balancing the volt-seconds. Diode D_3 will prevent i_μ from actually going negative, so the minimum flux level in the transformer is $B_{\min} = 0$. When D_3 turns off, v_P steps to zero, where it stays until the transistor turns on again.

The duty ratio is often a control variable, so we must choose V_c to guarantee that i_μ will be reset to zero in the worst case. For example, the larger the duty ratio in the converter of Fig. 7.1, the higher will be the positive volt-seconds seen by the transformer and the larger V_c must be. Therefore we must choose V_c on the basis of the expected maximum duty ratio. For duty ratios below this maximum value, i_μ will simply return to zero more quickly than necessary, and a greater amount of the transistor's off time will be spent with $v_P = 0$.

There are many different ways of implementing the clamp function, only some of which we discuss. In practice, we frequently use a breakover device—such as a Zener diode or a metal-oxide varistor—as a direct replacement for the diode and clamp voltage source of Fig. 7.1(a) if the energy involved is within the ratings of the breakover device. However, the principles involved are independent of the specifics of implementation. For this reason, we use easily understood clamp circuits to illustrate the important issues of the clamping function.

EXAMPLE 7.1

Choosing a Clamp Voltage

Assume that the circuit of Fig. 7.1(a) has an input voltage $V_1 = 50$ V and that the transformer has a turns ratio $N = 5$. If the transistor's duty ratio is D, the output voltage will be

$$V_2 = D\left(\frac{V_1}{N}\right) = D(10) \text{ V} \tag{7.1}$$

If the requirement on V_2 is $0 < V_2 < 8$ V, we need a maximum duty ratio of 80%. We can find the minimum necessary value of V_c by equating the positive and negative volt-seconds on the primary:

$$V_1 D_{max} T = (V_c - V_1)(1 - D_{max})T \tag{7.2}$$

or

$$V_c = \frac{V_1}{1 - D_{max}} = \frac{50}{0.2} = 250 \text{ V} \tag{7.3}$$

The clamp voltage should be somewhat greater than this minimum value. How much greater depends on how sure we are that the maximum duty ratio will not exceed 80% or how carefully we control transients that can cause flux offsets. Note that we must choose a transistor rated to withstand V_c, which is high relative to other voltages in the circuit.

Energy is absorbed by the clamping circuit every time it operates, that is, once a cycle. This energy is often difficult and uneconomical to recover from the source V_c, and hence it represents a loss. We can calculate the amount of this energy by integrating $V_c i_c$ over the time that D_3 is on. The voltage V_c is a constant, and the current i_c decreases linearly from its peak value of:

$$I_{\mu p} = \frac{V_1 DT}{L_\mu} \tag{7.4}$$

to zero over a time,

$$\Delta t = \frac{L_\mu I_{\mu p}}{V_c - V_1} \tag{7.5}$$

The energy flowing into the source V_c every cycle, E_c, therefore is

$$E_c = \int_0^{\Delta t} V_c i_c \, dt = V_c \frac{I_{\mu p}}{2} \Delta t = \frac{1}{2} L_\mu I_{\mu p}^2 \left(\frac{V_c}{V_c - V_1}\right) \tag{7.6}$$

which is larger than $L_\mu I_{\mu p}^2 / 2$, the energy stored in L_μ.

We usually develop the clamp voltage V_c by charging a relatively large capacitor. In order to prevent V_c from increasing every cycle, we must provide a means to remove E_c. One simple solution is to place a resistor in parallel with the capacitor—R in Fig. 7.1(b). Because both the energy flowing into and out of the capacitor depend on V_c, we must choose exactly the right value of R to get the desired clamping voltage. To keep V_c constant, this resistor value has to change with a change in either V_1 or D. Therefore we generally use a linear network that senses and regulates V_c instead of a simple resistor of fixed value. If the application is a very high-power one, we might justifiably use an auxiliary dc/dc converter to recover E_c by transferring it to either the source or load.

EXAMPLE 7.2

A More Efficient Placement of the Clamp Discharge Resistor

By placing the clamp circuit regulator or discharge resistor in parallel with the clamp capacitor (assumed to be infinite), we dissipate an amount of energy given by *(7.6)*. If we instead connect the resistor between the capacitor and the input voltage source, as shown in Fig. 7.2, part of the energy removed from the capacitor returns to the source. The ratio of the energy dissipated in the resistor every cycle, E_R, to that absorbed by the clamp, E_c, is equal to the ratio of the resistor voltage, $V_c - V_1$, to the clamp voltage, V_c. The energy dissipated is thus:

$$E_R = E_c \left(\frac{V_c - V_1}{V_c} \right) = \frac{1}{2} L_\mu I_{\mu p}^2 \left(\frac{V_c}{V_c - V_1} \right) \left(\frac{V_c - V_1}{V_c} \right) = \frac{1}{2} L_\mu I_{\mu p}^2 \qquad (7.7)$$

Because the energy given by *(7.7)* is less than that given by *(7.6)*, the resistor placement in Fig. 7.2 is more efficient than that in Fig. 7.1(b).

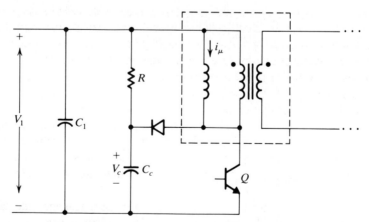

Figure 7.2   An alternative to the discharge resistor placement in Fig. 7.1(b).

7.1.2 A Transformer-Coupled Clamp

We show another way to provide the clamp function in Fig. 7.3, where a third winding of N_T turns has been added to the transformer. This third winding, called

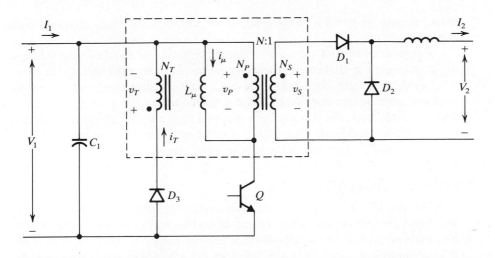

Figure 7.3   Clamp function provided by a tertiary transformer winding.

a *tertiary* or *clamp winding*, permits i_μ to circulate through the primary and clamp windings when Q turns off. The current i_T therefore is

$$i_T = \frac{N_P}{N_T} i_\mu$$

Note that when Q turns off, i_μ will flow out of the dot on N_P, causing i_T to flow into the dot on N_T and D_3 to turn on. Because $v_P = v_T(N_P/N_T)$ and $v_T = -V_1$ during the time when D_3 is on, L_μ discharges, as desired. However, i_T cannot go negative, and therefore we can still reduce the transformer flux only to zero. The advantage of this approach is that the magnetizing energy returns directly to the source, instead of to a separate clamp circuit.

We can adjust the effective clamp voltage (the voltage across L_μ when Q is off) by changing the ratio N_P/N_T. For a maximum duty ratio of 50%, N_P/N_T must have a minimum value of 1. For a maximum duty ratio of 75%, N_P/N_T must have a minimum value of 3. Note that this turns ratio affects the off-state voltage of the transistor. For the 1:1 ratio, the transistor's off-state voltage is $2V_1$, and for the 3:1 ratio it is $4V_1$.

A problem with the circuit of Fig. 7.3 is that any leakage inductance between the primary and tertiary windings keeps the magnetizing current from commutating immediately to the tertiary winding when the transistor turns off. An additional clamp, or a transient suppressing circuit called a *snubber*, must be placed across the transistor to keep its voltage from rising too much. (We discuss snubber circuits in detail in Chapter 24.) The energy associated with this leakage inductance typically is small compared to that of the magnetizing inductance, so the dissipation in this clamp or snubber is not a problem unless the circuit is switching at a very high frequency.

Instead of connecting the clamp winding across V_1, we could connect it across V_2, transferring the magnetizing energy to the output rather than back to the input. The result is a slight improvement in the circuit's efficiency. Owing to safety isolation specifications that often call for a minimum spacing between source and load windings, however, this approach makes a tight coupling of the primary and tertiary windings difficult. The greater dissipation caused by the larger leakage inductance easily offsets the expected gain in efficiency.

7.1.3 The Isolated Hybrid Bridge

A technique frequently used to maintain current continuity in inductive loads switched by transistors is to place a diode around the load, as shown in Fig. 7.4(a). The problem with this technique for our present purposes, however, is that the discharging voltage is only a diode drop plus the drop across the resistance of the winding. But the circuit of Fig. 7.4(a) suggests another possibility. When Q_1 is off, the bottom terminal of the load is connected to the positive rail through D_1. Thus, if we could connect the top terminal of the load to the negative rail, we could apply $-V_1$ across the inductor to discharge it. We can do so by adding Q_2 and D_2, as shown in Fig. 7.4(b). The transistors Q_1 and Q_2 turn on and off simultaneously. Note that again, because the diodes do not permit i_μ to reverse, the transformer core resets to a minimum flux value of only $B_{min} = 0$. Because only two of the four bridge switches are controllable, this circuit is still only single-ended and is called a *single-ended isolated hybrid-bridge*.

Although this converter has twice as many primary side switches as that of Fig. 7.3, they all have a smaller required voltage rating. And as the magnetizing current does not need to commutate from the primary winding to a clamp winding, we avoid the leakage inductance in the circuit of Fig. 7.3. We do need to drive two transistors whose emitters are at different potentials, however, and the circuit is limited to a maximum duty ratio of 50%. We can obtain this duty-ratio limit from the current waveforms shown in Fig. 7.4(c). These waveforms are based on the assumption that the transformer leakage inductance is zero. This is not true in a practical transformer: A commutation period results, during which both D_3 and D_4 are on, as we will discuss in Section 7.4.

7.1.4 Switch Stresses in the Single-Ended Isolated Converter

One drawback to the single-ended isolated converter is that power flows through the transformer only during the period DT. As the circuit is idle for the period $(1 - D)T$, the voltage and current stresses on the transformer, switches, and filter elements are higher than they would be if the circuit were transferring energy from input to output continuously. You can see the reason for these effects by considering the voltage and current stresses on the transistor in the circuit of Fig. 7.1.

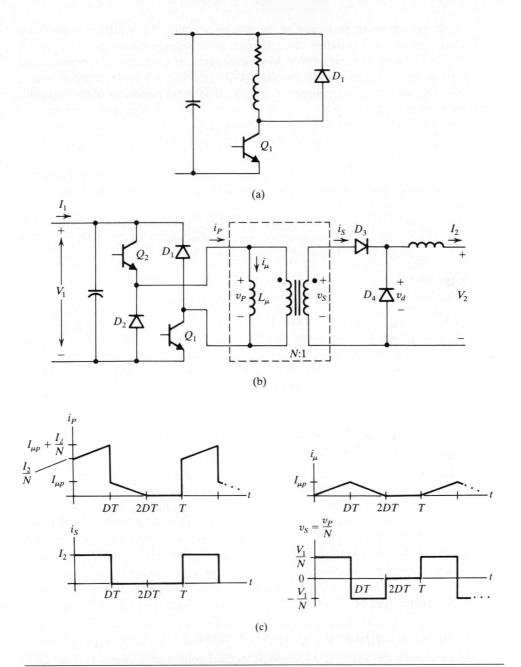

Figure 7.4   (a) Conventional application of a clamp diode to an inductive load. (b) Extension of the circuit of (a) to provide an L_μ discharge voltage equal to $-V_1$.   (c) Waveforms for the circuit of (b).

If the maximum duty ratio of the converter in Fig. 7.1 is 50%, the minimum clamp voltage, and therefore the minimum peak voltage across the transistor, is $2V_1$. If P_o is the average power flowing through the converter, the primary side current, which is the current carried by Q (ignoring i_μ), is rectangular with a 50% duty ratio and an amplitude of $2P_o/V_1$. The stress parameter of the transistor therefore is

$$V_{Qp}I_{Qp} = (2V_1)\left(\frac{2P_o}{V_1}\right) = 4P_o \tag{7.8}$$

showing that it operates with a stress factor of 4 in this circuit. Similarly, the two output diodes carry the output current and withstand twice the output voltage, so their stress factors are 2.

In general, if the maximum duty ratio is D_{max}, the clamp voltage must be at least $V_1/(1 - D_{max})$, and the current carried by the transistor is $P_o/(D_{max}V_1)$. The stress parameter of the transistor therefore is

$$V_{Qp}I_{Qp} = P_o\frac{1}{D_{max}(1 - D_{max})} \tag{7.9}$$

This parameter is minimal when $D_{max} = 0.5$. Even at the minimum value of its stress parameter, the switch must be rated at a stress factor of 4. Therefore, if we do not need isolation, we would not replace a conventional, nonisolated direct converter with one containing a transformer and a single transistor unless the output/input conversion ratio V_2/V_1 was less than 0.25 (see Fig. 6.23).

7.2 THE DOUBLE-ENDED ISOLATED FORWARD CONVERTER

Even though the secondary voltage in the circuit of Fig. 7.4(b) is a symmetric ac waveform, we cannot use a full-wave rectifier on the output. The reason is that diodes D_1 and D_2 prevent i_P from going negative, which would be necessary if we used a full-wave rectifier circuit on the secondary voltage. The result is that the frequency of the ripple in v_d equals the switching frequency of the transistors. *Double-ended* converters permit full-wave rectification and a doubling of the fundamental ripple frequency—and also result in better utilization of transformer core material. Their disadvantage is that they are more complicated than their single-ended counterparts.

7.2.1 The Double-Ended Bridge Converter

If we replace the diodes and transistors in the hybrid-bridge circuit of Fig. 7.4(b) with bilaterally conducting switches, we can use a full-wave rectifier on the output. This approach has two advantages. First, we could make the output inductor smaller, because the ripple frequency would be twice that of the single-ended circuits.

Second, the magnetizing current i_μ could change sign, thereby permitting the core material to be better used, because the flux could now vary between $\pm B_s$ instead of between 0 and $+B_s$, where B_s is the saturation flux density of the material. Such a double-ended converter is shown in Fig. 7.5(a).

The waveform of v_d for the double-ended converter is shown in Fig. 7.5(b). Here, we define the duty ratio D in terms of the period of the full-wave rectified voltage instead of the switching period of the primary side switches. If $D = 1$, each of the primary side switches would be on 50% of the time and $\langle v_d \rangle = V_1/N$. We can create the waveform of v_d by several different primary side switching sequences. They differ in the path provided for i_μ during the time between DT and T, and their applicability is constrained by the magnitude of i_μ relative to the load current reflected to the primary, I_2/N. They also result in different switch implementations. But in every case the core resets during the next interval, DT, when v_P reverses.

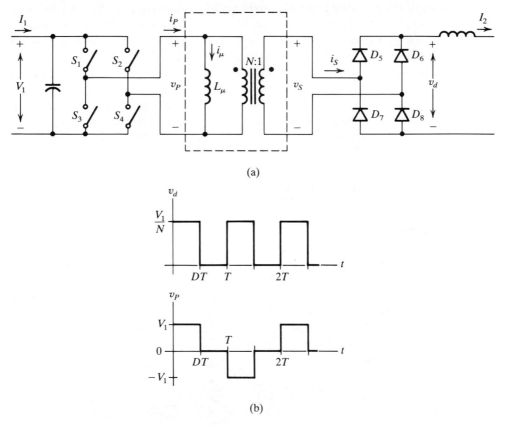

(a)

(b)

Figure 7.5   (a) Double-ended converter with primary side switches arranged in a bridge.   (b) Rectifier output voltage v_d and primary voltage v_P.

The only function that the switches must perform during time $(1 - D)T$ is to maintain the continuity of i_μ. We do not have to apply a resetting voltage to the transformer during this time because the core resets during the next period DT when v_P reverses. We can maintain the continuity of i_μ by using any one of three switch combinations. First, we can turn off all the switches, forcing the magnetizing current to flow in the secondary winding, that is, $i_S = -Ni_\mu$. The second combination is S_1 and S_2 on, and the third combination is S_3 and S_4 on.

The first option, all the primary side switches off, is possible only if $Ni_\mu < I_2$, so that at least three secondary diodes can be on simultaneously to allow $i_S \neq I_2$. For instance, if $0 < Ni_\mu < I_2$, the diodes D_5, D_6, and D_7 (or, alternatively, D_6, D_7, and D_8) would have to be on, with I_2 circulating through D_5 and D_7, and $i_S = -Ni_\mu$ circulating in the negative direction through D_5 and in the positive direction through D_6. If $Ni_\mu > I_2$, then D_5 would be conducting a net negative current, which would be inconsistent with its on state. (Note that the state of D_8 during this time is ambiguous but makes no difference to the result.) The advantage of this option is that the primary switches carry only a unipolar current. But its disadvantage is that the continuity of energy stored in any transformer leakage inductance (which is not included in the figure) cannot be maintained.

The second and third options are functionally indistinguishable. In both cases the switches carry bilateral current, a path is provided to maintain continuity of primary leakage energy, and the magnetizing current i_μ will split between the primary and secondary windings. Exactly how i_μ distributes itself between the primary and secondary windings is a function of second-order parameters, such as switch and diode drops, and the relative sizes of the primary and secondary leakage inductances. In practice, we utilize both options by alternating them each cycle, as in the sequence $S_{1,4}$, $S_{4,3}$, $S_{3,2}$, and $S_{2,1}$. This ensures that each switch is subjected to the same current stress. A switch implementation suitable for this operating mode is shown in Fig. 7.6.

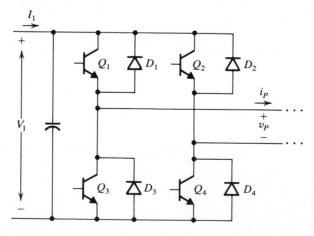

Figure 7.6   Switch implementation for the double-ended converter of Fig. 7.5(a).

EXAMPLE 7.3

Transformer Primary Current in the Double-Ended Full-Bridge Converter

Let's determine the primary current i_P in the double-ended converter of Fig. 7.5. We use the switching sequence $S_{1,4}$, $S_{4,3}$, $S_{3,2}$, and $S_{2,1}$, and we assume that i_μ circulates only on the primary side during the period $(1 - D)T$, when S_1 and S_2 or S_3 and S_4 are on. (Recall that T is the period of v_d, as shown in Fig. 7.5b.) During this time, i_μ remains constant at $I_{\mu p}$ because $v_P = 0$. A possible path for I_2 while $v_P = 0$ is the freewheeling path created if D_5 and D_7 or D_6 and D_8 are on.

At the start of the period DT at $t = 0$, we assume that $i_\mu(0) = -I_{\mu p}$. During DT, S_1, S_4, D_5, and D_8 are conducting, and the primary current is

$$i_P = \frac{1}{N} I_2 - I_{\mu p} + \frac{1}{L_\mu} \int_0^t V_1 \, dt \qquad (7.10)$$

At $t = DT$, $i_\mu = +I_{\mu p}$ (we assume operation in the periodic steady state), and S_1 turns off and S_3 turns on, maintaining a path for i_μ. With $v_P = 0$ during $(1 - D)T$, $i_\mu = I_{\mu p}$ until S_4 turns off and S_2 turns on at $t = T$. The resulting waveforms of i_μ, i_P, and i_S/N are shown in Fig. 7.7(a).

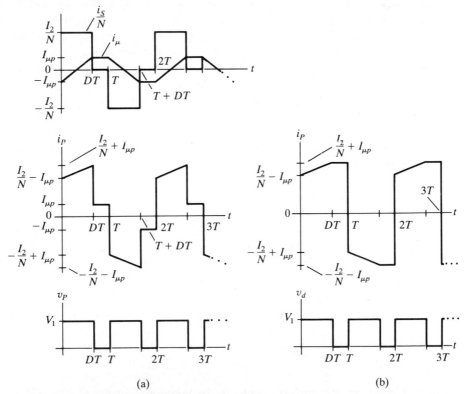

Figure 7.7   Waveforms as derived in Example 7.3 for the circuit of Fig. 7.5 or 7.6: (a) I_2 freewheeling during $(1-D)T$; (b) D_5 and D_8 or D_6 and D_7 remaining off during $(1 - D)T$.

Although we assumed that I_2 freewheeled through D_5 and D_7, or D_6 and D_8, during $(1 - D)T$, the possibility exists that D_5 and D_8 remain on and that D_6 and D_7 remain off during this time, causing an additional current I_2/N to circulate through S_3 and S_4. In this case, i_P is as shown in Fig. 7.7(b). The ambiguity present in this example disappears if the transformer model includes some leakage, for then i_S could not change instantaneously at $t = DT$, and D_5 and D_6 would remain on.

7.2.2 The Double-Ended Isolated Half-Bridge Converter

The fact that the emitters of the transistors are at three different potentials complicates the base-drive circuit for the bridge converter of Fig. 7.6. We use a center-tapped primary winding in the circuit shown in Fig. 7.8(a) to solve this problem. As a result, the emitters of both transistors connect to the same point, the negative rail of the input supply in this case. We call this converter a *push–pull* converter because of the alternating action of the two transistors. Although we show a full-wave centertapped rectifier on the secondary, a full-bridge rectifier circuit works just as well.

We control output voltage in the same way as in the full-bridge converter. But we can create the zero-voltage period $(1 - D)T$ only by turning off both primary side switches. If we ignore the magnetizing current, and the transformer has no leakage inductance, both primary winding currents are zero during this time. The transformer forces $i'_1 = i'_2$, splitting I_2 evenly between the two secondary windings, which are now shorted because both D_3 and D_4 are on. Under this condition, all the winding voltages must be zero, and therefore $v_d = 0$.

We now consider the effects of magnetizing current by modeling L_μ as an inductor across the lower primary winding of the ideal four-winding transformer. If S_1 is on, the voltage across L_μ is equal to $-V_1$, and i_μ ramps down linearly from its peak positive value $I_{\mu p}$. Because the switches are controlled to make v_P a symmetric waveform, $i_\mu = -I_{\mu p}$ at the end of the S_1 conduction period. During the S_1 conduction period, v_P is negative and D_3 is on, so $i'_1 = I_2$. As $i_2 = -i_\mu$ during this time, $i_1 = i_\mu - I_2/N$, because the ideal transformer requires that the sum of all the Ni products of the windings be zero. When S_1 turns off at DT, i_1 becomes zero, and both $i_\mu (= -I_{\mu p})$ and I_2 must be continuous. The only way for this to happen is for both D_3 and D_4 to be on. The result is that I_2 splits between the two diodes so that the difference between the two secondary currents is $i'_1 - i'_2 = Ni_\mu = -NI_{\mu p}$. Combining this expression with the constraint $i'_1 + i'_2 = I_2$ yields

$$i'_1 = (I_2 - NI_{\mu p})/2 \qquad\qquad (7.11)$$

and

$$i'_2 = (I_2 + NI_{\mu p})/2 \qquad\qquad (7.12)$$

Continuing this analysis results in the waveforms of Fig. 7.8(b), which were drawn for $I_{\mu p} < I_2/N$. In this case, the switch currents are unilateral, and the diodes are not required in the implementation of S_1 and S_2. However, if $I_{\mu p} >$

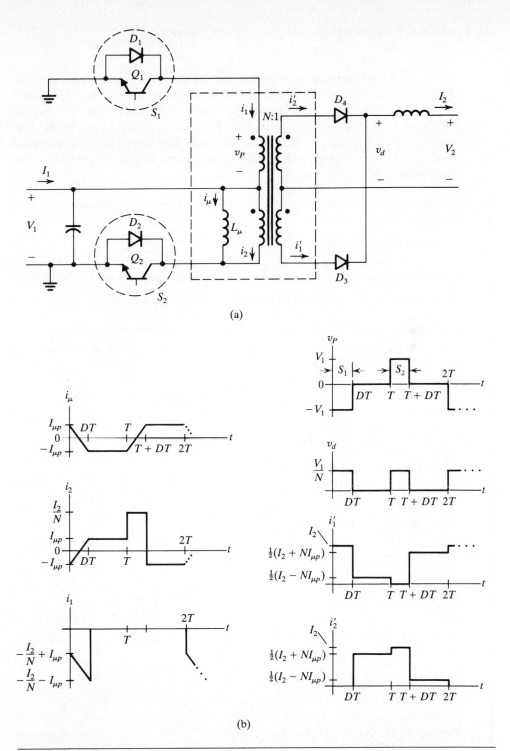

(a)

(b)

Figure 7.8 (a) A transformer coupled half-bridge, or push–pull, converter. (b) Waveforms of branch variables.

I_2/N, the switch currents are bilateral and the diodes are necessary. Note that $i_\mu(t) + i_2(t) = -i_1(t - T)$; that is, the switch currents are identical but shifted in time by T. When we recognize that L_μ is part of the transformer, we see that the circuit is symmetrical. We could model L_μ as being across any of the four windings (with a change in value to L_μ/N^2 when it is placed across a secondary winding) or even across the two primary or secondary windings. In the latter case the symmetry is more obvious.

EXAMPLE 7.4

A Current-Fed Half-Bridge Converter

So far in this chapter, we have only discussed power circuits in which the power flows from the voltage source side to the current source side of the canonical cell. In Chapter 6 we

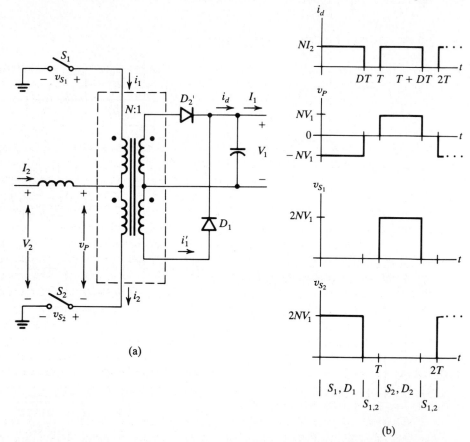

(a)

(b)

Figure 7.9 (a) Current-fed half-bridge converter of Fig. 7.8 configured to provide power flow from V_2 to V_1 ($L_\mu = \infty$). (b) Branch variables in the circuit of (a).

showed that power can flow in the other direction if we control the switches differently. To do so, in practical terms, we have to rearrange the transistors and diodes. For example, Fig. 7.9 shows the push–pull converter of Fig. 7.8(a) configured to provide for power flow from V_2 to V_1. In this example, we assume L_μ to be infinite. This circuit is called a *current-fed half-bridge converter*. As for the up converter on which it is based, the output voltage V_1 of this converter can range from a minimum value of V_2/N (the input voltage transformed by the turns ratio) to infinity.

The operation of a current-fed converter is slightly different from the operation of a voltage-fed converter. When we turn on only one of the switches, the input current flows through the transformer to the load, and the *load voltage* determines the voltage across the transformer. To obtain an interval when power does not flow to the load (the period $(1 - D)T$), we turn both S_1 and S_2 *on*. The input current then splits evenly between the two primary windings, just as the load current did in the secondary windings of the voltage-fed converter when both switches were off, and the transformer voltage is forced to be zero. Branch variables for the current-fed converter are shown in Fig. 7.9(b).

7.3 THE FLYBACK CONVERTER

We can easily transform the indirect converter into a circuit that provides electrical isolation because the voltage across its filter inductor has no dc component. For example, we can provide isolation in the up/down circuit of Fig. 6.18(b). We substitute a transformer having a magnetizing inductance of value $L_\mu = L$ for the filter inductor L in the circuit, as shown in Fig. 7.10(a). When Q turns on, D turns off and energy flows into L_μ. When Q turns off, D turns on and the stored magnetizing energy flows to the load through the secondary winding. Television receivers have a converter of this type with $N \ll 1$ to generate a very high voltage that drives the retrace of the beam in the picture tube. What happens visually is that the spot on the screen "flies back" to start another horizontal line. Thus the isolated indirect converter is commonly known as a *flyback converter*.

If it were not for leakage inductance, the operation of the flyback converter would be identical to that of the nonisolated indirect converter, except for the step-up or step-down created by the transformer. Primary side leakage stores energy that, in the circuit of Fig. 7.10(a), has no place to go when the transistor turns off. We usually place a snubber across the transistor to solve this problem.

Unlike the transformers used to provide isolation in the direct converter topologies discussed previously—which are transformers designed for minimum magnetizing current—the magnetizing inductance of the transformer used in a flyback converter must have a specified value in order to store energy. Such transformers are known as *energy-storage transformers*, and in the case of the flyback converter, they are called *flyback transformers*.

The Ćuk version of the indirect converter can also be isolated, as shown in Fig. 7.11. In this circuit, however, the transformer does not replace the filter inductor. Instead it operates the same as in the forward converter—it provides an impedance transformation but stores no energy.

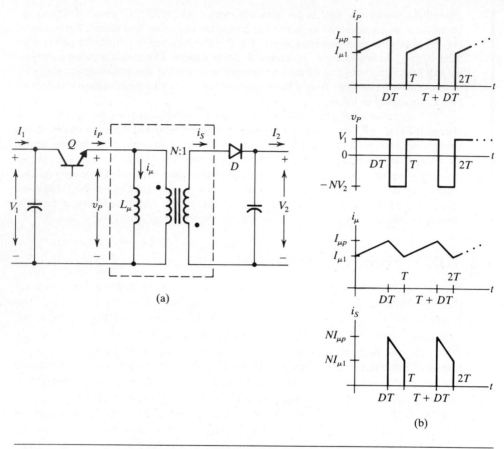

(a)

(b)

Figure 7.10 (a) The flyback converter, an isolated version of the indirect converter of Fig. 6.18(b). (b) Branch variables for the converter of (a).

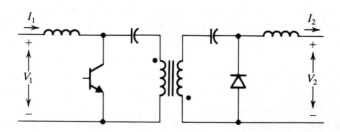

Figure 7.11 The isolated Ćuk converter.

7.4 EFFECTS OF TRANSFORMER LEAKAGE INDUCTANCE

So far in this chapter we have not considered the detailed effects of transformer leakage inductance. In this section we describe how the primary winding to secondary winding leakage inductance affects the operation of isolated converters.

7.4.1 Leakage Effects in the Single-Ended Converter

In the single-ended converter without leakage shown in Fig. 7.1, the load current immediately commutates from D_2 to D_1 when Q turns on and the voltage v_d makes a step change from zero to V_1/N. Figure 7.12(a) shows the same converter with leakage modeled as an inductor L_ℓ in series with the secondary winding. Now when the transistor turns on, the current in D_1 cannot make a step change. Instead, there is a commutation period of duration τ_{u_1} during which both diodes are on and the commutating voltage is $v_x = V_1/N$. The current in L_ℓ then increases linearly with slope V_1/NL_ℓ until it reaches the full load current I_2, as shown in the waveforms of Fig. 7.12(b). The time τ_{u_1} is

$$\tau_{u_1} = \frac{NL_\ell I_2}{V_1} \tag{7.13}$$

Only after τ_{u_1} does D_2 turn off and allow the voltage v_d to make its step change to V_1/N.

Because $v_d = 0$ instead of V_1/N during the commutation period, the output voltage $V_2 = \langle v_d \rangle$ is less than DV_1/N, its value without leakage. Therefore, to maintain a desired output voltage over a specific range of input voltage, we must either decrease the turns ratio N or increase the maximum duty ratio. For the

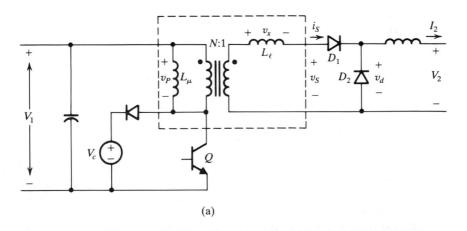

(a)

Figure 7.12   (a) A single-ended forward converter with leakage inductance.

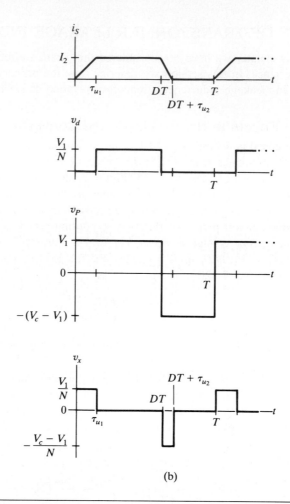

(b)

Figure 7.12 (cont.) (b) The behavior of i_S, v_d, v_P, and the commutating voltage v_x in the converter of (a).

same values of V_2 and I_2, decreasing N increases the peak and rms values of the transistor current, and the reverse voltage at the diodes. Increasing the maximum duty ratio increases the transistor's required voltage rating.

Turning the transistor off transfers the energy stored in L_ℓ to the clamp voltage source V_c in the following manner. The transformer primary voltage v_P immediately steps from $+V_1$ to $-(V_c - V_1)$, initiating commutation of I_2 from D_1 to D_2. The commutating voltage is now $-(V_c - V_1)/N$, so this commutation period, τ_{u_2}, is different from τ_{u_1}. During this time, the direction of energy flow is from the secondary winding to the primary winding, as the polarities of v_P and i_S show.

EXAMPLE 7.5

Determination of D for a Single-Ended Forward Converter with Leakage

The single-ended forward converter of Fig. 7.12(a) has the following parameters:

$$V_1 = 50 \text{ V} \qquad V_2 = 5 \text{ V} \qquad I_2 = 40 \text{ A} \qquad N = 5$$

$$L_\ell = 100 \text{ nH} \qquad L_\mu = \infty \quad \text{and} \quad f_s = 200 \text{ kHz}$$

What must D be to give the specified 5-V output?

With leakage inductance, we need to increase the duty ratio to account for the commutation interval τ_{u_1}. From (7.13), we determine this time to be

$$\tau_{u_1} = \frac{(5)(100 \times 10^{-9})(40)}{50} = 0.4 \ \mu s \tag{7.14}$$

This time corresponds to 8% of the switching period. Therefore $D = 58\%$. The clamp voltage must be high enough to satisfy the condition that the core be reset at this duty ratio, or:

$$V_1 D < (V_c - V_1)(1 - D) \tag{7.15}$$

From this condition, we determine that $V_c > 120$ V.

If this converter had no leakage, the duty ratio would be 50%, and V_c could be as low as 100 V. Because the transistor voltage rating must be at least as high as V_c, another result of leakage in this circuit is to increase the transistor voltage rating.

The peak stored energy E_ℓ in L_ℓ is

$$E_\ell = \frac{1}{2} L_\ell I_2^2 = 80 \ \mu J \tag{7.16}$$

However, because the input source V_1 is also supplying energy while L_ℓ is discharging, more than E_ℓ is absorbed by the clamp. We leave the calculation of this energy to an end-of-chapter problem, but the result in this case is that $E_c = 137 \ \mu J$, which is approximately 70% greater than E_ℓ. If this energy were dissipated every cycle, the power lost from leakage inductance would be 27.5 W, or about 14% of the output power. If instead we use the more efficient placement of the discharge resistor discussed in Example 7.2, we lose only E_ℓ per cycle, or 8% of the output power.

Note from Example 7.5 that the energy lost from leakage inductance in a single-ended forward converter is significant. For this reason, we want to return the clamp energy to the input voltage source using the methods shown in Figs. 7.3 and 7.4. Double-ended circuits automatically return the clamp energy to the source.

The presence of leakage also increases the required voltage rating of the transistor in the flyback converter—but for a different reason. Figure 7.13 shows a flyback converter from which the ideal part of the transformer has been removed, which does not change its operation in any way that is significant to this discussion. For simplicity we also assume that L_μ is large enough that i_μ is constant at I_μ. If

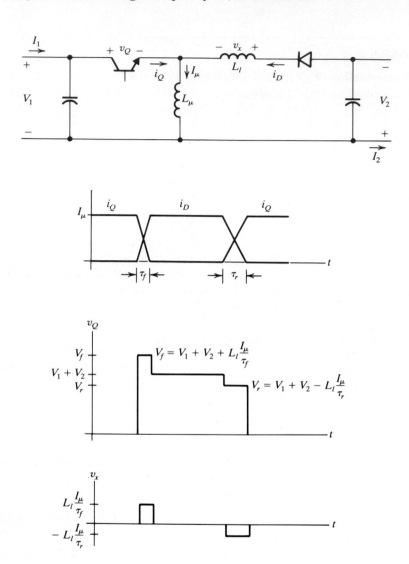

Figure 7.13   A simplified flyback converter analyzed to show the effect of leakage inductance. The magnetizing current is assumed constant at I_μ.

we now turn off Q so that i_Q falls linearly to zero in time τ_f, then during this fall time $v_x = L_\ell(I_\mu/\tau_f)$, and the transistor voltage is

$$v_Q = V_1 + V_2 + L_\ell\left(\frac{I_\mu}{\tau_f}\right) \tag{7.17}$$

This voltage is greater than the transistor voltage in the absence of leakage $(V_1 + V_2)$ by the amount necessary to force the current to commutate from Q to L_ℓ. If we were to clamp the transistor voltage at $V_1 + V_2$ while it was turning off, the current would never commutate to L_ℓ. Therefore leakage increases the required voltage rating of the transistor by an amount determined by the desired switching speed. This characteristic often makes the flyback topology less desirable than the forward converter topology, particularly at power levels high enough to justify the cost of the additional parts for the forward converter.

7.5 CONVERTERS WITH MULTIPLE OUTPUTS

Isolated converters supplying power to electronic equipment, such as computers, must usually provide several outputs at different voltages. Rather than make a separate supply for each output, it is often better to add additional secondary windings to the transformer, each with its own rectifier and output filter. The problem created by this approach is that, invariably, each output requires a slightly different value of duty ratio D because of different load regulation characteristics of the different secondary windings and rectifiers.

One approach to solving this problem is to control the highest power output directly with the duty ratio and to adjust the turns ratios of the other outputs to ensure that they will always be slightly high. We then use a linear regulator on each of these lower power outputs to give the desired voltage.

Notes and Bibliography

Some of the practical problems of using transformers in dc/dc converters are described nicely in Section 4.4 of [1]. The use of a transformer in a flyback converter and solutions to the problem of resetting the core are detailed in [2]. Harada's paper [3] discusses the effects of capacitive coupling between transformer windings in a dc/dc converter. Detailed models that include leakage are developed and analyzed for both two and three winding transformers.

The design of transformers for switching frequencies in excess of 100 kHz is discussed in [4]. This is an excellent reference, but you should understand Chapter 20 before attempting it.

1. R. P. Severns and G. Bloom, *Modern DC-to-DC Switchmode Power Converter Circuits* (New York: Van Nostrand Reinhold Co., 1985).
2. J. N. Park and T. R. Zaloum, "A Dual Mode Forward/Flyback Converter," in *IEEE Power Electronics Specialists Conference (PESC) Record* (1982), 3–13.
3. K. Harada, T. Ninomiya, and H. Kakihara, "Effects of Stray Capacitances Between Transformer Windings in the Noise Characteristics in Switching Power Converters," in *IEEE Power Electronics Specialists Conference (PESC) Record* (1981), 112–123.
4. N. R. Coonrod, "Transformer Computer Design Aid for Higher Frequency Switching Power Supplies," *IEEE Trans. Power Electronics*, 1 (4): 248–256 (October 1990).

Problems

7.1   Reconsider Example 7.1, but assume now that the output voltage is held constant at 5 V, while the input voltage varies between 50 and 100 V, that is, $50 < V_1 < 100$ V. What is the minimum value for V_c?

7.2   We stated in Section 7.2.1 that the option of opening all the switches in the double-ended bridge converter of Fig. 7.5(a) during the period $(1 - D)T$ created a problem if the transformer contained leakage. Describe this problem, and argue that it makes no difference whether the leakage is on the primary or secondary side of the transformer.

7.3   Determine and sketch the waveforms of i_μ, i_P, and i_S in Fig. 7.4(b) when leakage inductance is present between the primary and secondary windings of the transformer. Assume that all the leakage is on the secondary side, as shown in Fig. 7.14.

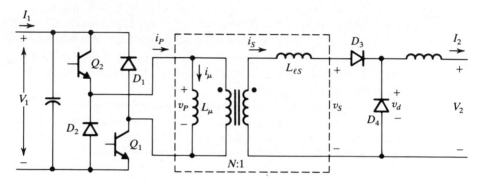

Figure 7.14   The single-ended converter of Fig. 7.4(b) with the addition of secondary leakage inductance $L_{\ell S}$. The effect of this leakage is the subject of Problem 7.3.

7.4   Repeat Example 7.3 under the assumption that the transformer has secondary side leakage $L_{\ell S}$, as shown in Fig. 7.15(a). How does the behavior of the circuit differ if the leakage is on the primary side, that is, $L_{\ell P} = N^2 L_{\ell S}$, as shown in Fig. 7.15(b)?

7.5   Repeat Example 7.3 under the condition that $I_{\mu p} > I_2/N$.

7.6   Compare the switch voltage stresses of the bridge converter of Fig. 7.6 and the push–pull converter of Fig. 7.8.

7.7   The waveforms of Fig. 7.8(b) were drawn for $I_{\mu p} < I_2$. Derive and draw these waveforms for the case $I_{\mu p} > I_2$.

7.8   Determine and sketch the branch variables i_1, i_μ, i_d, and v_P for the current-fed push–pull converter of Fig. 7.9(a) under the assumption that the magnetizing current is no longer negligible but that $0 < |I_{\mu p}| < I_2$. Model the magnetizing inductance by placing it across the lower primary winding. How would you implement S_1 and S_2 for this circuit?

7.9   Repeat Problem 7.8 under the assumption that $I_2 < |I_{\mu p}|$.

7.10 In many respects, the circuit of Fig. 7.9(a) is the dual of the circuit of Fig. 7.8(a). With this in mind, we might expect that the switch implementations, for nonzero magnetizing currents, would be duals of each other. As Problems 7.8 and 7.9 showed, this is not so. Where does the duality between these two circuits break down?

7.11 How do the device stresses in the isolated flyback converter of Fig. 7.10(a) compare with those in the nonisolated version having the same input voltage and the same output voltage?

7.12 Determine and sketch the branch variables shown in Fig. 7.10(b) assuming secondary side leakage.

7.13 When the transistor turns off in the circuit of Fig. 7.12(a), the energy absorbed by the clamp source V_c is greater than the energy stored in the transformer leakage L_ℓ. Determine and plot the energy absorbed by the clamp as a function of the ratio V_1/V_c. Interpret the result for $V_c = V_1$. What is the disadvantage of making V_c very large?

7.14 The centertapped converter topology of Fig. 7.8(a) can also be used to provide power flow from right to left, that is, from V_2 to V_1. How would you implement and control the switches in this case?

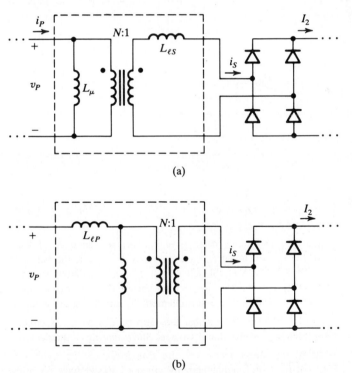

(a)

(b)

Figure 7.15 The double-ended converter of Fig. 7.5 with the addition of transformer leakage inductance: (a) all the leakage referred to the secondary winding; (b) all the leakage referred to the primary winding. This circuit is the subject of Problem 7.4.

7.15 The magnetizing branch of the transformer model in the push–pull converter of Fig. 7.8(a) was arbitrarily placed across the lower primary winding in the transformer model. However, the magnetizing branch may be connected in several places in the model. Figure 7.16 shows an alternative placement, where L'_μ is across the entire secondary winding. Show, by deriving and plotting the real transformer's terminal currents i_1, i_2, i'_1, and i'_2, that the behavior of this circuit is identical to that of Fig. 7.8. How is L'_μ related to L_μ in Fig. 7.8(a)?

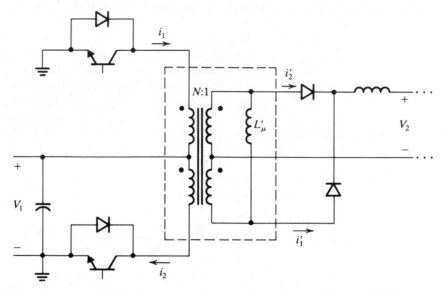

Figure 7.16 Push–pull converter of Fig. 7.8(a) redrawn with the magnetizing inductance L'_μ across the secondary winding. This circuit is analyzed in Problem 7.15.

7.16 Determine the conversion ratio V_1/V_2 as a function of D for the current-fed half-bridge circuit of Fig. 7.9(a).

7.17 Figure 7.17 shows the current-fed half-bridge converter of Example 7.4 with the magnetizing inductance L_μ now included. Determine the behavior of this circuit for the case $I_{\mu p} < NI_2$. Sketch i_1, i_2, i_μ, and i_d. What is the maximum possible value of i_μ? Does the bridge rectifier change the behavior of this circuit from that of Fig. 7.9(a)? How would you implement S_1 and S_2 in the circuit of Fig. 7.17?

7.18 We stated in Example 7.2 that the ratio of E_R to E_c is equal to the ratio of the resistor voltage to the clamp voltage. Show that this statement is true.

7.19 What is the stress factor for the transistors in the hybrid-bridge converter of Fig. 7.4(b)? (Neglect magnetizing current.) Based only on minimizing the switch stress factor, at what conversion ratio V_2/V_1 would you choose this isolated circuit over the nonisolated direct converter?

7.20 Show that the switch currents in the circuit of Fig. 7.8(a) are symmetrical.

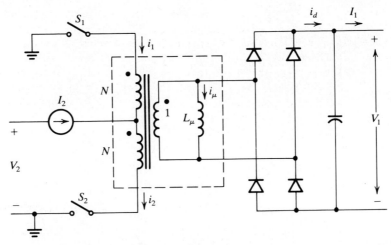

Figure 7.17   Current-fed half-bridge converter with magnetizing inductance. This circuit is the subject of Problem 7.17.

Chapter 8

Variable-Frequency dc/ac Converters

THE phase-controlled dc/ac converter introduced in Chapter 5 requires that the external ac system be a voltage source, typically the ac utility line. This condition is necessary because the phase-controlled converter uses the reversal of the ac voltage to drive the commutation process. Therefore the ac frequency in these circuits is constrained to be that of the ac source. In this chapter we remove the restriction that the ac system be a voltage source, realizing that by doing so we must use means other than line commutation to turn devices off. Alternatives might be the use of auxilary circuits to *force* the commutation of thyristors (we discuss these commutation circuits in Chapter 22) or the use of fully controllable switches, such as transistors or *gate-turn-off* (GTO) thyristors. (The GTO is a member of the thyristor family that can be turned off from its gate. We discuss it in Chapters 15 and 19.) However, the benefits are that we can control both the ac voltage and frequency. These circuits, which we call *variable-frequency dc/ac converters*, are sometimes referred to as *stand-alone converters*, because they do not need the presence of the ac line voltage to function.

The most common application of variable frequency dc/ac converters is to drive ac motors at varying speeds. These *variable-speed drives* are used to control the speed of electrically driven vehicles such as trains; to vary the speed of pumps and compressors so that they operate at maximum efficiency under varying loads; to control conveyer speeds; to control and coordinate the speed of sequential rollers in manufacturing operations such as found in steel, paper, and textile mills; and to control the speed and positioning of machine tools. Other applications of variable-frequency converters include uninterruptible power supplies (which use batteries to provide standby ac power), frequency changers, mobile power supplies, and systems that match ac loads to alternative energy sources that produce dc, such as photovoltaic arrays.

In the discussion of dc/dc converters in Chapter 6, we emphasized that inputs and outputs remain undefined until we specify the switches and external networks. The same is true of dc/ac converters. Until we specify the types of switches to be

167

utilized and the networks connected to the converter ports, we cannot tell whether power is flowing from the dc side to the ac side (inversion), or vice-versa (rectification).

8.1 THE BASIC VARIABLE-FREQUENCY BRIDGE CONVERTER

The basic bridge converter circuit is shown in Fig. 8.1(a). It is nothing more than a bridge connection of switches controlled so as to periodically reverse the polarity of the voltage applied across the ac system. Here, the ac system is a resistive load, R (so the circuit is an inverter), and in Fig. 8.1(b) the ac voltage shown is a square wave. The switches are controlled so that either switches S_1 and S_4 or S_2 and S_3 are on; that is, the source is always connected to the load. The ac frequency is controlled by the rate at which the switches open and close. An inverter circuit such as this, in which the switches create an ac *voltage* from a dc voltage source, is called a *voltage-source inverter*. Its complement, the *current-source inverter* creates an ac *current* from a dc current source.

We can control parameters of the ac voltage (its rms value or the amplitude of its fundamental component, for instance) by varying the dc port voltage. This requires a complicated dc system that might, for instance, use a phase-controlled

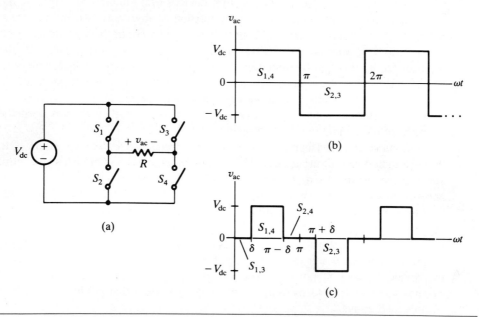

(a)

(b)

(c)

Figure 8.1　(a) A bridge inverter driving a resistive load. (b) Waveform of the ac voltage when diagonal switches open and close simultaneously. (c) Waveform of the ac voltage when the switches are controlled to provide a variable ac voltage.

rectifier or a dc/dc converter. This approach is relatively straightforward and we do not discuss it further. An alternative technique is to use a third switch state during which $v_{ac} = 0$ to create the waveform of Fig. 8.1(c). In the third switch state, switches S_1 and S_3 or S_2 and S_4, close for a time $2\delta/\omega$, shorting the ac system. A bridge converter capable of providing a zero voltage state at its output is known as a *tristate inverter*, with the states 1, 0, and -1 denoting the amplitude of the output relative to the dc input voltage. Which parameters of the ac voltage we control depends on the specific requirements of the load. Here, we have modeled the load as a resistor, so we might want to control the rms value of the output voltage. As a function of our controlling variable δ, $V_{ac\,rms}$ is

$$V_{ac\,rms} = \sqrt{\frac{1}{\pi} \int_{\delta}^{\pi-\delta} V_{dc}^2 \, d(\omega t)} = V_{dc}\sqrt{1 - \frac{2\delta}{\pi}} \qquad (8.1)$$

In general the ac loads for inverters are not as simple as the resistor of Fig. 8.1(a). Almost invariably the power factor of the load is not unity, and in many cases average power is transferred at only one frequency, generally the fundamental. For instance, an ac rotating machine (a machine without a commutator, mechanical or electronic) accepts or supplies average power only at that electrical frequency corresponding to the machine's mechanical speed.[†] Although the converters discussed in this chapter have the ability to produce variable-frequency ac, they are often connected to ac sources (or sinks) of constant frequency, such as a utility line. In these applications the need for a high power factor and low distortion interface to the line makes the simpler phase-controlled rectifier/inverter unsuitable, because the THD of its line current is high and its power factor is low at large values of α. Instead we can use a variable-frequency converter operated as described in Section 8.2.

8.1.1 Bridge Converters with Nonunity Power Factor Loads

Figure 8.2 shows a voltage-source inverter supplying a reactive load. If $L/R > \pi/\omega$, the third harmonic component of i_a is on the order of 10% of its fundamental, allowing us to approximate i_a as:

$$i_a(t) \approx I_{a_1}\sin(\omega t - \theta) \qquad (8.2)$$

where

$$\theta = \tan^{-1}\left(\frac{\omega L}{R}\right) \qquad (8.3)$$

$$I_{a_1} = \frac{V_{a_1}}{\sqrt{(\omega L)^2 + R^2}} \qquad (8.4)$$

[†]The relationships between power and frequency in multiport systems are described by the *Manley–Rowe relations*. See Penfield, *Frequency-Power Formulas*, MIT Press, 1960.

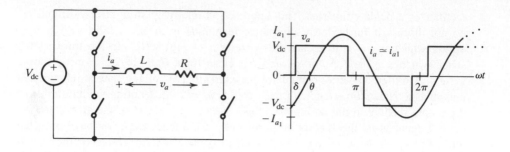

Figure 8.2   A variable-frequency bridge converter supplying a load with a nonunity power factor.

and

$$V_{a_1} = \frac{2V_{dc}}{\pi} \int_{\delta}^{\pi-\delta} \sin(\omega t)\, d(\omega t) = \frac{4V_{dc}}{\pi} \cos \delta \qquad (8.5)$$

The average power P delivered to R is

$$P = I_{a_1 \text{rms}} V_{a_1 \text{rms}} \cos \theta = \frac{I_{a_1} V_{a_1}}{2} \cos \theta \qquad (8.6)$$

Substituting (8.4) and (8.5) into (8.6) yields:

$$P = \frac{8 V_{dc}^2}{\pi^2 \sqrt{(\omega L)^2 + R^2}} \cos^2 \delta \cos \theta \qquad (8.7)$$

The conclusion we draw from this analysis is that the functional dependence of power on δ, as given by (8.7) for a reactive load, is different than it is for a resistive load. If we were to consider additional harmonic contributions to the power, the relationship would be more complicated, but would still depend on δ and the fundamental and harmonic displacement power factors. For the passive load of Fig. 8.2, the power factor is fixed, and we can control the power only by varying δ or V_{dc}.

8.1.2 Power Control for a Load Containing an ac Voltage Source

If we replaced the resistive load in Fig. 8.2 with an ac voltage source, we could use θ in addition to δ as a control variable. Figure 8.3 shows a full-bridge inverter connecting a dc voltage source, V_{dc}, to a load modeled by an inductance in series with a sinusoidal voltage source, v_{ac}. This model might represent, for instance, a single-phase synchronous motor under certain operating conditions (in which case the ac source is a model for the *back-EMF* of the motor), or a utility grid (in which case the dc source might represent a photovoltaic array). The inverter produces a tristate output as shown in Fig. 8.3. As we can now control the angle ϕ between v_a

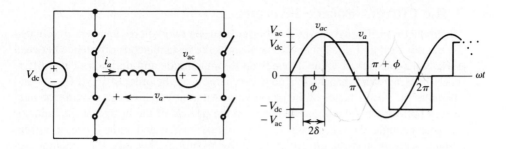

Figure 8.3 A full bridge inverter feeding an ac load with a source of EMF.

and v_{ac}, we can also control the phase angle between the fundamental component of i_a and v_{ac}, giving us another handle on power.

EXAMPLE 8.1

Control of an Inverter Feeding an ac Voltage Source

Our problem is to specify values of δ and ϕ for the inverter of Fig. 8.3 so that the power delivered to v_{ac} is 10 kW. The circuit parameters are

$$v_{ac} = 400 \sin(377t) \text{ V} \qquad V_{dc} = 350 \text{ V} \quad \text{and} \quad L = 10 \text{ mH}$$

No unique combination of δ and ϕ yields a power of 10 kW, unless we add a requirement to the performance of the circuit. The additional constraint we choose is that the ac source operate at unity power factor; that is, v_{ac} is in phase with i_{a_1}.

The complex amplitude of the fundamental component of the current i_a, $\hat{I}_{a_1}$, is

$$\hat{I}_{a_1} = \frac{\hat{V}_{a_1} - \hat{V}_{ac}}{j\omega L} \tag{8.8}$$

As v_{ac} and i_{a_1} are in phase, we can express the average power as:

$$\langle p(t) \rangle = P = \frac{I_{a_1} V_{ac}}{2} = 10^4 \text{ W} \tag{8.9}$$

which gives $I_{a_1} = 50$ A. Choosing $\hat{V}_{ac}$ as our reference for angular measurement (thus $\hat{V}_{ac} = V_{ac}$), we can express $\hat{V}_{a_1}$ in terms of ϕ and δ and use (8.5) and (8.8) to determine these angles, that is,

$$\frac{((4V_{dc}/\pi) \cos \delta)e^{-j\phi} - V_{ac}}{j\omega L} = 50 \text{ A} \tag{8.10}$$

By equating the real and imaginary parts on the two sides of (8.10), we obtain

$$\phi = -25.2° \quad \text{and} \quad \delta = 7.1°$$

8.1.3 The Current-Source Inverter

The inductor in Fig. 8.3 serves as a buffer between two voltage sources, v_a (created by the inverter) and v_{ac}. This inductor absorbs the instantaneous difference between these voltages, and its value depends on the magnitude and duration of the difference. In many cases its actual value is very large—on the order of $V_a/\omega I_{a\max}$—and degrades the power factor of the ac network. Alternatively, we can provide the necessary buffering by placing an inductor on the dc side of the bridge. If the inductor is large enough, we can model the dc voltage source and inductor as a current source. This is a frequently used buffering technique, because the inductor can be made arbitrarily large without degrading the power factor of the ac network. But it does degrade the dynamic performance of the converter, because more time is required for the load current to change in response to a control command. A current-source converter of this type is shown in Figure 8.4, and we now analyze its performance.

The circuit diagram of Fig. 8.4 alone contains insufficient information for us to know whether the function performed is inversion or rectification. Only when we specify the switch control can we determine the direction of power flow. Let's assume that it flows from the dc side to the ac side (inversion). The waveforms of Fig. 8.4 reflect this assumption.

The average power P delivered to the ac source is

$$P = \frac{1}{T} \int_0^T v_{ac} i_a \, dt = \frac{V_{ac} I_{a_1}}{2} \cos \theta \qquad (8.11)$$

where $I_{a_1} \cos \theta$ is the amplitude of the fundamental component of i_a that is in phase with the ac source v_{ac}. Thus for a fixed δ, we may control the average power delivered to the load by varying θ through switch timing. But because the amplitudes of all the frequency components of i_a are dependent on δ, we may also control the power by varying δ. Calculating I_{a_1} in terms of δ gives

$$I_{a_1} = \frac{2I_{dc}}{\pi} \int_\delta^{\pi-\delta} \sin(\omega t) \, d(\omega t) = \frac{4I_{dc}}{\pi} \cos \delta \qquad (8.12)$$

so the average power is

$$P = \frac{2V_{ac} I_{dc}}{\pi} \cos \delta \cos \theta \qquad (8.13)$$

In certain situations, controlling power by varying δ rather than θ may be more desirable because θ control generally requires bidirectional switches. You can see the need for switches with bidirectional blocking capability in the current-source inverter by studying the waveforms of Fig 8.4. Note that the voltage across the open switches is v_{ac}, which changes sign during the S_1, S_4 and S_2, S_3 conduction periods if δ is zero and θ varies. Thus each switch blocks both voltage polarities. Similarly, we can show that the switches in the voltage-source converter must carry bidirectional current. If θ is zero, δ control does not require bidirectional switches.

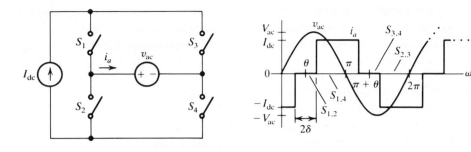

Figure 8.4 A bridge inverter with the dc-side network modeled as a current source. This circuit is known as a current-source inverter.

Another disadvantage of θ control is that it reduces power by decreasing the power factor at the load terminals. For instance, if we decrease the output power in the circuit of Fig. 8.4 to zero by setting $\theta = \pi/2$, we still have an output current with a peak of I_{dc}, and many of the inverter losses remain unchanged from their full-load ($\theta = 0$) values.

8.2 HARMONIC REDUCTION

Reducing the harmonic content of the ac port voltage or current is one of the most difficult challenges of dc/ac converter design. Harmonics not only reduce the power factor of the ac port, but also interfere with the proper operation of the converter or other equipment by appearing as noise in control circuits. Furthermore, if the converter drives an electromechanical load, such as an ac motor, harmonics in the ac waveform can excite mechanical resonances causing the load to emit acoustic noise.

You have already seen in Chapter 6 that low-pass filters consisting of a shunt capacitor and a series inductor were required to eliminate the ac components caused by switching from appearing at the terminals of a dc/dc converter. However, a low-pass filter is not nearly so effective when applied to dc/ac converters. The reason is that the ratio of the switching frequency to the input or output frequency is infinite for a dc/dc converter but finite for a dc/ac converter. For a dc/dc converter, therefore, the amount of ripple attenuation is limited only by the physical size of the inductor or capacitor or, perhaps, the desired control bandwidth. However, for the dc/ac converter the size and effectiveness of the filter elements are determined by factors such as how much attenuation or phase shift can be tolerated in the fundamental. As a result, we often control the switches in a dc/ac converter to achieve *active harmonic reduction*. In this section we discuss two ways of doing this. The first is *harmonic elimination*, in which we control the switches to eliminate certain harmonics. The second is *harmonic cancellation*, in which we add the outputs of two or more converters so as to cancel certain harmonics.

An alternative approach to harmonic reduction is to move the harmonics to frequencies high enough to make filtering possible with smaller components. This is called *pulse-width modulation* (PWM), which we discuss separately in Section 8.3.

8.2.1 Harmonic Elimination

An important potential benefit of controlling δ in the converter of Fig. 8.3 is that the amplitude of the third harmonic of v_a may be controlled. In fact, by appropriate choice of δ we can completely eliminate the third harmonic. The size of passive filters is generally determined by the lowest frequency to be eliminated. Thus elimination of the third harmonic by δ control has a major beneficial effect on the size of the ac-side filter components, because the lowest harmonic present in v_a will then be the fifth harmonic.

If the converter of Fig. 8.3 is controlled so that $\theta = 0$ (which is not necessary to the result, but makes the math a bit simpler), the third harmonic amplitude in v_a is

$$V_{a_3} = \frac{2V_{dc}}{\pi} \int_{\delta}^{\pi-\delta} \sin 3\omega t \, d(\omega t) = \frac{4V_{dc}}{3\pi} \cos 3\delta \qquad (8.14)$$

Therefore the third harmonic is eliminated if we control the switches so that $\delta = \pi/6$. In fact, all harmonics of order $3n$ are eliminated, and the waveform is said to be free of *triple-n* harmonics. Of course, if we fix δ, we no longer use this parameter to control v_{a_1} or power.

So far we have assumed that the 0-state of our tristate inverter occurs as a step between the 1 and -1 states. This is not necessary, nor is it necessary that there be only two 0-states per cycle. We can obtain more sophisticated harmonic control by creating 0-state regions, called *notches*, during the 1 and -1 states of the output waveform.

EXAMPLE 8.2

Simultaneous Elimination of Third and Fifth Harmonics

By placing a 0 state of width 2δ between the 1 and -1 states, and additional notches of width γ in the 1 and -1 states, we can simultaneously eliminate the third and fifth harmonics from a square wave. A waveforem of v_a containing a single notch in each half cycle is shown in Fig. 8.5. If we introduce two notches in each half cycle, it is easier to eliminate both the third and fifth harmonics. Instead of calculating the third- and fifth-harmonic components of such a waveform as functions of δ and the widths and locations of the notches we use a more insightful graphic technique to determine the position and widths of the additional notches.

We first consider the graphic representation of *(8.14)* shown in Fig. 8.6(a), focusing on only a half cycle of the fundamental because v_a is an odd function. The product of v_a

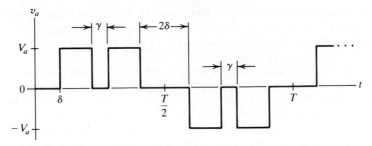

Figure 8.5   A tristate waveform containing notches of width γ.

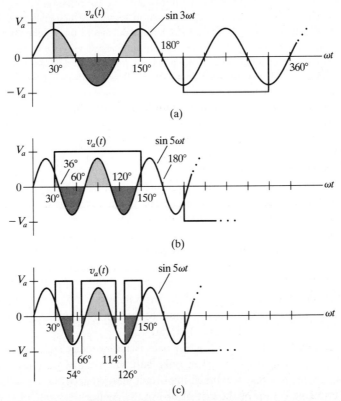

Figure 8.6   (a) Graphic representation of the integrand of (8.14). The positive (lightly shaded) and negative (heavily shaded) areas are equal, canceling the third harmonic of the square wave.   (b) The third-harmonic free waveform of (a) superimposed on a fifth-harmonic sine wave. (c) The square wave notched so that it is free from both third and fifth harmonics. A fifth-harmonic sine wave illustrates that the product of it and $v_a(t)$ has zero net area.

and a third-harmonic sinusoid is shown shaded. The integral of the shaded region is given by *(8.14)*. The positive lightly shaded area is equal to the negative heavily shaded area, which is the result given by *(8.14)* for $\delta = 30°$, the value for which Fig. 8.6(a) was drawn.

If we are to introduce notches into v_a while maintaining the cancellation of the third harmonic, we must position these notches so as to eliminate equal amounts of the positive and negative areas of Fig. 8.6(a). Notches placed symmetrically about 60° and 120°, the zero-crossings of the third-harmonic sinusoid, meet this criterion. Of course, these notches must also meet the other criterion of eliminating the fifth harmonic.

A fifth-harmonic sine wave and the waveform of v_a are superimposed in Fig. 8.6(b). Again, the result of integrating their product (the process whereby we calculate the fifth-harmonic amplitude) is shown as lightly (positive) and heavily (negative) shaded regions. The net result of the integration is negative, which is what we expect with the fifth harmonic of v_a being 180° out of phase with the sine wave shown. The points around which we can place notches without disturbing the third harmonic cancellation ($\omega t = 60°$ and 120°) are also shown in Fig. 8.6(b). These points are in regions of negative area, which is what we want to eliminate. The question now is how wide should the notches be?

Considering Fig. 8.6(b) and visualizing notches of varying width around the 60° and 120° points, we conclude that, if the notch has a width of 12°, we have achieved our goal of creating a product with a net area of zero. The resulting v_a, which is free of both the third and fifth harmonics, is shown in Fig. 8.6(c). The positive and negative small triangular segments cancel each other, and the two negative quarter cycles sum to cancel the single positive half cycle.

8.2.2 Harmonic Cancellation

We can derive an alternative harmonic reduction technique by recognizing that we can create the tristate waveform of v_a in Fig. 8.6(a) by adding two square waves of amplitude $V_{dc}/2$, shifted by 60° with respect to each other. Two voltage-source square-wave inverters configured this way are shown in Fig. 8.7. If the switching of one circuit is delayed by 60° with respect to the other, then v_a will have no third harmonic. We have already considered this cancellation in the graphic integration, which gave the coefficents of the terms in the Fourier series. However, we may interpret the lack of a third harmonic in v_a of Fig. 8.7 as resulting from the addition of two harmonic components of the same frequency (third) and amplitude, but differing in phase by $3(60°) = 180°$.

Although the circuit of Fig. 8.7 uses twice as many switches as that of Fig. 8.3, each of the switches is rated at only half the voltage of those of Fig. 8.3. This reduced voltage stress can be an important advantage if the ac voltage is very high.

Implementation of the circuit of Fig. 8.7 is not very practical because the two dc sources must be separate; that is, they cannot share a common terminal. A more practical circuit is shown in Fig. 8.8. The dc sides of the two bridges share a common terminal, but the ac sides are isolated by transformers whose secondaries are connected in series to sum the ac-side voltages.

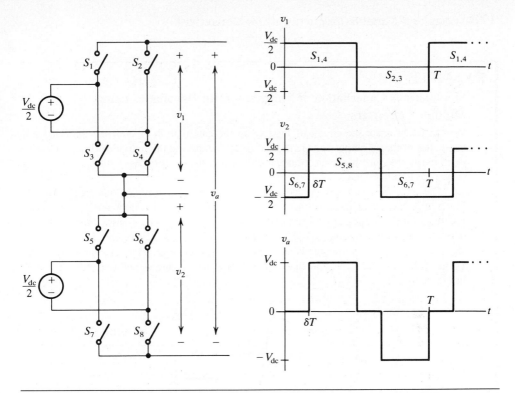

Figure 8.7 Harmonic cancellation achieved by adding the out-of-phase voltages of two similar bridge converters.

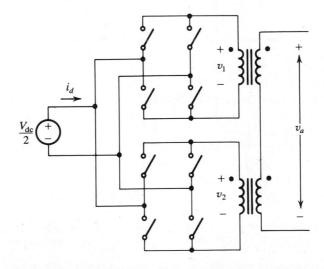

Figure 8.8 Two harmonic canceling bridge converters with a common dc-side voltage and transformer coupled ac-side voltages.

EXAMPLE 8.3

Simultaneous Cancellation of Third and Fifth Harmonics Using Multiple Converters

We can easily adapt the circuit of Fig. 8.8 to the cancellation of both the third and fifth harmonics of the ac output voltage. We do so by controlling the individual bridge switches to eliminate the third harmonic from v_1 and v_2 and then shifting v_2 with respect to v_1 to cancel the fifth harmonic.

The individual voltages v_1 and v_2 are similar to the third-harmonic free voltage of Fig. 8.6(a). Neither of these voltages contains a third-harmonic component, so we cannot reintroduce it by creating any linear combination of v_1 and v_2. Therefore we are free to shift v_2 so that its fifth harmonic is 180° out of phase with the fifth harmonic in v_1. This is a shift of $180°/5 = 36°$ of the fundamental period. Now when we add v_1 to the phase-shifted v_2, the fifth-harmonic terms will cancel. The appropriate v_1 and v_2 and their sum are shown in Fig. 8.9.

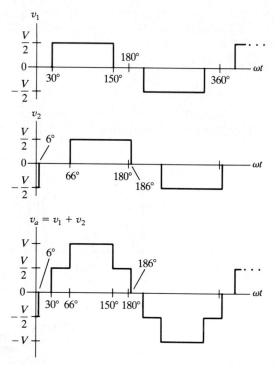

Figure 8.9   Waveforms for the circuit of Fig. 8.8, showing how the third and fifth harmonics can be simultaneously eliminated from the ac voltage.

We accomplish the phase shift of v_2 simply by delaying the switching sequence of the lower bridge by 36° relative to the upper bridge. The switching patterns of both bridges are identical. The resulting ac voltage waveform is called a *stepped waveform*, whereas the ac voltage of Fig. 8.5 is called a *notched waveform*.

The generation of a stepped waveform, such as v_a in Fig. 8.9, requires considerably more circuit complexity than the creation of a notched waveform. But the stepped waveform has a much lower THD. For instance, although the notched waveform of Fig. 8.6(c) and the stepped waveform of Fig. 8.9 are both free from third- and fifth-harmonic components, the notched waveform has a THD of 63%, compared to 17.5% for the stepped waveform. The notches in Fig. 8.6(c) eliminate only the fifth harmonic, while the addition of the two phase-shifted voltages to create the stepped waveform of Fig. 8.9 cancels more than just the fifth harmonic.

8.3 PULSE-WIDTH MODULATED DC/AC CONVERTERS

In the discussion of dc/dc converters in Chapter 6, we pointed out that the output voltage of a down converter could vary from zero to the input voltage V_i, depending upon the duty ratio D. If we were to vary D slowly, relative to the switching frequency, we could synthesize a waveform whose "average" value varied with time and was given by $d(t)V_i$. The averaging time must be long relative to the switching frequency but short relative to the rate of change of $d(t)$. We refer to the result of this averaging process on some quantity $x(t)$ as the *local average* of $x(t)$, and denote it by an overbar, or $\bar{x}(t)$.

An example of a down converter with a modulated $d(t)$ is shown in Fig. 8.10. Here $d(t) = 0.5 + 0.25 \sin \omega_a t$, and $T \ll L/R \ll 2\pi/\omega_a$. The voltage v_d is a *pulse-width modulated* (PWM) waveform with a dc component, a fundamental component of frequency ω_a, and additional, unwanted components at and above the switching frequency, $1/T$. The load voltage v_2 is the local average $\bar{v}_d$, which results from putting v_d through the low-pass filter composed of L and R. The local average contains a dc component of $V_1/2$.

In this section we explore the use of this high-frequency PWM, or *waveshaping*, technique in the control and construction of variable-frequency dc/ac converters. The advantage that high-frequency PWM techniques have over the relatively low-frequency switching techniques already discussed is that the undesirable harmonics in the output are at a much higher frequency and thus easier to filter.

8.3.1 Waveshaping and Unfolding

If we modulate $d(t)$ in Fig. 8.10 to produce a waveform $v_2 = |V_2 \sin \omega_a t|$, that is, the same waveform produced by a full-wave rectifier, we can then use a bridge of switches to "unfold" this waveform across the load resistor. A converter functioning in this way is shown in Fig. 8.11. The bridge transistors are switched at the cusps of v_2. The time-dependent duty ratio necessary to create v_2 is $d(t) = k|\sin \omega_a t|$, where k is a constant between zero and 1 and is known as the *depth of modulation*. The amplitude of the resulting sinusoidal load voltage is $V_2 = kV_1$. This circuit, although shown generating a sinusoidal output, actually is a more general *switching power amplifier*. If $f(t)$ is any arbitrary waveform with a bandwidth less than R/L

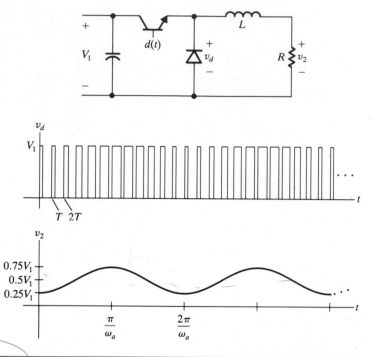

Figure 8.10 A down converter whose duty ratio is modulated to produce a load voltage consisting of a dc value and a sinusoid of frequency w_a much lower than the switching frequency.

(a Mozart sonata, for instance) normalized so that $|f(t)| < 1$, we could amplify this signal by setting $d(t) = |f(t)|$ and controlling the bridge switches according to the polarity of $f(t)$.

A problem with the circuit of Fig. 8.11 is that, contrary to the way the waveform appears, the wave-shaped voltage v_2 does not approach zero sinusoidally. Because the current i_2 falls at a rate equal to v_2/L and $v_2 = Ri_2$, both the current and v_2 approach zero asymptotically. Figure 8.12(a) shows the v_d and v_2 of Fig. 8.11 expanded around $w_a t = \pi$ to illustrate the problem. The fact that v_2 never quite reaches zero at times $wt = n\pi$ results in a step at these times in the unfolded waveform v_{ac}, as shown in Fig. 8.12(b). This *crossover distortion* results in the generation of harmonics of the output frequency w_a, which, if they are large enough, can compromise the advantage of using high-frequency PWM.

Reversing the voltage across the inductor L in Fig. 8.11 could force the current i_2 to zero, as we desire. We can do so by placing L inside the bridge, as shown in Fig. 8.13. The desired output, $v_{ac} = \bar{v}_a$ and the current i_a are also shown in the figure. Because the effective load now has an inductive component, the current lags the voltage. Because of this current lag, i_2 is negative for short periods. Therefore

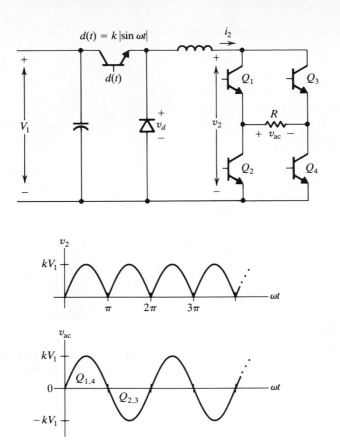

Figure 8.11 A waveshaping down converter and an unfolding bridge controlled to produce a sinusoidal load voltage.

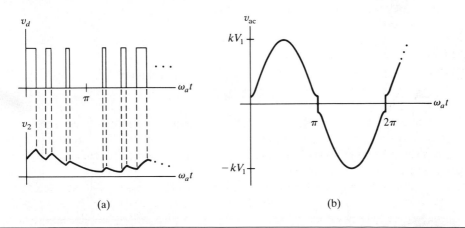

Figure 8.12 (a) An expansion of v_d and v_2 of Fig. 8.11 around $\omega_a t = n\pi$. (b) The output voltage v_{ac}, which exhibits crossover distortion.

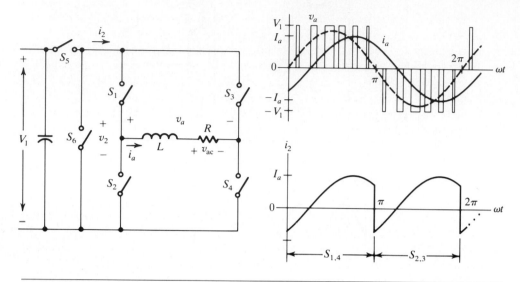

Figure 8.13   A waveshaping down converter and an unfolding bridge with the inductor placed inside the bridge to reduce crossover distortion.

we must configure the waveshaping down converter for two-quadrant operation: positive v_2 and positive or negative i_2. The bridge switches also must carry bipolar current.

A practical implementation of the dc/ac converter of Fig. 8.13 is shown in Fig. 8.14. Although this circuit reduces the crossover distortion present in v_{ac} in the circuit of Fig. 8.11, circuit control is more complicated. Not only must we now control the down converter as a function of the polarity of i_2, but also the potential

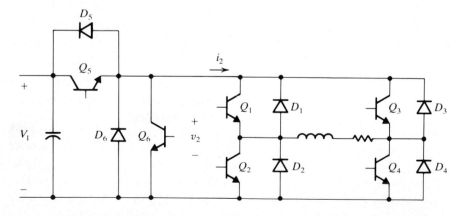

Figure 8.14   A practical implementation of the dc/ac converter of Fig. 8.13.

short-circuit paths that we have introduced require more precise switch timing. For example, any overlap in the times when Q_1 and Q_2 are conducting causes Q_5 to turn on into a short circuit. The inductor on the dc side of the bridge of Fig. 8.11 provided some buffering, but in its present position it does not. Furthermore, we added four low-frequency (bridge) diodes, one high-frequency diode (D_5), and one high-frequency transistor (Q_6). We can achieve the same result with a simpler circuit by incorporating the high-frequency waveshaping function in the bridge. We discuss this approach next.

8.3.2 The High-Frequency Bridge Converter

If S_5 remains on and S_6 remains off, we can create the waveform v_a in Fig. 8.13 by controlling the bridge switches only, thus eliminating the need for the waveshaping converter switches S_5 and S_6. When $v_a = V_1$, S_1 and S_4 are on; when $v_a = 0$, S_1 and S_3, or S_2 and S_4 are on; and when $v_a = -V_1$, S_2 and S_3 are on. Moreover, careful consideration of this switching sequence shows that when $v_a \geq 0$, S_4 can remain on and v_a can be created by switching S_1 and S_2 on and off in a complementary fashion. Similarly, when $v_a \leq 0$, S_3 remains on while S_1 and S_2 alternate. That is, S_1 and S_2 are switching at the high frequency of the carrier $1/T$, while S_3 and S_4 unfold the modulated carrier by switching at the much lower bridge frequency ω_a. We now have a circuit performing the same function as that of Fig. 8.13, with four switches instead of six, and yet only two of them are high-frequency switches.

A practical high-frequency bridge converter is shown in Fig. 8.15, along with a sketch of v_a showing the switching sequence. The low-frequency variation of v_a, $v_{ac} = \bar{v}_a$, is the load voltage we desire. Although in this case we have again assumed $\bar{v}_a$ to be a sinusoid, it can have any arbitrary time variation so long as the variation is slow enough to permit proper filtering. We choose the value of L to filter the high switching frequency from the load current, that is,

$$\frac{2\pi}{\omega_a} \gg \frac{L}{R} \gg T \qquad (8.15)$$

In addition to the desired low-frequency fundamental, v_a contains harmonics centered around the high switching frequency $1/T$. The advantage of PWM over the harmonic reduction schemes presented in Section 8.2 is that the harmonic distortion is moved to higher frequencies, making filtering easier. The THD of the PWM waveform, however, is greater than that of a square wave, as we show in Example 8.4. But the important harmonic issue is almost always the spectrum of the load current, which can be limited most easily in the case of a high frequency PWM voltage waveform. Furthermore, controlling the amplitude of the fundamental in a notched or stepped waveform is difficult, while simultaneously trying to control harmonics.

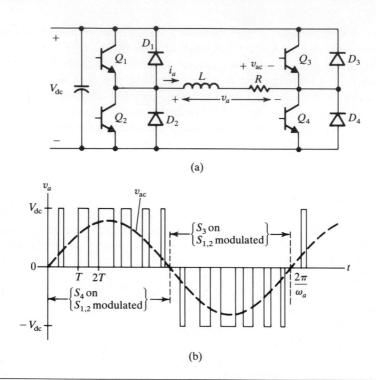

Figure 8.15   (a) The high-frequency pulse-width modulated (PWM) bridge converter, showing a practical switch implementation.   (b) The PWM voltage v_a and the output voltage $v_{ac} = \bar{v}_a$.

EXAMPLE 8.4

THD of a Sinusoidal PWM Waveform

In this example we calculate the THD of a sinusoidal PWM waveform, such as that of v_a in Fig. 8.15(b), and compare it to the THD of a square wave. We create the PWM waveform by varying $d(t)$ sinusoidally, so that:

$$d(t) = |\sin \omega_a t| \qquad (8.16)$$

If the switching frequency is much greater than the ac output frequency, we can assume that $\sin \omega_a t$ is constant over each switching period. If the amplitude of the dc voltage is V_{dc}, the height of each pulse is V_{dc}, and the rms value squared of a pulse occurring at time t_o is

$$v_{arms}^2(t_o) = V_{dc}^2 d(t_o) = V_{dc}^2 \sin \omega_a t_o \qquad (8.17)$$

We can now determine the rms value squared of the PWM pulse train, V_{arms}^2, by averaging (8.17) over half a cycle of v_a:

$$V_{arms}^2 = \frac{1}{\pi} \int_0^\pi V_{dc}^2 \sin \omega_a t \, d(\omega_a t) = \frac{2V_{dc}^2}{\pi} \qquad (8.18)$$

Because we assumed that $d(t)$ varies from 0 to 1, v_{ac} has a peak value of V_{dc} and an rms value of $V_{acrms} = V_{dc}/\sqrt{2}$. We can now calculate the THD using the definition given by (3.45):

$$\text{THD} = \sqrt{\frac{V_{a\,rms}^2 - V_{ac\,rms}^2}{V_{ac\,rms}^2}} = 52\% \tag{8.19}$$

To put this THD value into perspective, let's compare it to that for a square wave of fundamental frequency ω_a. If the square wave has an amplitude V, its rms value is also V, and its fundamental has an amplitude of $4V/\pi$ with an rms value of $2\sqrt{2}\,V/\pi$. The THD for this square wave is

$$\text{THD} = \sqrt{\frac{V^2 - (2\sqrt{2}V/\pi)^2}{(2\sqrt{2}V/\pi)^2}} = 48\% \tag{8.20}$$

Thus $v_a(t)$ and a square wave have essentially the same THD, although the significant harmonics of the PWM waveform are at much higher frequencies than those of the square wave. If we supplied the load through a simple inductive filter, as shown in Fig. 8.15, the required value of the inductor to obtain a given THD for the current would be much smaller for the PWM converter than for a square-wave converter.

8.3.3 Generation of d(t) for the PWM Inverter

The conventional way of generating a sinusoidal $d(t)$ for a PWM inverter is to use the *sine-triangle intercept* technique. Its essential features are shown in Fig. 8.16. A full-wave rectified sine wave v_S, with peak amplitude k, and a unipolar triangle

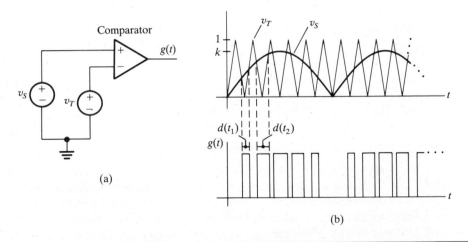

(a)

(b)

Figure 8.16   The sine-triangle intercept technique for generating a pulse train $g(t)$, having a sinusoidally varying duty ratio $d(t)$.   (a) A circuit that compares a sine waveform v_S to a triangle waveform v_T to produce the pulse train $g(t)$.   (b) The waveforms of v_S, v_T, and $g(t)$.

wave v_T, are fed into a comparator whose output $g(t)$ is then high when the value of the sine wave exceeds that of the triangle wave. The duration of each output pulse is therefore weighted by the value of the sine wave at that time; that is, $d(t) \approx k|\sin \omega_a t|$, which is what we desire. These pulses then drive the high-frequency transistors in the circuit of Fig. 8.15(a), alternating between Q_1 and Q_2 each half cycle.

We control the amplitude of the fundamental by varying the depth of modulation k. This, in turn, varies the absolute widths of the pulses comprising $g(t)$, while retaining their relative widths. However, decreasing the amplitude of the fundamental by decreasing the depth of modulation increases the THD of v_a.

8.4 TRANSFORMER-COUPLED CONVERTERS

In principle, we can couple any dc/ac converter to a load through a properly designed transformer. For instance, we can couple both the PWM bridge converter of Fig. 8.15 and the square-wave converter of Fig. 8.1 through a transformer by replacing the resistor with the transformer primary and loading the secondary with the resistor. In both cases the transformer must have sufficient core cross-sectional area to carry the fundamental frequency magnetic flux without saturating. (We discuss this requirement in detail in Chapter 20.) The important constraint is that the product of the core cross section A_c, the maximum permissible flux density B_s, and the number of turns on one of the windings N must exceed the maximum value of the integral with respect to time of the voltage across that winding. That is,

$$NA_cB_s \geq \int v_{ac}(t)dt \qquad (8.21)$$

The maximum value of the volt–time integral is about the same for the output voltage of either the square-wave or PWM converter. The PWM converter, however, generates a voltage v_a containing components of much higher frequencies than those produced by the square-wave converter. Consequently, we must use a magnetic material with better high frequency characteristics in the transformer for the PWM circuit, unless these high frequencies are filtered before the transformer.

8.4.1 High-Frequency Transformer Isolation

The PWM approaches that we have discussed so far result in a component of v_a at the low frequency ω_a. By utilizing a modulating process that moves all components of the modulated waveform to frequencies in the vicinity of the switching frequency, we can use a much smaller transformer to provide the requisite isolation. Unfortunately, a demodulating circuit (instead of a simple low-pass filter) is now required to recover the low frequency ac waveform.

A modulation technique that moves the lowest frequency component of the transformer voltage to the switching frequency is shown in Fig. 8.17(a). We con-

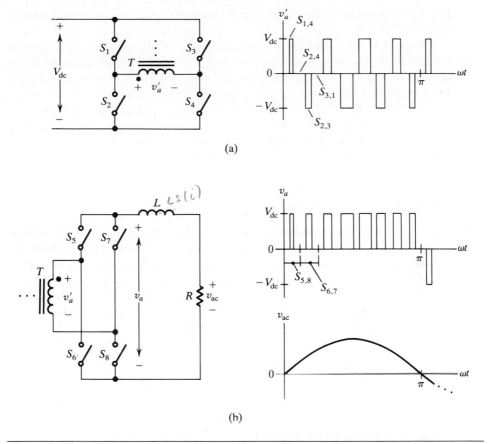

(a)

(b)

Figure 8.17   A method of PWM modulation that translates all frequency components of v_a' to the region above the switching frequency, permitting the use of a small transformer: (a) the modulator and transformer primary voltage;   (b) the demodulator and its waveforms on the transformer secondary. Transformer T has a 1:1 turns ratio.

trol switches S_1–S_4 to create pulses of alternate polarity across the transformer primary. The result is a waveform v_a', whose maximum volt–time integral value is approximately the area of the largest pulse. Contrast this result to the sum of the areas of all the pulses between $\omega_a t = 0$ and π for the PWM inverter of Fig. 8.15. Therefore the transformer linking the modulator and demodulator in the converter of Fig. 8.17 is considerably smaller than the one in the converter of Fig. 8.15. Although *(8.21)* shows that we can reduce the transformer size (implicit in A_c) in direct proportion to the reduction in the volt–time integral, the size reduction is less than this in practice. The reason is the higher frequency waveform (that of

Fig. 8.17a) generally requires the use of a magnetic core material with a lower value of B_s.

Figure 8.17(b) is one version of a demodulator that we can use to recover the desired low frequency waveform from v_a'. The circuit shown is similar to the unfolder of Fig. 8.11, except that it unfolds each pulse instead of each half cycle of the desired output waveform. The voltage v_a of Fig. 8.17(b) is simply the PWM waveform of Fig. 8.15, which we can low-pass filter to give the desired voltage v_{ac}.

The price we pay for the relatively small transformer in the circuit of Fig. 8.17 is greater control complexity and the need for all the switches in the circuit to operate at the switching frequency.

8.5 3-PHASE CONVERTERS

Many applications, particularly the control of rotating machinery, require 3-phase ac sources of varying frequency. We can use together three of any of the single-phase inverters discussed so far to generate a 3-phase set of waveforms. We do so by controlling the switches of each inverter to produce an output phase shifted by $\pm 2\pi/3$ with respect to the outputs of the other two inverters. However, six of the 12 bridge or unfolding switches are redundant; a practical 3-phase inverter circuit contains only six switches, configured as a 3-phase bridge.

8.5.1 Evolution of the 3-Phase Inverter Circuit

Figure 8.18(a) illustrates the connection of three inverters to create a symmetrical 3-phase output. However, 3-phase loads are seldom three separate loads isolated from one another; they are generally connected in a Y or Δ configuration. In the Y connection, one output terminal of each inverter is attached to the common point of the Y. In the Δ connection, the inverter outputs are connected in series. Because all three inverters share a common dc bus, we cannot control their switches arbitrarily but must coordinate them to prevent shorting the dc source.

If we consider the Δ connected load and the connection of two of the converter bridges, the circuit of Fig. 8.19(a) results. Here, it is clear that S_{a_3} and S_{b_1} are redundant, as are S_{a_4} and S_{b_2}. Note that this redundancy also eliminates the independent operation of the two switches and thus the arbitrary control of the individual converters. If the third converter were added to the figure, the switches S_{a_1}, S_{a_2}, S_{b_3}, and S_{b_4} would each have a redundant counterpart. Consequently, we can eliminate six switches from the individual converter version. The resulting 3-phase–bridge converter circuit is shown in Fig. 8.19(b).

The vast majority of 3-phase loads are balanced (the phases are loaded by equal impedances) and are supplied from a balanced set of voltages. The result is an absence of neutral current if the load is Y connected. In this case the converter of Fig. 8.19(b) could just as well drive a Y load.

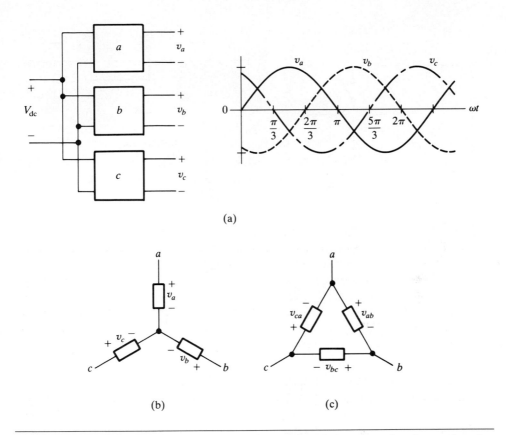

Figure 8.18   (a) A 3-phase converter consisting of three single-phase circuits.
(b) A Y connected 3-phase load.   (c) A Δ connected 3-phase load.

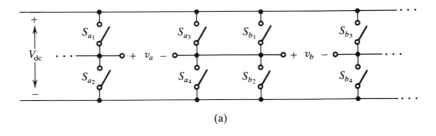

Figure 8.19   Simplification of the independent converter circuit for the case of a
Δ connected load.   (a) Connection between two of the independent
converters.

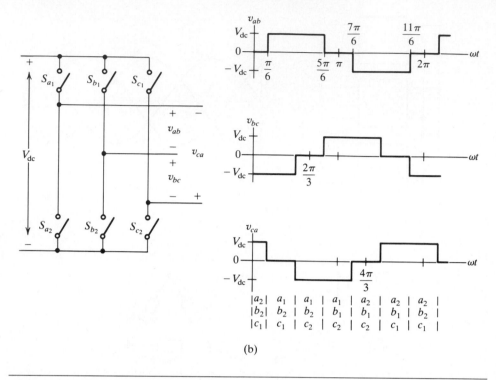

(b)

Figure 8.19 (cont.)   (b) The simplified 6-switch 3-phase converter and its phase-to-phase waveforms.

EXAMPLE 8.5

A 3-Phase Inverter with a Centertapped dc Voltage Source

A 3-phase inverter feeding a Y connected load from a centertapped dc source is shown in Fig. 8.20. In this circuit a separate half-bridge inverter supplies each phase, making independent control of the phase voltages possible. Therefore the conduction angle of each bridge leg may exceed 120°, as shown in the waveforms of Fig 8.20, where each switch is conducting for 180°. However, the circuit is capable only of bistate operation, the zero volt state is not available.

The benefits of this circuit are that the centerpoint of the Y load can be grounded to the source, better use is made of the switches because they conduct for 180° instead of 120°, and the circuit ensures that the Y connected load has a balanced 3-phase voltage, even if the load itself is not balanced. The disadvantages are that the circuit cannot produce a zero volt state, and the load *must* be Y connected with an accessible centerpoint. With no centerpoint connection, the circuits of Figs. 8.20 and 8.19(b) are identical.

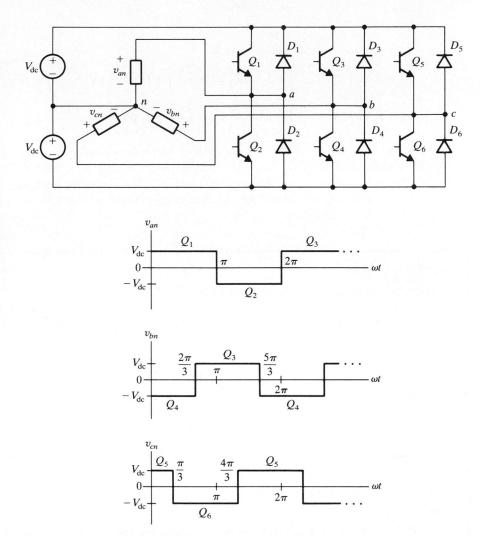

Figure 8.20 A 3-phase inverter with a centertapped dc source and its phase-to-neutral waveforms.

8.5.2 Harmonics in 3-Phase Inverter-Driven Loads

Three-wire 3-phase loads, that is, loads without a centerpoint connection, impose important constraints on the harmonic content of the phase voltages and currents. For instance, if we apply KVL and KCL to the output terminals of the inverter of Fig. 8.19(b), both the voltages and the currents must sum to zero. If we express these currents and voltages as Fourier series, the three current or voltage series cannot contain terms that add to nonzero values. For example, if the three phase currents are

$$i_a = I_1 \sin \omega t + I_2 \sin 2\omega t + I_3 \sin 3\omega t + \cdots$$

$$i_b = I_1 \sin\left(\omega t + \frac{2\pi}{3}\right) + I_2 \sin 2\left(\omega t + \frac{2\pi}{3}\right) + I_3 \sin 3\left(\omega t + \frac{2\pi}{3}\right) + \cdots$$

$$i_c = I_1 \sin\left(\omega t - \frac{2\pi}{3}\right) + I_2 \sin 2\left(\omega t - \frac{2\pi}{3}\right) + I_3 \sin 3\left(\omega t - \frac{2\pi}{3}\right) + \cdots$$

the sum $i_a + i_b + i_c$ would consist only of those harmonics of order $3n$ (called the *triple-n* harmonics). But KCL precludes existence of these harmonics. We can show the same to be true of the voltages v_{ab}, v_{bc}, and v_{ca}. The interesting consequence is that it is impossible to control the switches of the inverter of Fig. 8.19 so as to generate triple-n harmonics.

EXAMPLE 8.6

Triple-n Harmonics in a Balanced 3-Phase Load

The inverter of Fig. 8.19(b) is supplying a Δ connected resistive load, as shown in Fig. 8.21(a). We show that the phase currents do not contain any triple-n harmonics.

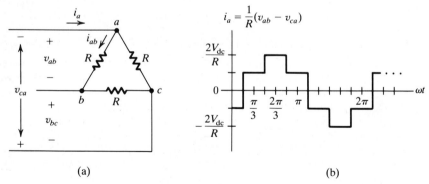

(a)   (b)

Figure 8.21   (a) A Δ connected 3-phase load supplied by the inverter of Fig. 8.19(b).
(b) The a-phase current.

The phase current i_a is the sum of the currents in the two branches connected to phase a, or:

$$i_a = \frac{v_{ab}}{R} - \frac{v_{ca}}{R}$$

This current waveform is shown in Fig. 8.21(b). The b- and c-phase currents are identical in form but shifted by $\pm 120°$ relative to the a-phase current. As the phase currents are the sum of two branch currents, they cannot contain any harmonics that are not already present in the branch currents. Therefore we show that there are no triple-n harmonics in the simpler branch current waveforms, which are the branch voltages of Fig. 8.19(b) divided by the branch resistance R. Arbitrarily picking the a–b branch, the harmonic amplitudes I_n of the branch current i_{ab} are

$$I_n = \frac{2V_{\text{dc}}}{R\pi} \int_{\pi/6}^{5\pi/6} \sin n\theta \, d\theta = \frac{2V_{\text{dc}}}{Rn\pi} \left[\cos \frac{n\pi}{6} - \cos \frac{5n\pi}{6} \right] \qquad (8.22)$$

Because I_n is zero for values of n that are multiples of 3, the phase currents contain no triple-n harmonics.

Notes and Bibliography

Harmonic elimination through proper choice of PWM patterns is the subject of [1]–[3]. In [1] Patel and Hoft derive patterns to eliminate five harmonics from the ac waveform, and in [2] they consider the result of varying these patterns to control voltage as well as harmonics. A hardware controller that does not employ a look-up table for pattern selection, and which allows continuous control of the fundamental amplitude, is the subject of [3].

The problem of avoiding transformer saturation due to asymmetry in the switching patterns of a PWM inverter is addressed in [4]. This problem is critical to any high-frequency converter using a transformer. The authors suggest a means for sensing impending saturation and for making corrections to the switching pattern to avoid saturation.

The use of high-frequency PWM to generate a sinusoidal current is the subject of [5]. The particular application is a utility-mains interface, such as might be required for a solar-photovoltaic source. The size of the filter elements required to produce a specified THD is determined. The crossover distortion depicted in Fig. 8.12 is clearly illustrated in [5].

1. H. S. Patel and R. G. Hoft, "Generalized Techniques of Harmonic Elimination and Voltage Control in Thyristor Inverters: Part I—Harmonic Elimination Techniques," *IEEE Trans. Industry Applications* IA–9 (3): 310–317 (May/June 1973).
2. H. S. Patel and R. G. Hoft, "Generalized Techniques of Harmonic Elimination and Voltage Control in Thyristor Inverters: Part II—Voltage Control Techniques," *IEEE Trans. Industry Applications* IA–10 (5): 666–673 (September/October 1974).
3. S. Bolognani, G. S. Buja, and D. Longo, "Hardware and Performance-Effective Microcomputer Control of a Three-Phase PWM Inverter," in *International Power Electronics Conference (IPEC) Record*, (Tokyo, 1983), 360–371.
4. H. R. Weischedel and G. R. Westerman, "A Symmetry Correcting Pulsewidth Modulator for Power Conditioning Applications," *IEEE Trans. Industry Applications* IA–9 (3): 318–322 (May/June 1973).
5. M. F. Schlecht, "Novel Topological Alternatives to the Design of a Harmonic-Free Utility/DC Interface," in *IEEE Power Electronics Specialists Conference (PESC) Record* (1983), pp 206–214.

Problems

8.1 Determine the average power delivered to the load in Fig. 8.2 if the third-harmonic component of i_a is not neglected. (*Hint:* do your analysis in the frequency domain.)

8.2   In Example 8.1 we determined values of ϕ and δ that would result in 10 kW being delivered to the source v_{ac} at unity power factor. Redo that example under the constraint that the *inverter* has a unity power factor at the fundamental frequency. That is, i_{a_1} and v_{a_1} are in phase. Which of these two control schemes results in the lowest value of rms load current?

8.3   Determine and sketch the load current i_a in Example 8.1. (*Hint:* Use circuit replacement by equivalent source to represent v_a and then use superposition of the responses to v_a and v_{ac}.) Can switches permitting only unilateral current flow be used in this circuit? Explain.

8.4   Determine graphically the switching sequence for the inverter of Fig. 8.1 that would result in elimination of the 7th harmonic of the load voltage. What has happened to the 3rd and 5th harmonics relative to their amplitudes for a true square wave, that is, a square wave having no 0 state?

8.5   The waveforms of Figs. 8.6(c) and 8.9 were constructed to be free of both the third and fifth harmonics. Compare these waveforms with respect to their THDs.

8.6   Example 8.2 showed how both the third and fifth harmonics could be eliminated by appropriately placed notches. However, doing so eliminates the possibility of controlling the fundamental of v_a. But we can use the notches in the waveform of Fig. 8.6(c) to control the fundamental of v_a if we are not required to eliminate the 5th harmonic. Determine the dependence of the fundamental on the width of the two notches in v_a of Fig. 8.6(c).

8.7   Calculate the THD of the dc source current i_d in Fig. 8.8 if the converters are controlled as described in Example 8.3. Assume that the converter has a resistive load at its ac terminals. How does the THD of this current compare to the THD of the ac terminal voltage v_a?

8.8   Show that decreasing the depth of modulation of a PWM-generated sinusoid, such as that shown in Fig. 8.11, increases the THD of the PWM waveform.

8.9   Example 8.3 showed how multiple converters could be used to eliminate the third and fifth harmonics from the ac voltage. How would you eliminate three harmonics (e.g., third, fifth, and seventh) using two bridge converters? Draw the resulting ac voltage waveform.

8.10   The low-pass filter used in the PWM inverter of Fig. 8.17 introduces a phase shift between the fundamental of v_a and v_{ac}. Express this phase shift in terms of L and R.

8.11   The resistive load in the PWM inverter of Fig. 8.17 is replaced by a sinusoidal voltage source having the same polarity as that of v_{ac} shown. How would the eight switches be implemented and controlled so that power flows from this ac source to the dc side of the circuit?

8.12   Figure 8.22 shows a half-bridge inverter driving a load consisting of a sinusoidal voltage source v_s in series with an inductor L. How should the switches be controlled to maximize the power *delivered to* the load? Express this maximum power in terms of the parameters of the circuit. How could the switches be implemented for this condition?

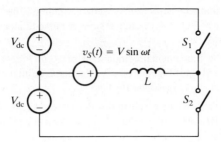

Figure 8.22 The half-bridge inverter discussed in Problem 8.12.

8.13 A balanced Y connected load is supplied by the inverter of Fig. 8.19(b). The switches are controlled so that their conduction angles are all 120°. Sketch the Y-branch voltages, that is, the voltages between the phase lines and the Y centerpoint. How does the harmonic content of these voltages compare with that of the phase voltages if the load were connected in Δ?

8.14 Is there any operating condition for the current-source inverter of Fig. 8.4 under which the switches are required to carry bilateral current?

8.15 Propose an implementation of the switches in the inverter of Example 8.1.

8.16 In Section 8.2 we discussed the elimination of harmonics from the inverter's ac terminal variables. It is almost always equally important to consider harmonics generated in the dc variables. Calculate, sketch, and dimension the dc terminal current for the inverter of Fig. 8.3 under the operating conditions determined in Example 8.1.

   One possible filter circuit designed to eliminate the ac components from the dc input current is shown in Fig. 8.23. Using intelligent approximations, determine values for L and C so that the peak-to-peak ripple current at the dc source is 5% of the dc current, and the peak-to-peak ripple voltage at the inverter input terminals is 5% of V_{dc}.

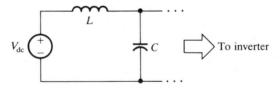

Figure 8.23 Input filter for the inverter of Fig. 8.3. Values for L and C are determined in Problem 8.16.

8.17 Reconsider the filter designed in Problem 8.16. Sketch and dimension the inverter input voltage if all four switches of the inverter are suddenly opened simultaneously. Does this behavior create a problem? If so, how would you solve it?

8.18 In the text we stated that more than just the third and fifth harmonics had been canceled in the waveform of v_a in Fig. 8.9. What are the other harmonics that had been canceled?

8.19   Assuming that a centertapped secondary winding is permissible in the high-frequency transformer of Fig. 8.17, design a simpler demodulator circuit to replace the 4-switch bridge shown.

8.20   Compare the fundamental components of v_a of Fig. 8.6(a), which contains no third harmonic, and v_a of Fig. 8.6(c), which contains neither the third nor the fifth harmonic. Assume that V_a is the same for both.

8.21   Compare the THD of a sinusoidal PWM waveform, as calculated in Example 8.4, with the THD of v_a in Fig 8.6(c), which is notched to eliminate both the third and fifth harmonics.

Chapter 9

Resonant Converters

RESONANT converters, like the phase-controlled converters of Chapter 5, connect a dc system to an ac system. They also share the feature that the switching frequency equals the fundamental ac frequency. But unlike phase-controlled converters, resonant converters usually control energy flow by varying their switching frequency, not their phase. When controlled in this manner, resonant converters can only be used if the exact frequency of the ac system does not matter or when the dc voltage can be varied to control power.

One application of resonant converters is induction heating, in which a coil driven by an ac current induces a current by transformer action in a conductive object. The dissipation caused by this induced current heats the object. Because this heating process works satisfactorily over a relatively broad range of frequency (at least two to one), the resonant converter driving the coil can make use of this frequency range to control the power. Another common application of the resonant converter is in very high-frequency dc/dc power supplies, where the converter's ac output is rectified and filtered to produce dc. This approach to dc/dc conversion is particularly useful when an isolation transformer is used, because an intermediate ac waveform must be generated anyway.

Although there are many topological variations of the resonant converter, our purpose in this chapter is to highlight their common features, rather than to analyze specific topologies. The following is a summary of these features.

First, the switches in a resonant converter create a square-wave ac waveform from the dc source. Inductors and capacitors then remove the unwanted harmonic components from this square wave. As the difference in frequency between the fundamental component and the lowest harmonic (the third) of the square wave is so small, we use a resonant LC circuit tuned to approximately the switching frequency, rather than a simple low-pass filter, to remove harmonics from the fundamental. Hence the name *resonant converter*. The tuned filter can be very selective if its Q is high enough to give good peaking in its impedance versus frequency characteristic. This selectivity, discussed in more detail in Section 9.1, also provides one of the means by which we control power.

Second, because the network composed of the resonant filter and the external ac system has a reactive impedance at all but its resonant frequency, the switches in a resonant converter must be able to transfer energy in both directions. We therefore implement each switch to either carry a bipolar current or block a bipolar voltage. Thus we can also use a resonant converter designed for average power flow from the dc system to the ac system to transfer energy in the other direction should the application require it.

Third, the resonant converter's semiconductor devices can have significantly lower switching losses than those of the semiconductor devices in a high-frequency dc/dc or dc/ac converter. The energy lost when a device Q switches on or off is

$$E_{\text{loss}} = \int_0^{t_{\text{on,off}}} v_Q i_Q \, dt \qquad (9.1)$$

where $t_{\text{on,off}}$ is the time it takes the semiconductor device to turn on or off, that is, the rise or fall time of its current and/or voltage. We can design a resonant converter to have one of the switch variables remain near zero during this time, resulting in low switching losses. Note the word *can*. Not all resonant topologies have this feature. For those that do, the reduced dissipation allows us to use certain devices (for example, bipolar transistors) in a resonant converter at a higher switching frequency than possible otherwise in a high-frequency switching converter. This advantage is often the basis for making a decision to use the resonant converter topology. Unfortunately, in return for the lower switching losses, the semiconductor devices are subjected to higher on-state currents and off-state voltages, giving them higher stress parameters than they would have in a nonresonant topology operating at the same power level. Therefore more expensive devices are often required, and higher conduction losses usually result.

Finally, there are two approaches to the resonant converter—one the dual of the other. In the first, the switches create a square wave of voltage that is applied to a series resonant circuit. This is called a *series resonant converter*. In the second, the switches create a square wave of current that is applied to a parallel resonant circuit, resulting in the *parallel resonant converter*.

In many resonant converter circuits the switch currents oscillate and would reverse direction if the switch could carry bilateral current. These circuits are especially well suited to the use of SCRs, as they can be turned off by the resonant action of the circuit trying to force the current to become negative. This process is called *resonant commutation*.

To simplify the discussion, we assume that the converter is designed to deliver ac power to a resistive load. If the load is reactive, we can use its reactive elements as part of the resonant filter, as in the case of induction heating. In Section 9.7 we discuss the use of a rectified load for dc/dc conversion. We start this chapter with a brief review of the behavior of second-order systems.

9.1 A REVIEW OF SECOND-ORDER SYSTEM BEHAVIOR

In each of its topological states, a resonant converter is generally a second-order system. To prepare you for studying these circuits, we present briefly the behavior of a second-order system in both the time and frequency domains.

9.1.1 The Time-Domain Response

Let's analyze the switched RLC network shown in Fig. 9.1 to determine v_C, assuming initial conditions $i_a(0) = 0$ and $v_C(0) = V_{Co}$. Recognizing that $i_a = C\, dv_C/dt$ and applying KVL around the circuit for $t > 0$, we can write the differential equation for v_C as follows:

$$\frac{d^2 v_C}{dt^2} + \frac{R}{L}\frac{dv_C}{dt} + \frac{1}{LC}v_C = \frac{V_{dc}}{LC} \tag{9.2}$$

Defining $\alpha = R/2L$ and $w_o = 1/\sqrt{LC}$, we can rewrite (9.2) as:

$$\frac{d^2 v_C}{dt^2} + 2\alpha\frac{dv_C}{dt} + w_o^2 v_C = w_o^2 V_{dc} \tag{9.3}$$

This equation has natural frequencies, s_1 and s_2, at:

$$s_{1,2} = -\alpha \pm \sqrt{\alpha^2 - w_o^2} \tag{9.4}$$

If $w_o > \alpha$, then $s_{1,2}$ are complex, and we can write them in terms of the *damped resonant frequency* of the circuit w_d:

$$s_{1,2} = -\alpha \pm jw_d \tag{9.5}$$

$$w_d = \sqrt{w_o^2 - \alpha^2} \tag{9.6}$$

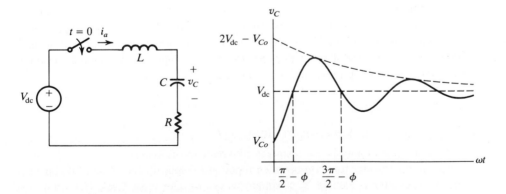

Figure 9.1 Switched RLC network for analysis in the review of second-order system behavior.

We can now write the general solution to *(9.3)* as:

$$v_C = e^{-\alpha t}(A \sin \omega_d t + B \cos \omega_d t) + V_{dc} \tag{9.7}$$

where V_{dc} is the particular solution, that is, the value of v_C after a long time. The initial conditions on v_C are

$$v_C(0) = V_{Co} \tag{9.8}$$

$$\left.\frac{dv_C}{dt}\right|_0 = 0 \tag{9.9}$$

Applying these conditions to *(9.7)*, we can determine A and B:

$$A = \frac{\alpha(V_{Co} - V_{dc})}{\omega_d} \tag{9.10}$$

$$B = V_{Co} - V_{dc} \tag{9.11}$$

The specific solution to our problem then is

$$v_C = (V_{Co} - V_{dc})e^{-\alpha t}\left(\frac{\alpha}{\omega_d}\sin \omega_d t + \cos \omega_d t\right) + V_{dc} \tag{9.12}$$

which we can simplify by combining the sine and cosine terms:

$$v_C = (V_{Co} - V_{dc})e^{-\alpha t}\left(\sqrt{1 + (\alpha/\omega_d)^2}\cos(\omega_d t + \phi)\right) + V_{dc} \tag{9.13}$$

Hence:

$$\phi = \tan^{-1}\left(\frac{\alpha}{\omega_d}\right)$$

The voltage v_C plotted in Fig. 9.1.

If there were no resistor in the circuit of Fig. 9.1, the capacitor voltage would oscillate forever around V_{dc} at the *undamped resonant frequency* $\omega_o = 1/\sqrt{LC}$. We always are careful to distinguish between undamped and damped resonant frequencies by using the symbols ω_o and ω_d, respectively.

9.1.2 The Frequency-Domain Response

We often obtain control of a resonant converter by utilizing the frequency dependence of the transfer function between input and output. Because the resonant filter of the converter is excited by a square wave, we can most easily determine the filter's effectiveness in eliminating harmonics by using frequency-domain analysis. Therefore understanding the frequency-domain behavior of a second-order system is important.

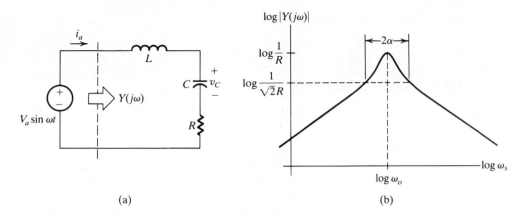

Figure 9.2 (a) A series resonant circuit. (b) The magnitude of the admittance $Y(j\omega)$ for the circuit of (a).

The admittance of the series RLC circuit of Fig 9.2, as a function of complex frequency $s = j\omega$, is

$$Y(s) = \frac{1}{sL + 1/sC + R} = \frac{sC}{s^2 LC + sRC + 1} \tag{9.14}$$

We can rewrite (9.14) as:

$$Y(s) = \left(\frac{1}{R}\right) \frac{2\alpha s}{s^2 + 2\alpha s + \omega_o^2} \tag{9.15}$$

The magnitude of this admittance is graphed as a function of frequency in Fig. 9.2. At resonance, $s = j\omega_o$, the impedances of the inductor and capacitor cancel, and the admittance is a pure conductance $1/R$.

A measure of the sharpness of the admittance function, (9.15), is its width at the points where its magnitude is less than its maximum by $1/\sqrt{2}$, called the *half-power* or *3 dB* points. These points occur at $s = j(\omega_o \pm \alpha)$, and the width is 2α. The ratio $\omega_o/2\alpha$ is a normalized measure of the filter's *selectivity*, that is, its ability to select a desired frequency and reject all others. This ratio is called the *quality factor*, or Q, of the filter, and the higher its value, the sharper the curve is. In terms of Q, (9.15) becomes

$$Y(s) = \left(\frac{1}{R}\right) \frac{(1/Q)(s/\omega_o)}{(s/\omega_o)^2 + (1/Q)(s/\omega_o) + 1} \tag{9.16}$$

One last observation worth making about the behavior of resonant circuits is the effect of Q on the magnitude of one or the other of the state variables. At resonance, the voltage across the resistor in Fig. 9.2 is equal to the source voltage. But if we calculate the capacitor voltage, we find that it can be substantially larger than the source. The transfer function between V_C and V_a is

$$\left|\frac{V_C}{V_a}\right| = \left|\frac{\omega_o^2}{s^2 + 2\alpha s + \omega_o^2}\right| \tag{9.17}$$

which, when evaluated at $\omega = \omega_o$, gives

$$\left.\left|\frac{V_C}{V_a}\right|\right|_{(\omega=\omega_o)} = \frac{\omega_o}{2\alpha} = Q \tag{9.18}$$

This result means that, if the source has an amplitude of 100 V and $Q = 10$, the capacitor would see a peak voltage of Q times the source voltage, or 1000 V! Such an excessive voltage (or current, in the case of a parallel resonant circuit) often is the most serious disadvantage of using resonant converters, especially under a light load when Q is high.

9.2 THE VOLTAGE-SOURCE SERIES RESONANT CONVERTER

The first form of the resonant converter that we discuss is one in which the switches create a square wave ac voltage waveform, and a series resonant filter is used to extract the fundamental. We assume that the dc system is a voltage source; that is, its impedance at the switching frequency ω_s is nearly zero. If the dc system does not have a low impedance at ω_s, we can place a large capacitor in parallel with it. Figure 9.3(a) shows the dc source and switches replaced by the equivalent square wave voltage source v_a.

9.2.1 The Filter

The series resonant LC filter of Fig. 9.3(a) placed in series with the load R makes the load current i_a nearly sinusoidal. At the resonant frequency ω_o, the impedance of the inductor and the capacitor exactly cancel, and the admittance of the RLC network equals $1/R$. Therefore, when the resonant converter is switched at $\omega_s = \omega_o$, the full fundamental component of the square wave, which has an amplitude $V_{a_1} = 4V_{dc}/\pi$, appears across the load resistor. If the Q of the filter is high, the harmonic content of i_a is low, and i_a is nearly sinusoidal. Thus:

$$i_a \approx |Y(j\omega_o)|V_{a_1}\sin\omega_o t = \left(\frac{1}{R}\right)\left(\frac{4V_{dc}}{\pi}\right)\sin\omega_o t \tag{9.19}$$

This current and its relationship to v_a are shown in Fig. 9.3(b). Note that v_a and i_a are in phase.

9.2.2 Control of the Output Waveform

It is not necessary that $\omega_s = \omega_o$ for the LC network to adequately filter harmonics from the current i_a. However, if ω_s is slightly higher or lower than ω_o, the LC

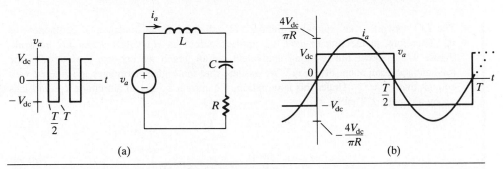

Figure 9.3 (a) The basic topology of the series resonant converter. The switches and dc sources have been replaced by an equivalent square-wave voltage source v_a. (b) The waveform of the load current i_a drawn for $\omega_o = \omega_s = 2\pi/T$.

filter looks like a small inductor or a large capacitor, respectively. This additional impedance reduces the magnitude of voltage across the load resistor. If the filter Q is high, we can achieve more than an order-of-magnitude change in the output power by shifting ω_s only a small amount. This shifting of the drive frequency, therefore, becomes a means by which we can control the output power or voltage.

If we achieve output voltage control by using a switching frequency lower than the resonant frequency, it is possible (if ω_s is sufficiently below ω_o) for the third harmonic of the square wave to be at a frequency for which the filter's transfer function is relatively high. Substantial third-harmonic currents would then flow into the load. This problem is particularly acute if the filter's Q is low and the admittance curve of Fig. 9.2 is broad. If we want an output waveform with low distortion, this problem imposes a lower limit on the drive frequency and/or the Q of the filter.

However, if the switching frequency is higher than ω_o, the harmonic components of the square wave are always adequately filtered because $|Y(n\omega_s)| \ll |Y(\omega_s)|$ for $\omega_s > \omega_o$. But this mode of control also has its limitations, because semiconductor devices have an upper limit to their switching frequency.

EXAMPLE 9.1

Load Current Determination When Switching Off Resonance

If the resonant converter of Fig. 9.3(a) is driven above resonance by a factor $\omega_s/\omega_o = \beta$, then at ω_s the combined impedance of the inductor and capacitor is

$$Z(j\omega_s) = \frac{(j\omega_s)^2 LC + 1}{j\omega_s C} = \frac{-(\beta^2 - 1)}{j\omega_s C} = j\omega_s L \left(\frac{\beta^2 - 1}{\beta^2} \right) = j\omega_s L_e \qquad (9.20)$$

where:

$$L_e = L \left(1 - \frac{1}{\beta^2} \right) \qquad (9.21)$$

The LC network is an equivalent inductor of value L_e at the fundamental component of v_a. Therefore, at the fundamental, v_a appears across an equivalent series LR network. Figure 9.4(a) shows the resulting equivalent circuit which we can use to determine the fundamental component of i_a: i_{a_1}. We assume that the higher harmonics are filtered well enough that $i_a \approx i_{a_1}$. Under this assumption, the current i_a has an amplitude I_{a_1} and lags v_a by an angle θ. That is,

$$i_a \approx I_{a_1} \sin(\omega_s t - \theta) \tag{9.22}$$

$$I_{a_1} = \frac{4V_{dc}}{\pi R} \left(\frac{1}{\sqrt{\omega_s^2 L_e^2 / R^2 + 1}} \right) \tag{9.23}$$

$$\theta = \tan^{-1} \frac{\omega_s L_e}{R} \tag{9.24}$$

Note that the amplitude of i_a is less than its amplitude when $\omega_s = \omega_o$, as given by (9.19).

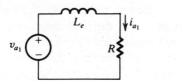

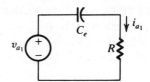

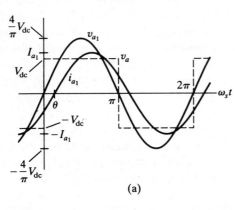

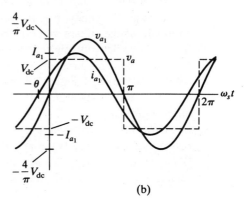

(a)   (b)

Figure 9.4   Behavior of the circuit of Fig. 9.3(a) for $\omega_s \neq \omega_o$:   (a) equivalent circuit and waveforms for $\omega_s > \omega_o$;   (b) equivalent circuit and waveforms for $\omega_s < \omega_o$.

When the converter is driven below resonance by a factor of $\omega_s/\omega_o = \gamma$, the LC circuit looks like an equivalent capacitor of value C_e. Thus:

$$Z(j\omega_s) = \frac{(j\omega_s)^2 LC + 1}{j\omega_s C} = \frac{(1 - \gamma^2)}{j\omega_s C} = \frac{1}{j\omega_s C_e} \tag{9.25}$$

and

$$C_e = \frac{C}{1 - \gamma^2} \tag{9.26}$$

The fundamental of v_a therefore sees a series RC load. If w_s is not too much lower than w_o, the harmonic content of i_a is low and again we assume that $i_a \approx i_{a_1}$. Thus:

$$i_a \approx I_{a_1} \sin(w_s t + \theta) \tag{9.27}$$

$$I_{a_1} = \frac{4V_{dc}}{\pi R} \left(\frac{1}{\sqrt{1/(w_s C_e R)^2 + 1}} \right) \tag{9.28}$$

$$\theta = \tan^{-1} \frac{1}{w_s R C_e} \tag{9.29}$$

The equivalent circuit and waveforms for $w_s < w_o$ are shown in Fig. 9.4(b).

9.2.3 Implementation of Switches

We can use any one of several ways to generate the square-wave voltage waveform of the source v_a in Fig. 9.3(a). For instance, if two dc voltage sources are available, we can use the half-bridge switch configuration shown in Fig. 9.5(a).

The switches of Fig. 9.5(a) must carry current in both directions if, to obtain control, w_s is to deviate from w_o. At w_o, the load current i_a is in phase with v_a, so it is positive when S_1 is on and negative when S_2 is on. Therefore both switches carry a current that is always positive. At any other frequency, however, i_a is either leading or lagging v_a, as shown in Fig. 9.4(a) or (b), so both switches must carry bilateral current. One possible switch implementation is shown in Fig. 9.5(b). By constraining the switching frequency so that $w_s < w_o$, we can use SCRs in place of the transistors in Fig. 9.5(b). The reason is that the conduction time of a diode allows its companion SCR to recover its forward blocking capabability. We explore the use of SCRs in this circuit more fully in Example 9.3.

9.2.4 Values for L and C

To attenuate the harmonic components of the square wave and to achieve a wide range of output voltage control with only a small change in switching frequency, we need to make the Q of the filter as high as possible. However, the higher the Q, the higher the peak energy storage requirements of the inductor and capacitor become. This relationship is shown in the definition of Q:

$$Q = \frac{2\pi \times \text{Peak stored energy}}{\text{Energy dissipated per cycle}} \tag{9.30}$$

The energy dissipated per cycle is proportional to the load power, which is specified by the application. Therefore the only way to increase Q is to increase peak energy storage.

Another way of looking at this relationship is to reduce the definition of Q for the series resonant filter:

$$Q = \frac{2\pi \left(\frac{1}{2} L I_p^2 \right)}{\frac{1}{2} R I_p^2 / (w_o / 2\pi)} = \frac{w_o L}{R} = \frac{1}{w_o R C} \tag{9.31}$$

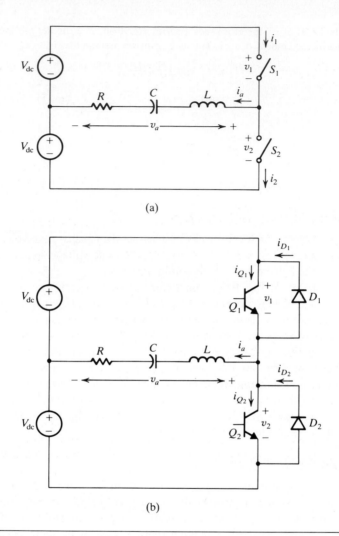

(a)

(b)

Figure 9.5 Generation of v_a in Fig. 9.3(a) from two dc voltage sources: (a) basic topology; (b) switch implementation.

where I_p is the peak filter current. From this expression you can see that L must be made larger and C must be made smaller to get a higher Q. Because the load current remains constant, both actions increase the peak energy storage of the two elements, so:

$$E_L = \frac{1}{2}LI_p^2 \tag{9.32}$$

$$E_C = \frac{1}{2}C\left(\frac{I_p}{\omega_o C}\right)^2 = E_L \tag{9.33}$$

EXAMPLE 9.2

Determining Values of L and C for the Series Resonant Converter

Suppose that we have a 10 Ω resistive heating load across which we want to apply a 40 kHz sinusoidal waveform having a peak voltage $V_p = 100$ V, using the topology of Fig. 9.5(b). The power delivered to the load is

$$P_{max} = \frac{V_p^2}{2R} = 500 \text{ W} \tag{9.34}$$

An additional requirement is that the power level be reducible to 125 W without decreasing the frequency by more than 20%. We need to determine the minimum amplitude of the dc voltage sources V_{dc}, and the values of L and C necessary to meet these requirements.

For any voltage source amplitude V_{dc}, the load voltage is largest when the converter is switched at resonance. All of the fundamental component of the square wave then appears across the load. The peak of this sinusoidal voltage V_p is

$$V_p = 100 = \frac{4}{\pi} V_{dc} \tag{9.35}$$

Substituting (9.35) into (9.34), we determine that $V_{dc} \approx 80$ V.

Because resonance is to occur at $\omega_o = 2\pi(40 \times 10^3)$ rad/s, we know that:

$$LC = \left(\frac{1}{(2\pi)(40 \times 10^3)} \right)^2 \tag{9.36}$$

We also want the magnitude of the admittance of the RLC network to decrease by a factor of 2 when the frequency decreases by 20% to $\omega_s = 0.8\omega_o = 2\pi(32 \times 10^3)$ rad/s. This decrease will give half the load current and therefore reduce the power delivered to the load by a factor of 4 to 125 W. From (9.15), when $s = j0.8\omega_o$, the magnitude of the admittance is

$$|Y(j0.8\omega_o)| = \left(\frac{1}{R} \right) \sqrt{\frac{1}{1 + Q^2(0.8 - 1/0.8)^2}} \tag{9.37}$$

As the admittance at $s = j0.8\omega_o$ must be $1/2R$, we can establish the relationship:

$$|Y(j0.8\omega_o)| = \frac{1}{2R} = \frac{1}{R}\sqrt{\frac{1}{1 + 0.2Q^2}} \tag{9.38}$$

We can solve (9.38) for Q:

$$Q = \frac{\omega_o L}{R} = 3.85 \tag{9.39}$$

We can now solve (9.36) and (9.39) to obtain values for L and C:

$$L = 153 \ \mu\text{H} \quad \text{and} \quad C = 0.1 \ \mu\text{F}$$

The peak energy storage requirement occurs at $\omega_s = \omega_o$ and is the same for both L and C.

At resonance, $I_p = V_p/R = 10$ A, and the value of the peak stored energy is

$$E_L = E_C = \frac{1}{2}(153 \times 10^{-6})(10)^2 = 7.65 \text{ mJ} \qquad (9.40)$$

Both the L and C carry a peak current I_p of 10 A, and their peak voltages V_{Lp} and V_{Cp} are

$$V_{Cp} = \frac{I_p}{\omega_o C} = QRI_p = 385 \text{ V} \qquad (9.41)$$

$$V_{Lp} = \omega_o L I_p = QRI_p = 385 \text{ V} \qquad (9.42)$$

Note how high these voltages are compared to the maximum load voltage of 100 V. This is the penalty paid for requiring a high value of Q.

To determine how large the third harmonic in the load current is, we evaluate the magnitude of the admittance at $\omega = 3\omega_s$ when the inverter is operating at minimum power, that is, $\omega_s = 0.8\omega_o$:

$$|Y(j2.4\omega_o)| = \left(\frac{1}{R}\right)\sqrt{\frac{1}{1 + Q^2(2.4 - 1/2.4)^2}} = \frac{0.13}{R} \qquad (9.43)$$

As the third-harmonic component of the square-wave voltage has only one third the amplitude of the fundamental ($V_{a_3} = V_{a_1}/3 = 33.3$ V), the maximum amplitude of the third-harmonic current is 0.43 A, or 4.3% of the full-load current and 8.7% of the quarter-load current. The power supplied by the third-harmonic current is never greater than 1% of the power in the fundamental, so our decision to neglect all but the fundamental component of load current is justified.

9.2.5 Switching Losses

In the introduction to this chapter we mentioned that one potential advantage of the resonant converter is the reduced switching losses in semiconductor devices. For the topology of Fig. 9.5 this condition is true only if the switches are driven at exactly the resonant frequency. Only at this frequency will the switch current, which is equal to the load current, be passing through zero at exactly the time when the switches change state, as we can infer from the waveforms of Fig. 9.3(b). Therefore at $\omega_s = \omega_o$, a step change occurs in the voltage of each switch at the transition. But its current is approximately zero, so no energy is lost in the semiconductor device while it changes state. However, when the converter is operated away from resonance, the switches are subjected to simultaneous step changes in voltage and current. Practical switches, of course, exhibit nonzero rise or fall times of their voltage and/or current, which create switching losses.

Figure 9.6(a) shows typical waveforms for the ideal switch voltages and currents in Fig. 9.5(a) when the switching frequency, ω_s, is higher than the resonant frequency, ω_o. The load is inductive at this frequency, so the load current lags v_a. Therefore each switch starts its conduction period with a negative current and ends

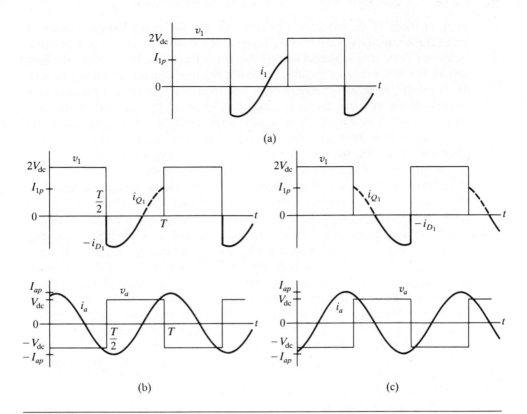

(a)

(b)   (c)

Figure 9.6 (a) Switch S_1 voltage and current waveforms for the circuit of Fig. 9.5(a) when $w_s > w_o$. (b) Variables in the circuit of Fig. 9.5(b) when $w_s > w_o$. (c) Same as (b) for $w_s < w_o$.

with a positive current. For the case $w_s < w_o$, the switches begin conduction with a positive current and end conduction with a negative current. The diode and transistor currents in the implemented switches of Fig. 9.5(b) are shown in Fig. 9.6(b) and (c) for these two cases.

Note that at both the turn-on and turn-off transitions, the switch current and voltage undergo simultaneous step changes, just as they do for the high-frequency dc/dc converters discussed in Chapter 6. Thus the topology of Fig. 9.5 does not offer the important attribute of low switching losses generally ascribed to resonant converters. In Section 9.4 we discuss a modified topology that does offer this advantage.

For the circuit of Fig. 9.5(b), there are actually four semiconductor device transitions per cycle of v_a. The two just discussed occur whenever the square wave of voltage goes through a step change. The other two occur when the load current changes direction. For these latter two transitions, the current simply commutates from the diode to the transistor, or vice versa, and the switch voltage remains at zero. These transitions therefore are lossless. As a result, each of the four semiconductor devices is subjected to only one lossy transition. If $\omega_s > \omega_o$, the load current first flows through the diode before it changes direction and flows through the transistor, as shown in Fig. 9.6(b). The diode turning on and the transistor turning off are therefore the lossy transitions. If $\omega_s < \omega_o$, the load current will lead v_a. It will flow through the transistor at the beginning of each half cycle and then commutate to the diode when it changes polarity, as shown in Fig. 9.6(c). The transistor turn on and the diode turn off are now the lossy transitions.

In the circuit of Fig. 9.5(b) we can create a short circuit across the dc sources if the two transistors are on simultaneously. This unpleasant condition is known as *shoot-through* and is caused by turning on one transistor before the other has turned off. To avoid shoot-through, we insert a delay between the turn off of one transistor and the turn on of the other. This delay does not keep a switch from conducting during the delay period—only the transistor part of the switch. For example, when $\omega_s > \omega_o$, the diode of the switch turning on conducts first. Thus if the top switch is turned off, the bottom diode will automatically turn on, even though the base drive to the bottom transistor is delayed. Alternatively, if $\omega_s < \omega_o$, the diode is the last part of the top switch to conduct. Thus the top transistor can be turned off before the top switch is forced off by the bottom transistor turning on.

One advantage of the operating region $\omega_s < \omega_o$ is that we can use an SCR in place of a transistor for the controllable switch. Because each SCR is naturally commutated when the load current changes direction, no special commutation circuitry is required. However, because of the turn-off time t_q required by an SCR, we must ensure that the diode is on for a time greater than t_q. That amount of time allows the antiparallel SCR to recover its ability to block forward voltage before the next SCR turns on. Otherwise, the SCR turning off would turn on again, creating a shoot-through condition. This requirement places an upper limit on ω_o and a lower limit on how close ω_s can be to ω_o.

EXAMPLE 9.3

A Resonant Converter Using SCRs

We designed the series resonant converter in Example 9.2 for full power to occur at the resonant frequency ω_o. Operation at ω_o is possible because we implicitly assumed switches that had no turn-off time constraint—a transistor, for instance. Instead, if SCRs having a turn-off time $t_q = 5$ μs are used, as shown in Fig. 9.7, the maximum switching frequency must be far enough below resonance to guarantee that the antiparallel diodes conduct for at least 5 μs. Using the element values derived in Example 9.2, which give $Q = 3.85$, we can determine the maximum permissible value of the switching frequency ω_s and its consequences for the harmonic content of the load voltage.

If, as shown in Fig. 9.4(b), θ is the phase of the admittance of the RLC network at

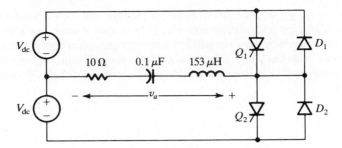

Figure 9.7 Series resonant converter using SCRs and diodes as the switches.

ω_s, then:

$$\frac{\theta}{\omega_s} \geq 5 \; \mu s \qquad (9.44)$$

The phase of the admittance at ω_s is

$$\theta = \angle Y(j\omega_s) = \angle \left[\frac{1}{jQ\left(\omega_s/\omega_o - \omega_o/\omega_s\right) + 1} \right]$$

$$= \tan^{-1}\left[-Q\left(\frac{\omega_s}{\omega_o} - \frac{\omega_o}{\omega_s} \right) \right] \qquad (9.45)$$

We then combine (9.44) and (9.45) to give a transcendental equation, which we solve iteratively to give $\omega_s = 0.81\omega_o$ and $\theta = 58.4°$. Note that this maximum switching frequency is close to the frequency for which we designed the filter to give one quarter the power in Example 9.2. The admittance magnitude is $|Y(j0.81\omega_o)| \approx 1/2R$, and as we now need to deliver the full load (500 W) at this frequency, the amplitude of the dc voltage sources must be twice the value determined in Example 9.2, that is, 160 V instead of 80 V. But doing so stresses the semiconductor devices to twice the voltage they are subjected to under the operating conditions determined in Example 9.2. Furthermore, to reduce the power to the minimum of 125 W, we must decrease the switching frequency to give an admittance magnitude of $1/4R$. Using (9.15) with $s = j\omega_s$, we determine that the new minimum power frequency is $\omega_s = 0.62\omega_o$. At this value of ω_s, the magnitude of the admittance at the third harmonic of ω_s, $1.86\omega_o$, is

$$|Y(j1.86\omega_o)| = \frac{1}{R}\sqrt{\frac{1}{1 + Q^2(1.86 - 1/1.86)^2}} = \frac{0.19}{R} \qquad (9.46)$$

Proceeding as we did in Example 9.2 to calculate the third-harmonic current, we find it to be 25% of the fundamental current at low power, which is three times its value in Example 9.2.

The higher source voltage requirement and the higher harmonic distortion in the output voltage are two consequences of using SCRs in the resonant converter of Fig. 9.5(a). We could have lowered the resonant and switching frequencies (assuming that the application would permit it), so that t_q would represent less of the switching cycle. But to maintain the same Q, we would have to make the inductor and capacitor larger. In general, the inability of many resonant converter circuits to operate at resonance with SCRs means that the circuit components are stressed by higher voltages and/or currents than they would be if transistors were used. But SCRs are available with much higher voltage and current ratings than transistors, so in many very high-power applications the SCR is still the device of choice.

9.3 THE CURRENT-SOURCE PARALLEL RESONANT CONVERTER

The current-source parallel resonant converter is the dual of the voltage-source converter. It requires that the dc system function like a dc current source, presenting a high impedance to switching frequency currents. We usually meet this requirement by placing an inductor in series with the dc system. The switches then create a square wave of current that passes through a parallel RLC circuit, as shown in Fig. 9.8(a). This circuit shows an equivalent square-wave source i_a for the dc current source and switches.

Figure 9.8(b) shows the magnitude of the impedance presented to the source i_a as a function of frequency; it has the same shape as the admittance curve in

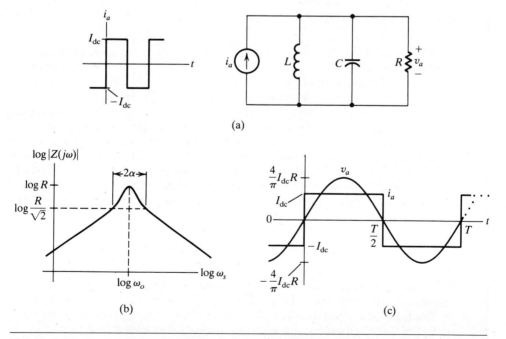

(a)

(b)   (c)

Figure 9.8 Current-source parallel resonant converter: (a) basic topology; (b) magnitude of the impedance presented to i_a; (c) waveform of the load voltage v_a when $\omega_o = \omega_s = 2\pi/T$.

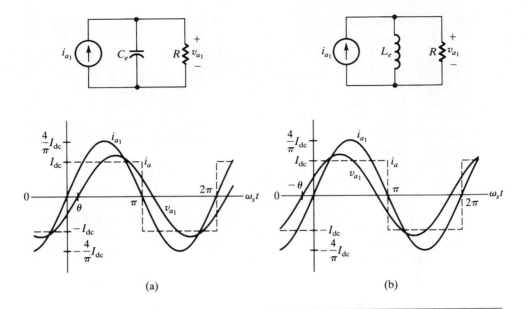

(a)   (b)

Figure 9.9   Equivalent circuit of the parallel resonant converter of Fig. 9.8 and
the fundamental load voltage v_{a_1}, when: (a) $\omega_s > \omega_o$;   (b) $\omega_s < \omega_o$.

Fig. 9.3(b). We can also describe this impedance using the expression for admittance, *(9.15)*, but with a coefficient of load resistance, R, instead of conductance, $1/R$, so that:

$$Z(s) = R\left[\frac{s/Q\omega_o}{s^2/\omega_o^2 + s/Q\omega_o + 1}\right] \qquad (9.47)$$

where $Q = \omega_o RC$ for the parallel resonant circuit. The voltage across the load resistor approximately equals the product of the fundamental component of the driving current and this impedance. Hence the curve of Fig. 9.8(b) also represents the converter's input current to output voltage transfer function.

When $\omega_s = \omega_o$, the impedance looks purely resistive to the fundamental component of i_a. Higher harmonics are subjected to a much lower impedance, so the load voltage v_a is nearly sinusoidal and in phase with i_a, as shown in Fig. 9.8(c).

As for the voltage-source series resonant converter, we can control the power delivered to the load of a parallel resonant converter by shifting ω_s relative to ω_o. For $\omega_s > \omega_o$, the parallel LC network has a finite impedance that is, in effect, capacitive (instead of the infinite impedance it has at ω_o). This effective capacitance C_e, which is in parallel with the load resistor, reduces the load voltage and makes it lag the square wave, as shown in Fig. 9.9(a). For $\omega_s < \omega_o$, the parallel LC network is, in effect, inductive. Again, shunting of some of the source current by this effective inductor L_e reduces the load current and voltage. In this case v_{a_1} leads i_a, as shown in Fig. 9.9(b).

For this converter we face the same problems with moving w_s too far away from resonance that we did for the series resonant converter. If we lower w_s, at some point the third-harmonic component of i_a will approach w_o and not be adequately attenuated. If we raise w_s, we have to consider carefully the rate at which the semiconductor devices are being switched. These constraints limit the range over which we can vary w_s and therefore the range of control that is possible using this technique.

Figure 9.10(a) shows a current-source parallel resonant converter that is the dual of the voltage-source series resonant converter of Fig. 9.5(a). Recall that, for the series resonant converter, we must make sure that both switches are not on at the same time in order to avoid a shoot-through. In the parallel resonant converter, we must make sure that both switches are not *off* at the same time in order to avoid opening the current source.

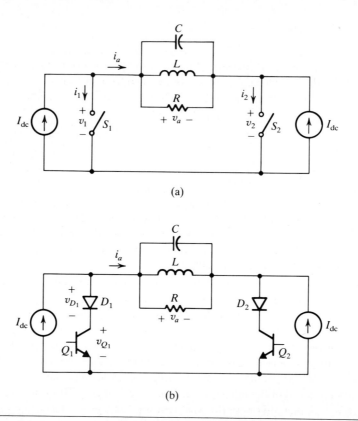

(a)

(b)

Figure 9.10   (a) A current-source parallel resonant converter using two current sources.   (b) A practical switch implementation.

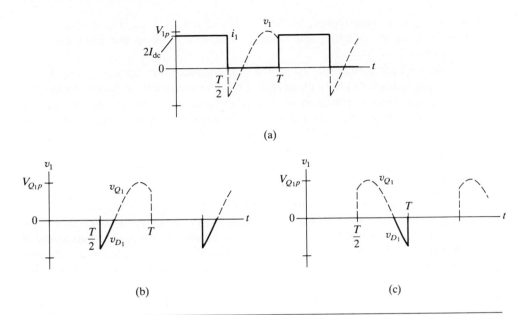

(a)

(b)   (c)

Figure 9.11   (a) Switch S_1 voltage and current waveforms for the circuit of
Fig. 9.10(b) for $\omega_s > \omega_o$.   (b) Diode and transistor voltages for the
circuit of Fig. 9.10(b) for $\omega_s > \omega_o$.   (c) Same as (b) for $\omega_s > \omega_o$.

Another difference between the two converters is the way in which we must
implement the switches. Recall that the series resonant converter needs a switch
that can carry current in both directions if we want to vary the switching frequency from resonance. But the parallel resonant converter needs a switch that can
block a bipolar voltage, as shown by the switch voltage and current waveforms of
Fig. 9.11(a). We can make an appropriate switch by connecting a transistor and
diode in series, as shown in Fig. 9.10(b). We could also use an SCR, but must
restrict circuit operation to the region $\omega_s > \omega_o$ if we want to force one SCR off
by turning on the other.

Figure 9.11(a) shows that when the parallel converter is operated off resonance,
a switch's voltage and current exhibit simultaneous step changes at both turn on
and turn off. This means that both transitions are lossy. As for the series converter
of Fig. 9.5(b), however, the diode and transistor comprising a switch in Fig. 9.10(b)
each experience only one lossy transition. If $\omega_s > \omega_o$, Fig. 9.11(b) shows that the
diode will experience switching loss at turn off, and the transistor will experience
loss at turn on. In this case it is interesting to note that since a switch is turned
off by reverse biasing its diode (when the other switch turns on) rather than by
turning off its transistor, the storage delay time of the transistor will not influence
the timing of the switch as long as the excess base charge is completely removed
from the transistor while the diode is blocking. (For explanations of storage delay

time and excess base charge, see Section 15.3.2.) If $w_s < w_o$, Fig. 9.11(c) shows that there will be switching loss when the transistor turns off and when the diode turns on.

As we have mentioned, the switch of a current-source parallel resonant converter can be implemented with an SCR. From the waveforms of Figs. 9.9 and 9.11, you can see that the condition $w_s > w_o$ is required in order for the SCRs to be commutated. When an SCR turns on, it imposes a negative voltage across the previously conducting SCR, forcing it to turn off.

EXAMPLE 9.4

A Practical Current-Source Implementation

Figure 9.12(a) illustrates a parallel resonant converter in which the current sources have been implemented with dc voltage sources and large inductors. Because the source ends of both inductors are at the same voltage, we can connect them to a single source, as shown in Fig. 9.12(b). We assume that the inductors are arbitrarily large. The circuit is to deliver

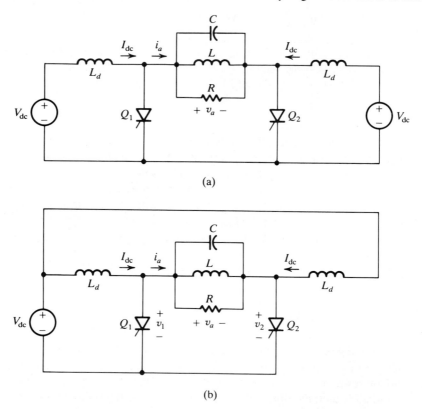

(a)

(b)

Figure 9.12 (a) A parallel converter, showing a current-source implementation. (b) The circuit of (a), simplified by combining the two voltage sources.

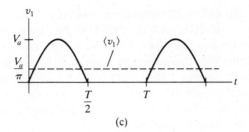

(c)

Figure 9.12 (cont.) (c) The voltage across Q_1 or Q_2, when $w_s \approx w_o$.

10 kW to the load resistor $R = 5\ \Omega$. What value of V_{dc} is required to meet this specification? We also assume that w_o is low enough that sufficient time is provided for the SCRs to turn off, even though w_s is close to, but greater than, w_o.

For $w_s \approx w_o$, v_a is a sinusoid having a peak voltage V_a, and v_1 or v_2 are as shown in Fig. 9.12(c). We can determine V_a from the power requirement:

$$\frac{\left(V_a^2/2\right)}{5} = 10^4 \tag{9.48}$$

$$V_a = 316\ V$$

Because the voltage across L_d can have no average value, $V_{dc} = \langle v_1 \rangle$. Approximating v_1 as a half-wave rectified voltage gives

$$V_{dc} = \frac{V_a}{\pi} \approx 100\ V \tag{9.49}$$

This current-source implementation without the use of feedback to control I_{dc} is far from ideal. For instance, if one of the SCRs were to remain on, I_{dc} would increase without bound. Also, with this implementation, I_{dc} is a function of R and w_s.

9.4 MODIFIED RESONANT CONVERTER TOPOLOGIES

For many applications, we make slight modifications to the two basic resonant converter topologies. In this section we discuss some of these modifications, the reasons for them, and the resulting changes in circuit operation.

9.4.1 Splitting the Inductor in the Voltage-Source Converter

If $w_s < w_o$ for the voltage-source converter of Fig. 9.5, the step change in v_a occurs when a transistor turns on, reverse biasing the previously conducting diode. But the diode actually exhibits a reverse recovery characteristic that lets it carry current in the negative direction for a brief period (see Chapter 17). Thus there is a time when both the upper and lower switches are conducting, creating a transient shoot-through condition. The circuit of Fig. 9.5 provides no impedance to limit the resulting shoot-through current.

To limit shoot-through current, we normally place a small inductor, called a *turn on snubber*, in series with the switches. This inductor limits the rate at which the switch current can rise. (In Chapter 24 we discuss turn-on snubbers in detail.) This need for limiting shoot-through current in the circuit of Fig. 9.5 leads to the resonant converter topology shown in Fig. 9.13, known as a *Mapham inverter*. Here, the inductor of the resonant circuit is divided between the two switch branches. In these positions, the inductors provide more than enough impedance to limit shoot-through current. Another consequence of this circuit modification is that a switch can no longer turn off until the current in its series inductor is zero.

Analysis of this modified circuit for $\omega_s < \omega_o = 1/\sqrt{LC}$ requires more care than before, because the switches no longer simply create a square-wave voltage that is filtered by a single series resonant circuit. During part of each half cycle

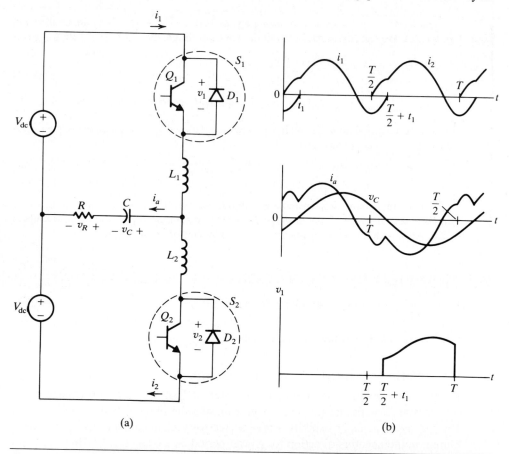

(a)   (b)

Figure 9.13   (a) The Mapham inverter: a series resonant converter with split inductor.   (b) Waveforms of variables in the circuit of (a).

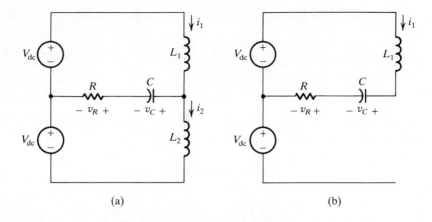

(a)   (b)

Figure 9.14   The two topological states of the converter of Fig. 9.13(a) during one half cycle, $0 < w_s t < \pi$:   (a) S_1 and S_2 both on $(0 < w_s t < w_s t_1)$; (b) S_1 on and S_2 off $(w_s t_1 < w_s t < \pi)$.

both switches are on, and the equivalent circuit is that shown in Fig. 9.14(a). A period then follows during which only one switch is on, resulting in the circuit of Fig. 9.14(b). The second half cycle of operation begins when the second switch is turned on again.

We now analyze this circuit by assuming that S_2 of Fig. 9.13(a) has been on and that at $t = 0$ switch S_1 turns on, so $i_1(0) = 0$. We also assume the initial conditions:

$$v_C(0) = V_o \qquad (9.50)$$

$$i_2(0) = -I_o \qquad (9.51)$$

The current i_2 is negative because, for $w_s < w_o$, D_2 is on at the end of the conduction period of S_2. So long as $i_2 < 0$, the equivalent circuit is that of Fig. 9.14(a). The following equations describe the behavior of i_1, i_2, and v_C during this time:

$$\frac{di_1}{dt} = \frac{1}{L_1}[V_{dc} - v_C - R(i_1 - i_2)] \qquad (9.52)$$

$$\frac{di_2}{dt} = \frac{1}{L_2}[V_{dc} + v_C + R(i_1 - i_2)] \qquad (9.53)$$

$$\frac{dv_C}{dt} = \frac{1}{C}(i_1 - i_2) \qquad (9.54)$$

At $t = t_1$, $i_2 = 0$ and S_2 turns off. (The diode turns off in the implemented switch. The transistor, which stopped carrying current before $t = 0$, could have turned off at any time between $t = 0$ and t_1.) At this point the boundary conditions are

$$v_C(t_1) = V_1 \tag{9.55}$$

$$i_1(t_1) = I_1 \tag{9.56}$$

$$i_2(t_1) = 0 \tag{9.57}$$

The following equations then describe the circuit behavior for $t_1 < t < T/2$, during which Fig. 9.14(b) is the equivalent circuit:

$$\frac{di_1}{dt} = \frac{1}{L_1}\left(V_{dc} - v_C - Ri_1\right) \tag{9.58}$$

$$\frac{di_2}{dt} = 0 \tag{9.59}$$

$$\frac{dv_C}{dt} = \frac{i_1}{C} \tag{9.60}$$

In the periodic steady state, the boundary conditions at $t = T/2$ are

$$i_1(T/2) = i_2(0) = -I_o \tag{9.61}$$

$$v_C(T/2) = -v_C(0) = -V_o \tag{9.62}$$

$$i_2(T/2) = i_1(0) = 0 \tag{9.63}$$

We can now solve (9.52)–(9.63) for the unknowns, i_1, i_2, v_C, I_o, I_1, V_o, V_1, and t_1. The iterative solution approach discussed in Example 3.1 is appropriate to this problem. The qualitative aspects of the resulting waveforms are shown in Figure 9.13(b). Note that we have eliminated the two lossy switch transitions present in the basic series converter of Fig. 9.5. When a switch turns on, a step change in switch voltage still occurs, but now the inductor constrains the current to rise with a finite derivative rather than a step as it did in Fig. 9.6(c). Similarly, the switch current at turn off approaches zero with a finite slope.

One disadvantage of this circuit is that the voltage across a nonconducting switch is higher than it was for the circuit of Fig. 9.5. The reason is that the off-state voltage, of S_1 for instance, is

$$v_1 = 2V_{dc} + L_2\frac{di_2}{dt} \tag{9.64}$$

The derivative is positive over some part of the cycle, so the maximum value of v_1 is greater than $2V_{dc}$, its value in the basic series resonant converter of Fig. 9.5.

The meaning of "resonance" is ambiguous for the circuit of Fig. 9.13(a). At times in a cycle, only one inductor is in the circuit; at other times, both inductors are present (as in Fig. 9.14b). In practical terms, $\omega_s < \omega_o$ means switching at a frequency low enough to ensure that the current in the *on* switch has reversed before the *off* switch turns on.

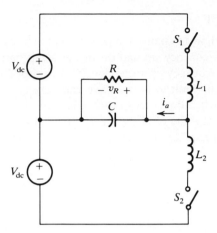

Figure 9.15   Repositioning the load in the split inductor converter of Fig. 9.13(a) to reduce load voltage distortion.

9.4.2 Repositioning the Load for Lower Distortion

One side effect of splitting a filter element is an increased distortion of the load waveform. The load-voltage distortion in the split inductor converter of Fig. 9.13 is apparent in the waveform of i_a. This distortion arises because the current is ringing with a frequency $\omega_1 = 1/\sqrt{LC}$ during the time when only S_1 is on and with a different frequency $\omega_2 = 1/\sqrt{(L/2)C}$ when both switches are on.

To reduce this load-voltage distortion, we can place the load resistor in parallel with the capacitor, as shown in Fig. 9.15. To first order, especially if the Q of the circuit is high, the load voltage will be

$$v_R \approx \frac{1}{C} \int i_a \, dt \qquad (9.65)$$

which reduces harmonics of order n by the factor $1/n$. A disadvantage of this circuit, however, is that under certain conditions the capacitor voltage may have a dc component, which is unacceptable to some loads, such as, for instance, a transformer-coupled load.

9.5 THE BRIDGE TOPOLOGY

The half-bridge of switches that we have been using to create a square wave of current or voltage has the disadvantages that two dc sources, V_{dc} or I_{dc}, are required and that each switch is subjected to a voltage stress equal to $2V_{dc}$ or a current stress of $2I_{dc}$. Using four switches in a full bridge, we can utilize a single dc source, reduce the individual switch stresses, and obtain an additional control method. Of course, we now have twice as many switches.

9.5.1 Power Control

We have discussed how to control the power level in a resonant converter by moving the switching frequency up and down the steep part of the resonant filter's impedance versus frequency characteristic. This control method has two disadvantages. The first is the limit to how far we can vary ω_s from ω_o—up or down—because of either switch limitations or the presence of the third harmonic in v_a or i_a. The second is that if the Q of the LRC circuit is not very high, the impedance curve is very broad near ω_o. This breadth requires a large change in frequency to

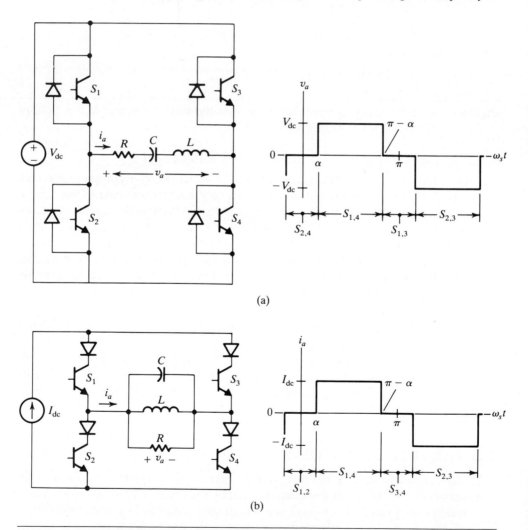

(a)

(b)

Figure 9.16   Full bridge with phase-angle adjustment for power-level control: (a) voltage-source case;   (b) current-source case.

achieve a wide range of control. The result is that controlling the power over the desired range is difficult at times.

A solution to this problem for the series resonant converter is shown in Fig. 9.16(a), where the load is placed inside a full bridge of switches. At $t = 0$, S_2 and S_4 are on, so $v_a = 0$. At $t = \alpha/\omega_s$, S_2 turns off and S_1 turns on, and $v_a = V_{dc}$. At $t = (\pi - \alpha)/\omega_s$, S_4 turns off and S_3 turns on, giving $v_a = 0$. This switching sequence continues and results in the waveform of v_a shown in Fig. 9.16(a). We can also view this waveform as resulting from the addition of two square waves of amplitude $V_{dc}/2$ that are phase shifted from each other by 2α. If V_{dc} is split so that the centerpoint $V_{dc}/2$ is accessible, we can visualize how each half-bridge $S_{1,2}$ and $S_{3,4}$ creates one of the square waves between the half-bridge and source centerpoints.

We call the waveform v_a in Fig. 9.16(a) a *quasi-square wave*. Its fundamental component is

$$V_{a_1} = \frac{4V_{dc}}{\pi} \cos \alpha \tag{9.66}$$

If $\omega_s = \omega_o$, this fundamental component appears across the load resistor. We can therefore control the load voltage by varying α from 0 to $\pi/2$, while maintaining $\omega_s = \omega_o$. Besides the broad range of control offered by this topology, there is the added benefit that the switching frequency remains constant for those situations in which this condition is desirable. (One advantage of constant-frequency control is that we can use high-Q tuned filters to remove switching frequency noise from sensitive equipment external to the power circuit.)

Splitting the inductor or the capacitor is still possible with the bridge circuit. We can also use the bridge topology for the parallel resonant converter, as shown in Fig. 9.16(b).

9.6 DISCONTINUOUS MODE CONTROL

So far in this chapter we have discussed resonant converters that are operated with what is called *continuous mode control*. In this mode, the switches are controlled so that the resonant current i_a, in the voltage-source converters, or the voltage v_a in the current source converters, is continuous. By continuous we mean that the oscillation is not interrupted. Another way to operate the resonant converter is with *discontinuous mode control*. In this mode the resonating current or voltage variable is interrupted every cycle or half cycle, and during these interruptions the variable usually has a value of zero. Power control is obtained by varying the duration of the interruption, much as in duty-ratio control of dc/dc converters.

9.6.1 Basic Discontinuous Mode Operation

The basic series resonant converter operating under discontinuous mode control is shown in Fig. 9.17. The transistor Q_1 is turned on and the load current rings first

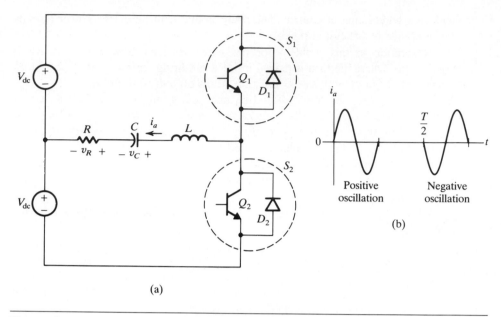

Figure 9.17 Discontinuous mode control of the series resonant converter: (a) the series resonant converter; (b) load-current waveform.

positively and then negatively back through D_1. While D_1 is on, Q_1 is off, but Q_2 is not on yet. When the load current returns to zero and D_1 turns off, all four semiconductor devices are off and i_a is zero. After a period of time during which no power flows to the load, Q_2 turns on and the switching sequence is repeated in the lower half of the circuit, which completes a cycle. Figure 9.17(b) shows the resulting waveform of the load current i_a.

Note that this mode of operation eliminates switching losses without splitting the filter elements. Whenever a switch turns on, its voltage makes a step change to zero, but its current starts at zero and rises with a limited di/dt because of the filter inductor. Similarly, when a switch turns off, it does so after its current has slowly returned to zero, reducing diode reverse recovery problems.

The negative aspect of discontinuous mode control is that the output waveform, which now has regions of zero value, is no longer a sinusoid. But there are many applications, such as high-frequency induction heating and solid-state ballasts for fluorescent lamps, for which a harmonic-free ac waveform is not necessary.

EXAMPLE 9.5

Resonant Conversion with Discontinuous Mode Control

We want to find the total energy delivered to the load during each oscillation of the series resonant converter of Fig. 9.17(a). Let's focus on the oscillation initiated by S_1 and find the resulting load-current waveform.

We must first specify the initial conditions for the oscillation. The initial inductor current $i_a(0)$ is zero by the definition of discontinuous mode control. The initial capacitor voltage $v_C(0)$ requires some analysis, however, because the positive excursion of the oscillation is not symmetrical with the negative excursion, owing to the damping introduced by R. As a result, the charge delivered to C during the positive excursion of i_a is greater than the charge removed during the negative excursion. Hence, at the end of the damped cycle of the oscillation, the capacitor is charged to a higher voltage than that with which it started. In the steady state, this increase in capacitor voltage is balanced by a corresponding decrease during the oscillation initiated by S_2, so the net charge transfer to C is zero over a complete cycle.

If we assume that $v_C(0) = -V_{Co}$ when the top switch is closed, the load current is

$$i_a = \left(\frac{V_{dc} + V_{Co}}{\sqrt{L/C}\sqrt{1 - (1/2Q)^2}} \right) e^{-\alpha t} \sin \omega_d t \qquad (9.67)$$

where $\alpha = R/2L$, $Q = \omega_o/2\alpha$, and $\omega_d = \sqrt{\omega_o^2 - \alpha^2}$. This current stops flowing at $t = 2\pi/\omega_d$, when it has completed a full cycle of oscillation. If we integrate the load current from $t = 0$ to $t = 2\pi/\omega_d$, we find that the net charge transferred to the capacitor is

$$\Delta q = \int_0^{2\pi/\omega_d} i_a \, dt = \frac{V_{dc} + V_{Co}}{\sqrt{L/C}\sqrt{1 - (1/2Q)^2}} \left(1 - e^{-(2\pi\alpha/\omega_d)} \right) \frac{\omega_d}{\omega_o^2} \qquad (9.68)$$

In the steady state, symmetry requires that the capacitor voltage corresponding to this change in charge be $v_C(T/2) = +V_{Co}$, the negative of $v_C(0)$. Therefore:

$$2V_{Co} = \frac{\Delta q}{C} = \frac{V_{dc} + V_{Co}}{\sqrt{1 - (1/2Q)^2}} \left(1 - e^{-(2\pi\alpha/\omega_d)} \right) \frac{\omega_d}{\omega_o} \qquad (9.69)$$

$$V_{Co} = V_{dc} \frac{1 - e^{-2\pi\alpha/\omega_d}}{\left(\frac{\omega_o}{\omega_d} \right) - \left(1 - e^{-(2\pi\alpha/\omega_d)} \right)} \qquad (9.70)$$

Figure 9.18 shows the resulting waveforms of i_a and v_C.

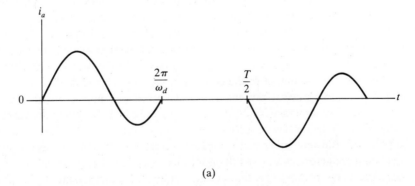

(a)

Figure 9.18   (a) Load current and capacitor voltage determined for discontinuous mode control of the converter of Fig. 9.17(a).

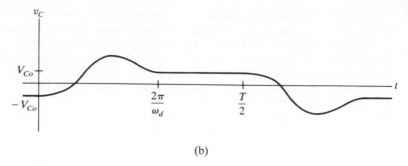

(b)

Figure 9.18 (cont.)   (b) Capacitor voltage determined for discontinuous mode control of the converter of Fig. 9.17(a).

The energy delivered to the load during one complete switching cycle is given by:

$$E = 2 \int_0^{2\pi/\omega_d} Ri_a^2 \, dt \qquad (9.71)$$

The power is

$$P = \frac{\omega_s}{2\pi} E \qquad (9.72)$$

which is a linear function of the switching frequency $\omega_s = 2\pi/T$, so long as ω_s is low enough to maintain discontinuous conduction.

9.7 RESONANT DC/DC CONVERTERS

An important use of the resonant converter is for dc/dc conversion. Although we described in Chapter 6 how to perform this function with the canonical switching cell, the semiconductor devices in those high-frequency dc/dc converters have relatively high switching losses. For each transition, both the switch voltage and the switch current change at the same time, and substantial power can be dissipated during the time required to complete the transition. We have shown in this chapter that the resonant converter can provide benign switch transitions during which either the voltage or the current of a device remains near zero. By rectifying and filtering the ac output of the converter, we can achieve dc/dc conversion without the switching losses present in the canonical cell implemented with practical switches. This approach gives us a way to raise the switching frequency above what semiconductor device technology permits for the nonresonant or *square-wave* topologies. But remember that resonant converters impose more severe stresses on the semiconductor devices and the filter elements than the canonical cell topologies do. And keep in mind that the rectified output of the resonant converter requires low-pass filter elements in addition to the energy-storage elements comprising its resonant filter.

9.7.1 Basic Topologies

To obtain a dc output, we must place a bridge rectifier and large dc filter element between the load and the filter of the resonant topologies discussed in Sections 9.2–9.4. The dc filter can be one of two types—either a large capacitor in parallel with the load, or a large inductor in series with it. The capacitor makes the load appear as a voltage sink at the switching frequency, and the inductor makes it appear as a current sink. Figure 9.19 shows a series resonant half-bridge converter with its dc filter configured both ways.

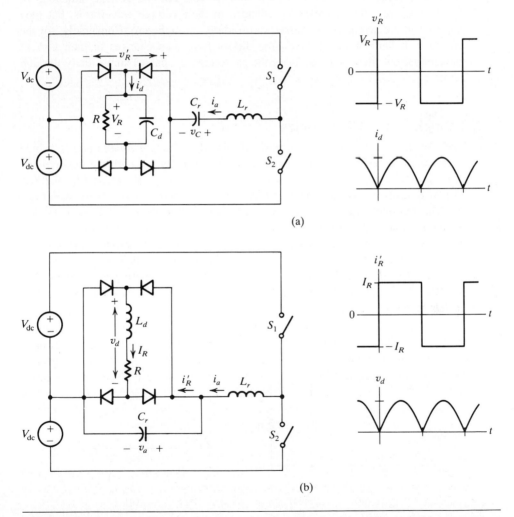

(a)

(b)

Figure 9.19   A series resonant converter used as the basis for a dc/dc converter: (a) voltage sink load created by filtering the dc voltage with a large capacitor, C_d;   (b) current sink load created by filtering the dc current with a large inductor, L_d.

From the ac side of the rectifier bridge, the inversion caused by the switching of the diode bridge makes the load act like a square-wave voltage sink if we use a capacitor on the dc side, and a square-wave current sink if we use an inductor. If we use a capacitor filter, we cannot place the load in parallel with the resonating capacitor C_r. Instead we must place the load in series with the resonating inductor L_r. Conversely, if the load resembles a current sink, we cannot place it in series with the inductor but must place it across the capacitor.

The diodes of the rectifier bridges in these topologies have low-loss switch transitions because they are rectifying sine waves. For the voltage sink load of Fig. 9.19(a), the diodes have step changes in their voltage waveforms, but they occur when the ac current i_a passes sinusoidally through zero. Conversely, for the current sink load of Fig. 9.19(b), the diodes have step changes in their current waveforms, but these steps occur as the ac voltage v_C passes sinusoidally through zero.

9.7.2 Single-Ended Topologies

The resonant converters that we have discussed so far are *double ended*, which means that the resonant filter is excited symmmetrically, on both the positive and negative half cycles. A *single-ended* resonant converter is shown in Fig. 9.20(a). The SCR turns on to start the ring and, when i_a returns to zero, the SCR turns off. The resonant capacitor C_r then discharges to zero through the load resistor. Because v_o has a dc component, the single-ended circuit has limited applications. For example, it is not generally suitable for supplying loads through a transformer. We can use it to create high-current or high-voltage pulses with little switching loss. We can also use it as the basis of a dc/dc converter by putting a low-pass filter between C_r and the load. In the latter case the circuit is known as a *quasi-resonant dc/dc converter*.

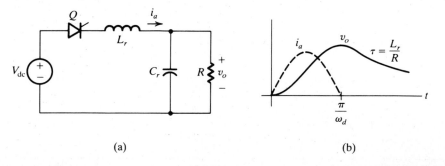

(a)   (b)

Figure 9.20 (a) A single-ended resonant converter. (b) Waveforms of inductor current i_a and output voltage v_o.

EXAMPLE 9.6

dc/dc Conversion with Discontinuous Mode Control of a Single-Ended Resonant Converter

Figure 9.21 shows the converter of Fig. 9.20 with a large inductor placed in series with the load resistor to form a low-pass filter. We added the clamp diode D to prevent I_d from discharging the capacitor voltage v_C to a large negative value while Q is off. What switching period T is required to give $V_o = 30$ V?

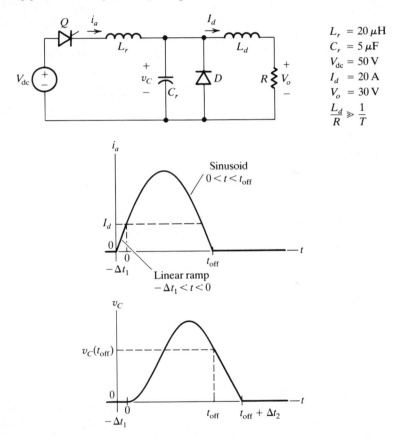

$L_r = 20\,\mu H$
$C_r = 5\,\mu F$
$V_{dc} = 50$ V
$I_d = 20$ A
$V_o = 30$ V
$\dfrac{L_d}{R} \gg \dfrac{1}{T}$

Figure 9.21   A single-ended resonant dc/dc converter. This topology is known as a quasi-resonant dc/dc converter.

The operation of this circuit is straightforward. When Q turns on, $i_a = 0$ and $v_C = 0$. The clamp diode remains on until $i_a = I_d$, at which time L_r and C_r begin to ring. Modeling the filter inductor and load resistor as a current source leaves the ringing frequency undamped, at $\omega_o = 1/\sqrt{L_r C_r}$. When $i_a = 0$ again, Q turns off, I_d discharges C_r to zero, and D turns on. The load current then freewheels through D until Q turns on again at T. If we establish our time reference to be when D turns off and ringing starts, then $v_C(t < 0) = 0$

and for $t > 0$,

$$i_a = I_d + \frac{V_{dc}}{\sqrt{L_r/C_r}} \sin \omega_o t \tag{9.73}$$

$$v_C = V_{dc}(1 - \cos \omega_o t) \tag{9.74}$$

Setting (9.73) to zero and solving for t_{off}, the time Q turns off, yields

$$t_{off} = \frac{1}{\omega_o} \sin^{-1} \frac{-I_d \sqrt{L_r/C_r}}{V_{dc}} = 40.7 \ \mu s \tag{9.75}$$

From (9.74), $v_C(t_{off}) = 80$ V, and it takes 20 μs for I_d to discharge C_r to zero. The waveforms of i_a and v_C are shown in Fig. 9.21.

We can now calculate V_o by taking the average value of v_C over the switching period T:

$$V_o = \langle v_C \rangle = \frac{1}{T} \left[\int_0^{t_{off}} V_{dc}(1 - \cos \omega_o t) \, dt + \left(\frac{1}{2}\right)(80)(20 \times 10^{-6}) \right] \tag{9.76}$$

$$= \frac{3.24 \times 10^{-3}}{T}$$

Setting $\langle v_C \rangle = 30$ V yields $T = 108 \ \mu s$.

The quasi-resonant converter of Fig. 9.21 resembles the direct converter connection of the canonical cell of Chapter 6. In fact, if the switching frequency $1/T$ is high enough, we cannot avoid the presence in the canonical cell of parasitic inductance comprising L_r and the junction capacitance of D comprising C_r. As with all the converters discussed in this chapter, the current in the dc source contains high-frequency components unless we add a low-pass filter between this source and the switches. For the circuit of Fig. 9.21, this filter would include a capacitor across V_{dc}, which is the capacitor of the canonical cell.

Notes and Bibliography

The literature is rich with papers on resonant converters. Below is a sample of good papers in several areas. The Mapham inverter, which was originally developed for SCR switches, is introduced in [1]. Papers [2]–[4] present innovative circuit designs. The application discussed in [2] is an induction cooking appliance, which you might find particularly interesting. The time-shared inverter introduced in [4] permits the use of thyristors at frequencies several times the limit imposed by t_q. Design and modeling are the subjects of [5] and [6]. The concept of the quasi-resonant converter was first discussed in [7].

As most new topologies are presented by their advocates, the methods of comparing and evaluating circuits described and used in [8] is instructive in helping you come to independent conclusions about a new topology's characteristics. Two good papers that carefully compare various circuits are [9] and [10].

1. N. Mapham, "An SCR Inverter with Good Regulation and Sine Wave Output," *IEEE Trans. Industry and General Applications* 3 (2): 176–187 (March/April 1967).

2. H. Omori, M. Nakaoka, H. Yamashita, and T. Maruhashi, "A Novel Type Induction-Heating Single-Ended Resonant Inverter Using New Bipolar Darlington Transistor," *IEEE Power Electronics Specialists Conference (PESC) Record* (1985) 590–599.

3. J. G. Kassakian, "A New Current Mode Sine Wave Inverter," *IEEE Trans. Industry Applications* 18 (3): 273–278 (May/June 1982).

4. M. Nakaoka, N. M. Vietson, T. Maruhashi, and M. Nishimura, "New Voltage-Fed Time-Sharing High Frequency Thyristor Inverter Circuits and Their Applications," in *Proc. IEEE IAS Annual Meeting* (1979) 399–413.

5. D. M. Divan, "Design Considerations for Very High Frequenncy Resonant Mode dc/dc Converters," *IEEE Trans. Power Electronics* 2 (1): 45–54 (January 1987).

6. R. J. King and T. A. Stuart, "Modeling the Full Bridge Series Resonant Power Converter," *IEEE Trans. Aerospace and Electronic Systems* 18 (4): 449–460 (July 1982).

7. K.-H. Lie, R. Oruganti, and F. C. Lee, "Resonant Switches—Topologies and Characteristics," in *IEEE Power Electronics Specialists Conference (PESC) Record* (1985) 106–116.

8. M. F. Schlecht and L. F. Casey, "Comparison of the Square-Wave and Quasi-Resonant Topologies," in *Proc. IEEE Applied Power Electronics Conference (APEC)* (San Diego, 1987), 124–134.

9. R. L. Steigerwald, "A Comparison of Half-Bridge Resonant Converter Topologies," in *Proc. IEEE Applied Power Electronics Conference (APEC)* (San Diego, 1987), 135–144.

10. S. D. Johnson, A. F. Witulski, and R. W. Erickson, "A Comparison of Resonant Topologies in High Voltage DC Applications," in *Proc. IEEE Applied Power Electronics Conference (APEC)* (San Diego, 1987), 145–156.

Problems

9.1   Determine and plot the current i_a in the switched RLC network of Fig. 9.1. If the switch opens at the first zero-crossing of i_a, how much energy is trapped in C?

9.2   Figure 9.22 shows two parallel RLC networks. The network of Fig. 9.22(a) is driven by a sinusoidal voltage source. The network of Fig. 9.22(b) is excited by a current source. Calculate the inductor current i_L in (a) and the capacitor current i_C in (b) when the circuits are driven at resonance, that is, $\omega = \omega_o$. Define α and Q for these circuits and express your answers in terms of α, Q, and the *characteristic impedance* of the tank $Z_o = \sqrt{L/C}$.

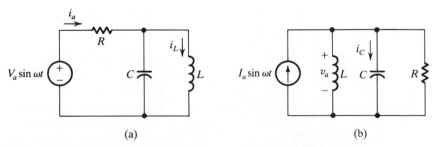

(a)   (b)

Figure 9.22   Two parallel resonant filters analyzed in Problem 9.2

9.3 Show that the definitions of Q given by *(9.18)* and *(9.30)* are equivalent for a parallel resonant circuit.

9.4 Determine values in terms of R, L, C, and Q for C_e and L_e in Fig. 9.9.

9.5 Calculate and plot the power delivered to the load resistor in the series resonant converter of Fig. 9.3(a) for $\omega_o < \omega_s < 2\omega_o$, and $Q = 5$. Make the same assumptions as in Example 9.1. What is the error resulting from ignoring the third harmonic at the extremes of ω_s?

9.6 The power to the load in Example 9.2 was controlled by *reducing* the the switching frequency from resonance. Redo that example, including the calculation of the third-harmonic power, if one of the requirements is that ω_s cannot be *increased* beyond $1.2\omega_o$.

9.7 What are the current-source inductor currents I_{dc} in the parallel converter of Fig. 9.12(b)? If the peak-to-peak ripple current in these inductors is to be less than 10% of I_{dc}, what value would you choose for L_d if $\omega_s = 2\pi(10^4)$ r/s?

9.8 Explain why the Mapham inverter of Fig. 9.13 cannot be operated in the region $\omega_s > \omega_o$ if the transistors are replaced by SCRs.

9.9 The parallel resonant converter of Fig. 9.12 is designed to have $Q = 2$ for the specified load resistor $R = 5\ \Omega$. We now increase the switching frequency to $\omega_s = 1.2\omega_o$. If V_{dc} remains unchanged from the value calculated in Example 9.4, that is, $V_{dc} = 100$ V, what is the new value of I_{dc}, and what is the power delivered to R? (*Hint:* express $Z(j\omega_s)$ in terms of Q and $j(\omega_s/\omega_o)$, then approximate v_a by its fundamental component, which can be determined by $Z(j\omega_s)$ and the fundamental component of i_a. Now use the constraint $\langle v_{1,2}\rangle = V_{dc}$ to determine I_{dc}.)

9.10 Figure 9.23 shows a single-ended resonant dc/dc converter similar to that of Fig. 9.21, except that the switch is implemented with a transistor and diode to carry bilateral current. At what value of the switching period T is $V_o = 30$ V?

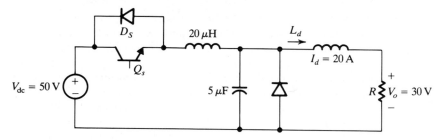

Figure 9.23 The single-ended resonant dc/dc converter analyzed in Problem 9.10.

9.11 What is the maximum allowable value of the circuit commutated turn-off time t_q for the SCR in the converter of Fig. 9.21? Compare this value to that for an SCR used in place of the transistor in the converter of Fig. 9.23.

9.12 The current-source inverter of Fig. 9.12(b) is designed to deliver a maximum power of 10 kW to an 8 Ω load. The tank circuit (RLC) has an undamped resonant frequency $\omega_o = 10^4$ r/s, and $Q(\omega_o) = 6$. As SCRs are used in this circuit, it must be operated in the region $\omega_s > \omega_o$.

(a) Assuming that $L_d \approx \infty$, determine and plot as a function of ω_s the turn-off time

t_{off} for the SCRs. If SCRs with $t_q = 50\ \mu s$ are used, what is the permissible range of w_s? (*Hint:* use only the fundamental component of i_a, and justify this approximation by calculating the amplitudes of the third-harmonic voltages at the extremes of w_s.)

(b) Sketch v_1 at the minimum value of w_s and determine a value for V_{dc}. If the peak-to-peak ripple in i_d is to be less than 10% of I_{dc}, what is the minimum value of L_d?

(c) What is the power delivered to R at the maximum value of w_s for the value of V_{dc} determined in (b)? Again assume that $L_d \approx \infty$.

(d) What value of V_{dc} would you choose, and over what range of w_s would you operate the converter, if the design requirements were a maximum power of 10 kW, a 10:1 power control range, and minimum device stress factors?

9.13 We can make the split source required in a half-bridge converter, using a single source and a pair of large capacitors, as shown in Fig. 9.24 for a series resonant topology, where $C_1 = C_2$. The inductor L_d together with C_1 and C_2 form a low-pass filter to keep the high-frequency currents away from the dc source V_{dc}. The two resistors shunting the capacitive voltage divider ensure that $\langle v_1 \rangle = \langle v_2 \rangle$. Sketch the capacitor currents i_1 and i_2 for the circuit parameters of Example 9.2 and operation at 500 W.

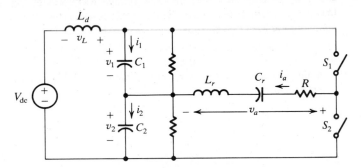

Figure 9.24 The half-bridge converter of Problem 9.13 in which the two required dc sources are approximated by C_1 and C_2.

9.14 Resonant converters are often used to generate very high-frequency waveforms. When w_s approaches the MHz range, the effects of circuit parasitic inductance and capacitance become important. One source of parasitic capacitance is the collector-emitter or drain-source capacitance of a bipolar junction transistor (BJT) or metal-oxide-semiconductor field effect transistor (MOSFET). Under certain conditions we can use this capacitance to advantage as all or part of the resonating capacitance in the converter. Figure 9.25 shows a series resonant converter using the MOSFET drain-source capacitance in this manner. Although shown separately, the antiparallel diode is an integral part of the power MOSFET, as you will see in Chapter 19.

Determine and sketch i_a for $T = 5 \times 10^{-7}$ s. Note that the two possible states of the circuit are both switches off or one off and one on. Start your analysis by assuming that Q_1 turns off at $t = 0$. At $t = t_1$, $v_1 = 2V_{\text{dc}}$ and D_2 turns on. Set up the equations for v_1 and i_a and use the periodicity condition, $i_a(T/2) = -i_a(0)$, to determine the initial condition on i_a. The Q of the resonant circuit is high when both switches are off, so $w_d \approx w_o$. One method of determining $i_a(0)$ and t_1 is to assume

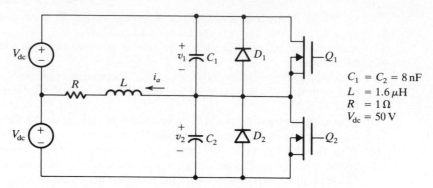

Figure 9.25 Series resonant converter using device junction capacitances as the resonating capacitors. This circuit is the subject of Problem 9.14.

a value for $i_a(0)$, calculate t_1 from the condition $v_1(t_1) = 2V_{dc}$, and then calculate a
new value of $i_a(0)$ from the condition $i_a(T/2) = -i_a(0)$. With a few iterations you
should converge on the correct initial condition.

9.15 In Section 9.5.1 we described how the voltage waveform of Fig. 9.16(a) can be
viewed as the sum of two square waves. Make a similar argument for Fig. 9.16(b),
identifying the location of the two sources.

9.16 We stated in Section 9.3 that in order to use SCRs for the switches in Fig. 9.10(a),
we must restrict operation of the circuit to the region $w_s > w_o$. Explain why this
constraint is necessary. Reference to the waveforms of Fig. 9.9 might prove useful
in your explanation.

Chapter 10

ac/ac Converters

AC/AC converters take power from one ac system and deliver it to another with waveforms of different amplitude, frequency, or phase. The ac systems can be single phase or polyphase, and reactive power flow can exist at the input, the output, or both, depending on how we configure the converter. The major application of ac/ac converters is variable-speed motor drives. These devices range in complexity from simple ac controllers found on products such as variable-speed electric drills or kitchen appliances to highly sophisticated four-quadrant pulse-width modulated drives used for traction applications.

The most common approaches to designing ac/ac converters are those that utilize a dc link between the two ac systems and those that provide direct conversion. The first is called a *dc-link converter*, and the second is called a *cycloconverter*. In the dc-link approach, we first rectify the input ac waveform to give a dc waveform, which in turn we invert to give the output ac waveform. A capacitor or inductor placed between the two converter stages stores the instantaneous difference between the input and output powers. This intermediate dc stage is known as a *dc link* or *dc bus*. We can control the ac/dc and dc/ac converters independently, so long as the *average* energy flows of the two are equal.

We can implement the input and output stages of the dc-link converter with a phase-controlled converter topology, a high-frequency dc/ac topology, or a resonant topology, depending on the application. You have already studied these basic topologies in Chapters 3–9, so we focus our discussion of dc-link converters in this chapter on the requirements of the dc energy storage element.

The cycloconverter avoids the intermediate dc bus by converting the input ac waveforms directly into the desired output waveforms. Figure 10.1 illustrates the basic principle of cycloconversion. The input voltage source v_i is sinusoidal with a higher frequency than that of the output (a factor of 3 in this case). We control the switches in the cycloconverter to synthesize a waveform having a fundamental equal to the desired output frequency. We then use filters to eliminate higher frequency components. Although the design shown here uses a single-phase

235

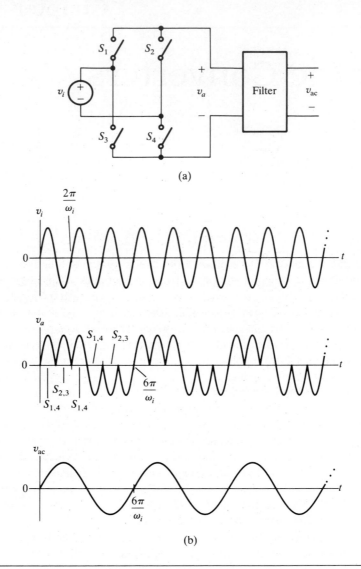

Figure 10.1   An illustration of the principle of operation of a cycloconverter: (a) an elementary cycloconverter circuit;   (b) waveforms of the ac source voltage v_i, the unfiltered output voltage v_a, and the filtered output voltage v_{ac}.

input source, a polyphase source is more common and produces a more flexible converter.

Generally speaking, we control the switches in a cycloconverter to connect the output terminals of the converter to the input voltage source whose value is closest to the desired output voltage at the time. The switches that make these connections usually must support bipolar voltage and carry bidirectional current, which is one of the reasons why cycloconverter circuits are frequently quite complex. We can consider the ac controller described in Chapter 2 as the simplest form of the cycloconverter, but we can use it only when frequency conversion is not required.

Very high-power applications, such as traction or steel-mill roll drives, require the use of SCRs because of the high voltages and currents at which these drives operate. A dc-link converter designed with SCRs would require the use of forced commutation at the inverter stage. An important advantage of the cycloconverter for these high-power applications is that it can use line commutation to turn off the SCRs. When operated this way, the cycloconverter is called a *naturally commutated cycloconverter*.

10.1 ENERGY STORAGE REQUIREMENTS IN A DC-LINK CONVERTER

In general, the instantaneous power flowing into a dc-link converter does not equal the instantaneous power flowing out of it. For instance, assume that the input and output waveforms have negligible distortion—and therefore are sinusoids—but have different amplitude, frequency, or phase, as shown in Fig. 10.2. The power waveforms at the two ports then each have a dc component and an ac component (at the second harmonic of the respective port's fundamental frequency). These instantaneous input and output powers, p_i and p_o, are also shown in Fig. 10.2.

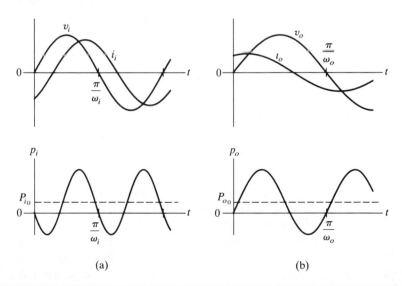

(a)   (b)

Figure 10.2 Voltage, current, and power waveforms at the input and output ports of a dc-link converter (neglecting harmonic distortion): (a) input voltage and current and input power $p_i(t)$; (b) output voltage and current and output power $p_o(t)$.

If the converter is 100% efficient, the average input power equals the average output power. Therefore the dc components of the power waveforms shown in Fig. 10.2(a) and (b) must be the same. But as the input and output can be at different frequencies, the ac components of the two power waveforms need not be the same. The difference between the instantaneous input and output powers must be absorbed or delivered by an energy storage element within the converter. For the dc-link converter, this *load-balancing* energy storage element is located at the dc bus.

The load-balancing energy storage element can be either a capacitor, an inductor, or both. The choice depends on the topologies used for the input and output conversion stages. For those topologies that require the dc system to resemble a voltage source, we need a capacitor; for those that require the dc system to resemble a current source, we need an inductor. For instance, in Section 8.3.2 you learned how to create an ac voltage from a dc voltage by modulating the duty ratio of bridge switches operated at a high frequency. So long as the switches have antiparallel diodes, we can make the average power flow in either direction. Hence we can use one of these converters for the input stage of our dc-link converter and another for the output stage. Both converters require a voltage source at their dc ports, so we place a large capacitor across the dc bus, as shown in Fig. 10.3(a).

We can also use a phase-controlled converter for the input and output stages of a dc-link converter. As you saw in Chapter 5, the phase-controlled converter needs a large inductor in series with its dc port to hold the dc current constant over a cycle of the ac waveform. These inductors serve as the load-balancing energy storage element of the dc-link converter, as shown in Fig. 10.3(b). The use of phase-controlled converters imposes the additional constraint that both the input and output be connected to ac voltage sources (lines). This is precisely the method used for high-voltage dc transmission, in which the dc bus is the dc transmission line.

A typical implementation of a dc-link converter would use a phase-controlled rectifier to take power from the utility and a voltage source dc/ac converter to deliver it to an ac load. The former requires a current source at the dc bus, the latter a voltage source. In this case, we must achieve the load-balancing energy storage with two components: a series inductor and a shunt capacitor, as shown in Fig. 10.3(c).

The size of a load-balancing energy storage element depends on the amount of ac energy it must absorb and the level of ripple that its state variable can tolerate. For instance, to determine the value of C_B in Fig. 10.3(a), we need to first find the difference between input and output power waveforms. The integral of this difference is the energy that flows in and out of C_B. We can then relate its peak-to-peak amplitude, ΔE_C, to the peak-to-peak ripple in the capacitor voltage Δv_{dc}. If $\langle v_{dc} \rangle = V_{dc}$, then:

$$\Delta E_C = \left(\frac{1}{2}\right) C_B \left[\left(V_{dc} + \frac{\Delta v_{dc}}{2}\right)^2 - \left(V_{dc} - \frac{\Delta v_{dc}}{2}\right)^2 \right] \qquad (10.1)$$

from which we can determine the value of C_B.

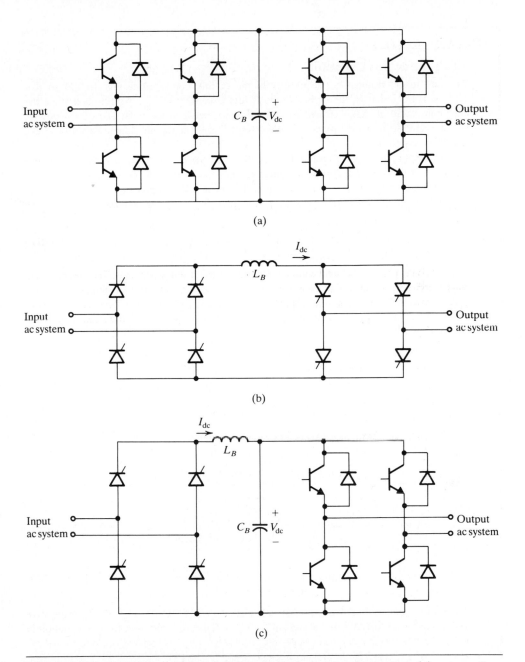

Figure 10.3 Possible choices of dc bus energy storage elements: (a) a capacitor, when both dc/ac converters require a dc voltage source; (b) an inductor, when both converters require a dc current source; (c) both an inductor and a capacitor, when one dc voltage source and one dc current source are needed.

EXAMPLE 10.1

Sizing the dc Bus Capacitor

We want to determine the size of C_B in the dc-link converter of Fig. 10.3(a) so that $\Delta v_{dc} < 0.1 V_{dc}$, that is, the dc bus ripple is less than 10% of V_{dc}. We first determine the maximum value of ΔE_C created by the difference between the input and output powers and then use *(10.1)* to determine C_B. We assume that the input and output waveforms have negligible distortion.

The input power waveform will have a dc component and a second-harmonic component, as shown in Fig. 10.2(a). If the input voltage and current are in phase, the amplitude of the second harmonic P_{i_2} equals the amplitude of the dc component P_{i_0}. If they are out of phase, meaning that there is reactive power flow, $P_{i_2} > P_{i_0}$. In general,

$$P_{i_2} = \frac{P_{i_0}}{\cos \theta_i} \tag{10.2}$$

where θ_i is the phase between the input voltage and the current waveforms. The output power waveform has a similar relationship between its dc and second-harmonic components.

The two dc components P_{i_0} and P_{o_0} are equal, and the difference between the two ac components flows into C_B. If the frequencies of the two power waveforms are ω_i and ω_o, these components are

$$p_{i_2}(t) = P_{i_2} \cos 2\omega_i t = \frac{P_{i_0}}{\cos \theta_i} \cos 2\omega_i t \tag{10.3}$$

$$p_{o_2}(t) = P_{o_2} \cos 2\omega_o t = \frac{P_{o_0}}{\cos \theta_o} \cos 2\omega_o t \tag{10.4}$$

As $P_{i_0} = P_{o_0}$, the difference between *(10.3)* and *(10.4)* is

$$p_{i_2}(t) - p_{o_2}(t) = \frac{P_{i_0}}{\cos \theta_i} \left(\cos 2\omega_i t - \beta \cos 2\omega_o t \right) \tag{10.5}$$

where $\beta = \cos \theta_i / \cos \theta_o$.

The dc bus capacitor C_B has a total energy, $E_C(t)$, composed of a dc component, E_{dc}, and an ac component, $E_{ac}(t)$. We can find the ac component by integrating *(10.5)*:

$$E_{ac}(t) = \int \left[p_{i_2}(t) - p_{o_2}(t) \right] dt = \frac{P_{i_0}}{\cos \theta_i} \left(\frac{\sin 2\omega_i t}{2\omega_i} - \beta \frac{\sin 2\omega_o t}{2\omega_o} \right) \tag{10.6}$$

The positive peak of this function occurs when $\sin 2\omega_i t = 1$ and $\sin 2\omega_o t = -1$; its negative peak occurs when $\sin 2\omega_i t = -1$ and $\sin 2\omega_o t = 1$. Therefore the peak-to-peak variation in E_C is

$$\Delta E_C = \frac{P_{i_0}}{\cos \theta_i} \left(\frac{1}{\omega_i} + \frac{\beta}{\omega_o} \right) \tag{10.7}$$

Equating ΔE_C to the change in energy of a capacitor whose voltage changes from $0.95 V_{dc}$

to $1.05V_{dc}$ (10%), we can find C_B, or:

$$C_B = \frac{\Delta E_C}{\frac{1}{2}\left[(1.05V_{dc})^2 - (0.95V_{dc})^2\right]} \qquad (10.8)$$

If we use the parameters $P_{i_o} = 10$ kW, $\omega_i = 2\pi \times 60$, $\omega_o = 2\pi \times 400$, $\theta_i = \pi/6$, $\theta_o = 0$, and $V_{dc} = 200$ V, then $\beta = 0.866$, $\Delta E_C = 34.6$ J, and $C_B = 8650$ μF.

10.2 THE NATURALLY COMMUTATED CYCLOCONVERTER

The cycloconverter is utilized primarily to provide ac/ac conversion at very high power levels, typically in excess of 100 kW. A thyristor is the only device that can meet the switch voltage and current ratings needed at these power levels. Naturally commutated cycloconverter circuits provide a means of using SCRs without requiring forced commutation. Although we can design cycloconverters for lower power levels using fully controlled switches, such as transistors, the number of required switches and the complexity of their control usually makes a dc-link converter a more attractive choice. Therefore we confine ourselves here to the naturally commutated cycloconverter, utilizing SCRs.

10.2.1 Principles of Operation

In Section 5.4 we discussed the operation of the 6-pulse phase-controlled converter shown in Fig. 10.4(a). For fixed α, the output voltage of this converter is

$$v_o = V_d = \langle v_d \rangle = \frac{3V_{\ell\ell}}{\pi} \cos\alpha = V_{do} \cos\alpha \qquad (10.9)$$

where $V_{\ell\ell}$ is the amplitude of the line-to-line voltage. Note that as α increases from 0 to π, the output voltage varies from V_{do} to $-V_{do}$. If we modulate α slowly compared to the input frequency ω_i, the short-time average of $v_d(t)$, $\overline{v}_d(t)$, yields an output voltage that varies with time, $v_o(t) = \overline{v}_d(t)$. This technique of synthesizing a waveform that varies with time by varying α is similar to the waveshaping technique discussed in Section 8.3. There, we used a duty ratio $d(t)$ that varied with time to modulate the output of a dc/dc converter.

The waveshaped output of a modulated dc/dc converter is directly proportional to $d(t)$. Shaping the output of a phase-controlled converter, however, is complicated by the nonlinear relationship between $\alpha(t)$ and $\overline{v}_d(t)$ given by (10.9). For example, suppose we want $\overline{v}_d(t) = v_o$ to be sinusoidal, or:

$$v_o = V_o \sin\omega_o t \qquad (10.10)$$

where $\omega_o \ll \omega_i$ and $V_o < V_{do}$. The required $\alpha(t)$ must be such that:

$$v_o = V_{do} \cos\alpha(t) = V_o \sin\omega_o t \qquad (10.11)$$

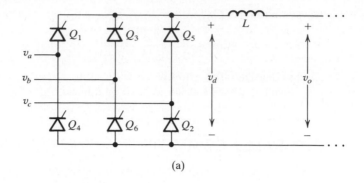

(a)

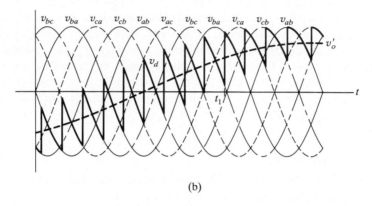

(b)

Figure 10.4   (a) A six-pulse phase-controlled, two quadrant (I and II) converter. (b) Output voltage, when the circuit of (a) is controlled to be a naturally commutated cycloconverter. The voltage v_o' is the desired waveform, and if L is properly sized $v_o \approx v_o'$.

which gives

$$\alpha(t) = \cos^{-1}\left(\frac{V_o}{V_{do}}\sin\omega_o t\right) \qquad (10.12)$$

Figure 10.4(b) shows the waveforms that result from this $\alpha(t)$. The waveform we are trying to synthesize is v_o', and if $\omega_o \ll \omega_i$ so that the inductor effectively removes the harmonics from v_d, then $v_o \approx v_o'$. However, as ω_o and ω_i get closer, the output filter will affect both the phase and amplitude of the fundamental of v_d.

In a practical control circuit, the function for α given in (10.12) is generally implemented indirectly by comparing the line-to-line input voltages with the desired output waveform v_o' and turning on the next SCR when the voltage v_d that would result is closer to v_o' than is the present value of v_d. For instance, just before t_1 in Fig. 10.4(b), v_{bc} is connected to the output through Q_3 and Q_2. The next commutation event is from Q_2 to Q_4, connecting v_{ba} to the output. At t_1, v_{bc} and

v_{ba} are equidistant from v_o', but v_{bc} is diverging while v_{ba} is converging on v_o'. At this time then, Q_4 is triggered and v_d assumes the value of v_{ba}. The next event is initiated based on a comparison of $v_{ca} - v_o'$ with $v_o' - v_{ba}$, and so on.

The problem with the ac/ac converter we have developed so far is that, although the output voltage can be either positive or negative, the output current can only be positive. To obtain bilateral load current, we need to place in parallel with this "positive" converter a "negative" converter, created by inverting its SCRs, relative to those in the positive converter, so that they can carry negative load current. A negative converter is shown in Fig. 10.5(a); its output voltage waveform is shown in Fig. 10.5(b). Note that, because the SCRs are inverted, the output voltage takes a negative step at commutation. The combined circuit, known as a *four-quadrant, naturally commutated cycloconverter*, is shown in Fig. 10.6(a). The waveforms of Fig. 10.6(b) illustrate how the two converters produce an ac output voltage for the output current shown. Note that the direction of the step change in v_d at commutation indicates which converter is operating at any point in the cycle. At t_o, operation shifts from the negative to the positive converter.

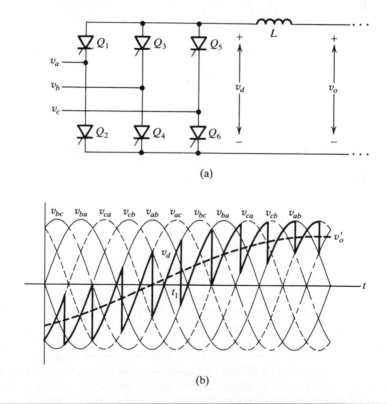

(a)

(b)

Figure 10.5   (a) A two-quadrant (III and IV) negative converter.   (b) The voltage v_d of the converter in (a) and the desired output waveform v_o'.

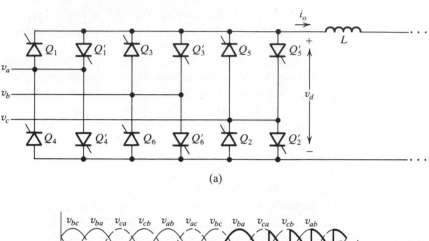

(a)

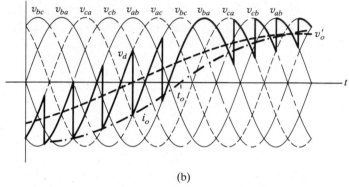

(b)

Figure 10.6   (a) A naturally commutated four-quadrant cycloconverter made by combining a positive and a negative phase-controlled converter. (b) Output voltage and current for the cycloconverter of (a).

EXAMPLE 10.2

A Control Circuit for the Naturally Commutated Cycloconverter

Figure 10.7 is a block diagram of a circuit suitable for controlling the six naturally commutated switches in the cycloconverter of Fig. 10.6. The finite-state machine has six states, one allocated to each of the six switches (12 SCRs). The output of analog multiplexor MUX_1, v_{d_1}, is a synthetic, signal-level version of the converter voltage v_d. The second analog multiplexor MUX_2 has an output v_{d_2} one commutation event (60°) ahead of v_d and v_{d_1}.

The essence of this control scheme is that commutation of the converter begins when v_{d_1} and v_{d_2} are equidistant from but on opposite sides of v_o'. At this point, the algebraic sum of $v_{d_1} - v_o'$ and $v_{d_2} - v_o'$ is zero. The voltages v_1 and v_2 are the differences between the multiplexor outputs and v_o'. For example, just before t_1 in Fig. 10.4(b), $v_{d_1} = v_d = v_{bc}$, and $v_{d_2} = v_{ba}$. At t_1, $v_1 < 0$ and $v_2 = -v_1$. At this time the zero-crossing detector clocks the finite-state machine and the commutation event at t_1 takes place, as well as the

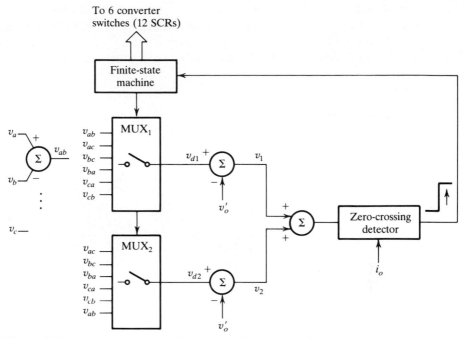

Figure 10.7   A circuit for controlling the cycloconverter of Fig. 10.6

switching of the multiplexors. This description is based on the assumption that the load current i_o is positive and therefore that the positive converter is functioning. In this case the zero-crossing is from positive to negative. If load current were negative, the zero-crossing would be oppositely directed, from negative to positive, as you can see from Fig. 10.5(b). The polarity of the load current is thus an input to the zero-crossing detector, determining which zero-crossing direction is valid.

10.2.2 Cycloconverters with Polyphase Outputs

Most high-power applications of cycloconverters require polyphase output waveforms. For instance, we may need to drive a large 3-phase machine at varying speeds from a 3-phase fixed-frequency service. We could duplicate the cyclocon-verter of Fig. 10.6 to create the three output waveforms required, but could not directly connect the outputs of the individual converters in a delta or wye configuration. The reason is that the output terminals of each converter are directly connected to the input voltages by the SCRs that are on, and if two such terminals were connected, the input voltage sources would be shorted. Therefore, if we use independent cycloconverters for each phase, we must incorporate into the circuit input or output isolation transformers to allow cycloconverter interconnection at the load.

Another way to create a 3-phase cycloconverter, which avoids the need for isolation transformers and halves the number of switches required, is shown in

Fig. 10.8(a). This approach uses three 3-pulse phase-controlled converters to create the three output voltages. Each converter has six SCRs, three to carry positive load current and three to carry negative load current. The input voltage sources are indicated as line-to-neutral sources to make it easier for you to see how the

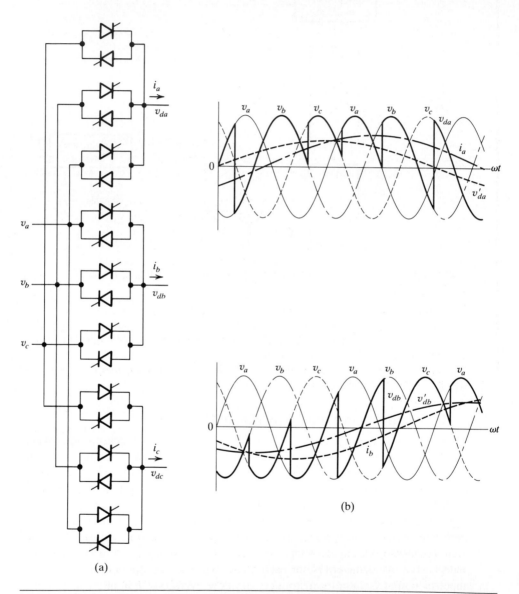

(a)

(b)

Figure 10.8 (a) A 3-phase in, 3-phase out cycloconverter using 3-pulse bridges.
(b) Waveforms at the a and b phases in the cycloconverter of (a).

output voltage waveforms are created. Figure 10.8(b) shows the line-to-neutral output voltages v_{da} and v_{db}. Note that as each 3-pulse converter has only three input voltages to choose from (the negative input voltages are not available), commutations occur half as often as with the 6-pulse circuit. The deviations of v_{da}, v_{db}, and v_{dc} from v'_{oa}, v'_{ob}, and v'_{oc} are therefore larger than the deviation of v_d from v'_o in Fig. 10.5(b). Other than that, v_{da}, v_{db}, and v_{dc} have the same features as v_d, including positive step changes when the load current is positive and negative step changes when the load current is negative.

Notes and Bibliography

An exhaustive and detailed treatment of cycloconverters can be found in [1]. The treatment tends to be more mathematical and theoretical than practical, but there is a good chapter on control of cycloconverters that includes a number of block diagrams of control system implementations.

An ac/ac converter of a type sometimes referred to as a *high-frequency link* converter is described in numerous papers by Fransisc Schwarz and J. Ben Klaassens. A good introduction to the concept is given in [2]. A more detailed application of the high-frequency link converter is to be found in [3]. The high-frequency link converter is especially suitable to situations where isolation is required between the two ac systems. However, its operation, and in particular its control, is substantially more complex than the two systems presented in this chapter.

1. L. Gyugyi and B. Pelly, *Static Power Frequency Changers*, (New York: Wiley Interscience, 1976).
2. F. C. Schwarz, "A Doublesided Cycloconverter," in *IEEE Power Electronics Specialists Conference (PESC) Record* (1979), 437–447.
3. J. B. Klaassens, "Dc-ac Series Resonant Converter System with High Internal Frequency Generating Multiphase ac Waveforms for Multikilowatt Power Levels," *IEEE Trans Power Electronics* 2 (3): 247–256 (July 1987).

Problems

10.1   The simple single-phase cycloconverter of Fig. 10.1 is designed to drive a resistive load through a filter consisting of an inductor, as shown in Fig. 10.9. Design a suitable implementation for the switches.

10.2   The cycloconverter of Fig. 10.9 is designed to produce an output v_{ac} with a fundamental frequency of $\omega_i/3$, as shown in Fig. 10.1. If $R/L = \omega_i/3$, what are the amplitudes of the fundamental and first nonzero harmonic of v_{ac} relative to the amplitude of the input voltage V_i? Sketch v_{ac}, considering only these two components.

10.3   Sketch the input current i_i for the cycloconverter of Fig. 10.9 with the values of L and R as specified in Problem 10.2. What is the power factor of the converter?

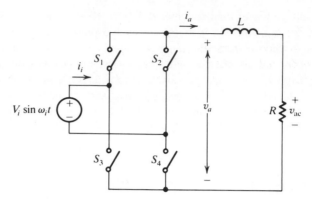

Figure 10.9 The cycloconverter of Fig. 10.1 with a simple low-pass filter. This circuit is the subject of Problems 10.1, 10.2, and 10.3.

10.4 The ac sources in a system containing a cycloconverter are seldom without some series impedance. Figure 10.10 shows inductance L_c in series with the source $V_i \sin \omega_i t$ of the converter operating as shown in Fig. 10.1. Assume that $L_c \ll L$ and make qualitative sketches of i_i and v_a.

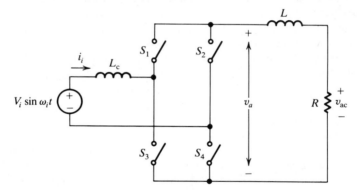

Figure 10.10 A single-phase cycloconverter with commutating inductance L_c. This circuit is the subject of Problem 10.4.

10.5 A HVDC transmission system consists of two-phase controlled converters connected by a dc link, as shown in Fig. 10.11 for a simplified single-phase ac system. It is often used to tie together ac utility systems that have different frequencies or are out of phase relative to each other. One of the advantages of a HVDC intertie between two ac systems is that the intertie can be used to control real and reactive power independently, enhancing the stability of the overall system. In answering the following questions, assume that the ac filters remove all but the fundamental components of the line current and do not affect the fundamental at all.

(a) How is the link current I_{dc} controlled?

(b) Sketch v_{d_1} and v_{d_2} if α_1 and α_2 are controlled so that the real power flow is zero, but the reactive power flowing into system 2 is maximized for a given I_{dc}, and the power factor for ac system 2 is leading—that is, the source of this reactive power acts like a capacitor.

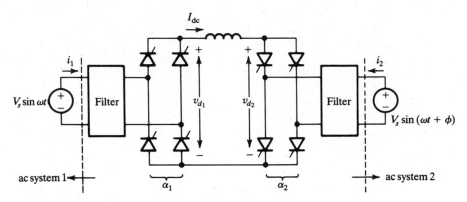

Figure 10.11   The HVDC intertie analyzed in Problem 10.5.

(c) Repeat (b) for zero reactive power and maximum real power.

(d) Can power (real or reactive) flow from ac system 2 to ac system 1 in the circuit of Fig. 10.11?

10.6   Figure 10.12 shows a HVDC intertie between two 3-phase ac systems.

(a) Sketch v_{d_1} and v_{d_2} if α_1 and α_2 are controlled so that the real power flow is zero, but the reactive power flowing into system 2 is maximized for a given I_{dc}, and the power factor for ac system 2 is lagging—that is, the source of this reactive power acts like an inductor.

(b) Repeat (b) for zero reactive power and maximum real power.

(c) For both parts (a) and (b), sketch the line–line voltages v_{ab_1}, and v_{ab_2} and the line currents i_{a_1} and i_{a_2} at the ac ports.

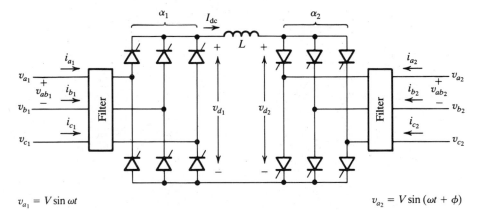

$$v_{a_1} = V \sin \omega t$$

$$v_{b_1} = V \sin \left(\omega t - \frac{2\pi}{3} \right)$$

$$v_{c_1} = V \sin \left(\omega t + \frac{2\pi}{3} \right)$$

$$v_{a_2} = V \sin (\omega t + \phi)$$

$$v_{b_2} = V \sin \left(\omega t + \phi - \frac{2\pi}{3} \right)$$

$$v_{c_2} = V \sin \left(\omega t + \phi + \frac{2\pi}{3} \right)$$

Figure 10.12   A dc-link converter connecting two three-phase ac networks. This dc intertie is considered in Problem 10.6.

Dynamics
and Control

Chapter 11

Dynamics and Control: An Overview

IN Part I we examined the form and function of the major families of power electronic circuits. Our goal was to show how the intended power conversion function is achieved in each case by appropriate configuration of the circuit components and by proper operation of the switches. Throughout those earlier chapters, our concern was with *nominal* operating conditions, that is, the ideal operating conditions in which a circuit is designed to perform its primary conversion function. As nominal operation in most power electronic circuits involves a *periodic steady state*, we focused on situations in which circuit operation and behavior are the same from cycle to cycle.

Now we need to deal with the consequences of the inevitable disturbances or errors that cause circuit operation to deviate from the nominal. These disturbances include variations and uncertainties in source, load, and circuit parameters, perturbations in switching times, and events such as startup and shutdown. We refer to the resulting evolution of the deviations from nominal behavior as the *dynamic behavior* of the circuit. If such deviations have a negligible effect on circuit operation, the user may be content to let the circuit run without corrective action. This is rarely the case, however.

Most often, departures from nominal conditions have to be counteracted through properly designed controls. We show several examples of circuits that do not recover, or whose recovery is incomplete or too slow, without such controls. A controller or "compensator" must first provide the user with a simple and convenient means of selecting the desired nominal operating condition. Second, it must automatically regulate the circuit at this operating condition by delaying or advancing the times at which switches are turned on and off. Hence in this chapter we begin to build the basis for analyzing the dynamic behavior of power circuits and for designing and implementing controls that regulate these dynamics, maintaining operation in the vicinity of the nominal, despite disturbances or errors.

The focus of Part II is on analysis and control design by means of appropriate dynamic models. This approach allows us to anticipate the behavior of the circuit under diverse operating conditions, generate candidate controller structures and parameters, plan simulation studies, understand experimental results, recognize which regimes of operation call for further investigation, and so on. Such an approach is especially critical in power electronics because of the effort and cost involved, even in breadboarding, and the expense (not to mention the distress and, sometimes, smoke!) associated with component failure. Analytical studies must, of course, be combined with engineering experience and intuition, experimentation, and other ingredients of the design process in order to be successful.

11.1 CONTROL SYSTEM CONFIGURATION

All control is based, explicitly or implicitly, on a model that describes how control actions and disturbances are expected to affect the future behavior of the system. Control of a power circuit entails specifying the desired nominal operating condition and then regulating the circuit so that it stays close to the nominal in the face of disturbances and modeling errors that cause its operation to deviate from the nominal.

In simple *open-loop* control, the controller is not given any information about the system during operation, although the open-loop controller may be constructed on the basis of prior information or models. In open-loop control with *feedforward*, the controller utilizes measurements of some of the disturbances affecting the system. Using feedforward, the controller can attempt to cancel the anticipated effects of measured disturbances. Feedforward alone is usually insufficient, however, to obtain satisfactory performance in power electronic circuits.

A better strategy is for the controller to also use measurements that reveal the circuit's present behavior. The controller can thereby assess the extent of departure from the desired behavior and choose control,actions aimed at restoring the system rapidly and safely to nominal operation. This strategy is the essence of *closed-loop* or *feedback control*. When the model that underlies the choice of control actions is itself updated on the basis of measurements, we refer to the controller as *adaptive*.

The block diagram in Fig. 11.1 represents the typical situation. Each connection between the blocks represents one or more actions, measurements, or information flows affecting the block at which the arrowhead terminates—and originating in the block at the other end of the connection. We frequently refer to the *feedback loop* in Fig. 11.1. The output measurements fed back to the controller provide information about the behavior of the system and about the variables that we want to control.

The quantities directly affecting the system, namely, the control inputs and disturbances, are shown on Fig. 11.1. It also shows the feedforward of those disturbances that can be measured. The unmeasured disturbances, as well as modeling errors, cause the controller's actions to have unanticipated effects. The output measurements available to the controller are corrupted by what is termed measurement noise, or sensor noise. This noise, along with the unmeasured disturbances and modeling errors, causes inaccuracies in evaluating measurements.

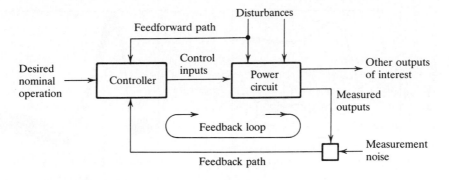

Figure 11.1   The typical control system configuration.

EXAMPLE 11.1

Open-Loop Control with Feedforward for an Up/Down Converter

Let's consider an up/down (or indirect, or buck/boost) dc/dc converter whose power circuit is built according to the circuit schematic in Fig. 11.2 and is operated at a frequency of 50 kHz, or with a switching period of $T = 20$ μs. Let $R = 2$ Ω, $C = 220$ μF, and $L = 0.25$ mH. (We discussed such circuits in Chapter 6.) We would like to maintain the average output voltage $\langle v_o \rangle$ within 5% of the nominal or reference value of $V_{ref} = -9$ V, despite step changes in the input voltage v_{in} from a nominal dc value of $V_{in} = 12$ V down to values as low as 8 V. For purposes of this example, let's assume that there are no other nonidealities or uncertainties in the circuit and, in particular, that the transistor and diode function as ideal switches. You will see that, even if the system behaves according to this idealized model, the circuit response can be unsatisfactory.

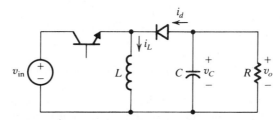

Figure 11.2   Circuit schematic of the power stage of an up/down converter.

   Recall from Chapter 6 that, if the transistor is turned on periodically and operated with a duty ratio D and if the converter is operating in continuous conduction mode with $v_{in} = V_{in}$, then $\langle v_o \rangle = -V_{in} D/D'$ to a good approximation, where $D' = 1 - D$. (Note the inversion of polarity between input and output.) The duty ratio must therefore be set at a nominal value of $D = V_{ref}/(V_{ref} - V_{in}) = 0.43$ in order to obtain the desired operation under nominal conditions.

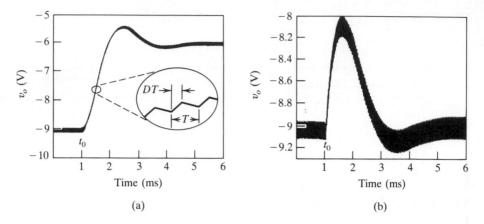

Figure 11.3 (a) Response of the ideal up/down converter circuit to a step from 12 V to 8 V in the input voltage v_{in}. (b) Response to the same step after incorporating feedforward.

Figure 11.3(a) shows the response of this idealized circuit to a step in the input voltage from 12 V to 8 V. The waveform before time t_0 corresponds to operation in a cyclic steady state with an input voltage of 12 V and a duty ratio of 0.43. Not unexpectedly, the ideal circuit produces the correct average output voltage under nominal conditions—before t_0. At t_0 the input voltage drops to 8 V. The circuit undergoes significant oscillatory transients and settles down to an incorrect average value of output voltage, namely, -6 V. We can explain completely the steady-state behavior with the static model that we already have, but the explanation of the oscillatory behavior must await the modeling results in Section 11.3.

One natural idea for obtaining better responses to changes in the input voltage v_{in} is to use feedforward. If D is made to vary in accordance with measured variations in v_{in} rather than being fixed by the nominal value V_{in} so that $-v_{in}D/D'$ is held constant at V_{ref}, then v_o will attain the correct steady-state average value $\langle v_o \rangle$ despite input-voltage variations. This *pulse-width modulation* (PWM) approach requires that we select $D = V_{ref}/(V_{ref} - v_{in})$. The resulting response to a step change in the input voltage is shown in Fig. 11.3(b). The ideal circuit with feedforward now settles down to the correct average output voltage, despite the step in input voltage. The peak excursions are considerably less, though still outside the allowed 5%. The transients, however, take as long to die out as they did without feedforward.

We can explain why the transients in Example 11.1 appear as they do after we obtain a dynamic model for the up/down converter in Section 11.3 (see especially Example 11.7). Feedforward in Example 11.1 compensated for the steady-state effects of disturbances in the input voltage but did nothing to modify the dynamics of the transient. In the presence of nonideal components, even the steady state would not be accurately restored with feedforward alone. More than feedforward is needed to obtain significantly better behavior. We return to the converter in Example 11.1 several times to illustrate various aspects of dynamic modeling and control design.

EXAMPLE 11.2

Closing the Loop on a Controlled Rectifier Drive for a dc Motor

The circuit schematic in Fig. 11.4(a) represents a separately excited dc motor, driven by a phase-controlled bridge rectifier supplied from a nominally sinusoidal voltage source, v_{ac}. (We discussed such controlled rectifiers in Chapter 5.) The voltage applied to the motor is denoted by v_d and the armature current by i_d. The waveforms expected on the basis of the analysis of nominal operation in Chapter 5 are displayed in Fig. 11.4(b), with the assumptions that the armature current remains positive throughout each cycle and that there is no commutating reactance. Figure 11.4(b) shows the role of the firing angle of the rectifier, denoted by α_k for the kth cycle, in specifying the voltage waveform.

Nominal operation of the drive system corresponds to a cyclic steady state in which the firing angle is maintained at a constant value $\alpha_k = \alpha$ and the armature current varies

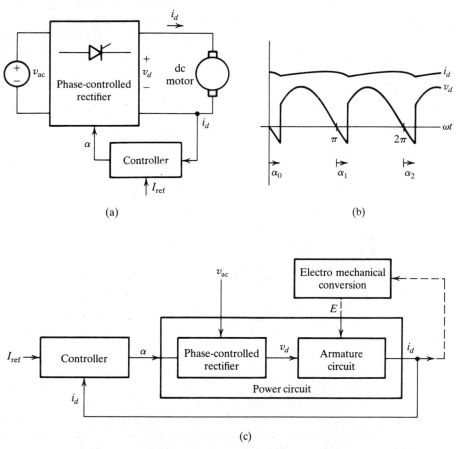

(a)   (b)

(c)

Figure 11.4   (a) Schematic diagram of a phase-controlled rectifier drive. (b) Waveforms of applied voltage v_d and current i_d, with a rectifier firing angle of α_k in the kth cycle. (c) Control block diagram of the drive.

periodically at twice the frequency of v_{ac}. Our objective is to use closed-loop control of the firing angle to regulate the average armature current $\langle i_d \rangle$ at any specified reference value in some range. This reference value, I_{ref}, is specified by a higher-level speed or torque controller for the motor. The average current $\langle i_d \rangle$ determines the average torque applied to the motor shaft. If the mechanical time constant of the motor is considerably larger than the period of the source, the speed of the motor displays very little ripple in the steady state and depends essentially on $\langle i_d \rangle$.

To represent the system entirely by a control block diagram—for comparison with the general control configuration presented in Fig. 11.1—we can transform Fig. 11.4(a) to Fig. 11.4(c). The presence of commutating reactance would introduce a feedback connection from i_d to the phase-controlled rectifier in this block diagram, indicating that the applied voltage v_d would then depend on i_d.

We have also chosen in Fig. 11.4(c) to represent the system at a finer level of modeling detail than in Fig. 11.4(a), with the dc motor model split into subsystems corresponding to the armature circuit and the electromechanical conversion process. The connections to and from the electromechanical subsystem are shown in dashed lines because the back emf, E, is most commonly treated as a slowly varying external disturbance acting on the armature circuit. The rationale for this treatment is that, generally, many cycles are needed for changes in the average current to cause changes in motor speed and hence changes in E.

In this case the problem becomes one of designing a controller to maintain the average armature current at a specified value, in the face of a slowly varying disturbance E. Note that questions of dynamic modeling are already tying in to questions of control design, even before we begin to develop quantitative or analytical models.

One fairly satisfactory and widely used control solution for such systems is based on a *proportional–integral (PI)* controller. Its heart is the proportional part, whose action is consistent with simple reasoning: It changes the firing angle from its nominal value by an amount proportional to the error $I_{ref} - \langle i_d \rangle$. The firing angle is reduced when the error is positive, because the average voltage and hence the average current are thereby increased, which decreases the error. When the error is negative, the proportional part of the controller does the opposite, increasing the firing angle in proportion to the error magnitude. The integral part of the controller acts on the integral of the error, and works on a slower time scale to correct for steady-state errors induced by parameter uncertainties and constant disturbances in the model.

It is also possible to use a proportional–integral–derivative (PID) controller instead, which can speed up the response by adding to the PI controller a contribution that depends on the derivative of the error or the derivative of the output (in which case it is called *rate feedback*). The PI, PID, and related control structures, such as lag, lead and lead–lag compensators, are ubiquitous in control.

Example 11.2 illustrates the form that the component blocks and signals in Fig. 11.1 may take in a power electronics application. We return to the system in this example later in this chapter.

Much of our attention in Part II centers on the feedback loop in Fig. 11.1. This loop is critical to a feedback-control solution. Feedback can speed up the response of a system to commanded changes, improve the recovery from unanticipated disturbances, and make performance less sensitive to system variations. However, a system that is stable and insensitive to perturbations in open loop could become sensitive, slow to recover, or even completely unstable in closed loop, if the controller reacts inappropriately to the feedback signals. The controller's actions could aggravate deviations from nominal operation instead of rapidly restoring nominal operation. In other words, to obtain the advantages of feedback control, we have to design and implement the feedback loop carefully.

11.2 MODEL SELECTION

The use of an appropriate model or set of models is central to the control design process. Different models may be needed for different stages or aspects of the control design. Even for a specific stage of the design, however, there are likely to be several possible models, differing in their explicitness, complexity, accuracy, domain of definition, flexibility, tractability, and so on. Obviously, several trade-offs are involved in selecting from among these models.

For instance, a model that functions well in simulating the open-loop behavior of a power circuit may not necessarily be a good basis for designing a closed-loop controller. A more complex model for a power circuit may predict observed open-loop behavior more accurately but may be less tractable analytically, less useful in generating candidate control designs, and perhaps less able to yield controllers that can withstand circuit variations. However, a simple model may miss crucial aspects of system behavior and therefore lead to unsatisfactory controllers.

In practice, you should work with several models, checking the predictions of one model against those of another—and against experimental observations. The insights from this process are used to iteratively refine the models and the control design. You must therefore become familiar with a variety of approaches to modeling power electronic circuits and their controllers. We begin our discussion of modeling approaches in Sections 11.3 and 11.4 by deriving circuit-averaged models that are useful in describing the average behavior of certain families of power circuits. In Chapters 12 and 13 we deal with state-space models, which embrace a much wider variety of modeling possibilities. We devote major attention to the special but important case of *linear, time invariant* (LTI) models.

Because we devoted so much effort in Part I to nominal operation in steady state, we need to emphasize here that, in general, we have to go beyond static or steady-state models to develop controllers for power circuits.

11.2.1 The Need for Dynamic Models

Elementary reasoning in Example 11.2 could have led us to at least the proportional part of the PI controller. With some tuning, a proportional controller alone could permit stable operation over a range of operating points (though probably with unacceptably high steady-state error). In this respect, the controlled rectifier drive is a benign system, because a reasonable controller can be derived from a model that is not much more sophisticated than the one in Example 11.2.

However, you can run into trouble by following common sense or using overly simplified analytical models when designing a feedback controller. For example, a controller based only on an understanding of static operating characteristics, ignoring dynamic effects, can fail badly.

EXAMPLE 11.3

Problems with Proportional Feedback Control of an Up/Down Converter

The desire for better dynamic performance than that obtained with open-loop and feedforward control of the up/down converter in Example 11.1 leads us to consider a feedback-control solution. We must now measure the deviation of the average output voltage from the desired value of $V_{ref} = -9$ V, and use the discrepancy to adjust the duty ratio from the nominal value D to $D + \tilde{d}$. The correction $\tilde{d}$ depends on the polarity and magnitude of the voltage deviation.

Examination of the (inverting) steady-state characteristic of the converter, $\langle v_o \rangle = -v_{in}D/D'$, suggests that when the error $\tilde{v}_o = \langle v_o \rangle - V_{ref}$ is negative, indicating that $\langle v_o \rangle$ is too negative, we should decrease the duty ratio. Similarly, we should increase the duty ratio when the error is positive. This is the "natural" pulse width modulation (PWM) control law suggested by the steady-state characteristic.

The *proportional* feedback-control system represented in the block diagram in Fig. 11.5(a) is one implementation of this control law. It gets its name from the fact that the

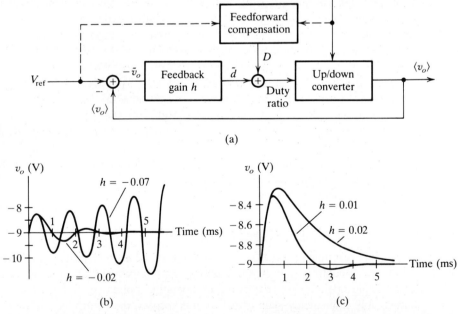

(a)

(b)

(c)

Figure 11.5 (a) Block diagram of a proportional feedback/feedforward-control system for an up/down converter. (b) Step response of the ideal circuit for increasingly negative gains. (c) Response for increasingly positive gains.

commanded duty ratio perturbations are proportional to the measured deviations of $\langle v_o \rangle$. The constant gain h has to be negative to provide the corrective action suggested by examination of the steady-state characteristic. The block diagram also indicates that the earlier feedforward compensation may be implemented along with the feedback.

Figure 11.5(b) presents the response of the ideal circuit to the same input-voltage step as in Example 11.1, for increasingly negative values of the proportional gain and with the previous feedforward in place. The response of the circuit clearly becomes *more* oscillatory, and the circuit actually goes *unstable* before h becomes very negative. The counter intuitive choice of a *positive* h does not immediately lead to the disaster that might be expected on the basis of the steady-state characteristic. As indicated by the waveforms in Fig. 11.5(c), the response can be stable for a range of positive h. Evidently our reasoning based on the steady-state characteristic has led us quite astray.

The waveforms presented in Example 11.3 result from subjecting the instantaneous switched waveforms to a moving short-term average or *local average*, computed over an interval equal to the switching period of the converter. (We give a formal definition later, in *(11.1)*, but you need not be concerned about it yet.) The effect of doing so is to remove the switching ripple from the waveforms, leaving only the smooth variations in local average over a time scale that is longer than the switching period. There are two reasons for displaying the averaged waveforms. First, this emphasizes that the quantity to be regulated in the up/down converter is the average value of the output voltage, not its instantaneous value; the details of the ripple are of little interest, provided the ripple is kept sufficiently small. Second, the averaged waveforms can be computed or simulated more simply than the instantaneous ones for this class of circuits by using the circuit-averaged models in Section 11.3.

Example 11.3 shows that a static model may be inadequate for the design of a feedback controller. This forces us to look at models that incorporate dynamic effects. In fact, a clearer picture of what is happening with the up/down converter in Fig. 11.5 emerges after we obtain a dynamic model for the response of the output voltage to variations in the duty ratio (see, especially, Example 11.9 in Section 11.4 and Example 11.11 in Section 11.5).

11.3 OBTAINING DYNAMIC MODELS BY CIRCUIT AVERAGING

A large majority of power circuit models have circuits with only LTI components and ideal switches. The analysis of such a circuit in each switch configuration (or topological state) is as simple as the analysis of an LTI circuit. Thus we can use various special techniques, such as impedance methods, in any switch configuration. The problem arises in piecing together the solutions from successive configurations, especially when the times of transition are controlled as functions of the circuit's behavior. We also need to model the dynamics of the controller along with that of the circuit in order to study the controlled, closed-loop system, but circuit methods are often not appropriate or convenient for controller modeling. These reasons motivate the study of more general models, most importantly state-space models.

However, we can analyze the dynamic behavior of many classes of power circuits using the notion of *averaged-circuit models*, which allows us to stay close to the circuit techniques that electrical engineers use so well. The ideas of circuit averaging are simple enough to develop concisely and yet yield very useful models for some important families of power circuits. We therefore use circuit averaging as a context for our discussion of dynamic modeling and control issues in this chapter. You should *not* infer from this decision that averaging is necessarily the main or best approach to constructing dynamic models for all types of power electronic circuits. We defer development of state-space models of the dynamics of general power circuits to Chapters 12 and 13.

Averaged-circuit models have traditionally been derived mainly for high-frequency switching dc/dc converters, usually through an averaging process applied to state-space models. Here we follow a more fundamental approach, which starts directly from the circuit diagram of the power circuit and can be applied more widely than the traditional approach. We show you how to obtain nonlinear circuits that describe the average behavior of various power circuits and how to derive associated linear circuits that approximately describe small-signal behavior.

11.3.1 Averaging a Variable

In many power electronic circuits, our interest is in the average values of voltages and currents rather than in their instantaneous values, provided the ripple or harmonics are sufficiently small. Thus, in the up/down converter of Examples 11.1 and 11.3, the control objective was to regulate the average output voltage at some fixed value, under the assumption that the voltage ripple was kept small. For the controlled rectifier drive of Example 11.2, our only interest was in controlling the average armature current in the dc motor, because the inherently low-pass nature of the mechanical dynamics causes the variations of the current around its mean to have relatively minor effects on the motion of the motor.

A slightly more elaborate example is provided by a typical high-frequency PWM waveshaping inverter for ac motor drives, of the kind discussed in Section 8.3. The objective in this case is regulation of the local average of the output current around a sinusoidal reference whose frequency is much lower than the switching frequency. The current ripple is assumed to be kept small.

Examples such as these lead to examination of the constraints governing average values of circuit variables. Our objective is to find a circuit approach to analyzing the local average behavior of circuit variables even during a transient, nonperiodic condition. The local average we use is a moving average defined by:

$$\bar{x}(t) = \frac{1}{T} \int_{t-T}^{t} x(\tau)\, d\tau \qquad (11.1)$$

for a fixed T. (A more complete notation would have been $\bar{x}_T(t)$, for example, but we rely on context to make clear what T is, so as to avoid notational clutter.) This average at any time is taken over the preceding interval of length T. Hence $\bar{x}(t)$ is a smoother function than $x(t)$ and is a continuous function of time unless $x(t)$ has impulses in it. An important consequence of *(11.1)* is that the derivative of the average of a variable equals the average of the derivative, as can be easily verified.

An appropriate choice of T is required in any specific application to obtain useful results. We almost invariably choose T to equal the shortest regular switching interval associated with the operation of a power circuit. In the special case where $x(\tau)$ is periodic, and we pick the averaging interval T to equal the period, $\bar{x}(t)$ is just the usual average—a constant that we have been denoting by $\langle x \rangle$. This special case is, of course, very important in power electronics, because the steady-state waveforms in typical power circuits are indeed periodic.

11.3.2 Averaging a Circuit

Averaging the constraint equations imposed on circuit variables by Kirchhoff's voltage and current laws (KVL and KCL), we find that the instantaneous and averaged variables satisfy identical constraints. The reason is that the KVL and KCL constraints are linear and time invariant (LTI), so their form is unchanged by averaging. Results that follow from KVL and KCL—such as power conservation and its generalization through Tellegen's theorem—therefore hold for averaged quantities as well as for instantaneous ones.

Similarly, averaging the constraint equations imposed by LTI components on their terminal voltages and currents, we find that the averaged terminal quantities are constrained in the same way as the instantaneous quantities. For example, averaging the equation $v_R(t) = R i_R(t)$ that governs an LTI resistor, we find that:

$$\bar{v}_R(t) = R \bar{i}_R(t) \qquad (11.2)$$

Averaging the equation $v_L(t) = L(di_L(t)/dt)$ that governs an LTI inductor, we obtain

$$\bar{v}_L(t) = L\frac{d\bar{i}_L(t)}{dt} \qquad (11.3)$$

These relationships allow us to construct an averaged circuit as follows. We replace all instantaneous voltages and currents in the circuit by their averages and keep all LTI components unchanged. Nonlinear or time-varying components in the original circuit, however, do not map into the same components in the averaged circuit. For example, switches in the original circuit become elements in the averaged circuit that simultaneously have a nonzero (average) voltage and nonzero (average) current—and so are no longer switches.

Despite the fact that only LTI components of the original circuit are preserved, this transformation to the averaged circuit can often be very useful. Parts of the averaged circuit, if not all of it, may be amenable to LTI analysis. For example, we can use impedance methods or superposition or Thévenin/Norton equivalents, which often is sufficient for a good understanding of circuit behavior.

Even when the exact averaged circuit is difficult or impossible to analyze, some approximations may yield useful insights. For example, it may be possible to approximately characterize an averaged nonlinear or time-varying component, such as a switch, in terms of the constraints it imposes on the averaged circuit variables at its terminals. In the case of a switch, the constraints will usually also depend on the control variables that govern it. Such a characterization produces an averaged component to substitute for the original one when we derive the averaged circuit. We illustrate this point with high-frequency switching dc/dc converters later.

EXAMPLE 11.4

Averaged Circuit for a Controlled Rectifier Drive

Figure 11.6(a) models the armature circuit of the controlled rectifier drive of Example 11.2. Here R and L represent the armature resistance and inductance, respectively, and E denotes the back emf of the motor. The waveform of the applied voltage v_d was shown in Fig. 11.4(b). The variable we want to control is the average armature current, which our new notation allows us to write as $\bar{i}_d(t)$. For this example, the natural choice of the averaging interval T is the period of v_d (which is one half the period of the sinusoidal source). With this choice, $\bar{v}_d$ and $\bar{i}_d$ in the steady state are both constant and equal to $\langle v_d \rangle$ and $\langle i_d \rangle$, respectively.

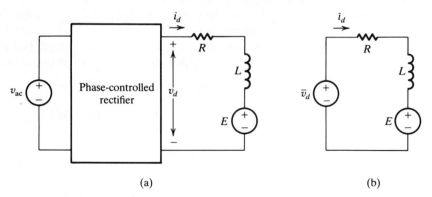

(a)   (b)

Figure 11.6 (a) Instantaneous and (b) averaged circuits for a controlled rectifier drive.

Figure 11.6(b) shows the result of averaging the circuit in Fig. 11.6(a). The averaged controlled rectifier is represented simply as a voltage source, because $\bar{v}_d$ here is completely defined by the control variable, namely, the firing angle α.

We can illustrate the usefulness of the averaged circuit by considering the open-loop step response of the system, from one steady state to another. A step change in the firing angle α from one cycle to the next leads to a transition in $\bar{v}_d$ in a single cycle, from its steady-state value before the step to its steady-state value after the step. The averaged circuit shows that, following this first cycle, the average current $\bar{i}_d$ approaches its new steady-state value exponentially, with a time constant of L/R. The first cycle is usually a small part of the transient. A typical value for L/R may be 40 ms, so the transient would last about 120 ms, whereas T for a 60-Hz supply is only 8.33 ms. (With a 6-pulse converter derived from a 3-phase supply, T would be just 2.78 ms.)

By working with the averaged circuit, we have been able to predict easily the overall form of the step response at a sufficient level of detail for many purposes. Examples 11.10 and 11.12 in Section 11.5 illustrate how we can use the averaged-circuit model as the basis for designing a simple feedback controller for the average armature current.

An exact analysis of average behavior becomes far more complicated if either of the initial assumptions in Example 11.2, namely, continuous conduction and no commutating reactance, is violated. The reason is that $\bar{v}_d$ then depends on the instantaneous values of waveforms in the original circuit and can no longer be represented as a voltage source in the averaged circuit. Nevertheless, an approximate analysis of average behavior is often still possible and useful.

We might have intuitively anticipated the overall features of the step response in Example 11.4. However, analysis gives our intuition a firm foundation.

EXAMPLE 11.5

Averaged Circuit for an Up/Down Converter in Discontinuous Conduction

Let's consider again the up/down converter used in Examples 11.1 and 11.3, but let's now assume that the resistance R of the load is high enough for the converter to be operating in discontinuous conduction mode. The corresponding inductor- and diode-current waveforms are shown in Fig. 11.7(a). We will construct an averaged model with the switching period T as the averaging interval.

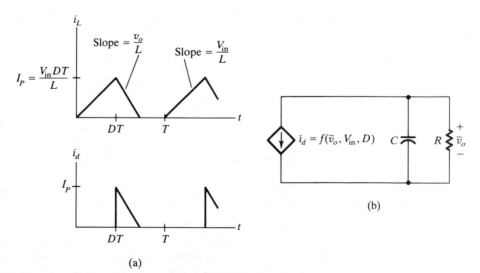

(a)

(b)

Figure 11.7 (a) Inductor- and diode-current waveforms for an up/down converter operating in discontinuous conduction mode. (b) Output portion of the averaged circuit.

Provided the input voltage is constant over a switching period, the inductor current rises linearly from 0 to $I_P = V_{in}DT/L$ while the transistor is conducting. The current decay to zero is shown in Fig. 11.7(a) as being linear, but it is actually a portion of the ringing waveform of the parallel RLC network formed when the transistor is turned off and the diode begins to conduct. Nevertheless, this portion of the ringing waveform is very well approximated by a linear segment, provided we assume that the output voltage does not vary significantly over a single cycle. More specifically, if the output voltage is well approximated by its local average (that is, if the ripple is small) and the average does not change significantly over a cycle, the inductor and diode currents decay essentially linearly, with a slope of $\bar{v}_o/L$.

With this linear approximation, we can easily compute the average current flowing through the diode over one switching period from the waveform in Fig. 11.7(a):

$$\bar{\imath}_d(t) = -\frac{V_{in}^2 T D^2}{2L\bar{v}_o(t)} = f(\bar{v}_o(t), V_{in}, D) \qquad (11.4)$$

(Recall that v_o is negative, so $-\bar{v}_o$, i_d, and $\bar{\imath}_d$ are all positive.) Assuming $\bar{v}_o$, V_{in}, and D are all slowly varying, (11.4) holds for all t, not just when the averaging interval is aligned with a switching cycle. Note the nonlinear dependence on $\bar{v}_o$, V_{in}, and D. We can now draw the output portion of the resulting averaged circuit as in Fig. 11.7(b), with the current through the diode represented by a voltage-controlled current source. This current is sometimes referred to as the *injected current*. We can use this nonlinear circuit to study the dynamics of $\bar{v}_o$ when V_{in} and D are constant or slowly varying.

The control variables in Examples 11.4 and 11.5 are the firing angle α and the duty ratio D, respectively. These control variables typically change from cycle to cycle, taking values α_k and d_k, respectively, in the kth cycle. However, in analysis and control design, working with models that involve both continuous-time and discrete-time quantities is awkward. Sampled-data models get around this obstacle by using samples of continuous-time waveforms in order to work entirely with discrete-time sequences. We discuss such models further in the remaining chapters of this part. In the context of averaged models, the opposite strategy is more natural, namely, representing the effects of discrete-time sequences such as α_k and d_k by continuous-time quantities.

11.3.3 Averaging a Switching Function

The schematic diagram in Fig. 11.8(a) covers such circuits as down or buck converters (Chapter 6) and PWM waveshaping inverters (Section 8.3), in which a controlled switching network is interposed between a dc voltage source V_{in} and a linear load, represented in the figure by its Norton equivalent. The voltage at the output of the switching network is $q(t)V_{in}$, where $q(t)$ is a *switching function* that constitutes the modulation of the source voltage; $q(t)$, for example, is determined by the duty ratios of switches in the switching network. The voltage applied to the load in the *averaged* circuit in Fig. 11.8(b) is then $\bar{q}(t)V_{in}$.

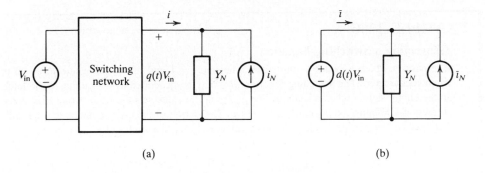

(a)   (b)

Figure 11.8 Averaging a modulated dc source to obtain a duty-ratio–controlled
dc or ac source: (a) switched circuit;   (b) averaged circuit with
$d(t) = \bar{q}(t)$.

We can carry out an exact analysis of average behavior when $\bar{q}(t)$ is determined
entirely by control variables and possibly by $\bar{\imath}(t)$, the average load current. Under
this condition, the voltage applied to the load in the averaged circuit may be
modeled by a (possibly current controlled) voltage source. Even if the condition
is violated, an approximate analysis of average behavior is often possible. An
approximate analysis is also possible if the source is not exactly dc but close to it,
in the sense that it varies little over any interval of length equal to the averaging
time T.

The function $q(t)$ is usually a waveform that switches in a controlled way
between a finite set of values: 1 and 0 for down converters, or 1, 0 and -1 for
PWM inverters. In such cases, we call the averaged switching function $\bar{q}(t)$ the
continuous duty ratio, for reasons that are suggested by Example 11.6, and denote
it by $d(t)$. The interval T over which the average is taken is some regular switching
interval defined by the clocking of the system. Note that $d(t)$ can be time varying
and can also go negative. However, if $q(t)$ is periodic and T is its period, then $d(t)$
is constant.

We can now label as $d(t)V_{in}$ the value of the voltage source in the averaged
circuit in Fig. 11.8(b). This expression shows that, by making $d(t)$ vary inversely
with any variations in V_{in}, we can essentially *eliminate* the effects of input voltage
variations on the averaged circuit. Such feedforward of the input voltage is therefore
commonly used in circuits of the type shown in Fig. 11.8. Note that this feedforward
eliminates both steady-state and transient effects, whereas our feedforward for the
up/down converter in Example 11.1 only dealt with steady-state errors.

The use of a switching function $q(t)$ and its average $\bar{q}(t) = d(t)$ is natural and
convenient in analysis and control design for many other types of power converters.
In most of these cases, the controller manipulates quantities closely related to $d(t)$
in order to control the average values of circuit waveforms.

EXAMPLE 11.6

Generating a Switching Function

The case in which a switch is turned on when $q(t) = 1$ and turned off when $q(t) = 0$ is often encountered. The switching function $q(t)$ may be generated as the output of the latch in the circuit shown schematically in Fig. 11.9(a). The clock sets the latch output to 1 every T seconds, defining the beginning of a cycle. The output of the comparator is low initially but switches later in the cycle to its high value, resetting the latch to 0.

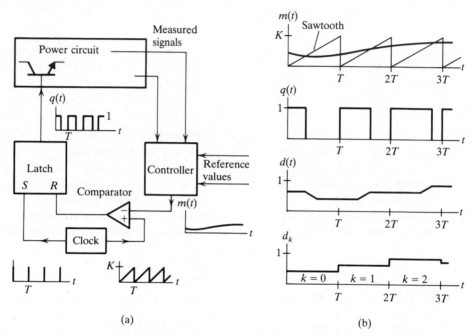

(a)	(b)

Figure 11.9   (a) Generating a switching function $q(t)$, with duty ratio determined by a modulating function $m(t)$. (b) Relationships among $m(t)$, $q(t)$, d_k, and $d(t)$.

The sawtooth waveform that is applied to the positive input of the comparator is synchronized to the clock, starting at 0 every T seconds and ramping up linearly to K. A modulating signal, $m(t)$, is fed to the negative input of the comparator and normally satisfies $0 \le m(t) \le K$. It follows that the output of the comparator is low at the start of each cycle and switches to its high value when the ramp crosses $m(t)$. The duty ratio d_k in the kth cycle therefore equals the ratio $m(t)/K$ at the instant that the sawtooth ramp in that cycle first crosses $m(t)$. Thus the modulating signal $m(t)$ controls the duty ratio.

We can obtain a constant duty ratio either by keeping $m(t)$ constant or by varying it at precisely the switching frequency. These two possibilities in turn suggest two ways to obtain slow changes in the duty ratio: by varying $m(t)$ slowly or by varying it at frequencies in the vicinity of the switching frequency. In the latter case, the duty ratio varies at the beat frequency.

Both methods are used in practice. In fact, for so-called current-mode control of high-frequency switching dc/dc converters, both methods are used simultaneously. One part of $m(t)$ is derived from the average output voltage, and varies at low frequencies; the other part is derived from the instantaneous inductor or switch current and therefore has components both at low frequencies and near the switching frequency. The periodic sawtooth waveform in the case of current-mode control is termed a stabilizing ramp (or compensation ramp), for reasons that we explain in Chapter 13. We assume in what follows that $m(t)$ varies slowly.

If $m(t)$ does not change significantly over the course of a cycle, that is, if significant variations in $m(t)$ occur at substantially less than one half the switching frequency, then $m(t)/K$ at any time closely approximates the prevailing duty ratio. The control circuit varies $m(t)$ around its nominal value in accordance with the need to increase or decrease the duty ratio. This determination in turn is made by feedback signals that measure how far the power circuit is from nominal operation. Evidently, the closed loop "bandwidth" of such a system, or the speed with which the controlled system returns to its nominal operation, is substantially less than one half the switching frequency under these conditions.

Figure 11.9(b) shows how $m(t)$, $q(t)$, the duty ratio d_k, and the continuous duty ratio $d(t)$ are interrelated. Note that, provided the change in duty ratio from cycle to cycle is sufficiently small, $d(t)$ also closely approximates the prevailing duty ratio d_k. Although we can choose to constrain $m(t)$ to have only slow variations, $d(t)$ is intrinsically constrained to vary slowly. The fastest possible variation in d_k occurs when the duty ratio alternates between high and low values in successive cycles. The corresponding period of $d(t)$ is $2T$. Thus $d(t)$ can never have a fundamental frequency higher than one half the switching frequency.

Note that variations in K could also be used to vary the duty ratio. Feedforward control that compensates for supply voltage variations is commonly implemented this way. With the up/down converter circuit in Examples 11.1 and 11.3, for instance, the required feedforward can be achieved by making K proportional to $V_{in} - V_{ref}$ (recall that V_{ref} is negative). For down converters and other circuits of the form shown in Fig. 11.8, we make K proportional to V_{in}. We leave you to verify these claims.

Our averaged models typically show the dynamic dependence of averaged waveforms in the power circuit on $d(t)$. However, of most interest in control design is the dependence on some real modulating control signal, such as $m(t)$ in Example 11.6. For slow variations in the modulating signal, we showed that $d(t) \approx m(t)/K$, so the averaged models are relevant. But the response of our averaged models to perturbations in $d(t)$ at frequencies that exceed, or even approach, one half the switching frequency do *not* approximate the response to perturbations in $m(t)$ at those frequencies.

11.3.4 Averaging a Switch

Although we have covered considerable ground without looking in detail at averaging the variables associated with a switch, it is now time to do so. Switch averaging is especially rewarding in the case of high-frequency switching or PWM converters. The reason is that we can approximately characterize the averaged switch in such circuits entirely in terms of the (control-dependent) constraints that

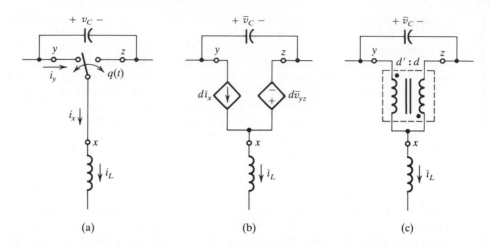

(a)   (b)   (c)

Figure 11.10   (a) The canonical switching cell for high-frequency switching converters. (b) Approximate averaged switching cell for continuous
conduction with duty ratio d, using controlled sources. (c) Approximate averaged switching cell, using ideal transformer; $d' = 1 - d$.

it imposes on the averaged circuit variables at its terminals. As a result, we can
replace the switch in the instantaneous circuit with a corresponding component
in the averaged circuit. The other components in the circuit are usually LTI, and
are unchanged by averaging. Hence circuit averaging for such a converter simply
involves replacement of the switch by its averaged model.

It is natural to start with the canonical cell of Fig. 6.7, reproduced in
Fig. 11.10(a) for convenience. The switching period T will again be chosen as
the averaging interval. The assumptions that we make in obtaining an averaged
model can be stated in different ways. We list them here as:

1. a *small ripple* assumption, namely, that the voltage v_{yz} $(=v_C)$ and current i_x
   $(=i_L)$ at time t are well approximated by their average values, $\bar{v}_{yz}(t)$ and $\bar{i}_x(t)$,
   respectively; and

2. a *slow variation* assumption, namely, that these two average values do not vary
   significantly over any averaging interval of length T, that is, they vary substantially more slowly than one half the switching frequency.

Both assumptions are generally satisfied in well-designed high-frequency switching
converters operating in continuous conduction.

Suppose that the switch in Fig. 11.10(a) is governed by a 0–1 switching function $q(t)$, such as that described in Example 11.6, and is in position y with duty
ratio $d(t)$. As $i_y = q i_x$, we can write $\bar{i}_y = \overline{q i_x}$. Now our assumptions allow us
to treat $i_x(\tau)$ as though it were approximately constant at $\bar{i}_x(t)$ over the averaging

interval $t - T \leq \tau \leq t$, so:

$$\bar{\imath}_y(t) = \overline{q\imath}_x \approx \bar{q}(t)\bar{\imath}_x(t) = d(t)\bar{\imath}_x(t) \tag{11.5}$$

Similarly, $v_{xz} = qv_{yz}$, so:

$$\bar{v}_{xz}(t) = \overline{qv}_{yz} \approx \bar{q}(t)\bar{v}_{yz}(t) = d(t)\bar{v}_{yz}(t) \tag{11.6}$$

The relationships obtained in *(11.5)* and *(11.6)* constitute the desired approximate characterization of the averaged switch in terms of the control-dependent constraints imposed on the averaged circuit variables at its terminals. (A three-terminal element is fully characterized by two constraint equations for its terminal variables.) Two circuit representations of this characterization are shown in Fig. 11.10(b) and (c). Although these two are equivalent, the representation in terms of controlled sources is simpler for such tasks as linearization, which we consider later. The representation involving the ideal transformer is useful for situations in which the duty ratio is held constant.

The high-frequency converter topologies in Chapters 6 and 7 had the canonical switching cell at their core, with all other elements being linear and time invariant. Hence approximate averaged circuits for all these converters are simply obtained by replacing the canonical cell by its approximate average.

EXAMPLE 11.7

Averaged Circuit for an Up/Down Converter in Continuous Conduction

Our development of the indirect switching converter in Section 6.4 identifies the canonical cell structure within the up/down converter topology in Fig. 11.2. Note that in this case v_{yz} of Fig. 11.10 is actually $v_{in} - v_C$ rather than v_C. To obtain the averaged model, we replace the canonical cell by its averaged version from Fig. 11.10(c), substituting the ideal transformer windings of $d'(t)$ and $d(t)$ turns, respectively, for the transistor and diode switches of Fig. 11.2. We also replace all instantaneous quantities by their averages. The result is in Fig. 11.11.

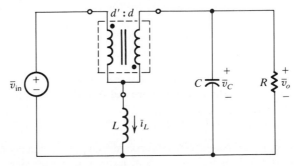

Figure 11.11 Averaged circuit for an up/down converter.

This averaged circuit can be used as the basis for generating other types of analytical models such as, for example, state-space models. The circuit can also be accepted directly by many standard circuit simulation packages. Note that the circuit depends nonlinearly on the control variable $d(t)$. The process of linearization that we examine in Section 11.4 provides one way of dealing with the nonlinearity.

When the duty ratio $d(t)$ is constant, the averaged circuit of Fig. 11.11 is LTI, so that analysis in this case is straightforward. For example, the circuit can immediately be used to read off average values of voltages and currents in the nominal steady state obtained with $\bar{v}_{\text{in}} = V_{\text{in}}$ and $d(t) = D$. In this steady state, the average inductor voltage and capacitor current are both zero, which leads to the following expressions for the average steady-state inductor current, denoted by I_L, and steady-state output voltage V_o (or capacitor voltage, V_C):

$$I_L = \langle i_L \rangle = -\frac{V_o}{RD'} \quad \text{and} \quad V_o = \langle v_o \rangle = -V_{\text{in}}\frac{D}{D'} \tag{11.7}$$

These expressions are consistent with the results in Section 6.4.

The averaged circuit also provides a basis for understanding the results in Example 11.1, where we examined the open-loop response of v_o ($= v_C$) to a step in the applied voltage v_{in}, with constant $d(t) = D$. Using the averaged circuit to compute the transfer function from the locally-averaged input voltage to the locally-averaged output voltage—referred to as the *audio susceptibility* transfer function of the circuit—we get

$$\frac{\bar{v}_o(s)}{\bar{v}_{\text{in}}(s)} = \frac{-D'D/LC}{s^2 + (1/RC)s + (D'^2/LC)} \tag{11.8}$$

With a step change in $v_{\text{in}}(t)$, the *averaged* input $\bar{v}_{\text{in}}(t)$ actually takes one averaging interval to cross between its initial and final values. Also, the averaging approximation in *(11.6)* is poor during this interval. However, this interval T is much smaller than the time constants of the averaged model, so we can consider $\bar{v}_{\text{in}}(t)$ itself to be a step function and can use the transfer function *(11.8)* of the average model to compute the response to this step.

The computations required to find the step response of the averaged model are outlined in Section 13.3, in the context of solving general LTI models. For now, you simply need to recall that the form of the response is largely determined by the *poles* of the transfer function *(11.8)*. For the underdamped case, which is what the parameter values in Example 11.1 correspond to, the poles λ_1 and λ_2 are complex and conjugate to each other:

$$\lambda_1 = \lambda_2^* = -\frac{1}{2RC} + j\omega_D \quad \text{and} \quad \omega_D = \sqrt{\frac{D'^2}{LC} - \frac{1}{4R^2C^2}} \tag{11.9}$$

where $*$ denotes complex conjugation. The transition from the steady state before the input step to the steady state after it is thus governed by a transient of the form:

$$c_1 e^{\lambda_1 t} + c_1^* e^{\lambda_1^* t} = c e^{-t/(2RC)} \sin(\omega_D t + \theta) \tag{11.10}$$

where c_1, c, and θ are constants fixed (upon invoking the continuity of the inductor current and capacitor voltage) by the initial conditions prior to the step. See Section 9.1 for a more detailed analysis of a very similar situation.

A detailed calculation based on *(11.8)* accounts very well—not only qualitatively but also quantitatively—for the average behavior of the open-loop step response of Fig. 11.3(a). The time constant $2RC$ is 880 μs or 44 switching cycles, and the period $2\pi/\omega_D$ is 2924 μs, or 146 cycles. A similar calculation can be carried out for the step response with feedforward, shown in Fig. 11.3(b). However, we now have to account for the fact that D has a step change as well, to a value determined by v_{in} after the step. Example 11.9 derives a transfer function that allows us to compute the response to (small) changes in D.

Sometimes the capacitor in the canonical cell has significant resistance, modeled as an equivalent series resistance (ESR). In this case, the voltage v_{yz} in Fig. 11.10(a) may no longer satisfy the assumption of small ripple that underlies the approximation in *(11.6)*. It is possible to refine the approximation and obtain an averaged-switch model that accounts for ESR. Alternatively, a state-space approach to averaging (Section 12.3) permits us to handle this situation systematically.

The switch averaging in Fig. 11.10 does not apply to *discontinuous* conduction (Section 6.7). The switch of the canonical cell in this case takes a third position, contacting neither y nor z. Using the fact that the average voltage across the inductor is now approximately zero, we can still obtain an averaged model of the switch. The style of analysis demonstrated in Example 11.5 is usually simpler, however.

Switch averaging is effective for a related family of dc/dc converters, namely quasi-resonant dc/dc converters, typified by the circuit in Example 9.6. The necessary computations are somewhat more involved than the preceding ones. The averaging exploits the fact that the average voltage of the resonant inductor and the average current of the resonant capacitor are both approximately zero. The resulting averaged-switch model again displays the control-dependent constraints that approximately constrain the average values of the switch variables.

11.3.5 A Generalization: The Local ω-Component

We began this discussion of circuit averaging by saying that for many power electronic circuits our interest is in the average values of circuit variables. Our definition of a local average then enabled us to study the dynamic behavior of the average. In circuits such as the resonant dc/ac converters of Chapter 9, however, we are interested in the *switching-frequency component* of the output. With resonant dc/dc converters, we are interested in the switching-frequency component of the resonant variables and the average value of the output.

The component at a frequency ω in a waveform $x(\tau)$ can be defined by the following generalization of the local average in *(11.1)*:

$$\bar{x}^{\omega}(t) = \frac{1}{T} \int_{t-T}^{t} e^{-j\omega\tau} x(\tau)\, d\tau \qquad (11.11)$$

We call this variable the *local ω-component* of x at time t. In terms of this notation, the local average is the local 0-component, given by $\overline{x}(t) = \overline{x}^0(t)$. The choices of T and ω are usually interrelated. For the resonant converter, we would pick T to be the switching period and set $\omega = 2\pi/T$. With this choice, $\overline{x}^\omega$ in the steady state is the (complex) amplitude of the fundamental Fourier series component for the periodic waveform $x(\tau)$. The expression in *(11.11)* extends the notion of a Fourier series component at the frequency ω to the nonperiodic case.

To study the dynamics of $\overline{x}^\omega(t)$, we need an expression for its derivative. It is

$$\frac{d}{dt}\overline{x}^\omega(t) = \overline{\dot{x}}^\omega(t) - j\omega\overline{x}^\omega(t) \qquad (11.12)$$

where $\overline{\dot{x}}^\omega$ is the ω-component of the derivative $dx(\tau)/d\tau$. We leave it to you to see what *(11.12)* implies for the construction of the ω-component circuit from the original switching circuit and to explore the use of this generalization in circuits such as resonant converters.

11.4 LINEARIZED MODELS

The static characteristics of power electronic circuits often depend nonlinearly on the control variables, and their dynamic characteristics are even more likely to do so. With linear or nonlinear feedback control, the closed-loop system is typically also nonlinear. Assessing stability and designing or evaluating controllers with nonlinear models are usually difficult. The most common, systematic and generally successful approach to these tasks is *linearization*. It yields *linear* models that approximately describe *small deviations* or perturbations from nominal operation of a system. Linear models are, of course, far easier to analyze than nonlinear ones.

Linearized models, also called small-signal models, are crucial to evaluating the stability of a nominal operating condition. Stability of the linearized model indicates that the nominal operating condition is stable for small perturbations at least. An initial goal for control design is therefore stabilization of the linearized model. This task is much easier than direct stabilization of the nonlinear model.

We outline the basis of linearization in this section and treat it in more detail in Chapter 13. We discuss the application of the resulting linear models in stability evaluation and control design in Section 11.5 and Chapters 13 and 14.

11.4.1 Linearization

The place to start when linearizing a continuous-time or discrete-time dynamic model is with a nominal solution, usually a steady-state solution. Even the steady state in a typical dynamic model for a power circuit may involve periodically-varying rather than constant quantities. For example, steady state in the instantaneous model of an up/down converter involves a periodic switching function and periodic waveforms. One way to get a constant steady state rather than a periodic steady state in power circuit models is by working with a discrete-time model that involves circuit variables sampled once per cycle. The reason becomes clear in Chapters 12 and 13.

Averaged models can also yield constant steady states. Steady state in the averaged-circuit model given in Fig. 11.11, when we impose a constant duty ratio $d(t)$, corresponds to the averaged variables taking constant values. However, with the high-frequency PWM inverters described in Section 8.3, even an averaged model has a steady state that varies periodically rather than remaining constant.

We can now represent small deviations from the nominal by expanding all the nonlinear terms of the model into Taylor series around the nominal values. Retaining only first-order terms results in a *linear* model that approximately governs small deviations; this is the linearized model. The parameters of the linearized model depend on the nominal operating condition, because the Taylor series coefficients depend on the nominal solution. If the original nonlinear model is time invariant and if the nominal solution corresponds to constant values of the variables, the linearized model turns out to always be LTI.

11.4.2 Linearizing a Circuit

For nonlinear models in circuit form, we can further simplify the process of linearization. Using simple operations on the nonlinear circuit, we can obtain the linearized model itself in circuit form. The procedure and justification are analogous to those we used in Section 11.3.2 to obtain an averaged-circuit model from an instantaneous-circuit model. The arguments should also be familiar from small-signal analysis of transistor amplifier circuits, for example.

We begin by replacing every voltage in the nonlinear circuit by its *deviation* from the nominal. This step results in voltage deviations that satisfy Kirchhoff's voltage law equations on the given circuit topology. The reason KVL is satisfied is that we obtain the deviations by taking the difference between two sets of voltages—the perturbed set and the nominal—that each satisfy the same linear equations. Similarly, we replace every current in the nonlinear circuit by its deviation from the nominal to obtain current deviations that satisfy Kirchhoff's current law on the given circuit topology.

The final step is to replace every nonlinear component in the circuit with its linearized version. (The linear components do not need replacement, because they impose the same constraints on the deviations as they do on the original variables.) The linearization of a nonlinear component is obtained by Taylor expansion of its characterizing equations up to first-order terms. The linearized component imposes linear constraints that approximately govern small deviations of the variables at the component's terminals and small deviations of any control variables that govern the component. The result of all these manipulations is a *linear circuit* that governs small deviations from the nominal.

The nonlinear averaged circuits that we described earlier are time invariant, and the nominal solution of interest is usually the constant steady state. The linearization in these cases is an LTI circuit. In the case of averaged-circuit models for the PWM waveshaping inverters in Section 8.3, the nominal steady state is periodically varying, and the corresponding linearization is a periodically varying circuit.

EXAMPLE 11.8

Linearized Circuit for an Up/Down Converter in Discontinuous Conduction

A nonlinear averaged-circuit model for the up/down converter in discontinuous conduction was obtained in Example 11.5, Fig. 11.7(b). Our derivation of the model there allows the duty ratio and input voltage to be slowly varying rather than constant. We can therefore rewrite (11.4) with $d(t)$ instead of D and $\overline{v}_{in}$ instead of V_{in}. We reserve D and V_{in} to denote the constant nominal values.

The nominal steady-state solution corresponds to constant values of the averaged quantities. In particular, $\overline{i}_d(t) = I_d$ and $\overline{v}_o(t) = V_o$. The capacitor in the averaged circuit acts as an open circuit in the steady state, so we see from Fig. 11.7(b) that:

$$V_o = -RI_d = -Rf(V_o, V_{in}, D) = R\frac{V_{in}^2 T D^2}{2LV_o} \tag{11.13}$$

Rearranging and taking the square root yields $V_o = -V_{in}D\sqrt{RT/2L}$. (With our sign convention, we need the negative square root.)

Now let the duty ratio be perturbed from D to $d(t) = D + \tilde{d}(t)$, but assume for simplicity that the input voltage is fixed at V_{in}. Correspondingly let $\overline{i}_d(t) = I_d + \tilde{i}_d(t)$ and $\overline{v}_o(t) = V_o + \tilde{v}_o(t)$. The superscript $\sim$ denotes a perturbation from the nominal. Linearizing the current source in the averaged circuit by a Taylor series expansion up to linear terms, we find

$$\tilde{i}_d(t) \approx \frac{\partial f}{\partial \overline{v}_o}\tilde{v}_o(t) + \frac{\partial f}{\partial d}\tilde{d}(t)$$

$$= \left[\frac{V_{in}^2 T D^2}{2LV_o^2}\right]\tilde{v}_o(t) - \left[\frac{V_{in}^2 T D}{LV_o}\right]\tilde{d}(t) \tag{11.14}$$

The partial derivatives in (11.14) are evaluated at the nominal solution. Using (11.13) to simplify (11.14), we have

$$\tilde{i}_d(t) \approx (1/R)\tilde{v}_o(t) + (V_{in}\sqrt{2T/RL})\tilde{d}(t) \tag{11.15}$$

The resulting linearized averaged circuit is shown in Fig. 11.12.

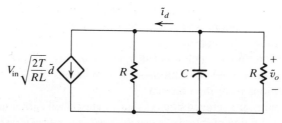

Figure 11.12 Linearized averaged circuit for an up/down converter in discontinuous conduction.

11.4.3 Linearizing the Averaged Switch

We have already shown how to average a high-frequency switched dc/dc converter by averaging the switch in the canonical cell. The nonswitch components are usually modeled as LTI and are therefore not changed for the averaged model. Similarly, the key step in linearizing the nonlinear averaged circuit for such a converter is to linearize the averaged model of the switch. If the remaining nonswitch components are LTI, they are preserved intact in the linearized circuit.

Linearization of the averaged-switch model in Fig. 11.10(b) is easily carried out. Let's denote nominal values by uppercase letters and deviations from the nominal by the superscript $\sim$, as in Example 11.8. Most commonly the nominal solution corresponds to a constant or periodic steady state, but we do not require this for the linearization that follows. We can write

$$d(t) = D + \tilde{d}(t) \quad \text{and} \quad d'(t) = D' - \tilde{d}(t) \tag{11.16}$$

and similarly for the other variables. We now expand the source terms in Fig. 11.10(b), namely, $d(t)\bar{i}_x(t)$ and $d(t)\bar{v}_{yz}(t)$, to first-order or linear terms in the perturbations. That is, we neglect terms that involve squares or products of the small perturbations. This linearized expansion yields the following first-order perturbations in the source terms:

$$d(t)\bar{i}_x(t) - DI_x \approx D\tilde{i}_x(t) + I_x\tilde{d}(t)$$
$$d(t)\bar{v}_{yz}(t) - DV_{yz} \approx D\tilde{v}_{yz}(t) + V_{yz}\tilde{d}(t) \tag{11.17}$$

The results of this calculation are represented in the linearized circuit in Fig. 11.13(a), with an equivalent representation in Fig 11.13(b).

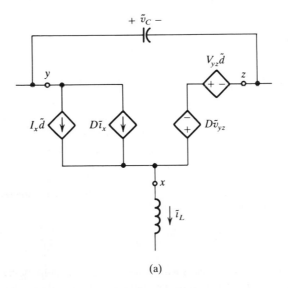

(a)

Figure 11.13 (a) Linearized averaged model of the canonical switching cell.

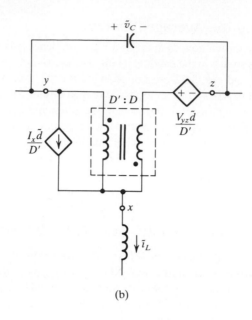

(b)

Figure 11.13 (cont.)   (b) Alternative representation.

EXAMPLE 11.9

Linearized Circuit for an Up/Down Converter in Continuous Conduction

We return once more to the up/down converter of Examples 11.1, 11.3, and 11.7 to obtain a linearized averaged model that describes small perturbations from the nominal steady state in the continuous conduction mode.

The linearized version of Fig. 11.11 is obtained by replacing all voltages and currents by their perturbations from the nominal and replacing the averaged canonical cell by its linearized version, using Fig. 11.13(b). All other converter elements are linear and are therefore preserved intact in the linearized circuit. The result is shown in Fig. 11.14, with $\tilde{v}_{\mathrm{in}}$ representing deviations of the averaged input source from its nominal dc value.

The linearized model corresponding to the constant nominal steady state is an LTI circuit. Hence solving for the values of all circuit variables is straightforward (for example, using impedance methods). For control design purposes, we must know the transfer function from perturbations $\tilde{d}$ in the duty cycle, which constitutes our control input, to perturbations $\tilde{v}_o$ in the output voltage. Setting $\tilde{v}_{\mathrm{in}} = 0$, a straightforward computation with the circuit in Fig. 11.14 shows the transfer function to be

$$\frac{\tilde{v}_o(s)}{\tilde{d}(s)} = \left(\frac{I_L}{C}\right) \frac{s - (V_{\mathrm{in}}/LI_L)}{s^2 + (1/RC)s + (D'^2/LC)} \tag{11.18}$$

(We have used the steady-state relationships in *(11.7)* to simplify the constant in the numerator.) This is the transfer function we referred to at the end of Example 11.7, in connection

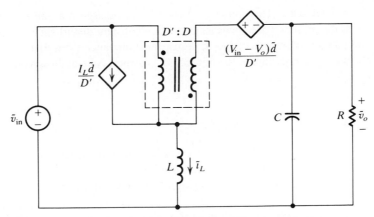

Figure 11.14 Linearized averaged model of an up/down converter.

with determining the effect of feedforward in Example 11.1. This transfer function also provides a good starting point for an explanation of why the proportional feedback scheme in Example 11.3 led to the behavior observed there. We pursue this explanation further in Example 11.11.

The transfer function computation in Example 11.9 illustrates the advantages of having an LTI circuit model for power circuit dynamics. Such a model allows us to apply the extensive array of concepts and techniques available for LTI circuits to the study of deviations from nominal operation in a power circuit. These concepts and techniques include superposition, impedance methods, Norton and Thévenin equivalents and their multiport generalizations, network interconnections, sensitivities, and so on. For instance, in addition to the transfer function computed in Example 11.9, we could determine input and output impedances or admittances, as well as various other relevant transfer ratios. These transfer ratios are important in describing how dynamic behavior depends on characteristics of the source, the load, the circuit itself, and the control design.

Apart from serving as a basis for computations that support the tasks of stability analysis and control design, a circuit model can sometimes directly suggest control designs. For example, consider again the circuit in Fig. 11.12, which we obtained by linearizing the averaged model of the up/down converter in discontinuous conduction. Its damping would evidently be increased and $\tilde{v}_o$ would decay to zero faster if additional conductance were placed in parallel with the capacitor. Adding a physical conductance is clearly ruled out on grounds of efficiency, but it is possible to reproduce the effect of additional conductance entirely by means of control.

Note that if we use *proportional* feedback to enforce $\tilde{d}(t) = h\tilde{v}_o(t)$, where h is a constant, the effect is the same as that of a conductance of value $hV_{in}\sqrt{2T/RL}$ replacing the current source in Fig. 11.12. *Proportional–integral* control has the effect of replacing the current source with a parallel combination of a conductance and an inductance. This inductance causes the steady-state value of $\tilde{v}_o$ to be zero even in the presence of parameter errors and constant disturbances (such as constant deviations of the input voltage or converter inductance from their nominal values). Thus using a circuit model to approach or interpret control design can often provide useful insights.

11.5 FEEDBACK CONTROL

We have shown how to obtain both nonlinear and linear circuit models for the dynamics of averaged variables in certain classes of power circuits. How do we use these models to analyze and design feedback controls?

It is generally hard to assess stability and to design or evaluate feedback control schemes directly with nonlinear models. In this section we focus on LTI models, such as those we obtain by linearization at a constant operating point. With LTI models, we can choose from a wide range of systematic analysis and design approaches to feedback control. The controllers designed using LTI models are usually LTI as well.

Controllers derived from linearized models cannot be guaranteed to provide satisfactory operation for large deviations from nominal operation, even if we expect them to function well for small perturbations. Also, the control design has to account for the fact that the linearized model will vary with the operating condition. Nevertheless, the majority of power circuit controllers are in fact designed on the basis of linearized models. A controller that performs well on the linearized model is likely to keep the circuit operating near the nominal and to make the operation relatively insensitive to small disturbances and errors. We can therefore reasonably expect that the nonlinearities have only secondary effects, except when there are major disturbances. We usually deal with nonlinear effects through refinements, modifications, or complements to a core LTI controller.

For these reasons, we confine ourselves in this overview to discussing feedback control design issues in the context of "classical" control with continuous-time LTI models. This allows us to use simple transfer function computations but still provides ample opportunity to appreciate the potential benefits and pitfalls of feedback control. Further consideration of control design using such models, as well as state-space models and discrete-time models, is left to Chapter 14. Most of Chapter 14 is accessible directly after Chapter 11.

11.5.1 The Classical LTI Control Configuration

Considerable insight into what feedback can accomplish—and an understanding of what the dangers are—may be obtained by considering the interconnection of single-input, single-output, continuous-time LTI subsystems in Fig. 11.15. The simple feedback configuration here is of the same form as Fig. 11.1, and lies at the heart of classical control design. The lessons you learn from this configuration are

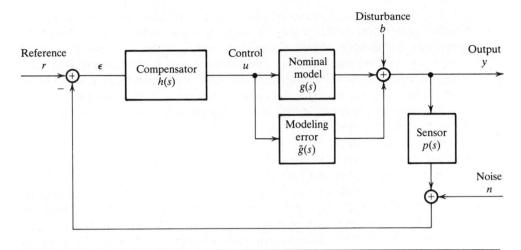

Figure 11.15   Block diagram of the classical LTI feedback configuration.

directly applicable to controlling the small-signal dynamic behavior of a wide range of power circuits.

Recent control approaches are rooted in the state-space methods developed since the early 1960s and are used to tackle much more general problems, with interconnected multi-input, multi-output systems. In Chapters 12, 13, and 14 we can do no more than hint at these relatively recent extensions. For the simple configuration in Fig. 11.15, however, the design approaches of classical control have withstood the test of time and provide a demanding standard for modern control approaches.

The quantity marked in each block of Fig. 11.15 is the transfer function that relates the output of the block to its input—in the Laplace transform domain (or frequency or impedance domain). Except where otherwise stated, we assume that all the transfer functions we work with are ratios of polynomials in s, or *rational* functions of s. This case is the most important one for control design. On occasion we must step outside this category of transfer functions, as when representing time delays; a time delay of η between an input and an output corresponds to the nonrational transfer function $e^{-\eta s}$.

To keep our notation streamlined, we denote time domain and transform domain representations of a signal such as u in Fig. 11.15 by $u(t)$ and $u(s)$, respectively, whenever necessary to make the domain explicit. However, this convention is an abuse of notation, because $u(s)$ is *not* obtained through replacement of t by s in the expression for $u(t)$. (An unambiguous notation would, for example, be $\hat{u}(s)$ for the transform, but such symbols are awkward to combine with our notation for averages, perturbations, and so on.) The context makes clear which domain we are working in. We always write transfer functions such as $g(s)$ with their arguments.

Figure 11.15 represents the controlled system or *plant* as having a transfer function $g(s) + \tilde{g}(s)$, which relates the output y to the control input u. Here $g(s)$ constitutes the *nominal* model of the plant, which is the basis for the control design, and $\tilde{g}(s)$ represents errors in the nominal model. These errors could, for example, reflect uncertainties regarding the load and other circuit parameters or simplifications, approximations, and compromises made during modeling.

The disturbances affecting the plant are represented in terms of their effects at the plant output, by means of the signal b. For simplicity, we assume that none of the disturbances can be measured, so no feedforward is possible. The feedback signal consists of a measurement of the system output, y, passed through a sensor whose transfer function is $p(s)$, and corrupted by measurement noise, n. This feedback signal is compared with the external signal, r, which represents the reference (or desired or commanded) value for y; we would ideally like to have $r = y$. The result of the comparison is the noise corrupted error signal, ϵ. To keep things simple, we shall omit consideration of sensor dynamics, so $p(s) = 1$. The control signal to the plant, namely u, is produced by an LTI controller or compensator with transfer function $h(s)$ that acts on ϵ.

EXAMPLE 11.10

Feedback Control of Armature Current in a Controlled Rectifier Drive

The control structure shown in Fig. 11.16 is based on a proportional–integral (PI) control design for the controlled rectifier drive treated in Examples 11.2 and 11.4. The plant transfer function $1/(sL + R)$, which relates the averaged signals in the armature circuit, is simply the input admittance of the averaged circuit in Fig. 11.6(b) of Example 11.4. We could have modeled the "disturbance" E as adding in directly at the plant input. Instead, we represent it by adding an equivalent signal at the plant output, to make the configurations in Figs. 11.15 and 11.16 look similar.

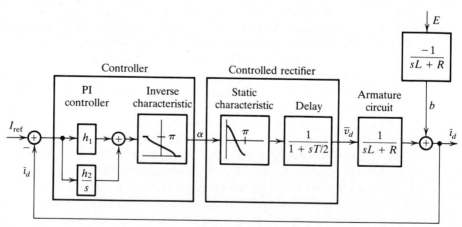

Figure 11.16   Feedback control for a phase-controlled rectifier drive.

The average current $\bar{i}_d(t)$ is usually obtained with sufficient accuracy in practice by filtering the instantaneous current $i_d(t)$ through a low-pass filter with transfer function $1/(1 + sT/2)$. This average current is compared with I_{ref}. Based on the discrepancy, the controller specifies the firing angle α for the phase-controlled rectifier. The output voltage waveform of the rectifier is thereby determined and hence so is its average value $\bar{v}_d(t)$.

We know from Chapter 5 that the steady-state relationship between the firing angle α and the average output voltage $\bar{v}_d(t)$ is $\langle v_d \rangle = (2V/\pi) \cos \alpha$, where V is the amplitude of the sinusoidal voltage source. Any transients in the average voltage, however, depend in a more complicated way on variations in the firing angle of the rectifier. For instance, we have already noted when discussing the open-loop response to a step in firing angle in Example 11.4, that it takes one cycle of duration T for the average output voltage to make the transition from its initial to its final value.

These facts lead to an approximate nominal model that is often used to represent the dynamics of the phase-controlled rectifier. The model comprises a cascade of the converter's steady-state characteristic and a dynamic block to represent a (mean) time delay of $T/2$ in the converter. This approximation of converter dynamics is reasonable when the signals in the system do not vary significantly during intervals of length T. Such models are often constructed in situations where sampling effects need to be approximately represented in a continuous time model. Sometimes, however, the approximation is too crude to permit a satisfactory analysis of the system, and there is then little choice but to go to a sampled-data model.

We have used the rational transfer function $1/(1 + sT/2)$ to approximate the nonrational transfer function $e^{-sT/2}$ of the time delay. Better approximations of this exponential, such as $(1 - sT/4)/(1 + sT/4)$, may be used, but the refinement is probably not justified unless matched by corresponding refinements elsewhere in the model.

The basic purpose of the controller is to vary $\bar{v}_d(t)$ in accordance with the error between the desired and actual average armature currents. In order to obtain a linear relationship from the error signal to the average voltage, the controller may be constructed as a cascade of a dynamic LTI section and a static nonlinearity that represents the *inverse* of the steady-state characteristic of the controlled rectifier. This approach, which was mentioned in Example 5.1, permits us to analyze and design the overall system using LTI models. The LTI portion of the controller shown in the block diagram is a PI controller, with h_1 representing the proportional gain and h_2 the integral gain. This controller is easily constructed by means of an operational amplifier circuit.

We can easily compute the transfer functions from the driving signals r, b, and n to the output y for the nominal system ($\tilde{g}(s) = 0$) in Fig. 11.15, using the relationships represented in it. We state the result in terms of the nominal *loop transfer function*, which is the product of the transfer functions around the feedback loop, namely, $\ell(s) = p(s)g(s)h(s)$ for the general case but $g(s)h(s)$ under our assumption here that $p(s) = 1$. The desired result then is

$$y = \frac{1}{1 + \ell(s)}b + \frac{\ell(s)}{1 + \ell(s)}(r - n) \quad \text{and} \quad \ell(s) = g(s)h(s) \qquad (11.19)$$

We can write *(11.19)* still more simply by defining what are known as the *sensitivity function* $\mu(s)$ and *complementary sensitivity function* $\mu'(s) = 1 - \mu(s)$:

$$\mu(s) = \frac{1}{1 + \ell(s)} \quad \text{and} \quad \mu'(s) = \frac{\ell(s)}{1 + \ell(s)} \qquad (11.20)$$

Note that $\mu'(s)$ is just the transfer function from r to y, so we also call it the *system transfer function*. One reason for giving $\mu(s)$ its particular name is that the fractional change in the system transfer function for a fractional change in the loop transfer function is given by:

$$\frac{d\mu'(s)}{\mu'(s)} = \mu(s)\frac{d\ell(s)}{\ell(s)} \qquad (11.21)$$

as you can verify quite simply. Now we can write *(11.19)* as:

$$y = \mu(s)b + \mu'(s)(r - n) \qquad (11.22)$$

We use these expressions to study three overlapping issues that are critical to assessing feedback control systems such as those in Figs. 11.15 and 11.16. These issues are *nominal stability*, *nominal performance*, and the *robustness* of both stability and performance to the presence of modeling errors. The nominal case corresponds to $\tilde{g}(s) = 0$, and robustness refers to the preservation of stability and performance when $\tilde{g}(s) \neq 0$.

Feedback control has the potential to turn an unstable open-loop system into a stable closed-loop system that performs significantly better than the open-loop system. It can also, unless care is taken, convert a stable open-loop system into a closed-loop system that performs poorly or becomes unstable. The design and implementation of feedback controls therefore require considerable care.

11.5.2 Nominal Stability

A single-input, single-output LTI system is called *bounded input, bounded output (BIBO) stable* if bounded signals applied at the input always produce bounded signals at the output, with the system initially at rest. A necessary and sufficient condition for BIBO stability is that the poles of the transfer function are strictly in the left half plane (which is equivalent to the impulse response decaying to zero). For brevity, we refer to the transfer function itself as stable in this case. A pole in the strict left half plane is also called a *stable pole*.

A system of interconnected LTI subsystems is called *internally stable* if bounded external signals added in at every subsystem input always produce bounded signals at every subsystem output, with the system initially at rest. A necessary and sufficient condition for internal stability is that *all* the corresponding transfer functions are stable. Internal stability is the appropriate notion of stability for application to the closed-loop system in Fig. 11.15 and is what we are referring to in the rest of this section when we talk of stability. (We defer to Section 13.3 a discussion of subtleties involved in relating internal stability, which is derived from an input–output approach, to asymptotic stability, which is defined by the response to initial conditions.)

Applying the preceding criterion to the nominal closed-loop system in Fig. 11.15 (with $\tilde{g}(s) = 0$ and $p(s) = 1$) shows that the system is stable if and only if $\mu(s)h(s)$, $\mu(s)g(s)$, and $\mu'(s)$ are all stable. We can easily verify that this condition is equivalent to the following two conditions together:

1. The sensitivity function $\mu(s)$ (or equivalently its complement $\mu'(s)$, the system transfer function) is stable.

2. No poles of the nominal plant $g(s)$ or compensator $h(s)$ in the closed right half plane (that is, including the imaginary axis) are canceled by zeros of $h(s)$ or $g(s)$, respectively.

Poles of $g(s)$ or $h(s)$ that are canceled by zeros of $h(s)$ or $g(s)$ are termed *hidden poles* of the nominal system. They are poles of the plant or compensator that do not appear as poles of the loop transfer function and are therefore unaffected by the feedback. Hence an equivalent statement of condition (2) is that there are no unstable hidden poles. The choice of $h(s)$ is up to us, so satisfying (2) is easy. We assume from now on that we have done so. The stability condition therefore reduces to condition (1), namely, that the sensitivity function (or the system transfer function) is stable.

Note that if the nominal plant in Fig. 11.15 were operated in an open-loop configuration instead of in the indicated closed-loop configuration, stability would be determined by the poles of $g(s)$. The nominal closed-loop system can be stable even if the plant is unstable, and vice versa, as you can verify with simple examples. This is one of the potential advantages, as well as dangers, of feedback control.

The *Routh–Hurwitz* criterion of classical control allows us to easily test for stability of a rational transfer function such as $g(s)$ or $\mu(s)$ without actually computing its poles. The test involves only simple computations with the coefficients of the denominator polynomial.

A better test for closed-loop stability is the *Nyquist criterion*, which we examine in Section 14.1. It is based on the *loop gain*, $\ell(j\omega)$, which is the frequency response of the loop transfer function. This test is especially suited to situations in which the plant is characterized by actual frequency response measurements, but its strengths go well beyond that. The Nyquist criterion underlies many strategies for control design, provides sound measures of nearness to instability, and is the basis for the results on robust stability noted later.

The pole locations provide important information about the degree of stability, in that they determine the speed of transients. The poles of $\mu(s)$, along with any hidden poles, constitute the system's *natural frequencies*. The number of natural frequencies is termed the *order* of the system. If there are no hidden poles, the denominator of $\mu(s)$ is the *characteristic polynomial* of the system, whose roots are the natural frequencies or *characteristic roots*.

A natural frequency at $\lambda_1 = \sigma_1 + j\omega_1$ contributes a term of the form $c_1 e^{\lambda_1 t} = c_1 e^{\sigma_1 t} e^{j\omega_1 t}$ to the system response. This term decays to zero when $\sigma_1 < 0$, that is, when λ_1 is in the left half plane. The more negative σ_1 is, the faster the decay or damping of this term will be. The larger the magnitude of ω_1 is, the more oscillatory this term will be.

Classical control has various rules that permit rapid graphic determination of how natural frequencies move in the complex plane in response to variations in the parameters of the controller or nominal plant model. Such *characteristic root loci* are a significant aid to understanding how system behavior is affected by these variations. They are also a valuable aid to synthesizing candidate control structures in the first place.

EXAMPLE 11.11

Stability of an Up/Down Converter under Proportional Feedback

We now have the information needed to explain the results of our attempt in Example 11.3 to regulate the up/down converter by proportional feedback. The feedforward compensation in the block diagram of Fig. 11.5(a) has no dynamics associated with it and does not affect the discussion of stability. If we ignore the feedforward, the remainder of the diagram can be redrawn as in Fig. 11.17(a). This is in the standard form of Fig. 11.15.

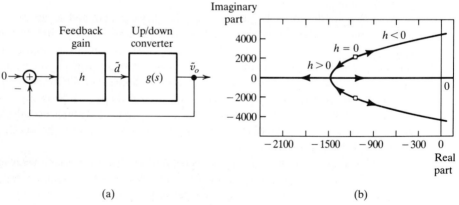

(a)   (b)

Figure 11.17   (a) Linearized model for an up/down converter with proportional feedback. (b) Locus of natural frequencies for $h < 0$ and $h > 0$.

The transfer function $g(s)$ from $\tilde{d}$ to $\tilde{v}_o$ is given by *(11.18)*, and $h(s)$ in this case is simply the gain h. The sensitivity function for this case therefore is

$$\mu(s) = \frac{s^2 + (1/RC)s + (D'^2/LC)}{\left[s^2 + (1/RC)s + (D'^2/LC)\right] + h(I_L/C)\left[s - (V_{in}/LI_L)\right]} \qquad (11.23)$$

The denominator has both its roots in the left half plane if and only if the coefficient of s and the constant term are both positive. The right half plane zero of the transfer function $g(s)$ prevents this condition from holding for all positive h. To satisfy the stability condition, we must constrain h to the interval:

$$-\frac{1}{I_L R} < h < \frac{D'^2}{V_{\text{in}}} \qquad (11.24)$$

An alternative expression for the lower bound is $-D'^2/V_{\text{in}}D$.

The roots of the denominator in *(11.23)* are also the natural frequencies of the system. Figure 11.17(b) shows the locus of the natural frequencies, as h takes both negative and positive values. Note that the damping *decreases* as h takes on increasingly negative values. The natural frequencies move closer to the imaginary axis and their imaginary parts increase. These variations correlate well with the time-domain responses shown in Fig. 11.5(b). For sufficiently small positive values of h, the natural frequencies move away from the imaginary axis and the damping *increases*. For more positive values of h, the roots become real, and one of them eventually crosses over to the right half plane. These movements again correlate well with the time-domain responses in Fig. 11.5(c).

11.5.3 Nominal Performance

Once stability of the nominal closed-loop system is ensured, other aspects of performance can be pursued. Evaluation of nominal performance is considerably more involved, however, than evaluation of nominal stability. Some desirable performance features are best stated and evaluated in the time domain; others are best adapted to frequency-domain specification and treatment. Necessary trade-offs can be difficult to resolve.

For time-domain performance, our first interest is in the transient response determined by the natural frequencies. The more negative the real part of a natural frequency λ_1 is, the faster the associated exponential term $c_1 e^{\lambda_1 t}$ decays. A quick response is often desirable. However, the locations of the natural frequencies alone do not tell us about such things as the peak values of system variables during a transient, which may be critical in evaluating performance, especially in the context of power circuits. Time-domain simulations of system behavior are valuable, even with linear models, in evaluating performance measures such as peak overshoot in the output and in internal variables.

Various connections between pole–zero locations and transient behavior are known for second-order systems. For example, fast responses with small overshoots are typically associated with complex poles whose real and imaginary parts are approximately equal; the presence of a zero tends to increase overshoots. To take advantage of such results, higher order systems are often controlled so as to make the dominant behavior essentially second order. This is done by placing a pair of closed-loop poles substantially closer to the imaginary axis than any of the other poles.

Performance specifications that are conveniently stated in the frequency domain can be studied using *(11.22)*. The nominal performance of the feedback control system in Fig. 11.15 is primarily measured by how closely the output y matches the reference signal r in the nominal system, despite the disturbance b and measurement noise n. Although r is commonly a constant or slowly varying value, the inputs b and n are best thought of as time-varying but bounded signals, with their power distributed over certain frequency ranges. Hence *(11.22)* suggests that the frequency response $\mu(j\omega)$ of the sensitivity function, or equivalently the loop gain $\ell(j\omega)$, plays a critical role.

Note first, for later comparison, that in the open-loop system the disturbance b appears without attenuation because $y = g(s)u + b$. Even if $g(s)$ were stable and u could be chosen to obtain $g(s)u = r$, we would have $y - r = b$. Measurement noise is not an issue in the open-loop case, because measurements are not used.

For the closed-loop system, *(11.22)* shows that we can make $y \approx r$, more or less uniformly over frequency, if:

1. $|\mu(j\omega)| \approx 0$, or equivalently $|\ell(j\omega)| \gg 1$, at frequencies ω where r and b have significant power compared to n; and

2. $|\mu(j\omega)| \approx 1$, or equivalently $|\ell(j\omega)| \ll 1$, at frequencies ω where n has significant power compared to r and b.

Condition (1) requires a large loop-gain magnitude in frequency ranges for which the reference signal and plant disturbance have relatively large power. Condition (2) requires a small loop-gain magnitude in frequency ranges for which the measurement noise has relatively large power. In frequency ranges for which r and b have power comparable to n, good performance may not be attainable, at least with the configuration in Fig. 11.15.

The loop-gain magnitude of a physical system naturally falls off to low values as the frequency increases. Therefore the best situation for control generally occurs when the reference signal r and disturbance b are low-frequency signals, and the measurement noise n is confined to higher frequencies. We would then try and choose the compensator transfer function $h(s)$ to obtain a large loop-gain magnitude at low frequencies and have the magnitude fall off (or "roll" off) at higher frequencies. Of course, this "shaping" of the loop gain is subject to the constraint that closed-loop stability is maintained. The Nyquist criterion is especially useful because it relates stability to the loop gain.

In Section 11.5.4 we discuss additional constraints on the loop gain. In Chapter 14 we indicate how the classical approach to control design tries to deal with all of these considerations. Related compensator design calculations are best carried out using *Bode plots* of the loop gain, which comprise a plot of its magnitude $|\ell(j\omega)|$ versus frequency on a log–log scale and a plot of its phase $\angle\ell(j\omega)$ versus frequency on a linear–log scale. The Bode magnitude plot traditionally uses the unit of *decibels* (dB). The loop-gain magnitude in dB is, by definition, $20\log_{10}|\ell(j\omega)|$.

The frequency at which the loop gain drops to a magnitude of 1 (or 0 dB) is termed the *unity-gain crossover frequency*, ω_c. Examination of *(11.20)* shows that the system frequency response at ω_c, namely, $\mu'(j\omega_c)$, has magnitude $\geq 1/2$. This magnitude is still comparable to the low-frequency value of approximately 1, so the system can still respond significantly to inputs at frequency ω_c. Hence ω_c provides some measure of the bandwidth of the system. (Strictly speaking, the bandwidth is defined as the frequency at which the system frequency response drops to $1/\sqrt{2}$ of its low-frequency value.)

Although we have summarized the traditional approach to control design, you should be alert to features of a problem that demand a different treatment. For example, many of the rules of thumb commonly used in control design apply only when behavior is dominated by a single pair of complex poles, and many do not apply when the plant has right–half-plane poles or zeros.

It should now be evident that feedback control can provide better performance than open-loop control. However, it should also be clear that a poor choice of compensator could increase the sensitivity to disturbances and measurement noise and make the performance worse than that of the open-loop system. The constraints on the loop transfer function imposed by disturbances, noise, and stability conditions may prevent a compensator from performing well for any choice of its parameters. In this case, an alternative compensator structure has to be sought. This is the situation with the up/down converter in Examples 11.3 and 11.11, for instance, where stability considerations limit the maximum loop gain magnitude to about 1. The feedforward in Example 11.3 helps to offset the effect of voltage-source changes, but other disturbances and errors can have severe consequences because of the low loop gain magnitude.

EXAMPLE 11.12
Nominal Performance of a Feedback Control
Design for a Controlled Rectifier Drive

The loop transfer function for the controlled rectifier drive model in Example 11.10 is

$$\ell(s) = \frac{(h_1 s + h_2)}{s(1 + sT/2)(sL + R)} \tag{11.25}$$

Let's assume that $T = (1/120)$ s $= 8.33$ ms, $R = 0.1\ \Omega$, and $L/R = 10T = 83.3$ ms. The magnitude and phase of $\ell(j\omega)$ are shown in the Bode plots of Fig. 11.18(a) for a candidate controller in which $h_1 = 0.35$ and $h_2 = 10.5$. The magnitude is high at low frequencies,

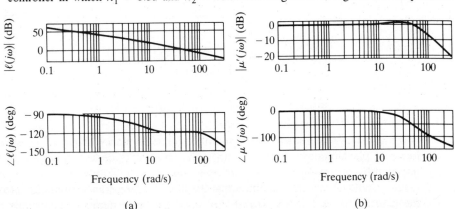

Figure 11.18 Bode plots of (a) loop gain and (b) system frequency response for a controlled rectifier drive.

thanks to the integrator, and crosses over at about 50 rad/s. The Bode plots in Fig. 11.18(b) are for $\mu'(j\omega)$, which is the closed-loop system frequency response from I_{ref} to $\bar{\imath}_d$.

The closed-loop pole locations are -193 and $-29.5 \pm j26.4$. The complex pair is dominant and the real pole is a reflection of our representation of time delay in the converter. The real and imaginary parts of the dominant pair are approximately equal, so we expect good transient responses. The time constant associated with these responses is on the order of (1/29.5=) 34 ms, or significantly larger than the period T. The design is therefore consistent with our modeling assumptions for the converter.

The simulation in Fig. 11.19, using the model of Fig. 11.16, shows the closed-loop response of the averaged armature current following a step in I_{ref}. Note that $\bar{\imath}_d(t)$ in the steady state equals I_{ref}, despite the presence of the unknown constant disturbance representing the back emf E. The reason for the zero steady-state error can be traced to the integrator in the PI controller. The integrator introduces a pole at $s = 0$ in the loop transfer function, which causes the loop gain at $\omega = 0$ to be infinite, and thereby makes the system insensitive to constant disturbances. Another way to view this result is to note that in the constant steady state obtained when this stable LTI system is driven by constant disturbances, the input to the integrator must be zero in order for its output to be constant. Hence $\bar{\imath}_d = I_{ref}$ in the steady state. More detailed simulations correlate well with predictions of the averaged model.

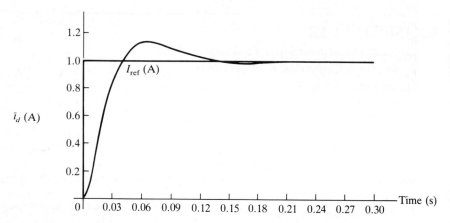

Figure 11.19 Closed-loop response of averaged armature current $\bar{\imath}_d$ to a step in the reference I_{ref}.

11.5.4 Robustness

Our robustness requirement is that stability and performance be preserved for some range of plant model uncertainty $\tilde{g}(s)$. The constraints on the loop gain that are imposed by our *stability robustness* requirement are highlighted in the following result. It holds under the condition that the nominal loop transfer function $\ell(s)$ and the actual transfer function $\ell(s) + \tilde{\ell}(s)$ have the same number of unstable poles (and no unstable *hidden* poles, as defined in Section 11.5.2). With this condition, the result is that stability of the actual system follows from stability of the nominal system if:

$$|\tilde{\ell}(j\omega)| < |1 + \ell(j\omega)| \qquad (11.26)$$

The proof of this result is based on the Nyquist criterion and is outlined Section 14.1.1. The constraint implies that we are especially vulnerable to errors or uncertainties in the loop gain at frequencies for which $\ell(j\omega) \approx -1$.

When the variation of the loop transfer function from its nominal is the result entirely of modeling errors, $\tilde{\ell}(s) = h(s)\tilde{g}(s)$. Relating (11.26) specifically to this case, we obtain the following sufficient condition for robust stability: If the nominal plant transfer function $g(s)$ and the actual transfer function $g(s) + \tilde{g}(s)$ have the same number of unstable poles, stability of the actual system follows from that of the nominal system if:

$$\frac{|\tilde{g}(j\omega)|}{|g(j\omega)|} < \frac{1}{|\mu'(j\omega)|} \qquad (11.27)$$

We usually cannot check such a condition, because the modeling error $\tilde{g}(s)$ is poorly characterized, except where it happens to represent portions of a more detailed model that were intentionally neglected to simplify the design of the nominal control. Often, in fact, we cannnot even restrict $\tilde{g}(s)$ to be rational. Some information about $\tilde{g}(j\omega)$ at low frequencies may be available from frequency-response measurements on the actual plant, if the open-loop plant is stable. However, this information too is subject to error, due to manufacturing variations or aging in the field, for example. For all these reasons, we have to design a controller with a range of possible $\tilde{g}(s)$ in mind. A reasonable assumption is that we have available an upper bound $\beta(\omega) \geq |\tilde{g}(j\omega)|$. Accordingly, we can replace the test in (11.27) with:

$$\frac{\beta(\omega)}{|g(j\omega)|} < \frac{1}{|\mu'(j\omega)|} \qquad (11.28)$$

but still with the requirement that the actual plant have the same number of unstable poles as the nominal. (The condition (11.28) is also *necessary* for robust stability if, rather than taking $\beta(\omega)$ as a bound on the frequency responses of a particular set of $\tilde{g}(s)$, we assume that *any* $\tilde{g}(s)$ whose frequency response is bounded by $\beta(\omega)$ can occur.)

The constraint in (11.28) requires that the frequency response $\mu'(j\omega)$ of the nominal closed-loop system have small magnitude in frequency ranges for which the fractional error in the plant model can be large. For example, at frequencies for which we are completely uncertain about the phase of the plant's frequency response, the left-hand side of (11.28) is ≥ 2. Thus we need $|\mu'(j\omega)| < 1/2$, which requires that we be beyond the crossover frequency. Resonance peaks in the system frequency response $\mu'(j\omega)$ mark regions that are particularly sensitive to modeling errors. Translating (11.28) to conditions on the nominal loop gain, we see that $|\ell(j\omega)|$ can be large at frequencies for which the fractional modeling error is small but must be small, and not close to -1, wherever the modeling error is large.

In classical control we commonly evaluate the distance of $\ell(j\omega)$ from -1 by means of a *phase margin*. This measures how much $\angle\ell(\omega_c)$ exceeds $-180°$, where ω_c is the crossover frequency, so $|\ell(j\omega_c)| = 1$. The phase margin tells us how much additional phase delay around the loop is needed to make $\ell(j\omega_c) = -1$. A phase margin of $45°$ is considered good if the modeling error around crossover is relatively small. For the candidate controller in Example 11.12, the phase margin is $60°$. Also of interest is the *gain margin*, defined as $1/|\ell(j\omega_o)|$, where ω_o is the *phase crossover frequency*, at which the phase of $\ell(j\omega)$ becomes $-180°$. The gain margin is therefore the factor by which the loop gain must be increased to make $\ell(j\omega_o) = -1$.

A major potential advantage of feedback is that it can robustly stabilize an unstable or insufficiently stable system. However, we evidently cannot consider stability robustness in isolation from good nominal performance. Both objectives impose constraints on the loop gain, and these constraints could very well conflict. The best situation generally occurs when significant modeling errors are confined to higher frequencies, the reference signal r and disturbance b are low-frequency signals, and the measurement noise n is at higher frequencies. In this case, we can make the loop gain high at low frequencies and roll it off at high frequencies. In other cases, the design of a good LTI feedback controller can be much more difficult.

We tend to have good low-frequency models in power electronics (for instance, averaged models). The modeling error is then indeed confined to higher frequencies, but often not as high as the noise frequencies. Modeling error—rather than measurement noise—then determines how soon the loop gain magnitude must be made small. For example, in the controlled rectifier circuit of Example 11.10, we constructed the rectifier model under the assumption that signals varied slowly relative to the switching period, so the model becomes less accurate as we approach one half the switching frequency. This limitation might cause us to reduce the loop gain magnitude well before measurement noise (such as the switching-frequency ripple in the current measurement) becomes a concern. The candidate controller in Example 11.12 may not be sufficiently conservative in this respect.

It is important to recognize that we cannot shape the loop gain at will. A practical compensator has only a limited number of parameters that we can vary. Also, what we do in one frequency range has consequences in other ranges. For instance, it can be shown that $\mu(s)$ for a stable closed-loop system is constrained by:

$$\int_0^\infty \ln|\mu(j\omega)|d\omega = \pi\rho \qquad (11.29)$$

Here ρ is a constant whose value depends on $\ell(s)$. If $\ell(s)$ has at least two more poles than zeros, then ρ is the sum of the real parts of the unstable poles of $\ell(s)$. If the number of poles of $\ell(s)$ is one more than the number of zeros and $\ell(s)$ is stable, then $\rho = -\ell_\infty/2$, where ℓ_∞ is the limit of $s\ell(s)$ as s goes to ∞.

The constraint in *(11.29)* forces trade-offs among the requirements in different frequency ranges. We want to keep $\mu(j\omega) \approx 0$ at frequencies for which good command following and disturbance rejection are needed. We want $\mu(j\omega) \approx 1$ at frequencies for which sensor noise and modeling errors are large. We are also forced to have $\mu(j\omega) \approx 1$ for high frequencies because the loop gain of physical systems must fall off. It follows from *(11.29)* that to maintain stability of our closed-loop system we must pay for these requirements by having $|\mu(j\omega)| \gg 1$, or equivalently, $|\mu'(j\omega)| \gg 1$, or equivalently, $\ell(j\omega) \approx -1$ in the remaining frequency ranges. The system could thus become vulnerable to even small modeling errors in these frequency ranges.

The severity of the constraint *(11.29)*—and of the trade-offs that it implies—increases with ρ. The more unstable the open-loop system, for example, the harder the control problem becomes. Zeros of $\ell(s)$ in the right half plane give rise to similar constraints, which complicate the trade-offs still further.

Ensuring *performance robustness* is more involved than ensuring stability robustness. The sensitivity relation in *(11.21)* suggests that, at least for small modeling errors, the closed-loop system frequency response does not change much at frequencies for which the sensitivity $\mu(j\omega)$ is small. We simply note here that, if nominal performance is satisfactory, actual performance is degraded but not destroyed, so long as the modeling errors satisfy *(11.26)* or *(11.28)* by an adequate margin.

We also are interested in studying the robustness of system stability and performance to *nonlinearities* in the model. One important nonlinearity in power electronics arises from the fact that control variables often have tight constraints. These constraints induce saturation nonlinearities in the feedback loop, the effects of which we can sometimes approximate in an LTI analysis by reducing loop gain.

More generally, the effects of nonlinearities on a control system designed by LTI methods are studied through simulations on more detailed models than the ones used for the initial controller design. The simulations should examine a variety of expected operating conditions, including those ignored during the initial design. Such simulations help to validate the control design or to expose problems stemming from the various modeling assumptions and the design approach used. The simulations can form the basis for refining the initial controller.

Notes and Bibliography

Averaging is a frequent theme in modeling and analysis of power electronic circuits, especially for high-frequency switching or PWM converters. Our treatment of circuit averaging in Section 11.3 was stimulated by the analysis of such converters in [1], though our development goes further. The averaging of the canonical-cell switch in Section 11.3.4 draws on [1] and [2, Part I]. The modifications needed to handle capacitor ESR are described in the latter reference. A more general perspective on circuit averaging for high-frequency switching converters is presented in [3].

See [2, Part II] for the extension of switch averaging to discontinuous conduction. The behavior of the averaged model in [2, Part II] at frequencies well below one half the switching frequency is essentially unchanged if the averaged inductor is replaced by a short circuit. In fact, certain computations in the derivation of the model effectively assume that the average voltage across the inductor in discontinuous conduction is approximately zero. The results in [2, Part II] suggest that retaining the inductor significantly improves the agreement with experimental data at frequencies up to one half the switching frequency. A convincing theoretical justification is still lacking, but perhaps the "hybrid" models mentioned in the Notes to Chapter 13 will provide a satisfactory explanation.

Switch averaging for quasi-resonant converters is developed in [4]. Again, replacing the averaged resonant inductor and resonant capacitor by a short circuit and open circuit, respectively, leaves the low-frequency behavior of the averaged model essentially intact.

The injected-current approach to averaging, illustrated in Example 11.5, is described in [5]. Further examples are presented in, for instance, [6], which also emphasizes the value of representing linearized converter models as controlled two-port networks. We provide additional references on averaging in Chapter 12, in connection with state-space averaging.

Applications of the local ω-component defined in Section 11.3.5 are explored in [7], which also mentions connections to describing function approaches. A related idea is described in [8].

Dynamic modeling and control design for dc and ac machine drives are covered very well in [9]. In particular, you will find a rather detailed treatment of the phase-controlled rectifier drive for a dc machine, which we have used in several examples in this chapter. The book also shows how to embed the current-control loop of the controlled rectifier in a larger control system that regulates machine speed or position. We look at this extension in Section 14.2. Other useful references on power electronics for electrical machine drives are [15] and [16].

There are many textbooks that are appropriate for a first course on control, but [10] is especially good. A rich context for studying feedback control is provided by op-amps, as [11] demonstrates. Our sketch in Section 11.5 of issues and objectives in control design is based on "neoclassical" or "postmodern" perspectives, as presented in [12], [13], and [14].

Many of the problems in Chapters 11–14 have a part that assumes access to appropriate computational tools. There are several public-domain and commercial programs for circuit and system analysis, some of them specifically aimed at power electronics. Since we have not carried out any serious comparison, we refrain from mentioning specific programs. However, it is appropriate to note that all the simulations and control design computations in Chapters 11–14 are done using MATRIX$_X$ and SystemBuild, which are packages for computer-aided control-system design, created by Integrated Systems, Inc., Santa Clara, California.

1. G. W. Wester and R. D. Middlebrook, "Low Frequency Characterization of Switched DC-DC Converters," in *IEEE Power Processing and Electronics Specialists Conference* (Atlantic City, May 1972).

2. V. Vorpérian, "Simplified Analysis of PWM Converters Using the Model of the PWM Switch: Parts I and II," *IEEE Trans. Aerospace and Electronic Systems* 26: 490–505 (May 1990).

3. S. R. Sanders and G. C. Verghese, "Synthesis of Averaged Circuit Models for Switched Power Converters," in *IEEE International Symposium on Circuits and Systems (ISCAS)*, (New Orleans, May 1990), 679–683.

4. V. Vorpérian, R. Tymerski and F. C. Lee, "Equivalent Circuit Models for Resonant and PWM Switches," *IEEE Trans. Power Electronics* 4: 205–214 (April 1989).

5. M. Clique and A. J. Fossard, "A General Model for Switching Converters," *IEEE Trans. Aerospace and Electronic Systems* 13: 397–400 (July 1977).

6. A. S. Kislovski, "Controlled-Quantity Concept in Small-Signal Analysis of Switching Power Cells," *IEEE Trans. Aerospace and Electronic Systems* 19: 438–446 (May 1983).
7. S. R. Sanders, J. M. Noworolski, X. Z. Liu and G. C. Verghese, "Generalized Averaging Method for Power Conversion Circuits," in *IEEE Power Electronics Specialists Conference (PESC)* (San Antonio, June 1990), 333–340.
8. C. T. Rim and G. H. Cho, "Phasor Transformation and its Application to the DC/AC Analyses of Frequency Phase-Controlled Series Resonant Converters (SRC)," *IEEE Trans. Power Electronics* 5: 201–211 (April 1990).
9. W. Leonhard, *Control of Electrical Drives* (Berlin: Springer-Verlag, 1985).
10. G. F. Franklin, J. D. Powell and A. Emami-Naeini, *Feedback Control of Dynamic Systems* (Reading, Massachusetts: Addison-Wesley, 1986).
11. J. K. Roberge, *Operational Amplifiers: Theory and Practice* (New York: John Wiley, 1975).
12. M. Morari and E. Zafiriou, *Robust Process Control* (Englewood Cliffs, New Jersey: Prentice-Hall, 1989).
13. J. M. Maciejowski, *Multivariable Feedback Design* (Wokingham, England: Addison-Wesley, 1989).
14. J. C. Doyle, B. A. Francis and A. Tannenbaum, *Feedback Control Theory* (New York: Macmillan, 1991).
15. J. M. D. Murphy and F. G. Turnbull, *Power Electronic Control of AC Motors* (Oxford: Pergamon, 1988).
16. B. K. Bose, *Power Electronics and AC Drives* (Englewood Cliffs, New Jersey: Prentice-Hall, 1986).

Problems

11.1 This problem deals with the derivatives of averages.

(a) For the local average $\overline{x}(t)$ defined in *(11.1)*, show that the derivative of the average equals the average of the derivative.

(b) Verify the more general result in *(11.12)*, relating the derivative of the ω-component of x to the ω-component of the derivative.

(c) How is the derivative of the ω-component related to the ω-component of the derivative when the averaging interval T is varied as a function of time? What does your result reduce to for $\omega = 0$?

11.2 We obtained an averaged circuit for a controlled rectifier drive in Fig. 11.6(b), Example 11.4, under the assumptions of continuous conduction and no commutating reactance.

(a) Use the results of Section 5.2 to show that the effect of a commutating reactance X_c on the averaged circuit can be approximately accounted for by incorporating an additional resistor of value $2X_c/\pi$ in series with the voltage source. Why is this only an approximate result?

(b) Suppose that there is no commutating reactance but that the drive goes into discontinuous conduction. Find an approximate averaged circuit for this case, making clear why the circuit is only approximate. (A good place to start is with a sketch of the rectifier voltage v_d for discontinuous conduction.)

11.3   This problem deals with an up/down converter in discontinuous conduction.

   (a) Extend the linearized model of the up/down converter in discontinuous conduction in Example 11.8 to include perturbations $\tilde{v}_{in}$ in the source value. Show that the effect is to add a $\tilde{v}_{in}$-dependent current source in parallel with the $\tilde{d}$-dependent current source.

   (b) Compute the transfer function from $\tilde{d}$ to $\tilde{v}_o$ in the linearized circuit of (a). Obtain Bode plots of the frequency response when $R = 300\ \Omega$ and the other parameter values are as in Example 11.1. (You should verify that the converter is indeed in discontinuous conduction for this R.) Compare these plots with Bode plots of the corresponding transfer function in *(11.18)*, for the up/down converter in continuous conduction, with $R = 2\ \Omega$.

   (c) Draw a block diagram of a closed-loop PI control system that could be used to regulate the up/down converter in discontinuous conduction, maintaining its output voltage at -9 V. Find the choice of the PI control gains that places the natural frequencies of the closed-loop system at $2p \pm j2p$, where p is the pole of the open-loop converter transfer function that you computed in (b). Obtain Bode plots of the loop gain for this closed-loop system.

   (d) Show that the PI control system in (c) can be interpreted as replacing the $\tilde{d}$ current source in the linearized circuit by a parallel resistor–inductor combination. Explain from the circuit diagram what this replacement does to the dynamic behavior of $\tilde{v}_o$ and to the steady state error in $\tilde{v}_o$ induced by parameter errors and constant disturbances, such as source deviations $\tilde{v}_{in} \neq 0$.

   (e) Compute the transfer functions from $\tilde{v}_{in}$ to $\tilde{v}_o$ (the so-called *audio susceptibility*) for the open-loop and closed-loop systems in (b) and (c), respectively. Compare their frequency-domain characteristics using Bode plots.

   (f) How would the controls in (c) perform if we actually had $R = 2\ \Omega$, with the converter in continuous conduction?

11.4   Following the analyses in Examples 11.5 and 11.8, obtain averaged models and their linearizations for the discontinuous conduction mode in the other high-frequency switching dc/dc converters in Chapters 6 and 7. Include at least the down (or buck), up (or boost), flyback, and forward converters. Use the linearized models to compute the transfer functions from $\tilde{d}$ and $\tilde{v}_{in}$ to $\tilde{v}_o$ in each case.

11.5   In *current-mode control* of high-frequency switching dc/dc converters, we specify the peak inductor current or switch current in each cycle, rather than specifying the duty ratio. This problem considers constant-frequency current-mode control, in which the switch is turned on every T seconds and turned off when its current reaches a specified threshold value, i_{th}. Develop averaged models and their linearizations for discontinuous conduction operation of the up/down converter and of the converters in Problem 11.4, under such a system. Find the transfer function from small perturbations $\tilde{i}_{th}$ in the threshold to perturbations $\tilde{v}_o$ in the output for each converter. Also find the audio susceptibility transfer function, from $\tilde{v}_{in}$ to $\tilde{v}_o$, in each case. Identify which converters can become unstable in an open-loop configuration for some operating condition.

11.6   To better understand the approximations in *(11.5)* and *(11.6)* that we used to average the switch in a high-frequency switching dc/dc converter, let's define the *ripple* at time τ in a variable w to be $\hat{w}(\tau) = w(\tau) - \overline{w}(\tau)$. If w is periodic and the averaging interval T equals the period, this expression reduces to our usual notion of ripple.

(a) Now show that:

$$\overline{qw} - \overline{q}\,\overline{w} = \overline{q\hat{w}} + \overline{q[\overline{w} - \overline{w}(t)]}$$

(All averages except the first one in the brackets are evaluated at time t.) Hence, with small ripple (or small $|\hat{w}|$) and slow variation in the local average (or small $|\overline{w} - \overline{w}(t)|$ over the averaging interval), we have $\overline{qw}(t) \approx \overline{q}(t)\overline{w}(t)$.

(b) Find an example to show that the term $\overline{q\hat{w}}(t)$ can be small even if the ripple is not small. Hence the small-ripple assumption is not always crucial to obtaining an averaged model.

11.7   Verify the analyses in Examples 11.7 and 11.9. Similarly obtain averaged models and their linearizations for the continuous conduction mode in the other high-frequency switching dc/dc converters in Chapters 6 and 7. Include at least the down (or buck), up (or boost), flyback, and forward converters. Use the linearized models to compute the transfer functions from $\tilde{d}$ and $\tilde{v}_{\text{in}}$ to $\tilde{v}_o$ in each case. Also characterize the portion of each linearized model that lies between the input and load as a *controlled two-port network*, taking the input to be a voltage source and the load to be a current source. (Do not forget to include the effects of the control variable $\tilde{d}$ in your two-port representation.)

11.8   This problem analyzes feedforward in the up/down converter.

(a) Use the transfer function of *(11.8)* in Example 11.7 to determine the response of the open-loop up/down converter in Example 11.1 to a step in input voltage from 12 V to 8 V, when feedforward is not used. How does this result relate to the waveform of Fig. 11.3(a)?

(b) To compute the response to the same step when the feedforward design in Example 11.1 is used, you have to be more careful. It is not enough to repeat the above calculation with D fixed at the value after the step, because now the response of the circuit is actually the combination of two effects: a step in the input voltage and an associated *step* in the *duty ratio*. Use the averaged circuit in Fig. 11.11 to help analyze this case, exploiting the fact that the inductor current and capacitor voltage cannot change suddenly (as no impulsive sources are present). How does your result relate to the waveform of Fig. 11.3(b)?

(c) Compute the audio susceptibility transfer function from small perturbations $\tilde{v}_{\text{in}}$ in supply voltage to perturbations $\tilde{v}_o$ in the response, when feedforward is present. You can do so by appropriately combining the transfer function of *(11.18)* in Example 11.9 (which gives the response to small perturbations $\tilde{d}$ in the duty ratio) with the transfer function of *(11.8)* (which gives the response to $\tilde{v}_{\text{in}}$ without feedforward). Use this transfer function to compute the approximate response to a supply drop from 12 V to 8 V and compare it to the results in (b).

(d) Compute the audio susceptibility transfer function for the *closed-loop* up/down converter in Example 11.3. (The results of Example 11.11 may be helpful here.) Use this transfer function to verify the responses shown in Fig. 11.5(b) and (c).

11.9   Use the ω-component introduced in Section 11.3.5 to analyze the dynamics of the series resonant converter in Example 9.1, picking ω equal to the switching frequency. You should be able to predict the form of the amplitude variations obtained in response to a step change in the switching frequency. Also verify that in the steady state you recover the results of Example 9.1.

11.10 Derive *(11.19)*, which relates the output y of the classical closed-loop control configuration in Fig. 11.15 to the driving signals r, b, and n. Derive the sensitivity result in *(11.21)*. Verify the conditions for closed-loop stability given in Section 11.5.2. Show that the robustness condition *(11.27)* follows from *(11.26)*.

11.11 We showed in Fig. 11.9 (Example 11.6) how we might generate a switching function to control a power circuit. The box labeled "controller" typically contains an analog circuit of the type shown in Fig. 11.20.

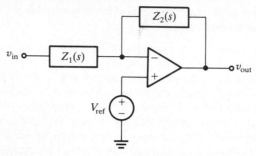

Figure 11.20 Op-amp realization of a PI controller.

(a) Assuming an ideal op-amp (infinite gain and input impedance), show that:

$$v_{out}(s) = -\frac{Z_2(s)}{Z_1(s)}[v_{in}(s) - V_{ref}] + V_{ref}$$

where $Z_1(s)$ and $Z_2(s)$ are the impedances in the forward and feedback paths, respectively. Suppose that $Z_1(s) = R_1$, a resistor. Find a simple choice of components for the feedback path to realize a PI controller.

(b) By incorporating appropriate circuitry in the box labeled "controller" in Fig. 11.9 and by modifying the sawtooth waveform, you can obtain the switching function needed to control a phase-controlled rectifier drive according to the configuration in Fig. 11.16. Specify what is needed, using Example 5.1 for guidance.

11.12 This problem relates to the closed-loop control design for the controlled rectifier drive in our examples.

(a) Determine the characteristic root loci for the closed-loop controlled rectifier system in Example 11.12, for the following parameter variations: (i) $-3 \le h_1 \le 3$, with $h_2 = 0$; (ii) $-2 \le h_2 \le 60$, first with $h_1 = 0.35$ and then with $h_1 = 1$.

(b) Verify the step response shown in Fig. 11.19. Also construct a simulation of the full nonlinear model of the closed-loop drive and verify that the step reponse of i_d is predicted well by our averaged analysis.

(c) Evaluate the closed-loop system obtained with $h_1 = 1$ and $h_2 = 60$. How does it compare with the system in Example 11.12?

11.13 Obtain an averaged-circuit model for the high-frequency PWM bridge inverter in Fig. 8.15 and use it to simulate the startup behavior of the system for different parameter values.

Chapter 12

State-Space
Models

$\mathbf{WE}$ introduced dynamic models for power electronic circuits in Chapter 11, using circuit averaging. We obtained circuit models to describe the average behavior of the variables in certain families of power circuits. When these circuit models were nonlinear, we showed how to linearize them to arrive at linear circuit models, which under certain conditions were time invariant as well. This provided a route for the application of the concepts and methods of linear, time invariant (LTI) circuit analysis and control system design to the case of power circuits.

While this approach can take us far, it does not provide a complete solution to the task of analyzing and controlling the dynamics of power circuits. Some of the reasons for this deficiency are worth noting. First, we are not always interested in the average behavior of circuit waveforms, so averaged circuits may be irrelevant. In some cases, using the ω-component models defined in Section 11.3.5 allows us to extend the notion of averaging to situations in which we are interested in the fundamental component at some frequency. This possibility deserves to be explored further, but it shares other limitations of the averaging approach.

Second, tractable averaged (or ω-component) models may not exist for many power circuits or may be obtainable only after various approximations. The assumptions or conditions underlying the approximations may fail in cases of interest to us. A common assumption in averaged models is that waveforms vary only slowly during an operating cycle, but this condition may not be satisfied in important or even critical cases. For instance, many power circuits can exhibit *ripple instability*, wherein the system variables alternate between low and high values from cycle to cycle. A model based on the assumption that variables change slowly may not be suited to analyzing such an instability.

Third, circuit models are sometimes inappropriate for describing the controllers used to regulate power circuits. This is especially true for digital or microprocessor based controllers, where we have to deal with clocked or sampled waveforms.

State-space models provide a much more general and powerful basis for dynamic modeling. They include switched and averaged circuit models as special cases but go considerably further. They are important in analyzing, simulating, and controlling both steady-state behavior and perturbations away from it. Our goal in this chapter and Chapter 13 is to show how to systematically develop and analyze dynamic models in state-space form for power electronic circuits and their controllers. (Part of Chapter 14 is devoted to feedback control design with state-space models, but the rest of that chapter is accessible without Chapters 12 and 13.)

In this chapter we describe nonlinear state-space models for both continuous-time and discrete-time variables and introduce compact matrix notation for these models. The discrete-time models are especially important for us, because the cyclic operation of power circuits makes it natural to model *samples* of the circuit variables taken once per cycle. We present several examples of state-space models in power electronics, including averaged models in state-space form, and discrete-time or sampled-data models.

In Chapter 13 we show how to linearize state-space models and how to analyze continuous-time and discrete-time LTI models in state-space form. We also introduce state-space descriptions for *piecewise LTI models*. These constitute the most common models in power electronics, because they describe circuits comprising LTI components and ideal switches.

12.1 FEATURES OF STATE-SPACE MODELS

There are several reasons for the growing use of state-space models in power electronics. State-space models focus on those variables that are central to describing the dynamic evolution of a system. By aiming for a state-space model of a power circuit, we impose a valuable discipline and direction on the modeling process. The variables to highlight, relationships to examine, and route the analysis should take become clearer. The state-space approach also allows us to handle nonlinear and time varying models in the same framework as LTI models. Furthermore, the formalism applies to both continuous-time and discrete-time or sampled-data systems. State-space models also provide a uniform and convenient starting point for such diverse tasks as steady-state computation, linearization, stability evaluation, control design, and simulation.

A state-space model of an LTI electrical circuit can be obtained systematically— and hence automatically by computer—from a circuit description. The resulting LTI description, as well as other LTI descriptions from, say, linearization, can be analyzed by familiar techniques, including impedance or transform methods. This treatment can then be extended to piecewise LTI circuits, consisting of LTI elements and ideal switches.

No other framework has all these features and advantages. In addition, certain disadvantages sometimes associated with state-space models are usually not serious in the context of power electronic circuits. Specifically, a state-space model may not reflect the "sparsity" of the interconnections among components in the system. Although that is of great concern in large power systems or complicated analog circuits—with hundreds or thousands of variables to deal with—it is not of concern in a typical power electronic circuit, which has far fewer variables. Moreover, the generalized state-space models introduced later can be used to reflect sparsity, if this is a concern.

12.1.1 State Variables, Inputs, and Outputs

The key variables in a state-space model are the *state variables*, whose values together define the *state* of the system. State variables summarize those aspects of the past that are relevant to the future. That is, they are the variables whose initial values are needed to determine future system behavior. Thus they are typically associated with a system's memory mechanisms or energy storage mechanisms. Natural state variables in electrical circuits are the currents or flux linkages in inductors and the voltages or charges on capacitors. For a digital controller, the natural state variables would be the contents of its registers.

In addition to the state variables, other variables are of interest in a description of a dynamic system. The *inputs* to the system are external signals, such as the waveforms of voltage and current sources that drive a power circuit and the signals that modulate the actions of controlled switches in it. Some of the inputs may be control variables that are under our command, whereas others may be disturbances that we have no control over. Specification of the inputs from the initial time onward—together with the initial values of the state variables—determines the future behavior of the state variables according to laws that govern their evolution. These governing laws are embodied in the *state-space model*.

The *outputs* of the system are either measured quantities or those whose values are of interest, even if not measured. The outputs that we consider have values at any instant that are functions of the system state and the inputs at that instant. In the context of electrical circuits, typical outputs may be the voltages, currents, or dissipation associated with selected elements.

12.2 CONTINUOUS-TIME MODELS

Suppose we have chosen to model the system of interest with n state variables x_i, $i = 1$ to n, and with m inputs u_j, $j = 1$ to m. A continuous-time state-space model of the system then takes the form of a set of coupled, nonlinear, time varying, first-order differential equations:

$$\frac{dx_1}{dt} = \dot{x}_1(t) = f_1\Big(x_1(t),\ x_2(t), \ldots,\ x_n(t),\ u_1(t), \ldots,\ u_m(t),\ t\Big)$$

$$\frac{dx_2}{dt} = \dot{x}_2(t) = f_2\Big(x_1(t),\ x_2(t), \ldots,\ x_n(t),\ u_1(t), \ldots,\ u_m(t),\ t\Big) \qquad (12.1)$$

$$\vdots$$

$$\frac{dx_n}{dt} = \dot{x}_n(t) = f_n\Big(x_1(t),\ x_2(t), \ldots,\ x_n(t),\ u_1(t), \ldots,\ u_m(t),\ t\Big)$$

They express the instantaneous rates of change of each of the n state variables as functions of the indicated arguments, namely, the instantaneous values of all the state variables and inputs and the time argument t itself. The model is said to be of nth *order*, because it involves n state variables. The functions $f_i(\cdot)$ can be fairly elaborate for even simple power circuits, as later examples show.

Also associated with the state-space model *(12.1)* are the output variables $y_\ell(t)$, $\ell = 1$ to p. We consider only models in which the outputs can be written in the form:

$$y_\ell(t) = g_\ell\left(x_1(t), x_2(t), \ldots, x_n(t), u_1(t), \ldots, u_m(t), t\right) \qquad (12.2)$$

so that the outputs are directly determined at any time by the state and inputs at that time. If an output variable is not of this form initially, we may be able to bring it to this form by appropriately defining additional state variables.

Figure 12.1 is a representation of the model in *(12.1)* and *(12.2)*. The outputs of the integrators represent the state variables, and the interconnection constraints ensure that the system is governed by *(12.1)* and *(12.2)*.

Linearity and Time Invariance   The description *(12.1)* is called *time invariant* if none of the functions $f_i(\cdot)$ explicitly involves t, so that:

$$f_i(\cdot) = f_i\left(x_1(t), x_2(t), \ldots, x_n(t), u_1(t), \ldots, u_m(t)\right) \qquad (12.3)$$

for all i. If all the $f_i(\cdot)$ in *(12.1)* are linear functions of the state variables and inputs, the model is termed *linear* and we have

$$
\begin{aligned}
f_i(\cdot) =\; & a_{i1}(t)x_1(t) + a_{i2}(t)x_2(t) + \cdots + a_{in}(t)x_n(t) \\
& + b_{i1}(t)u_1(t) + \cdots + b_{im}(t)u_m(t)
\end{aligned}
\qquad (12.4)
$$

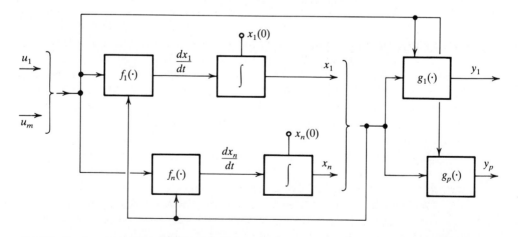

Figure 12.1   Representation of the continuous-time state-space model in *(12.1)* and *(12.2)*.

A most important special case is a linear *and* time invariant description, for which the coefficients $a_{ij}(t)$ and $b_{ik}(t)$ are constant. For such a *linear, time invariant (LTI)* model, we have

$$
\begin{aligned}
f_i(\cdot) =& a_{i1}x_1(t) + a_{i2}x_2(t) + \cdots + a_{in}x_n(t) \\
& + b_{i1}u_1(t) + \cdots + b_{im}u_m(t)
\end{aligned}
\tag{12.5}
$$

The output equations *(12.2)* are classified as linear and/or time invariant as done for the state-space model *(12.1)*. It is quite possible for the state-space model to be LTI while the output equations are not, or conversely.

Before highlighting certain important properties that distinguish state-space models, we present some examples of such models in the context of electrical circuits—and power circuits in particular.

12.3 STATE-SPACE MODELS FOR ELECTRICAL CIRCUITS

We begin by outlining a systematic way to obtain a state-space description of an electrical circuit model. The model is allowed to contain nonlinear and/or time varying inductors (with mutual couplings permitted), capacitors, nondynamic components (such as resistors, ideal transformers, and ideal switches), and sources (independent and dependent). The word *nondynamic* refers to components governed by constraints that involve only the instantaneous values of the voltages and currents at their terminals, not the derivatives or integrals or time shifts of these quantities.

In the case of LTI circuits (namely, fixed interconnections of LTI components), standard computer routines can use the procedure described here to produce a state-space description from a circuit specification. The resulting state-space model, not surprisingly, is also LTI. The same procedure can be used to automatically generate state-space models for certain classes of nonlinear and time varying circuits.

Natural state variables for a circuit, as already noted, are the inductor currents or flux linkages and capacitor voltages or charges. The state-space description requires expressions for the derivatives of these quantities, expressed entirely as functions of themselves and of the sources and control inputs driving the circuit. Outputs must also be expressed in this form.

Let's begin by considering the case where there is no mutual coupling between inductors. We know that, if the current through an inductor L is i_L, the inductor voltage is $d(Li_L)/dt$; and, if the voltage across a capacitor C is v_C, the capacitor current is $d(Cv_C)/dt$. These relationships suggest a way to express the state variable derivatives in the required form: Determine the inductor voltages and capacitor currents as functions of the inductor currents, capacitor voltages, source values, and control inputs. We can then readily obtain the state-space description. For convenience we break the procedure into the following three steps.

First, think of the inductors replaced by current sources and the capacitors replaced by voltage sources. The original circuit and the changes are illustrated schematically in Fig. 12.2. The effect of control inputs on a switch is represented by the associated switching function $q(t)$.

Second, solve the resulting nondynamic circuit for the complementary variables of these replacement sources. In other words, find the voltages across the inductor "current sources" and the currents through the capacitor "voltage sources," expressing them entirely in terms of all the control inputs and sources, including replacement sources. Any other circuit variables of interest, such as the output variables y_1 and y_2 in Fig. 12.2, are also solved for at this step.

An explicit solution can always be obtained at this second step for well-behaved LTI circuit models and for many nonlinear and time varying circuit models as well. With power circuit models, the solution must reflect the nonlinear, time varying nature of the switches. A natural way to obtain analytic expressions is to introduce appropriate switching functions, as suggested by the schematic diagram in Fig. 12.2. For that case, the result of the second step is

$$v_L = f_1'(i_L, v_C, v_S, i_S, q, t)$$
$$i_C = f_2'(i_L, v_C, v_S, i_S, q, t)$$
$$y_1 = g_1(i_L, v_C, v_S, i_S, q, t)$$
$$y_2 = g_2(i_L, v_C, v_S, i_S, q, t)$$

(12.6)

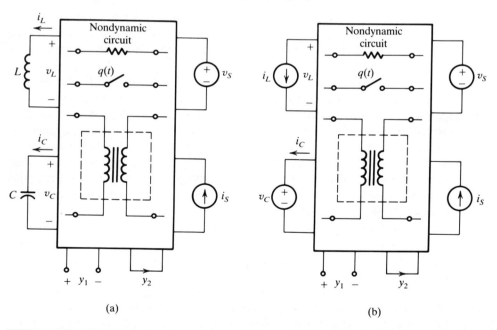

(a)   (b)

Figure 12.2 Obtaining a state-space description of a circuit: (a) schematic representation of original circuit; (b) inductors and capacitors replaced by current and voltage sources, respectively.

The necessary computations for this second step are straightforward for power circuit models that consist of LTI components and ideal switches. Such a model is LTI in any given switch configuration, so we refer to it as a *piecewise LTI* model.

Third, express the inductor voltages in terms of the derivatives of their currents and the capacitor currents in terms of the derivatives of their voltages. Using these expressions in the result of the second step, and rearranging the equations, we get the desired expressions for the derivatives of the state variables. When the inductors and capacitors are LTI, this third step is simple. For instance, in the notation of the circuit in Fig. 12.2, we substitute $v_L = L(di_L/dt)$ and $i_C = C(dv_C/dt)$ in (12.6), and rearrange the equations to obtain the desired expressions for di_L/dt and dv_C/dt:

$$\frac{di_L}{dt} = \frac{1}{L}f_1'(\cdot) = f_1(i_L, v_C, v_S, i_S, q, t)$$

$$\frac{dv_C}{dt} = \frac{1}{C}f_2'(\cdot) = f_2(i_L, v_C, v_S, i_S, q, t)$$

(12.7)

For the more general case of nonlinear, time varying inductors $L(i_L, t)$ and capacitors $C(v_C, t)$, we would have to make the substitutions:

$$v_L = \left(L + i_L \frac{\partial L}{\partial i_L}\right)\frac{di_L}{dt} + i_L \frac{\partial L}{\partial t}$$

$$i_C = \left(C + v_C \frac{\partial C}{\partial v_C}\right)\frac{dv_C}{dt} + v_C \frac{\partial C}{\partial t}$$

(12.8)

When mutual inductance is present, the voltage across an inductor also depends on the derivatives of the currents in all the inductors coupled to it. This introduces an additional step in the derivation of the state-space model. After rewriting each inductor voltage in terms of the derivatives of all the relevant inductor currents, you have to solve for the derivatives individually. We omit the details.

There are some exceptions to choosing all the inductor currents and capacitor voltages as state variables. If c capacitors form a loop that contains no other elements besides independent voltage sources, Kirchhoff's voltage law shows that only $c - 1$ of the capacitor voltages can have independent initial conditions and hence only $c - 1$ capacitor voltages should be chosen as state variables. A similar statement holds for the dual situation, in which a cutset of only inductors and current sources is constrained by Kirchhoff's current law to have currents that sum to zero.

Although we have stated several times that inductor currents and capacitor voltages are the natural choices of state variables for electrical circuits, they are not the only ones. We could pick any other set of variables that provides equivalent information. For example, we could have chosen $Ki_L - v_C$ and $Ki_L + v_C$ for the circuit in Fig. 12.2, with any nonzero constant K.

EXAMPLE 12.1

State-Space Model for an Up/Down Converter

Consider the up/down (buck/boost) converter shown in Fig. 12.3, which we used extensively as an example in Chapter 11. The only difference is that we have included an equivalent series resistance (ESR) of value R_C with the capacitor. We assume for simplicity that the converter is operated so that either the transistor or the diode is always conducting; this assumption rules out the discontinuous conduction mode, in which both devices are off during some part of each cycle. Let the switching function $q(t)$ represent the switch status, with $q(t) = 1$ when the transistor is on and $q(t) = 0$ when it is off; let $q'(t)$ denote $1 - q(t)$.

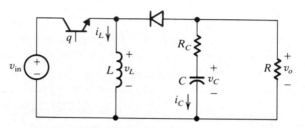

Figure 12.3 Up/down converter circuit.

To obtain a state-space model of the circuit, we focus on those variables whose initial values are essential to determining the future behavior of the system for a specified source value and switching pattern. The natural choice of state variables, as noted earlier, is the inductor current i_L and capacitor voltage v_C. The natural inputs are the source voltage $v_{in}(t)$ and the control signal $q(t)$.

When $q(t) = 1$,

$$v_L(t) = L\frac{di_L}{dt} = v_{in}(t)$$

$$i_C(t) = C\frac{dv_C}{dt} = \frac{-1}{R + R_C}v_C(t)$$

(12.9)

and when $q(t) = 0$,

$$v_L(t) = L\frac{di_L}{dt} = \frac{R}{R + R_C}\left[-R_C i_L(t) + v_C(t)\right]$$

$$i_C(t) = C\frac{dv_C}{dt} = \frac{-1}{(R + R_C)}\left[R i_L(t) + v_C(t)\right]$$

(12.10)

Combining *(12.9)* and *(12.10)* by introducing $q(t)$, and rearranging them into the form *(12.1)*, we obtain the desired state-space model:

$$\frac{di_L}{dt} = \frac{R}{L(R + R_C)}\left[-R_C q'(t)i_L(t) + q'(t)v_C(t)\right] + \frac{1}{L}q(t)v_{in}(t)$$

$$\frac{dv_C}{dt} = \frac{-1}{C(R + R_C)}\left[Rq'(t)i_L(t) + v_C(t)\right]$$

(12.11)

We leave it to you to derive an output equation of the form *(12.2)*, expressing the output voltage v_o as a function of the state variables and inputs.

Note that if $q(t)$ were fixed at either 0 or 1, the description in *(12.11)* would be LTI, justifying the label of "piecewise LTI." When $q(t)$ is considered as a control input, the description is nonlinear but time invariant. However, if $q(t)$ is a prespecified function of time, as happens when both the closing and opening of the switch are put under the open-loop control of a clocking waveform, then $q(t)$ behaves simply as a time varying parameter of the system, not as an input. The state-space model in this case is linear and time varying. Finally, in a feedback controlled system like that illustrated in Fig. 11.9, $q(t)$ is determined by the state variables as well as by a clocking waveform. The resulting description is nonlinear and time varying.

State-Space Averaging We can construct state-space models for the averaged circuits developed in Chapter 11 by following the procedure just described. An alternative route to a state-space model for the average behavior of a power circuit is to average a state-space description of the switched circuit. Such *state-space averaging* can be easier than direct circuit averaging, because the process of obtaining the state-space model of the circuit serves to organize some of the necessary computations.

EXAMPLE 12.2

State-Space Averaging for an Up/Down Converter

Averaging both sides of each expression in *(12.11)*, with temporary suppression of the time index to keep the notation simple, we find

$$\frac{d\bar{i}_L}{dt} = \frac{R}{L(R + R_C)}\left(-R_C\overline{q'i}_L + \overline{q'v}_C\right) + \frac{1}{L}\overline{qv}_{in}$$

$$\frac{d\bar{v}_C}{dt} = \frac{-1}{C(R + R_C)}\left(R\overline{q'i}_L + \bar{v}_C\right) \qquad (12.12)$$

These two equations do not yet constitute a state-space model, because the right-hand sides are not expressed entirely as functions of the averaged state variables $\bar{i}_L$ and $\bar{v}_C$ and of the averaged inputs $\bar{q} = d$ and $\bar{v}_{in}$. However, let's assume that i_L is well approximated during any interval of length T by its local average over that interval, as are v_C and v_{in}. (These assumptions are equivalent to those we made in Section 11.3.4 when averaging a switch.) Then we can express the averages of the products $q'v_C$, $q'i_L$, and qv_{in} as the products of the averages of the individual terms and approximate *(12.12)* by:

$$\frac{d\bar{i}_L}{dt} \approx \frac{R}{L(R + R_C)}\left[-R_C d'(t)\bar{i}_L(t) + d'(t)\bar{v}_C(t)\right] + \frac{1}{L}d(t)\bar{v}_{in}$$

$$\frac{d\bar{v}_C}{dt} \approx \frac{-1}{C(R + R_C)}\left[Rd'(t)\bar{i}_L(t) + \bar{v}_C(t)\right] \qquad (12.13)$$

which *is* in state-space form. This pair of equations is the desired state-space averaged model. The assumptions that permitted us to obtain this model are generally satisfied in well-designed switching converters. Note that the only difference between the averaged model and the switched model is that instantaneous quantities have been replaced by averages. Any output equations associated with the switched model in *(12.11)* can be averaged similarly.

When $R_C = 0$, the circuit of Fig. 12.3 is exactly the circuit of Fig. 11.2, which we averaged in Example 11.7. Not surprisingly, therefore, setting $R_C = 0$ in the preceding state-space averaged model turns out to yield a state-space description of the averaged circuit in Fig. 11.11 of Example 11.7.

Direct circuit averaging, as in Chapter 11, is usually more straightforward and revealing than state-space averaging. However, state-space averaging can sometimes be valuable. For example, when the capacitor in the up/down converter has significant ESR, the voltage across the R_C, C pair has significant ripple in it (even though the voltage v_C across C alone has small ripple). This condition violates the small-ripple assumption underlying the canonical cell averaging in Fig. 11.10 and prevents us from direct circuit averaging without some refinement of Fig. 11.10. The ESR poses no problem for state-space averaging, however, as Example 12.2 shows. A circuit that corresponds to the state-space averaged model in *(12.13)* can also be obtained, if desired.

12.4 PROPERTIES OF SOLUTIONS

We now examine two important properties of the solutions of continuous-time state-space models: the state property and the continuity property. This examination also gives us a chance to discuss some aspects of numerical solution for such models.

12.4.1 The State Property

The significance of the special form of description in *(12.1)* is that, knowing the values $x_i(t_0)$, $i = 1$ to n, of all the state variables at time t_0 and knowing the inputs $u_j(t)$ for t in some interval $t_0 \leq t < t_f$, we can determine $x_i(t)$ for t in the interval $t_0 < t \leq t_f$. For convenience, we refer to this property as the *state property* of the description *(12.1)*. Note that this property is consistent with the qualitative definitions of *state* and *inputs* mentioned in Section 12.1.1.

The state property can be made plausible through examination of the representation of *(12.1)* in Fig. 12.1. This schematic representation can form the basis for *solutions* or *simulations* of the state-space model. A continuous-time, analog simulation based on Fig. 12.1 would require an appropriate physical realization of the integrators (typically, using operational amplifiers), as well as some means of evaluating the functions $f_i(\cdot)$ and $g_\ell(\cdot)$. When we have set the initial conditions on the integrators, the system variables will evolve in response to the inputs. The state variables at any time will then be the outputs of the integrators at that time, which shows that the state property holds. Appropriate time and amplitude scalings are used in practice to give a convenient simulation, whose results can then be scaled inversely to recover the behavior of the underlying model.

Alternatively, a numerical solution or simulation can be carried out. This method would involve using some numerical integration routine instead of analog integrators in Fig. 12.1. The state variable waveforms or "trajectories" are determined at discrete points in time. We examine this approach in more detail, first because it makes the origin of the state property clearer, and second because numerical solutions of state-space models are of great practical importance.

12.4.2 Numerical Solution

Take the starting time t_0 to be 0 for notational simplicity. Assume that we know the initial values of the state variables $x_i(0)$, $i = 1$ to n, as well as the inputs $u_j(t)$ for $t \geq 0$. To approximately evaluate $x_1(\epsilon)$ in *(12.1)* for a small time step ϵ, we can use the expression:

$$x_1(\epsilon) \approx x_1(0) + \epsilon \frac{dx_1(0)}{dt} \qquad (12.14)$$

which is simply a Taylor series expansion truncated after the linear term. The symbol $dx_1(0)/dt$ stands for $dx_1(t)/dt$ evaluated at $t = 0$. The error in the approximation is of order ϵ^2, so the smaller ϵ is, the better the approximation will be. A graphic interpretation of *(12.14)* is given in Fig. 12.4. The dot on the heavy line at time ϵ represents the true $x_1(\epsilon)$. The square at time ϵ is the approximation given by the right-hand side of *(12.14)*, namely, the value at time ϵ of the *tangent* to the true solution at time 0.

Now the first expression in *(12.1)* shows that the derivative needed in *(12.14)* can be obtained by evaluating $f_1(\cdot)$ at $t = 0$. For this evaluation, all we need are the values of the state variables and inputs at $t = 0$, which we know. Hence we can find $x_1(\epsilon)$ as accurately as desired by picking a small enough value for ϵ. The same procedure can be used to compute the approximate values of all the other state variables at time ϵ, namely, $x_i(\epsilon)$, $i = 2$ to n.

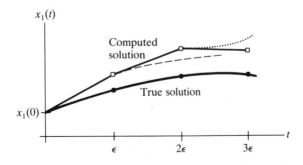

Figure 12.4   Linear, "forward Euler" approximation of the state trajectory. The heavy line denotes the true solution; the dashed line denotes the solution using the approximate $x_i(\epsilon)$; and the dotted line denotes the solution using the approximate $x_i(2\epsilon)$.

Beginning with these approximate values of all the state variables at $t = \epsilon$, and knowing the values of all the inputs at $t = \epsilon$, we can similarly step forward in time once more to evaluate $x_i(2\epsilon)$ approximately. The dashed line in Fig. 12.4 shows what the true solution for x_1 would be from time ϵ onward if we started at the *approximate* values of $x_i(\epsilon)$ computed at the previous step. Again, however, we approximate the dashed line by means of a tangent.

Iterating this process enables us, in principle, to approximately solve *(12.1)* forward in time. A rigorous analysis of the accumulated error is rather involved. It requires the functions $f_i(\cdot)$ in *(12.1)* to be sufficiently well behaved that trajectories such as those represented by the heavy, dashed, and dotted lines in Fig. 12.4 are close to each other when the errors in the computed values of $x_i(\epsilon)$, $x_i(2\epsilon)$, ..., are small.

For a less rigorous analysis, note that the error per step is of order ϵ^2 and the number of steps required to arrive at any specified time is of order ϵ^{-1}. The accumulated error in the computed value at the specified time must therefore be of order ϵ. Hence, by making ϵ small enough, we can compute the solution to any desired degree of accuracy. This establishes the state property of *(12.1)*, namely, that knowledge of the initial state and input trajectory is sufficient to determine the state trajectory.

This solution procedure is referred to in the numerical analysis literature as the *forward Euler* method. Although this method clarifies the origin of the state property, we would choose more sophisticated methods for actual numerical solution of *(12.1)*. For example, the popular "fourth-order Runge–Kutta" algorithm yields an accumulated error of order ϵ^4, but at the cost of evaluating the $f_i(\cdot)$ for four different sets of arguments at each time step.

A limitation of the Runge–Kutta and forward Euler methods is that the time step ϵ has to be kept relatively small compared to the *smallest* time constant in the solution. This condition must be enforced even though the solution over most of the interval is dominated by terms with much larger time constants. Thus we need other approaches for efficient solution of systems with widely separated time constants (*stiff* systems).

The *backward Euler* method is one such approach. The solution involves replacing $dx_1(0)/dt$ in *(12.14)* by $dx_1(\epsilon)/dt$, and similarly replacing the derivatives of the other state variables. Using *(12.1)* to substitute for $dx_i(\epsilon)/dt$, we obtain a set of *implicit*, nonlinear equations—the "corrector" equations—for $x_i(\epsilon)$. We can solve them by iterative methods, with initial guesses provided by explicit "predictor" equations, such as the forward Euler. We can also vary the step size ϵ as we go along, so as to maintain a balance between accuracy and computational efficiency. Many other possibilities exist, with their respective advantages and disadvantages, but are beyond the scope of this book.

In the case of LTI state-space models, you should expect to find analytical solutions, expressed in terms of sinusoids and exponentials. In Chapter 13 we show you how to find them. Piecewise LTI models have analytical solutions as well, and they are of special interest to us, as discussed in Chapter 13. The solution of piecewise LTI models is based on the following important property of continuous-time state-space models.

12.4.3 The Continuity Property

The state variables are continuous functions of time, as you can see by examining *(12.1)*. If any state variables had discontinuities, the derivatives on the left-hand side of *(12.1)* would have impulses in them, which would have to be balanced by impulses in the inputs $u_j(t)$ on the right-hand side. Hence, in the absence of impulsive inputs, state variables are continuous. For piecewise LTI models, we can invoke this continuity to piece together the analytical solutions obtained in each LTI regime and thus obtain the overall solution.

EXAMPLE 12.3

State Trajectories of a Resonant dc/dc Converter

The circuit in Fig. 12.5(a) represents a series resonant dc/dc converter of the type discussed in Chapter 9. We have omitted both the input dc source and the switching circuit that modulates it. Here we need only to indicate the resulting square-wave voltage v_{in}, of fixed amplitude V_1

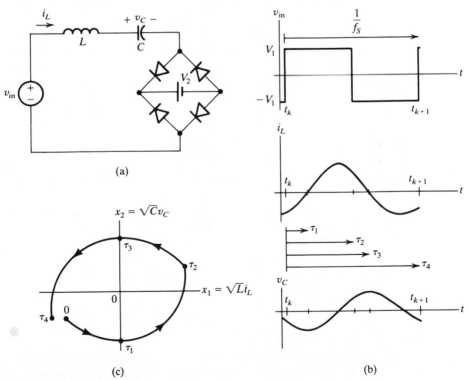

Figure 12.5 (a) Series resonant dc/dc converter circuit. (b) Associated waveforms. (c) State-plane representation.

and controllable frequency f_S, that is applied to the resonant circuit. This voltage produces a nearly sinusoidal inductor current i_L, which is rectified by the diode bridge and passed through the load. By controlling the frequency of the square wave relative to the resonant frequency f_o of the LC circuit, we control the amplitude of the resonant current and thereby the average current delivered to the load. The load in our example is a dc voltage source of value V_2. The circuit response, even for fixed f_S, is complicated by the action of the bridge rectifier, which subtracts the load voltage from v_{in} when $i_L > 0$ and adds it to v_{in} when $i_L < 0$.

Natural state variables for this circuit are again the inductor current and capacitor voltage. To be specific, let's take the nominal f_S to be above the resonant frequency. The current through the LC pair then lags the voltage across it in the steady state. We assume for simplicity that the transients we are modeling are not large enough to change this feature of the steady-state response, so that $i_L < 0$ at the instants that v_{in} goes from $-V_1$ to $+V_1$ (and $i_L > 0$ when v_{in} goes from $+V_1$ to $-V_1$). We consider a cycle of operation beginning at such an instant.

Typical transient waveforms during the kth cycle of operation, which extends from $t = t_k$ to $t = t_{k+1}$, are shown in Fig. 12.5(b). Note that f_S and the indicated transition times τ_ℓ, $\ell = 1, 2, 3$, and 4, can in general vary from cycle to cycle. Thus they should actually carry the argument k, but we drop the k for notational simplicity, relying on you to supply it as needed.

With these assumptions, the voltage applied to the resonant LC circuit at the start of every cycle is $V_1 + V_2$ and remains constant at this value until i_L goes positive. Hence, v_C during this interval is a portion of a sinusoid, with frequency f_o and mean $V_1 + V_2$. At the same time i_L is a portion of a sinusoid with the same frequency but with zero mean.

When i_L passes upward through zero at time τ_1 after the start of the kth cycle, v_C becomes a portion of another sinusoid whose frequency is still f_o but whose mean value is now $V_1 - V_2$. The current i_L correspondingly becomes a portion of another sinusoid, but with the same frequency and the same mean value of zero that it had before the transition.

At time τ_2 after the start of the kth cycle, the input voltage changes from V_1 to $-V_1$. The analysis during the second half of the cycle parallels that of the first half, except that all waveforms have the opposite polarity. At time τ_3, i_L passes downward through zero, and at time τ_4 the input voltage changes from $-V_1$ to V_1 again, signifying the start of the next cycle. During each cycle, therefore, the circuit goes through four configurations.

An illuminating companion to the waveform representation in Fig. 12.5(b) is the representation of state variable trajectories on the *state plane* (or "phase plane"), as shown in Fig. 12.5(c). The coordinates in such a representation are the state variables. The piecewise sinusoidal waveforms of i_L and v_C for the resonant converter would map into piecewise elliptical trajectories on the state plane. When we use the scaled quantities $x_1 = \sqrt{L}i_L$ and $x_2 = \sqrt{C}v_C$ as state variables, instead of i_L and v_C, the trajectories turn out to be piecewise circular, as you can easily verify. This scaling simplifies the analysis, so we use scaled quantities as state variables for the rest of this example and in Example 12.6, where we continue the analysis. The state-plane diagram in Fig. 12.5(c) already reflects this choice.

The circular arcs are centered at points determined by the mean values of the sinusoids that make up i_L and v_C during the intervals corresponding to these arcs. The center of each arc therefore has an x_1 coordinate of 0, and its x_2 coordinate takes one of the four values $\sqrt{C}(\pm V_1 \pm V_2)$. Because v_C increases when i_L is positive, the arcs are all traversed counterclockwise, with angular velocity $\omega_o = 2\pi f_o$. For clarity, only one cycle of operation is represented in Fig. 12.5(c). In the nominal periodic steady state, this trajectory would be a closed curve, because the state at the end of the cycle equals the state at the beginning.

State-plane representations often provide valuable insight into the dynamics of second-order circuits and can be a fruitful source of ideas for control design. We say more about this in Chapter 14.

Example 12.3 described the detailed behavior of the circuit waveforms of a converter during one cycle of operation. However, a higher level model that summarizes or condenses these intracycle details and highlights the intercycle behavior is often more useful. Averaged models are of this type, but have the limitations noted at the beginning of this chapter. We turn now to a class of models that are very useful in describing the cycle-to-cycle behavior of power circuits, namely, discrete-time or sampled-data models in state-space form. These models form the basis for evaluating the *stability* of cyclic steady-state operation of power converters, as we show in Chapter 13. The models are also the starting point for digital control design, as we illustrate in Section 14.4.

12.5 DISCRETE-TIME OR SAMPLED-DATA MODELS

A discrete-time state-space description most commonly comes from regular sampling of the state variables of a continuous-time description. Because the majority of power electronic circuits operate cyclically, working with models that involve quantities sampled once per cycle is especially natural.

Suppose that the underlying continuous-time description is given by *(12.1)* and is regularly sampled every T seconds. We often write $x_i[k]$ to denote $x_i(kT)$, where k is an integer that indexes the samples. To obtain useful sampled-data models, we need to assume that the inputs $u_j(t)$ in the continuous-time model *(12.1)* are determined by a finite set of variables in each sampling interval, with the precise form of dependence being allowed to vary from cycle to cycle. These *determining variables* are denoted by $p_1[k], \ldots, p_r[k]$ in the kth interval. The brackets around k serve to remind us that these quantities are discrete-time sequences.

If, for example, the input $u_1(t)$ changes only at sample times (remaining constant over the interval between successive sampling instants), we could use $p_1[k]$ to denote the value of $u_1(t)$ in the interval $kT \leq t < kT + T$. This situation is represented schematically in Fig. 12.6. As another example, suppose that the continuous-time input $u_2(t)$ in the kth sampling interval is a portion of a sinusoid, but with parameters that can vary from interval to interval. We could then represent $u_2(t)$ in the kth interval by $p_2[k] \sin(p_3[k]t + p_4[k])$. We present further examples when we examine piecewise LTI models for power electronic circuits in Chapter 13, where the controlled switching times of a circuit are determining variables for the functions $f_i(\cdot)$.

The state property of *(12.1)* shows that specifying the state variables $x_i[k]$ at time kT and the inputs $u_j(t)$ over the intervening interval $kT \leq t < kT + T$

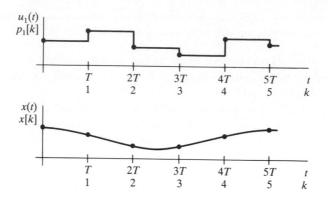

Figure 12.6   Sampling a continuous-time system whose inputs are piecewise constant.

completely specifies the state variables $x_i[k+1]$ at time $kT+T$. The evolution of the sampled state therefore is governed by some model of the form:

$$x_1[k+1] = \phi_1\Big(x_1[k], x_2[k], \ldots, x_n[k], p_1[k], \ldots, p_r[k], k\Big)$$

$$x_2[k+1] = \phi_2\Big(x_1[k], x_2[k], \ldots, x_n[k], p_1[k], \ldots, p_r[k], k\Big)$$

$$\vdots$$ *(12.15)*

$$x_n[k+1] = \phi_n\Big(x_1[k], x_2[k], \ldots, x_n[k], p_1[k], \ldots, p_r[k], k\Big)$$

which is what we term a discrete-time or sampled-data state-space model. The x_i are the state variables and the p_j are the inputs to the model.

The functions $\phi_i(\cdot)$ in *(12.15)* cannot in general be related in any simple, analytically expressible way to the underlying continuous-time model. Exceptions occur in some special but important cases. One such case is when the underlying continuous-time system is LTI or can be described by a succession of LTI models, which is the piecewise LTI case treated in Chapter 13. Another case occurs when the sampling interval is very short compared to the times over which the right-hand sides of *(12.1)* vary significantly; in this case we obtain an approximate sampled-data model by using the same kind of approximation as in *(12.14)*.

We can also obtain a discrete-time state-space model of the form *(12.15)* in ways that need not directly involve sampling an underlying continuous-time description. Whatever the origin of the discrete-time model, its key feature is that it expresses the state at the next discrete-time instant in terms of the state at the present instant and the inputs associated with the present instant. The description *(12.15)* therefore has a "state property" exactly like that of the continuous-time case: For a given sequence of inputs over some interval and a specified state at the beginning of the interval, the model allows us to compute the sequence of states,

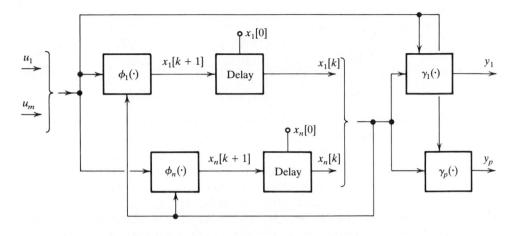

Figure 12.7   Representation of the discrete-time state-space model in *(12.15)* and *(12.16)*.

or the state trajectory, over the entire interval. This computation involves nothing more than iterative use of the state-space model.

The definitions of linearity and time invariance are the same as in the continuous-time case, with the functions $\phi_i(\cdot)$ of *(12.15)* used instead of the functions $f_i(\cdot)$ of *(12.1)*. Also, each output associated with a discrete-time state-space model is given, as in the continuous-time case, by a possibly time-dependent function of the current state and inputs:

$$y_\ell[k] = \gamma_\ell\Big(x_1[k],\, x_2[k],\ldots,\, x_n[k],\, p_1[k],\ldots,\, p_r[k],\, k\Big) \qquad (12.16)$$

A representation analogous to Fig. 12.1 for the discrete-time model *(12.15)* and *(12.16)* is shown in Fig. 12.7. We use *delay elements* instead of integrators and make other obvious changes, replacing $x(t)$ with $x[k]$, dx/dt with $x[k+1]$, and $f_i(\cdot)$ and $g_\ell(\cdot)$ with $\phi_i(\cdot)$ and $\gamma_\ell(\cdot)$, respectively. A delay element is characterized by the property that the signal present at its input at time k appears at its output at time $k+1$. Hence, when $x_i[k]$ is at the output of a delay element, its input must be $x_i[k+1]$.

Before giving an example of a discrete-time model for power electronic circuits, we introduce some streamlined notation.

12.6 NOTATION

Having compact notation to represent systems of equations such as *(12.1)*, *(12.2)*, *(12.15)*, and *(12.16)* is extremely useful. With it we do not have to write out the equations in detail until needed. The language of vector and matrix algebra provides this compact notation.

We define the state vector $x(t)$, input vector $u(t)$, and output vector $y(t)$ by:

$$x(t) = \begin{bmatrix} x_1(t) \\ x_2(t) \\ \vdots \\ x_n(t) \end{bmatrix} \qquad u(t) = \begin{bmatrix} u_1(t) \\ u_2(t) \\ \vdots \\ u_m(t) \end{bmatrix} \qquad y(t) = \begin{bmatrix} y_1(t) \\ y_2(t) \\ \vdots \\ y_p(t) \end{bmatrix} \qquad (12.17)$$

We do *not* use boldface or any other notational device to denote vectors such as $x(t)$, relying instead on the context of a discussion for clarity. We restrict ourselves to a few letters for vectors, such as x, u, y, and a handful more that we introduce later. Thus you should soon begin automatically to interpret these symbols as vectors, without the aid (or burden!) of additional notational frills. We usually denote the components of a vector by subscripting the symbol used for the vector, as in x_i for the ith component of the vector x. The symbol dx/dt, or sometimes $\dot{x}(t)$, denotes the vector whose entries are the derivatives dx_i/dt, $i = 1$ to n.

Now we can write *(12.1)* and *(12.2)* in far more compact notation:

$$\frac{dx}{dt} = \dot{x}(t) = f\Big(x(t), u(t), t\Big)$$

$$y(t) = g\Big(x(t), u(t), t\Big) \qquad (12.18)$$

where the vector functions $f(\cdot)$ and $g(\cdot)$ are given by:

$$f(\cdot) = \begin{bmatrix} f_1(\cdot) \\ f_2(\cdot) \\ \vdots \\ f_n(\cdot) \end{bmatrix} \qquad g(\cdot) = \begin{bmatrix} g_1(\cdot) \\ g_2(\cdot) \\ \vdots \\ g_p(\cdot) \end{bmatrix} \qquad (12.19)$$

Similarly, the discrete-time model *(12.15)* can be written as:

$$x[k + 1] = \phi\Big(x[k], p[k], k\Big) \qquad (12.20)$$

where the vector function $\phi(\cdot)$ is defined analogously to $f(\cdot)$, and the r-component vector of discrete-time inputs $p[k]$ is analogous to $u(t)$. The associated vector of outputs can be written as:

$$y[k] = \gamma\Big(x[k], p[k], k\Big) \qquad (12.21)$$

When the description *(12.1)* is linear, the use of matrix notation allows us to write it in the following more explicit and convenient form:

$$\frac{dx}{dt} = A(t)x(t) + B(t)u(t) \qquad (12.22)$$

where $A(t)$ and $B(t)$ are $n \times n$ and $n \times m$ matrices whose entries are functions of t only:

$$A(t) = \begin{bmatrix} a_{11}(t) & \cdots & a_{1n}(t) \\ \vdots & \cdots & \vdots \\ a_{n1}(t) & \cdots & a_{nn}(t) \end{bmatrix} \qquad B(t) = \begin{bmatrix} b_{11}(t) & \cdots & b_{1m}(t) \\ \vdots & \cdots & \vdots \\ b_{n1}(t) & \cdots & b_{nm}(t) \end{bmatrix} \qquad (12.23)$$

The $a_{ij}(t)$ and $b_{i\ell}(t)$ are the same as in (12.4). Although we generally denote matrices by capital letters, again we do *not* use any other special notational device, such as boldfacing, to distinguish them. As with vectors, we use only a handful of symbols for matrices, and you should soon begin to recognize them as matrices.

If the output equations (12.2) are also linear, they take the form:

$$y(t) = E(t)x(t) + F(t)u(t) \qquad (12.24)$$

where $E(t)$ and $F(t)$ are $p \times n$ and $p \times m$ matrices, respectively. For a continuous-time LTI model, the description is still of the form (12.22) and (12.24), except that now the coefficient matrices A, B, E, and F are *constant*:

$$\frac{dx}{dt} = Ax(t) + Bu(t) \qquad (12.25)$$

and

$$y(t) = Ex(t) + Fu(t) \qquad (12.26)$$

The appropriate notational modifications for the discrete-time case should be obvious. For now, we simply write the LTI case, which we examine in more detail later:

$$x[k+1] = \mathcal{A}x[k] + \mathcal{B}p[k] \qquad (12.27)$$

and

$$y[k] = \mathcal{E}x[k] + \mathcal{F}p[k] \qquad (12.28)$$

To illustrate how the notation defined here helps to express ideas and procedures concisely, we can use it to rewrite the forward Euler approximation introduced earlier in connection with the state property and numerical solutions. The following equations pertain, respectively, to the general case, the linear case, and the LTI case (with the identity matrix denoted by I):

$$\begin{aligned} x(t+\epsilon) &\approx x(t) + \epsilon f\Big(x(t), u(t), t\Big) \\ &= [I + \epsilon A(t)]x(t) + \epsilon B(t)u(t) \qquad (12.29) \\ &= (I + \epsilon A)x(t) + \epsilon Bu(t) \end{aligned}$$

EXAMPLE 12.4

Approximate Sampled-Data Model for an Up/Down Converter

We begin by expressing the continuous-time state-space model (12.11) for the up/down converter of Example 12.1 in matrix form. We let $x_1 = i_L$ and $x_2 = v_C$. Then, for continuous conduction, and neglecting the capacitor ESR for simplicity (so $R_C = 0$), we have

$$\frac{dx(t)}{dt} = \begin{bmatrix} 0 & q'(t)/L \\ -q'(t)/C & -1/RC \end{bmatrix} x(t) + \begin{bmatrix} q(t)/L \\ 0 \end{bmatrix} v_{\text{in}}(t)$$

$$= A_{q(t)} x(t) + B_{q(t)} v_{\text{in}}(t) \tag{12.30}$$

Everything in the rest of this example also applies to other high-frequency switching converters in continuous conduction, with the exception that the specific matrices $A_{q(t)}$ and $B_{q(t)}$ for other switching converters are different.

Now the switching period T in a well-designed up/down converter is much smaller than the time constants associated with the circuit in its different switch configurations. As a result, the inductor current and capacitor voltage waveforms in each switch configuration are essentially straight-line segments. Using this assumption to make approximations similar to *(12.29)*, we can obtain a sampled-data model for the up/down converter, starting with the continuous-time model *(12.30)*.

Suppose that the transistor is turned on every T seconds and turned off a time $d_k T$ later in the kth cycle, so that d_k is the duty ratio in the kth cycle. (For consistency with our notation for discrete-time sequences, we should actually write $d[k]$ instead of d_k, but the latter notation is less cluttered and is unambiguous because we do not attach any other subscripts to d.) Assume also that the source voltage v_{in} takes the constant value $v_{\text{in}}[k]$ in the kth cycle. Setting $q(t) = 1$ in *(12.30)* to represent the interval when the transistor is on, we obtain an expression similar to *(12.29)*:

$$x(kT + d_k T) \approx (I + d_k T A_1) x(kT) + d_k T B_1 v_{\text{in}}[k] \tag{12.31}$$

Now, setting $q(t) = 0$ in *(12.30)* for the interval when the transistor is off and letting $d_k' = 1 - d_k$, we get

$$x(kT + T) \approx (I + d_k' T A_0) x(kT + d_k T) + d_k' T B_0 v_{\text{in}}[k] \tag{12.32}$$

Using *(12.31)* to substitute for $x(kT + d_k T)$ in *(12.32)*, we get the desired approximate sampled-data model.

To streamline the description of this model, we write $x[k] = x(kT)$ and use the natural symbols A_{d_k} and B_{d_k} for the "averages" of $A_{q(t)}$ and $B_{q(t)}$, respectively, taken with duty ratio d_k:

$$A_{d_k} = d_k A_1 + d_k' A_0 \qquad B_{d_k} = d_k B_1 + d_k' B_0 \tag{12.33}$$

We obtain these matrices simply by replacing $q(t)$ with d_k in the expressions for $A_{q(t)}$ and $B_{q(t)}$ in *(12.30)*. The result of using *(12.31)* in *(12.32)* is then:

$$x[k+1] \approx \left(I + T A_{d_k} \right) x[k] + T B_{d_k} v_{\text{in}}[k]$$

$$+ T^2 d_k' d_k A_0 (A_1 x[k] + B_1 v_{\text{in}}[k]) \tag{12.34}$$

$$= \phi \left(x[k], v_{\text{in}}[k], d_k \right)$$

This is a nonlinear, time invariant, sampled-data model in state-space form, with inputs d_k and $v_{\text{in}}[k]$. Generally, the term involving T^2 can be omitted without significant loss of accuracy when the ripple in the state waveforms is small, or equivalently when T is small relative to the time constants in each switch configuration.

Figure 12.8 represents the relationships among the state variable trajectories of the sampled-data model in *(12.34)*, the underlying continuous-time switched model in *(12.30)*, and the continuous-time averaged model in *(12.13)* of Example 12.2. We could also have constructed another sampled-data model by starting with the averaged model instead of the switched model, and making approximations similar to *(12.29)*. The result differs only in that the terms involving T^2 in *(12.34)* are absent. The sample values in this case would lie on the dashed line in Fig. 12.8.

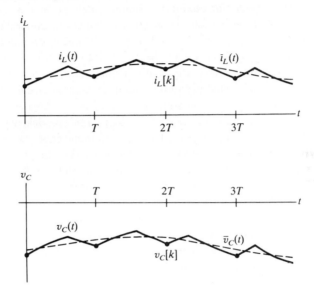

Figure 12.8 Relationships among the state variable trajectories of the sampled-data model, marked •; the underlying continuous-time switched model, marked —; and the continuous-time averaged model, marked – – –.

12.7 GENERALIZED STATE-SPACE MODELS

In Example 12.1 we wrote a state-space model directly in terms of the variables that are allowed in a state-space description. However, introducing auxiliary variables during the derivation is often convenient or necessary. In the cases that we deal with, we can explicitly or implicitly determine these auxiliary variables as functions of the state variables and inputs, using associated nondynamic constraints. When the auxiliary variables are explicitly determined, we can eventually eliminate them and get a state-space model in terms of the allowed variables. However, when they are only implicitly determined by the constraints, or when an explicit determination is complicated, we carry the constraints along with the state-space model. The combination of a state-space model that contains auxiliary variables, with associated nondynamic constraints that determine these variables, is what we term a *generalized state-space model*.

The most important case for us occurs in the context of sampled-data models for power circuits. The generalized state-space model in this case takes the form:

$$x[k+1] = \phi\Big(x[k], p[k], w[k], k\Big)$$

$$0 = \sigma\Big(x[k], p[k], w[k], k\Big)$$

(12.35)

Here $w[k]$ is a vector of auxiliary variables, defined analogously to the vectors in (12.17), and $\sigma(\cdot)$ is a vector of *constraint functions*, defined analogously to the functions in (12.19). Given $x[k]$ and $p[k]$, we could in principle solve the second equation for $w[k]$ and substitute the result into the first equation to determine $x[k+1]$. However, as $\sigma(\cdot)$ is typically a nonlinear function of $w[k]$, solution of the second equation usually requires iterative numerical calculations, as well as some good initial guess of the solution.

The auxiliary variables in $w[k]$ for power circuit models are primarily the time instants at which the topological state of the circuit changes in response to those switch transitions that are *not* directly controlled. (The times of directly controlled switch transitions appear as entries of the input vector $p[k]$.) Diodes turning on or off correspond to such indirectly controlled transitions, because they occur at times that are functions of the state as well as of the control inputs. Transitions in fully controllable switches such as transistors also become indirectly controlled when their operation has been made state dependent through feedback control.

The classification of a generalized state-space model as linear or time invariant follows the pattern already indicated for ordinary state-space descriptions, except that both $\phi(\cdot)$ and $\sigma(\cdot)$ must have the appropriate form. The outputs associated with (12.35) can be written as:

$$y[k] = \gamma\Big(x[k], p[k], w[k], k\Big)$$

(12.36)

There is an obvious continuous-time analog of (12.35), sometimes referred to as a differential/algebraic equation (DAE) description. We can also consider hybrid representations, where the state equation acts in continuous time but the constraints hold at discrete instants. We do not work with either of these representations here, so when we talk of generalized state-space models from now on, we mean the discrete-time case unless otherwise specified.

EXAMPLE 12.5

Generalized State-Space Model for an Up/Down Converter under Current-Mode Control

In (12.34), Example 12.4, we obtained an approximate sampled-data model for an up/down converter, with the duty ratio as the control variable. In *current-mode control*, however, the controller specifies a peak switch current in each cycle, or equivalently a peak inductor current, rather than the duty ratio. The switch may be turned on regularly every T seconds, as in the implementations discussed earlier, but is turned off when the transistor current or inductor current reaches the specified upper threshold value i_{th}. This threshold value is now the primary control variable; the duty ratio becomes an indirectly determined auxiliary variable. (Several modifications are possible. For example, a *hysteretic* current-mode controller uses a lower threshold also on the inductor current, to determine the time at which the transistor is turned on again. We focus here on implementations in which the turn-on occurs at constant frequency.) We discuss current-mode control further in Chapters 13 and 14.

The constraint that determines d_k in terms of i_{th} can be written as:

$$0 = i_L(kT + d_kT) - i_{th}(kT + d_kT) \qquad (12.37)$$

The threshold i_{th} is usually chosen as the sum of two signals: a slowly varying signal i_P determined by the controller on the basis of the discrepancy between the actual and nominal average output voltages; and a regular sawtooth ramp of slope $-S$ at the switching frequency, termed a stabilizing ramp for reasons that we explain when examining the stability aspects of this model in Chapter 13. The resulting waveforms are shown in Fig. 12.9. Note the connection with Example 11.6, where we would choose $K = ST$ and $m = i_P - i_L$ in order to obtain current-mode control.

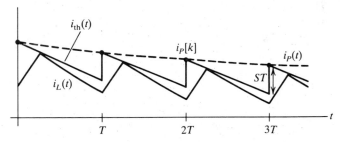

Figure 12.9 Inductor current waveform under current-mode control.

With this choice, $i_{th}(kT + d_kT) = i_P[k] - Sd_kT$, where $i_P[k]$ is the value of i_P in the kth cycle. This expression can be substituted into *(12.37)*, along with the approximate expression for $i_L(kT + d_kT)$ obtained from the first row of the matrix equation in *(12.31)*. In the case of an up/down converter, the substitution yields

$$0 \approx i_L[k] + (d_kTv_{in}[k]/L) - i_P[k] + Sd_kT$$
$$= \sigma\Big(x[k], v_{in}[k], i_P[k], d_k\Big) \qquad (12.38)$$

Together, *(12.34)* and *(12.38)* constitute a sampled-data model in generalized state-space form for an up/down converter under current-mode control. The appropriate expressions for other switching converters are obtained similarly.

Example 12.5 is representative of the ways in which you can obtain generalized state-space descriptions. In this particular case, solving explicitly for d_k from the constraint *(12.38)* actually is easy. Substituting the result into *(12.34)* yields an ordinary state-space model. Nonetheless, you may find it easier to work with the generalized form, even for this simple case. For example, we show in Chapter 13 that the task of linearization can be carried out completely on a generalized state-space model, and does not require reduction to an ordinary state-space model. In any case, with more complicated circuits or control laws and more accurate sampled-data models, carrying out such a reduction would be hard or impossible, and you would have to work with the generalized state-space model.

EXAMPLE 12.6

Generalized State-Space Model for a Resonant dc/dc Converter

Choosing the normalized state variables $x_1 = \sqrt{L}i_L$ and $x_2 = \sqrt{C}v_C$ for the resonant converter circuit in Fig. 12.5(a), Example 12.3, we know that the state trajectory in each cycle consists of four circular arcs in the state plane. With the help of Fig. 12.10, we can easily verify that, if $x(t_\alpha)$ and $x(t_\beta)$ are the state vectors at two points on the same circular arc centered at x_c and if $t_\beta > t_\alpha$, then:

$$x(t_\beta) - x_c = \Theta(t_\beta - t_\alpha)\Big[x(t_\alpha) - x_c\Big] \tag{12.39}$$

where $\Theta(t_\beta - t_\alpha)$ is the "rotation matrix" defined by:

$$\Theta(t_\beta - t_\alpha) = \begin{bmatrix} \cos\omega_o(t_\beta - t_\alpha) & -\sin\omega_o(t_\beta - t_\alpha) \\ \sin\omega_o(t_\beta - t_\alpha) & \cos\omega_o(t_\beta - t_\alpha) \end{bmatrix} \tag{12.40}$$

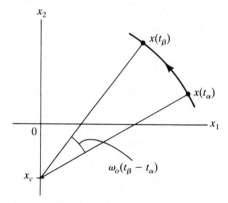

Figure 12.10 Relationship between two points on a circular trajectory in the state plane.

Using (12.40), we can translate the description of circuit behavior in Example 12.3 into the following succint, quantitative relationships for the first half of the kth cycle of operation (with the argument k again dropped from the times $\tau_1[k]$ and $\tau_2[k]$, for notational simplicity):

$$x(t_k + \tau_1) = \Theta(\tau_1)\Big[x(t_k) - x_{c_1}\Big] + x_{c_1} \qquad x_{c_1} = \begin{bmatrix} 0 \\ \sqrt{C}(V_1 + V_2) \end{bmatrix}$$

$$x(t_k + \tau_2) = \Theta(\tau_2 - \tau_1)\Big[x(t_k + \tau_1) - x_{c_2}\Big] + x_{c_2} \qquad x_{c_2} = \begin{bmatrix} 0 \\ \sqrt{C}(V_1 - V_2) \end{bmatrix} \tag{12.41}$$

We now use the first expression in *(12.41)* to eliminate $x(t_k + \tau_1)$ from the second, noting that $\Theta(\tau_2 - \tau_1)\Theta(\tau_1) = \Theta(\tau_2)$, as rotation over an interval τ_1 followed by rotation for $\tau_2 - \tau_1$ is equivalent to rotation over an interval τ_2. The result is

$$x(t_k + \tau_2) = \Theta(\tau_2)x(t_k) - \Theta(\tau_2)x_{c1} + \Theta(\tau_2 - \tau_1)\left[x_{c_1} - x_{c_2}\right] + x_{c_2} \qquad (12.42)$$

The time τ_2 in *(12.42)* is a control variable, whereas τ_1 is an auxiliary variable. Recall from Fig. 12.5(b) that τ_1 is determined by the constraint $i_L(t_k + \tau_1) = 0$ or, equivalently, $x_1(t_k + \tau_1) = 0$. Using the first row of the first matrix equation in *(12.41)*, we can express this constraint in the more detailed form:

$$x_1(t_k)\cos\omega_o\tau_1 - \left[x_2(t_k) - \sqrt{C}(V_1 + V_2)\right]\sin\omega_o\tau_1 = 0 \qquad (12.43)$$

Together, *(12.42)* and *(12.43)* constitute a sampled-data model in the generalized state-space form *(12.35)*, describing the evolution of the state over the first half of the kth cycle.

The symmetry of the circuit and its operation allows us easily to write a similar description for the evolution over the second half and to combine it with the model in *(12.42)* and *(12.43)* to obtain a generalized state-space model over the entire cycle. The control variables for this model are τ_2 and τ_4; in practice, the controller typically sets $\tau_4 = 2\tau_2$. The auxiliary variables are τ_1 and τ_3. With the obvious notation, namely, $x[k] = x(t_k)$ and so on, the result is exactly of the form *(12.35)*.

Note that the constraint *(12.43)* in Example 12.6 can actually be solved for $\tau_1[k]$ in this case, expressing it as the arctangent of a quantity determined by the state $x(t_k)$ at the beginning of the cycle. A similar solution can be obtained for $\tau_3[k]$ in the second half cycle. As with our model for current-mode control in Example 12.5, therefore, we can reduce the generalized state-space model to an ordinary state-space model. The result is messy, however, and not convenient for such tasks as linearization. We would thus be likely to retain the generalized form for many purposes. Furthermore, for slightly more involved circuits, no such reduction would be possible.

Examples 12.5 and 12.6 involve circuits with special features that we exploited to simplify the analysis. The up/down converter has (approximately) piecewise linear waveforms; the resonant converter has piecewise circular trajectories in the state plane. Nevertheless, as we show in Chapter 13, these examples actually contain the essential ingredients of dynamic modeling for much more general power electronic circuits. Using the *state-transition matrix* introduced in Chapter 13, we can conceptually deal with more complicated circuits in the same straightforward manner.

12.8 MODELS FOR CONTROLLERS AND INTERCONNECTED SYSTEMS

Our focus so far has been on developing state-space models for power circuits. State-space models can also be used to represent controllers, as well as interconnections of controllers with power circuits. Most continuous-time controllers that are encountered in practice, and certainly all those considered in this book, can be represented by a block diagram similar to Fig. 12.1 and therefore have a continuous-time state-space description. This result holds even if the control design was carried out and implemented, as it often is, without any reference to state-space models or methods. Similarly, all the discrete-time controllers of interest to us can be represented by a block diagram similar to Fig. 12.7, and therefore have a discrete-time state-space description.

Now consider an interconnection of a power circuit and a controller, as in Fig. 12.11. (We are representing any sensor dynamics as part of the controller.) A state-space description of the interconnection is simply obtained from the state-space models of the individual subsystems. We illustrate the procedure here for a continuous-time case, assuming time invariant models. You can easily extend the procedure to other situations, including time varying and discrete-time cases.

Assume that the power circuit is modeled by the state-space description:

$$\dot{x} = f(x, u, v)$$
$$y = g(x, v)$$

(12.44)

where the vectors u and v represent control and disturbance inputs, respectively. Suppose the controller can be modeled by a state-space description of the form:

$$\dot{x}_c = f_c(x_c, y, r, n)$$
$$u = g_c(x_c, r)$$

(12.45)

where the vectors r and n represent reference signals and measurement noise, respectively. The output of the controller is the control vector u that is applied to the power circuit.

The state variables of the combined system are those of the two subsystems, taken together. We can now combine their descriptions to find a state-space model of the interconnected system. This step simply entails merging the state equations

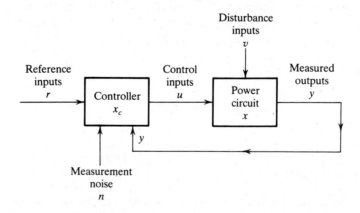

Figure 12.11   Interconnection of a power circuit and its controller.

in *(12.44)* and *(12.45)* and making the substitutions necessary for the model to contain only those variables allowed in a state-space description, namely, the state variables and external inputs. The result is

$$\begin{bmatrix} \dot{x} \\ \dot{x}_c \end{bmatrix} = \begin{bmatrix} f\Big(x, g_c(x_c, r), v\Big) \\ f_c\Big(x_c, g(x, v), r, n\Big) \end{bmatrix} \qquad (12.46)$$

In more complicated situations, we may be able to obtain only a generalized state-space description rather than an ordinary description.

Notes and Bibliography

The procedure in Section 12.3 for obtaining a state-space model for an electrical circuit is described in more detail in [1]. This imposing classic contains a great deal of useful material, including a comprehensive treatment of numerical algorithms for solving state equations. A more accessible reference for computer methods in circuit analysis is [2].

The approach to state-space averaging taken in [3] for high-frequency switching or PWM converters has been widely applied. Essentially the same idea is evident in [4] and [5]. (The approximate sampled-data model in Example 12.4 is also in [5].) State-space averaging for a broader class of converters that includes quasi-resonant converters is elegantly developed in [6].

The relevance to high-frequency switching converters of the classical averaging techniques developed by applied mathematicians in the 1940s is nicely demonstrated in [7]. This approach yields a sequence of approximations of increasing orders of accuracy, typically starting with the usual state-space averaged model. Apart from the asymptotic results in [7], the state space formulation has not yet yielded useful descriptions of the error incurred by averaging.

Much has been written on continuous-time generalized state-space systems, or differential/algebraic equations (DAEs). In discrete time, only the LTI case has received attention. However, the nonlinear, discrete-time case is what arises most commonly in power electronics. We defer references on generalized state-space models, current-mode control, and resonant converters to Chapters 13 and 14, except to note that state-plane analysis of resonant converters is treated in [8].

1. L. O. Chua and P.-M. Lin, *Computer-Aided Analysis of Electronic Circuits: Algorithms and Computational Techniques* (Englewood Cliffs, New Jersey: Prentice-Hall, 1975).
2. J. Vlach and K. Singhal, *Computer Methods for Circuit Analysis and Design* (New York: Van Nostrand, 1983).
3. R. D. Middlebrook and S. Ćuk, "A General Unified Approach to Modeling Switching Converter Power Stages," in *IEEE Power Electronics Specialists Conference (PESC)* 18–34 (Cleveland, June 1976), also *Intl. J. of Electronics* 42: 521–550 (June 1977).
4. R. W. Brockett and J. R. Wood, "Electrical Networks Containing Controlled Switches," in *IEEE Symposium on Circuit Theory* Addendum: 1–11 (April 1974).
5. H. A. Owen, A. Capel and J. G. Ferrante, "Simulation and Analysis Methods for Sampled Power Electronic Systems," in *IEEE Power Electronics Specialists Conference (PESC)* (Cleveland, June 1976), 44–55.

6. A. F. Witulski and R. W. Erickson, "Extension of State-Space Averaging to Resonant Switches and Beyond," *IEEE Trans. Power Electronics* 5:98–109 (January 1990).

7. P. T. Krein, J. Bentsman, R. M. Bass and B. L. Lesieutre, "On the Use of Averaging for the Analysis of Power Electronic Systems," *IEEE Trans. Power Electronics* 5:182–190 (April 1990).

8. R. Oruganti and F. C. Lee, "Resonant Power Processors, Part I—State Plane Analysis," *IEEE Trans. Industry Applications* 21:1453–1460 (November/December 1985).

Problems

12.1 The circuit of Fig. 12.12 is sometimes used to study the effect of an input filter on the dynamics of a controlled converter. The R and L represent, respectively, the combination of internal resistance and inductance of the source and externally added resistance and inductance in the filter. The block marked P is a nonlinear resistor that absorbs constant power P and is used as a simplified model for the controlled converter. This choice reflects the fact that a well-regulated, high-efficiency converter with a constant load absorbs essentially constant power from its input source. The purpose of the input filter is to reduce undesired current harmonics in the source.

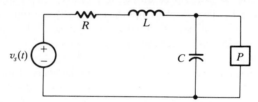

Figure 12.12   Model for a regulated converter with an input filter, analyzed in Problem 12.1.

(a) Obtain a state-space model of the circuit. Show that you can determine all the voltages and currents in the circuit at time t from knowledge of the state variables and source value at t.

(b) Use the state-space description in (a) to obtain a single second-order differential equation that describes the system. Explain how you would determine all voltages and currents in the circuit from knowledge of the solution of this equation.

(c) Suppose the capacitor has some equivalent series resistance (ESR) R_C. Obtain a continuous-time generalized state-space description for the resulting circuit. Can you obtain an ordinary state-space model from it?

(d) Suppose that the inductor is modeled as nonlinear, with its flux linkage λ_L being some nonlinear function of its current i_L, say, $\lambda_L = \ell(i_L)$. Find a state-space model of the circuit in the figure (assuming no ESR).

12.2 In Example 12.1, we obtained a state-space model *(12.11)* for an up/down converter in continuous conduction. Complete the model by deriving an equation for the output voltage v_o as a function of the state variables and inputs. Then repeat the derivation of state and output equations for at least the following switching converters described in Chapters 6 and 7: down (or buck), up (or boost), flyback, and forward converters. Include the ESR of the capacitor in your model.

12.3 This problem deals with state-space averaged models for the up/down converter and the converters in Problem 12.2.

    (a) Verify that the state-space averaged model for the up/down converter in *(12.13)*, with $R_C = 0$ (no ESR), describes the averaged circuit in Fig. 11.11 (Example 11.7).

    (b) For $R_C \neq 0$ in the up/down converter, find a circuit representation of the state-space averaged description *(12.13)* by modifying the averaged circuit in Fig. 11.11. Also express the average output voltage v_o as a function of the state variables and inputs of the averaged model. You can do this from the averaged-circuit representation or by averaging your expression for v_o in Problem 12.2.

    (c) Find state-space averaged models for the converters in Problem 12.2. Repeat (a) and (b) for these converters, with the necessary modifications.

    (d) Repeat your derivation of averaged models for all the preceding converters when the switches are no longer modeled as ideal. Specifically, instead of representing a conducting switch as a short circuit, model it as a (small) resistor in series with a (small) dc voltage source.

12.4 This problem compares some numerical properties of the forward and backward Euler algorithms by using them to solve (or simulate) a particular state-space model. Part (a) is devoted to generating a model that is simple enough to be transparent, yet sufficiently complicated to be interesting.

    (a) Suppose that $\dot{w}_1(t) = -10^2 w_1(t)$ and $\dot{w}_2(t) = -w_2(t)$. Show that the variables $x_1(t) = w_1(t) + w_2(t)$ and $x_2(t) = w_1(t) - w_2(t)$ are governed by an LTI state-space description. Find an analytical solution for $x_1(t)$ and $x_2(t)$ in terms of the initial conditions $x_1(0)$ and $x_2(0)$.

    (b) Suppose that you use the forward Euler method to compute a numerical solution for the state-space model in (a), using a step size ϵ. Let $x_1[k]$ and $x_2[k]$ denote the numerically computed values of $x_1(t)$ and $x_2(t)$, respectively, at time $t = k\epsilon$. Show that these variables are governed by a discrete-time LTI state-space equation.

    (c) Compare the analytical solution in (a) for $0 \leq t \leq 0.5$ with the numerically computed solutions using the algorithm in (b). Choose step sizes of 10^{-3}, 10^{-2}, and 10^{-1}, taking $x_1(0) = 3$ and $x_2(0) = 1$ in each case. Also examine the effect of using a step size of 10^{-3} for the first 30 steps and a step size of 2.1×10^{-2} thereafter.

    (d) Repeat (b) and (c) with the backward Euler algorithm instead of the forward.

12.5 Figure 11.7(b), Example 11.5, shows the averaged output circuit for an up/down converter in discontinuous conduction under duty-ratio control. Find state-space descriptions for the circuit, first using $\bar{v}_o$ as the state variable and then using $\bar{v}_o^2$ as the state variable. Can you see why using $\bar{v}_o^2$ may be preferable? How do your descriptions change if the circuit is operating under current-mode control?

12.6 This problem deals with the resonant converter in Examples 12.3 and 12.6.

    (a) Obtain a state-space model for each of the four configurations encountered during a cycle of operation; use the scaled quantities $x_1 = \sqrt{L}i_L$ and $x_2 = \sqrt{C}v_C$ as state variables.

    (b) Verify that the state-plane trajectories in each configuration are indeed circular arcs and that their centers are as claimed in Example 12.6.

(c) Use the simple geometry of the state-plane trajectories to obtain a closed-form solution for the steady-state radii of the circular arcs in (b).

(d) Verify the properties of the "rotation matrix" that were used in Example 12.6.

(e) Even though the resonant LC pair actually sees square-wave voltages presented by the input and load sources, replace the square waves by their fundamental (switching frequency) components. Also take the inductor current to be essentially a sinusoid at the switching frequency. Now determine the steady-state behavior of the resulting approximate circuit. Compare your result with that in (c).

(f) Verify that the analysis in (e) corresponds to a steady-state analysis of the ω_S-component defined in Section 11.3.5, where ω_S is the switching frequency. Try and deduce a model for the nonsteady-state behavior of the ω_S-component in this case.

12.7 Our analytical description of the operation of the resonant converter in Example 12.6 only went through one half-cycle of operation. Using the symmetry of the circuit and its operation, extend the description to the full cycle. Write the result in the standard generalized state-space form of *(12.35)*.

12.8 Obtain an output equation of the form *(12.16)* for the sampled-data model in *(12.34)*, Example 12.4, with the output $y[k]$ being the *average* value of $v_o(t)$ in the kth cycle. (You can make approximations similar to those in Example 12.4.)

12.9 Obtain an approximate sampled-data model for an up/down converter in discontinuous conduction. Do so for (i) duty-ratio control and (ii) current-mode control.

12.10 Obtain approximate sampled-data models for the converters in Problem 12.2 for (a) continuous conduction and (b) discontinuous conduction. Again, do so for (i) duty-ratio control and (ii) current-mode control.

12.11 We made extensive use in Chapter 11 of a phase-controlled rectifier drive for a dc motor. Let $i_d[k]$ denote the armature current at the beginning of the kth cycle and α_k be the firing angle in this cycle. Find a sampled-data description for the open-loop drive.

Chapter 13

Linear and Piecewise Linear Models

IN Chapter 11 we introduced the process of linearization, for nonlinear averaged-circuit models. This process allowed us to obtain linear, time invariant (LTI) circuit models for small perturbations of average values from constant nominal operating conditions. These LTI models then served as the basis for stability evaluation and control design in the examples considered in Chapter 11.

The results of Chapter 12 have expanded our modeling options to include state-space models. These models can describe continuous-time waveforms, their averages, or their discrete-time samples in open- or closed-loop systems comprising power circuits and controllers. In this chapter we show how to linearize continuous-time and discrete-time models that are given in state-space form rather than circuit form. We also develop some of the key concepts and results for the analysis of LTI state-space models. We then examine applications of these results for LTI models to analyzing piecewise LTI models and evaluating the stability of cyclic nominal operation in power circuits. We defer applications to control design to Chapter 14.

13.1 LINEARIZATION

Power circuit models are typically nonlinear and therefore difficult to work with. However, the process of linearization introduced in Chapter 11 allows us to obtain linear models that approximately govern small deviations from some nominal solution. These linearized models, also called small-signal models, are far more tractable than nonlinear models.

Linearized models are valuable for several tasks of analysis and design. In stability evaluation, we want to know whether a circuit can return to nominal operation after small perturbations away from it. Stability of the linearized model implies that nominal operation will not be destroyed by sufficiently small disturbances; instability of the linearized model implies that the nominal operating condition cannot be maintained without further control action. (Stability of the linearized model does not, of course, tell us about recovery from large disturbances.)

In controller design, we start by examining behavior under normal operating conditions, when the power circuit deviates only slightly from nominal. Models for small deviations are evidently central to this task. Linearized models are also useful in iterative numerical determination of steady-state behavior, because they allow us to predict the effect of small corrections to computed estimates at each stage of the iteration.

We showed in Chapter 11 that linearization of a circuit model is easy. All that is required is to replace the voltages and currents in the circuit model by their deviations from nominal and to replace each nonlinear component in the circuit by its linearized version. Just as simple is linearizing models that are represented by block diagrams, such as the state-space models in Figs. 12.1 and 12.7. Again, the only requirements are to replace each input and output variable in the block diagram by its deviation from nominal and to replace each block by its linearized version. The linearized version of any block represents the linear constraints that approximately govern small deviations of its inputs and outputs. These constraints are obtained from first-order Taylor series approximations of the nonlinear characteristics of the block.

In Section 13.2 we describe the details of the linearization process for continuous-time state-space models and in Section 13.5 treat the discrete-time case. The results in Section 13.5 are actually given for generalized state-space descriptions, which are more important for power circuits. The results for discrete-time systems in ordinary state-space form, as in the block diagram of Fig. 12.7, follow easily.

13.2 LINEARIZING CONTINUOUS-TIME MODELS

Here we show how to linearize the continuous-time state-space model given in *(12.18)* and repeated here for convenience:

$$\frac{dx}{dt} = f\Big(x(t), u(t), t\Big)$$
$$y(t) = g\Big(x(t), u(t), t\Big) \tag{13.1}$$

We use capital letters to denote the nominal values of the variables in *(13.1)*, so that in the nominal operating condition $x(t) = X(t)$, $u(t) = U(t)$, and $y(t) = Y(t)$. The linearization process is not predicated on these nominal values being constant or periodic; they can be any values that together satisfy *(13.1)*. However, the most important cases for our purposes are when the nominal solutions are constant or periodic.

Let's now consider deviations from nominal and mark these quantities with a $\sim$, so that away from the nominal solution we have

$$x = X + \tilde{x} \qquad u = U + \tilde{u} \quad \text{and} \quad y = Y + \tilde{y} \tag{13.2}$$

Substituting the relationships from *(13.2)* into *(13.1)* and expanding the resulting nonlinear terms in multivariable Taylor series around the nominal solution, we find that, to a first-order approximation:

$$\frac{dX}{dt} + \frac{d\tilde{x}}{dt} \approx f\Big(X(t), U(t), t\Big) + \frac{\partial f}{\partial x}\tilde{x}(t) + \frac{\partial f}{\partial u}\tilde{u}(t)$$

$$Y(t) + \tilde{y}(t) \approx g\Big(X(t), U(t), t\Big) + \frac{\partial g}{\partial x}\tilde{x}(t) + \frac{\partial g}{\partial u}\tilde{u}(t)$$

(13.3)

We drop the higher order terms in the perturbations $\tilde{x}$, $\tilde{u}$, and $\tilde{y}$, assuming that the perturbations are sufficiently small for these terms to be negligible in comparison with the first-order terms. Now using the fact that the nominal solution itself satisfies *(13.1)*, we obtain the desired linearized model:

$$\frac{d\tilde{x}}{dt} \approx \frac{\partial f}{\partial x}\tilde{x}(t) + \frac{\partial f}{\partial u}\tilde{u}(t)$$

$$\tilde{y}(t) \approx \frac{\partial g}{\partial x}\tilde{x}(t) + \frac{\partial g}{\partial u}\tilde{u}(t)$$

(13.4)

The notation in *(13.3)* and *(13.4)*, although designed to be simple and suggestive, needs definition and interpretation. The idea of a partial derivative of a vector function such as $f(x, u, t)$ with respect to a vector argument such as x may not be familiar. The symbol $\partial f/\partial x$ denotes a *matrix* (termed the *Jacobian* matrix), whose entry in the ith row and jth column is the partial derivative of the ith component of $f(\cdot)$ with respect to the jth component of x, namely, $\partial f_i/\partial x_j$. If, for instance, $f(x, u, t) = M(u, t)x$ for some matrix $M(u, t)$ that does not depend on x, we can easily show that $\partial f/\partial x = M(u, t)$. A similar definition holds for the other partial derivatives in *(13.3)* and *(13.4)*. In general, these partial derivatives are all functions of x, u, and t but must be evaluated at the nominal solution, with $x = X$ and $u = U$, when they are used in *(13.3)* and *(13.4)*.

Note that the linearized model in *(13.4)* is itself in state-space form, with the deviations $\tilde{x}$, $\tilde{u}$, and $\tilde{y}$ now constituting the state variables, inputs, and outputs, respectively. We can easily represent the model by a block diagram of the form in Fig. 12.1. The block diagram of the linearized model is in fact a direct linearization of the one in Fig. 12.1. The major advantage of *(13.4)* over *(13.1)* is, of course, that *(13.4)* is linear, because the partial derivatives that make up the coefficient matrices are functions only of the known nominal quantities X and U and do not depend on $\tilde{x}$ or $\tilde{u}$. We can write *(13.4)* in the form of *(12.22)–(12.24)*, namely:

$$\frac{d\tilde{x}}{dt} \approx A(t)\tilde{x}(t) + B(t)\tilde{u}(t)$$

$$\tilde{y}(t) \approx E(t)\tilde{x}(t) + F(t)\tilde{u}(t)$$

(13.5)

where the coefficient matrices are the partial derivative matrices in *(13.4)*.

If the original nonlinear model is time invariant, so that $f(\cdot)$ and $g(\cdot)$ are not explicitly dependent on time, and if the nominal solution is constant, then all the partial derivatives that define the linearized model are constant. The linearized model therefore is LTI in this case. If the nominal solution is periodically varying, the linearized model will also be periodically varying rather than LTI, even if the underlying nonlinear model is time invariant.

EXAMPLE 13.1

Linearized Switched and Averaged Models for an Up/Down Converter

Let's return to the switched and averaged models of an up/down converter in *(12.11)* and *(12.13)* of Examples 12.1 and 12.2, respectively. These two models have the same structure: Replacing all the instantaneous variables of the switched model by their local averages produces the averaged model and vice versa. We therefore focus first on linearization of the averaged model and then make a few remarks about the switched case.

We can conveniently rewrite the averaged model *(12.13)* in the matrix notation introduced in *(12.30)* of Example 12.4. With capacitor ESR neglected, so $R_C = 0$, we write

$$
\begin{aligned}
\frac{dx(t)}{dt} &\approx \begin{bmatrix} 0 & d'(t)/L \\ -d'(t)/C & -1/RC \end{bmatrix} x(t) + \begin{bmatrix} d(t)/L \\ 0 \end{bmatrix} \overline{v}_{\text{in}}(t) \\
&= A_{d(t)} x(t) + B_{d(t)} \overline{v}_{\text{in}}(t) \\
&= f\Big(x(t), \overline{v}_{\text{in}}(t), d(t) \Big)
\end{aligned}
\tag{13.6}
$$

where the components of $x(t)$ now denote averaged quantities: $x_1(t) = \overline{i}_L(t)$ and $x_2(t) = \overline{v}_C(t)$. If $R_C \neq 0$, we can still represent the averaged model in the form *(13.6)*, except that the particular entries of $A_{d(t)}$ and $B_{d(t)}$ are different. Furthermore, there are similar representations for other switching converters. We leave pursuit of these other cases to you.

With a constant duty ratio and a constant supply voltage, we have $d(t) = D$ and $\overline{v}_{\text{in}}(t) = V_{\text{in}}$. We find the corresponding constant nominal steady-state solution $x(t) = X$ by setting the derivative in *(13.6)* to zero and solving for X. This calculation yields

$$
X = \begin{bmatrix} I_L \\ V_C \end{bmatrix} = -A_D^{-1} B_D V_{\text{in}} = \begin{bmatrix} D/RD'^2 \\ -D/D' \end{bmatrix} V_{\text{in}}
\tag{13.7}
$$

These are the same formulas we obtained in *(11.7)* of Example 11.7 by direct analysis of the averaged circuit.

To compute the linearized model, we use *(13.4)*, with $f(\cdot)$ as in *(13.6)* and with $\sim$ now denoting deviations of averaged quantities from their nominal values. The linearized model then takes the form:

$$
\begin{aligned}
\frac{d\tilde{x}}{dt} &\approx \frac{\partial f}{\partial x} \tilde{x}(t) + \frac{\partial f}{\partial \overline{v}_{\text{in}}} \tilde{v}_{\text{in}}(t) + \frac{\partial f}{\partial d} \tilde{d}(t) \\
&= A_D \tilde{x}(t) + B_D \tilde{v}_{\text{in}}(t) + J\tilde{d}(t)
\end{aligned}
\tag{13.8}
$$

where the vector J is given by:

$$
J = (A_1 - A_0)X + (B_1 - B_0)V_{\text{in}}
\tag{13.9}
$$

(To derive *(13.9)*, you should note, as we did in *(12.33)*, that $A_d = dA_1 + d'A_0$ and $B_d = dB_1 + d'B_0$.) We leave to you verification that the linearized state-space description in *(13.8)* actually describes the linearized circuit obtained in Fig. 11.14 of Example 11.9.

Linearization of the switched model *(12.11)* results in a description like *(13.8)*, except that instantaneous quantities instead of averaged quantities appear. The nominal duty ratio D is replaced by the periodic nominal switching function $q(t) = Q(t)$; the nominal solution $X(t)$ in this case is periodically varying; $\tilde{d}(t)$ is replaced by $\tilde{q}(t)$; $\tilde{v}_{\text{in}}$ now denotes the deviation of the instantaneous rather than averaged supply voltage; and the entries of $\tilde{x}(t)$ now denote perturbations of the instantaneous waveforms from their periodic steady states. Figure 13.1 schematically represents the differences between the switched and averaged models. (The fact that $\tilde{q}$ takes only the values 0 and 1 introduces a subtle twist to the linearization argument. A "small" $\tilde{q}$ is one that has small *area* rather than small magnitude. We leave it to you to verify that the Taylor series linearization is still valid.)

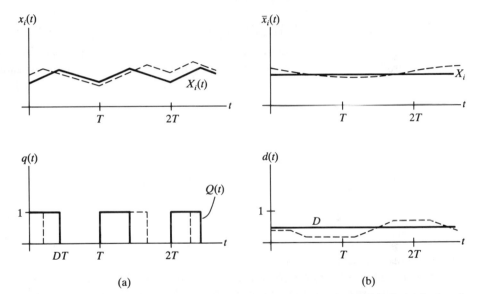

Figure 13.1 Representation of differences between the linearizations of switched and averaged models: (a) switched model; (b) averaged model.

13.3 ANALYSIS OF CONTINUOUS-TIME LTI MODELS

We have shown how to develop and apply continuous-time LTI models in several ways in power electronics. We use them to describe a fixed topological state of a power circuit model that is made up of LTI components and ideal switches. We obtain them when we linearize a nonlinear, time invariant, continuous-time model at a constant nominal operating point, as with the averaged model in Example 13.1. We also use them to describe some of the most common controllers or compensators, such as the PI controller for the phase-controlled drive in Examples 11.10 and 11.12.

In this section we develop some of the basic tools and results used to analyze continuous-time LTI models. The results are central to stability evaluation and control design. They also permit us, in Section 13.4, to describe the general solution of piecewise LTI models for power circuits, from which we can obtain useful sampled-data models.

Our starting point is an LTI model of the form *(12.25)* and *(12.26)*, repeated here for convenience:

$$\frac{dx}{dt} = Ax(t) + Bu(t)$$

$$y(t) = Ex(t) + Fu(t)$$

(13.10)

The results in this section can, of course, also be applied to a linearized model of the form *(13.5)* when its coefficient matrices are constant.

13.3.1 Transform-Domain Solution

The easiest way to solve *(13.10)* for specific initial conditions and inputs is to use Laplace transforms to convert the differential equations into algebraic equations. This procedure corresponds to using impedance methods for an LTI circuit. To avoid excessively cluttered notation, we denote the Laplace transform of a time function $h(t)$ simply by $h(s)$. (We have already noted this particular abuse of notation in Chapter 11.) It is not likely to cause confusion so long as we reserve the argument s for Laplace transforms and always display this argument. When more careful notation is called for, we write $\mathcal{L}\{h(t)\}$ instead of $h(s)$.

Recall the definition of the (one-sided) Laplace transform:

$$h(s) = \mathcal{L}\{h(t)\} = \int_0^\infty h(t)e^{-st}dt$$

(13.11)

Let's define the Laplace transform of a vector function of time $x(t)$ to be its component-wise Laplace transform, so that the ith entry of $x(s)$ is $x_i(s)$. With this definition, we can easily show that the following key property of the Laplace transform carries over from the scalar case to the vector case:

$$\mathcal{L}\left\{\frac{dx(t)}{dt}\right\} = sx(s) - x(0)$$

(13.12)

where $x(0)$ is the value of the *time* function $x(t)$ at $t = 0$. This result is usually referred to as the differentiation theorem for Laplace transforms and is the key to converting differential equations in the time domain to algebraic equations in the transform domain (or "frequency" domain). Many other results from the scalar case also have easy vector extensions. In particular, if A is a constant matrix, then $\mathcal{L}\{Ax(t)\} = Ax(s)$.

Taking transforms of the expressions in *(13.10)* and using the properties just discussed, we can write

$$sx(s) - x(0) = Ax(s) + Bu(s)$$

(13.13)

and

$$y(s) = Ex(s) + Fu(s) \qquad (13.14)$$

Our aim is to solve for $x(s)$ from *(13.13)*, given the initial condition $x(0)$ and the Laplace transform $u(s)$ of the input. We can then use the resulting expression for $x(s)$ in *(13.14)* to solve for $y(s)$. Gathering terms involving $x(s)$ to one side of *(13.13)*, we get

$$sx(s) - Ax(s) = (sI - A)x(s) = x(0) + Bu(s) \qquad (13.15)$$

(We introduced the identity matrix I in *(13.15)* to make the grouping $(sI - A)x(s)$ dimensionally correct. It is tempting to write $(s - A)x(s)$, but $(s - A)$ is not defined because s is a scalar and A is an $n \times n$ matrix.)

Solving *(13.15)* for $x(s)$ requires inversion of $(sI - A)$. The inverse exists if and only if the determinant of this matrix is not identically zero. Now, $\det(sI - A)$ is always an nth-degree polynomial in s, so:

$$\det(sI - A) = s^n + a_{n-1}s^{n-1} + \cdots + a_0 = a(s) \qquad (13.16)$$

for some coefficients $\{a_j\}$ determined by A. This polynomial $a(s)$ is termed the *characteristic polynomial* of A or of the system *(13.10)*, and its roots $\lambda_1, \ldots, \lambda_n$ are called the *characteristic roots*, or the *eigenvalues*, of A. Because these roots have units of reciprocal time, they are also referred to as *characteristic frequencies*, or *natural frequencies*, of the system *(13.10)*. The characteristic polynomial is bound to appear in any framework for analyzing an LTI system. In the state-space approach, the characteristic polynomial makes its appearance as $\det(sI - A)$.

As the characteristic polynomial is not identically zero, the inverse of $(sI - A)$ exists. The entries of the inverse are rationals in s, with the nth-degree characteristic polynomial $a(s)$ appearing as the denominator of every term and with every numerator being of degree less than n. (If a numerator root happens to coincide with a characteristic root, this root cancels out of both the numerator and denominator, yielding a denominator of reduced degree. We can show, however, that every characteristic root will be represented in the denominator of at least one entry of $(sI - A)^{-1}$.)

We now obtain the solution $x(s)$ of *(13.15)* by premultiplying that equation by $(sI - A)^{-1}$ to get

$$x(s) = (sI - A)^{-1}x(0) + (sI - A)^{-1}Bu(s) \qquad (13.17)$$

The first term constitutes the response to initial conditions alone, in the absence of external inputs, and is called the *natural response* or *free response* or *zero-input response* of the system. The second term, similarly, is the *zero-state response*, which we also refer to as the *forced response*. The full solution is thus the superposition of these two components.

To complete the solution of the LTI system, we substitute for $x(s)$ in the output equation *(13.14)* to obtain

$$y(s) = E(sI - A)^{-1}[x(0) + Bu(s)] + Fu(s) \qquad (13.18)$$

Transfer Functions With zero initial conditions, that is, $x(0) = 0$, we have

$$y(s) = [E(sI - A)^{-1}B + F]u(s) = G(s)u(s) \qquad (13.19)$$

where $G(s)$ denotes the $p \times m$ matrix in brackets in *(13.19)*. This matrix is the *transfer function* matrix of the system from the input u to the output y. Its entry in the ith row and jth column gives the transfer function from the jth component of the input to the ith component of the output.

The way in which $(sI - A)^{-1}$ enters the expression for $G(s)$ shows that each entry of the transfer function matrix will also be rational in s, with a denominator equal to the characteristic polynomial $a(s)$. (Again, if a root of the numerator in some entry coincides with a characteristic root, the resulting pole–zero cancellation leads to a denominator of reduced degree in that entry. We can show that, under so-called *controllability* and *observability* conditions, every characteristic root is represented in the denominator of at least one entry of $G(s)$. However, such refinements are beyond the scope of this book.) The numerator degree equals the denominator degree if the corresponding entry of F is nonzero and otherwise is less than the denominator degree. Also of interest for many computations is the transfer function matrix from u to x, which *(13.17)* shows to be $(sI - A)^{-1}B$.

Of particular significance in control design using LTI models is the *frequency response* of the system. It is simply the transfer function evaluated along the imaginary axis, that is, $G(s)$ evaluated for $s = j\omega$, $j = \sqrt{-1}$. We remind you shortly of the role of the frequency response in describing input–output behavior.

EXAMPLE 13.2

Transfer Functions for the Linearized Averaged Model of an Up/Down Converter

We return now to the linearized averaged model in *(13.8)* and examine it in the transform domain. Taking Laplace transforms and solving for $\tilde{x}(s)$ yields

$$\tilde{x}(s) = (sI - A_D)^{-1}[\tilde{x}(0) + B_D\tilde{v}_{\text{in}}(s) + J\tilde{d}(s)] \qquad (13.20)$$

The transfer function matrix between, for example, $\tilde{d}(s)$ and $\tilde{x}(s)$ can be obtained by setting $\tilde{x}(0) = 0$ and $\tilde{v}_{\text{in}}(s) = 0$:

$$\tilde{x}(s) = \begin{bmatrix} \tilde{\imath}_L(s) \\ \tilde{v}_C(s) \end{bmatrix} = (sI - A_D)^{-1}J\tilde{d}(s) \qquad (13.21)$$

For the case where the capacitor ESR is zero ($R_C = 0$):

$$\tilde{x}(s) = \begin{bmatrix} s & -D'/L \\ D'/C & s+(1/RC) \end{bmatrix}^{-1} \begin{bmatrix} (V_{\text{in}} - V_C)/L \\ I_L/C \end{bmatrix} \tilde{d}(s)$$

$$= \frac{1}{a_D(s)} \begin{bmatrix} s+(1/RC) & D'/L \\ -D'/C & s \end{bmatrix} \begin{bmatrix} (V_{\text{in}} - V_C)/L \\ I_L/C \end{bmatrix} \tilde{d}(s)$$

(13.22)

where

$$a_D(s) = s^2 + \frac{1}{RC}s + \frac{D'^2}{LC}$$

(13.23)

The transfer function from the controlling input (duty-ratio perturbations $\tilde{d}$) to the controlled output (perturbations $\tilde{v}_o$ in the output voltage) is of special interest in the design of a feedback controller and may be derived from (13.21). When $R_C = 0$, $v_o = v_C$, so the desired transfer function can be read from (13.22). Using (13.7) to simplify the result, we find

$$\tilde{v}_o(s) = \tilde{v}_C(s) = \left(\frac{I_L}{C}\right) \frac{s - (V_{\text{in}}/LI_L)}{s^2 + (1/RC)s + (D'^2/LC)} \tilde{d}(s)$$

(13.24)

which is exactly what we obtained from the linearized averaged circuit of Example 11.9. Figure 13.2(a) shows the locations of the poles and zero of this transfer function $g(s)$ for the parameter values in Example 11.1 and Fig. 13.2(b) a plot of its frequency response. Although the frequency response is shown up to one half the switching frequency ($\pi/T = 1.57 \times 10^5$ rad/s), you should keep in mind that the predictions of the averaged model become progressively less accurate as this frequency is approached.

Using (13.20) to compute the transfer function from $\tilde{v}_{\text{in}}(s)$ to the output leads similarly to the transfer function already obtained from the averaged circuit in Example 11.7. The polynomial appearing in the denominators of these transfer functions is indeed the characteristic polynomial of the matrix A_D in (13.8), as expected.

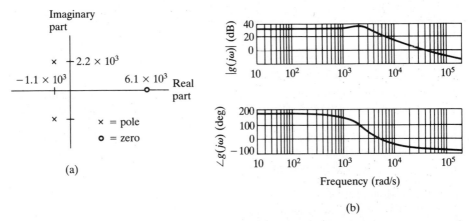

Figure 13.2 (a) Location of poles (**x**) and zero (**o**) of the transfer function $g(s)$ from perturbations in duty ratio to perturbations in output voltage. (b) Frequency response corresponding to this transfer function.

The results derived in Example 13.2 can also be obtained rather directly from the averaged circuits developed in Section 11.4, which provide a useful check on the computations. The example is an important way of building confidence in matrix computations and their results, for those of you who are unused to working with matrices. This confidence becomes important when you confront situations in which there is no direct, circuit-based route to the desired results.

13.3.2 Time-Domain Solution

To obtain the time-domain version of the transform-domain expression for $\tilde{v}_C$ in Example 13.2, or more generally for the vectors x or y in (13.17), (13.18), or (13.19), we must compute inverse Laplace transforms. We can do so easily when each entry of the vector whose inverse transform we want is rational in s. We compute a partial fraction expansion of each entry to obtain a sum of elementary terms whose inverse transforms we can write from inspection.

We begin this section with the time-domain response in two important special cases: (1) the natural response to initial conditions; and (2) the response to sinusoidal inputs. The first allows us to identify the condition for asymptotic stability, and the second brings out the role of the frequency response. We then introduce the so-called matrix exponential, which allows us to write the general solution of an LTI system.

Natural Response   The transform-domain expression for the natural response, from (13.17), is $(sI - A)^{-1}x(0)$. We start with an example, to remind you of what is involved in computing the inverse transform of an expression such as this.

EXAMPLE 13.3
Natural Response of the Linearized Averaged
Model of an Up/Down Converter

Let's continue with the circuit of Example 13.2. Now, suppose that we want to determine the output response $\tilde{v}_o(t) = \tilde{v}_C(t)$, assuming an initial deviation $\tilde{x}(0)$ from the nominal steady state but with $\tilde{d} = 0$ and $\tilde{v}_{in} = 0$. This formulation could be used, for example, to analyze how the circuit approaches the steady state corresponding to a new operating condition, if d and $\overline{v}_{in}$ have been set at their new values but $x(0)$ is still at a value determined by the old operating condition. The open-loop response of the circuit in Example 11.1 to a step in the source voltage could have been analyzed this way, rather than by the approach outlined in Example 11.7.

From (13.20), we determine that:

$$\tilde{v}_C(s) = \frac{1}{a_D(s)}\left[\tilde{v}_C(0)s - \tilde{\imath}_L(0)\frac{D'}{C}\right] \tag{13.25}$$

The characteristic polynomial $a_D(s)$ is given in *(13.23)*. We can rewrite it as:

$$a_D(s) = \left(s + \frac{1}{2RC} \right)^2 + \omega_D^2 \quad \text{and} \quad \omega_D = \sqrt{\frac{D'^2}{LC} - \frac{1}{4R^2C^2}} \qquad (13.26)$$

which identifies its roots as $\lambda_1 = -(1/2RC) + j\omega_D$, $\lambda_2 = \lambda_1^*$. Now, rewriting *(13.25)* as:

$$\tilde{v}_C(s) = \frac{1}{a_D(s)} \left[\tilde{v}_C(0) \left(s + \frac{1}{2RC} \right) - \frac{1}{\omega_D} \left(\tilde{i}_L(0) \frac{D'}{C} + \tilde{v}_C(0) \frac{1}{2RC} \right) \omega_D \right] \qquad (13.27)$$

allows us, using elementary Laplace transform results, to conclude that:

$$\tilde{v}_C(t) = e^{-t/2RC} \left\{ \tilde{v}_C(0) \cos(\omega_D t) - \frac{1}{\omega_D} \left[\tilde{i}_L(0) \frac{D'}{C} + \tilde{v}_C(0) \frac{1}{2RC} \right] \sin(\omega_D t) \right\} \qquad (13.28)$$

The solution is thus a sinusoidal oscillation at a frequency determined by the imaginary part of the characteristic roots but exponentially damped at a rate determined by the real part of the characteristic roots. The larger the imaginary part, the higher the frequency of oscillation will be; the more negative the real part, the greater the damping will be.

As suggested at the beginning of this example, we can use *(13.28)* to explain the waveforms in Fig. 11.3 of Example 11.1, where we obtained the open-loop response to a step in the source voltage. We fix the time origin to be the instant of the step and choose the initial conditions for the perturbed state variables in *(13.28)* to be the difference between their *actual* values before the step and their *nominal* steady-state values after the step. In other words, we set $\tilde{x}(0) = x(0) - X_{\text{new}}$, where we use *(13.7)* to compute the new steady-state vector X_{new}, utilizing the values that apply after the step. Also, the values of D and ω_D in *(13.28)* must be those after the step.

Figure 13.3 shows the result of computing $\tilde{v}_C(t)$ this way for the open-loop system with and without feedforward. In each case, we assumed that the circuit was in steady state prior to the step, so that $x(0) = X_{\text{old}}$. For example, in the case with feedforward, we

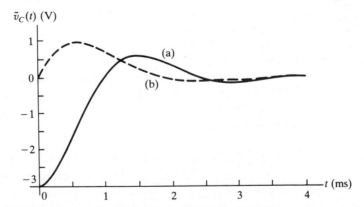

Figure 13.3 Output of the linearized model for two different initial conditions, representing the open-loop response of the up/down converter to a step in source voltage: (a) without feedforward; (b) with feedforward.

have $\tilde{v}_C(0) = 0$ and $\tilde{i}_L(0) = I_{L.\text{old}} - I_{L.\text{new}} = (-V_{\text{ref}}/R)[(1/D'_{\text{old}}) - (1/D'_{\text{new}})]$, where D_{new} is the duty ratio needed to maintain the output at -9 V when the input drops to 8 V, so $D_{\text{new}} = 9/(9 + 8) = 0.53$. Also, D'_{new} is used in *(13.28)* to compute ω_D. The correspondence with the waveforms in Fig. 11.3 should be evident.

Since the perturbations in Fig. 13.3 are not small, you might expect some discrepancies between the predictions of the nonlinear model and the linearized model. However, we already know from Examples 11.7 and 12.2 that the averaged model is linear when the duty ratio is fixed, so the transients in this particular example actually coincide with those predicted by the nonlinear model.

We can generalize the approach used in Example 13.3 and obtain a time-domain expression for the natural response of the state vector in any LTI system. As the transform-domain expression for the natural response is $(sI - A)^{-1}x(0)$, a useful first step in finding the corresponding time function is to compute the inverse Laplace transform of $(sI - A)^{-1}$, that is, to determine the $n \times n$ matrix whose entries are the inverse Laplace transforms of the corresponding entries of $(sI - A)^{-1}$.

Computation of the inverse transform of an entry of $(sI - A)^{-1}$ is as straightforward as the computations in Example 13.3. Because the denominator of each entry is $a(s) = \det(sI - A)$, inverse transformation yields a sum of exponentials and exponentially weighted sinusoids, corresponding, respectively, to the real and complex characteristic roots $\{\lambda_i\}$ of the system. The exponential $e^{\lambda_i t}$ decays or grows, depending on whether λ_i is in the left or right half of the complex plane, respectively. (The way that the characteristic roots enter the description of the time-domain solution gives rise to yet another name for them—the characteristic *exponents*.)

The inverse transform of $(sI - A)^{-1}x(0)$, namely, the natural response, is therefore also a combination of these same exponentials and exponentially weighted sinusoids. The precise combination of terms in the natural response is determined by the initial condition vector $x(0)$.

Asymptotic Stability From the discussion so far, you can see that the natural response decays asymptotically to zero if and only if all the natural frequencies are strictly in the left half of the complex plane. We term the system *asymptotically stable* under this condition, or *exponentially stable* because the asymptotic decay is bounded by some decaying exponential.

Frequency Response The frequency response $G(j\omega)$ of an LTI system with transfer function $G(s)$ determines how the system responds to sinusoidal inputs or, more generally, to combinations of sinusoids. This response is important for two reasons. First, reference signals, persistent disturbances, and measurement noise can often be represented as combinations of sinusoids (though we may require Fourier transforms or power spectral densities to determine the combinations). Second, sinusoidal signals in LTI systems represent the boundary between stability and instability, so the frequency response of the loop transfer function—the *loop gain*—is sufficient to determine whether a closed-loop system is stable or unstable. We mentioned the corresponding test, the Nyquist criterion, in Chapter 11 and we discuss it further in Chapter 14. For both these reasons, the frequency response is central to control design.

Suppose that the input to a single-input, single-output (SISO, or scalar) system with transfer function $G(s)$ is a sinusoid at a frequency ω_0, so that:

$$u(t) = u_0 \sin(\omega_0 t + \theta) \tag{13.29}$$

for some constants u_0 and θ. Carrying out a partial fraction expansion of $y(s) = G(s)u(s)$ and inverse transforming the result, we can show that the response at the output contains a sinusoid at the same frequency. If we assume that $G(s)$ does not have a pole at $j\omega_0$, the amplitude of the output sinusoid is $|G(j\omega_0)|$ times the input sinusoid, and its phase is retarded by $\angle G(j\omega_0)$:

$$y(t) = |G(j\omega_0)|u_0 \sin\left(\omega_0 t + \theta + \angle G(j\omega_0)\right) + \cdots \tag{13.30}$$

The terms that are represented by $\cdots$ correspond to exponentials at the natural frequencies of the system. If the system is asymptotically stable, these terms die out, and only the sinusoidal response persists. For stable systems, therefore, we can experimentally determine the frequency response on the basis of *(13.30)*. The magnitude is determined by the amplification of the sinusoid between the input and the output, and the angle is determined by the phase retardation.

We can extend the preceding interpretation of the frequency response to the transfer function matrix of a multi-input, multi-output (MIMO, or multivariable) system, by using it for each entry of the transfer matrix. The following alternative viewpoint will be useful when we deal with sampled-data systems later, and it holds for both SISO and MIMO systems. An input of the form:

$$u(t) = u_0 e^{j(\omega_0 t + \theta)} \tag{13.31}$$

produces an output of the form:

$$y(t) = G(j\omega_0)u_0 e^{j(\omega_0 t + \theta)} + \cdots \tag{13.32}$$

which can be verified using transform methods. The response to real sinusoids can easily be deduced from *(13.32)*, because a real sinusoid can be written as a superposition of complex exponentials.

As we note in Chapter 14, control design for MIMO systems on the basis of frequency-response information is significantly more intricate than the corresponding classical approach to SISO systems. In fact, it is beyond the scope of what we attempt in this book. We thus focus on frequency responses associated with scalar transfer functions.

We turn now from the special cases we have been considering, and develop an expression for the general time-domain solution of an LTI system. We also use this solution to describe the general solution of piecewise LTI models in Section 13.4. You can skip over this material and go directly to Section 13.5 or even to Chapter 14, coming back to it later.

The Matrix Exponential   Because *(13.17)* shows that $(sI - A)^{-1}$ is the key to representing the general transform-domain solution, we would expect that the inverse transform of $(sI - A)^{-1}$ is the key to the time-domain solution. The inverse transform of the scalar $(s - a)^{-1}$ is the exponential e^{at}, so we refer to the inverse transform of the matrix $(sI - A)^{-1}$ as the *matrix exponential* and denote it by e^{At}:

$$e^{At} = \mathcal{L}^{-1}\{(sI - A)^{-1}\} \tag{13.33}$$

In connection with our earlier determination of the natural response, we deduced that the entries of this matrix are made up of sums of (scalar) exponentials and exponentially weighted sinusoids, corresponding to characteristic roots or eigenvalues of A.

The matrix exponential has properties similar to the scalar exponential. For instance, we can write it as the following infinite series, which converges for all t:

$$e^{At} = I + At + \frac{1}{2!}A^2t^2 + \frac{1}{3!}A^3t^3 + \cdots \tag{13.34}$$

For small enough t, the first two terms provide a good approximation:

$$e^{At} \approx I + At \tag{13.35}$$

More precisely, for this to be a good approximation, we require $|\lambda_{max}t| \ll 1$, where λ_{max} is the eigenvalue of maximum absolute value among the eigenvalues of A.

The derivative of the matrix exponential, defined as the matrix whose entries are the derivatives of the corresponding entries of e^{At}, satisfies

$$\frac{de^{At}}{dt} = Ae^{At} = e^{At}A \tag{13.36}$$

This result can be obtained by differentiating *(13.34)* term by term, an operation that is justified because the series is sufficiently well behaved. The main difference from the scalar case occurs when two matrix exponentials (of the same dimensions, of course) are multiplied together:

$$e^{A_1t}e^{A_2t} \neq e^{(A_1+A_2)t} \neq e^{A_2t}e^{A_1t} \tag{13.37}$$

except when $A_1A_2 = A_2A_1$, in which case we have equalities in *(13.37)*.

The General Solution   The matrix exponential allows us to write an expression for the general solution of an LTI system. Note first that we can now write the natural response in the time domain as:

$$\mathcal{L}^{-1}\{(sI - A)^{-1}x(0)\} = e^{At}x(0) \tag{13.38}$$

The transform-domain representation of the forced response—the term $(sI - A)^{-1}Bu(s)$ in *(13.17)*—is the product of two transforms. Recall that multiplication of transforms corresponds to convolution of the associated time functions, in this case convolution of e^{At} with $Bu(t)$. Superposing the natural response and forced response leads us to the following expression for the complete time-domain equivalent of the frequency-domain solution in *(13.17)*:

$$x(t) = e^{At}x(0) + \int_0^t e^{A(t-\xi)}Bu(\xi)d\xi \tag{13.39}$$

The second term on the right is the forced response. The vector under the integral sign is integrated entry by entry. Also, as we are dealing with matrix products, the order of the factors must be respected. The forced response, like the natural response, contains exponentials at the natural frequencies of the system, but it also contains terms displaying the characteristics of the input. The choice of 0 as the time origin is arbitrary for a time invariant system, so we can rewrite *(13.39)* as:

$$x(t) = e^{A(t-t_0)}x(t_0) + \int_{t_0}^t e^{A(t-\xi)}Bu(\xi)d\xi \tag{13.40}$$

In Section 12.4.1 we made plausible the state property of a state-space description but did not prove it. For LTI systems, however, the explicit analytical solution in *(13.40) is* the proof of the state property. The role of the matrix exponential in *(13.40)* leads to its other name, the *state-transition matrix* of the system.

You have already encountered matrix exponentials acting as state-transition matrices. The "rotation matrix" $\Theta(t - t_0)$ used to analyze the resonant converter of Example 12.6 is the matrix exponential associated with the state-space description of resonant converter dynamics. Also, the matrices $(I + d_k T A_1)$ and $(I + d'_k T A_0)$ that were used to analyze the switched up/down converter in Example 12.4 are the appropriate matrix exponentials for that converter model, approximated according to *(13.35)*.

The analytical solution in *(13.40)* is the basis for various numerical solution algorithms, built on numerical approximations of the matrix exponential. For example, if $t - t_0$ is a sufficiently small number ϵ, we can use *(13.35)* to write $e^{A\epsilon} \approx I + A\epsilon$ in *(13.40)*. This substitution leads essentially to the forward Euler method of Section 12.4.2. Approximating the infinite series *(13.34)* by a partial sum such as *(13.35)* is *not* a good approach for stiff systems. Other choices, such as the approximation $e^{A\epsilon} \approx (I - A\epsilon)^{-1}$ on which the backward Euler algorithm is based, give much better results for stiff systems.

BIBO Stability   A notion of stability that is useful when inputs are present is that of *bounded input, bounded output (BIBO)* stability, where "output" broadly means any response of interest. For BIBO stability, the response to any bounded input must also be bounded, with the system initially at rest, that is, with $x(0) = 0$. As noted in Section 11.5.2, a necessary and sufficient condition is that the transfer function matrix between the input and output has the poles of all its entries strictly in the left half plane.

We can show from *(13.39)* that exponential stability of the zero-input response is sufficient to guarantee BIBO stability. The reason that the entries of the matrix exponential all decay exponentially under this condition, so the integral that gives the forced response is bounded. Whenever we refer to stability of an LTI system from now on, we mean asymptotic or exponential stability.

It *is* possible for a system to be BIBO stable even if it is not asymptotically stable. This condition occurs when the system's unstable characteristic roots do not appear in the transfer matrix, as a result of each unstable root being canceled by a corresponding numerator zero in every entry of the transfer function matrix. The possibility of such cancellations was mentioned earlier in connection with the structure of the transfer function matrix. In the absence of complete cancellation of unstable roots, BIBO stability and asymptotic stability are identical. Given an interconnection of subsystems where none of the subsystems has an unstable canceled root, internal stability of the interconnection, in the sense of Section 11.5.2, is identical with asymptotic stability of the interconnection.

13.4 PIECEWISE LTI MODELS

We now show how to obtain useful analytical descriptions for the most common circuit models used in power electronics, namely, interconnections of LTI components and ideal switches. You have already encountered several instances of such *piecewise LTI* models. The procedure we outline here generalizes Examples 12.5 and 12.6. Those examples perhaps are sufficient for you to proceed to Section 13.5 or even Chapter 14 at this point, returning to this section later.

A piecewise LTI circuit is LTI if the switch positions are frozen in any given configuration. Hence an LTI description can be associated with each switch configuration or topological state. We can therefore use expressions similar to *(13.40)* to solve for the behavior of the state variables in each of the circuit's successive configurations. We can then piece together these solutions from configuration to configuration by invoking the continuity of state variables. The final state in one configuration becomes the initial state for the next configuration. This procedure leads us naturally to discrete-time generalized state-space models.

Suppose that our piecewise LTI circuit is operated cyclically, taking a succession of N configurations in each cycle, with the kth cycle extending from time t_k to t_{k+1}. We number the configurations $0, 1, \ldots, N-1$ and let the state vector of the circuit be governed by the LTI description:

$$\frac{dx}{dt} = A_\ell x(t) + B_\ell u(t) \qquad (13.41)$$

in configuration ℓ, which in the kth cycle occurs from $t = t_k + \tau_\ell[k]$ to $t = t_k + \tau_{\ell+1}[k]$. Hence $\tau_0[k] = 0$ and $\tau_N[k] = t_{k+1} - t_k$. The *transition time* $\tau_\ell[k]$ denotes the time, relative to the start of the kth cycle, when the circuit enters configuration ℓ, and $\tau_N[k]$ is the duration of the kth cycle. Figure 13.4 is a schematic representation of the situation. Sometimes, considering still more general cases is useful—for example, where the number and identity of the state variables can change from configuration to configuration or where the number of configurations can vary from cycle to cycle—but we leave such extensions to you.

The constant matrices A_ℓ and B_ℓ in *(13.41)* are determined by the circuit topology and parameter values in configuration ℓ. As in the discussion preceding the sampled-data model *(12.15)*, we assume that the input vector $u(t)$ in this configuration of the kth cycle is completely specified by a vector of determining variables, denoted by $p_\ell[k]$.

By now applying *(13.40)* to the model *(13.41)*, we can solve for the state at the end of configuration ℓ in terms of the state at the beginning of this configuration and the inputs acting during the intervening interval. To simplify notation, we omit the argument k from the transition times τ_ℓ and $\tau_{\ell+1}$ and we initially write the resulting expression only for $k = 0$, assuming $t_0 = 0$. The resulting expression is

$$x(\tau_{\ell+1}) = e^{A_\ell(\tau_{\ell+1}-\tau_\ell)}x(\tau_\ell) + \int_{\tau_\ell}^{\tau_{\ell+1}} e^{A_\ell(\tau_{\ell+1}-\xi)}B_\ell u(\xi)d\xi \qquad (13.42)$$

To obtain the general, detailed expression for the kth cycle from *(13.42)*, we only need to attach the argument k to each transition time and to increment each transition time by t_k.

The input vector $u(\xi)$ in *(13.42)* is determined by $p_\ell[k]$, so we can write the integral in *(13.42)* as a function of $p_\ell[k]$, $\tau_\ell[k]$, and $\tau_{\ell+1}[k]$. We therefore find that:

$$x(t_k + \tau_{\ell+1}) = e^{A_\ell(\tau_{\ell+1}-\tau_\ell)}x(t_k + \tau_\ell) + \zeta_\ell\left(p_\ell[k], \tau_\ell, \tau_{\ell+1}\right) \qquad (13.43)$$

where we have continued to suppress the argument k on the transition times to

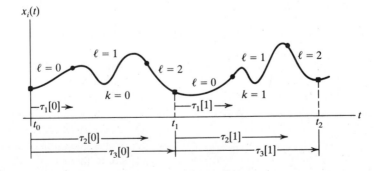

Figure 13.4   Schematic representation of waveforms in a cyclically operated piecewise LTI model.

simplify notation. The vector function $\zeta_\ell(\cdot)$ would depend explicitly on k if the form of dependence of $u(t)$ on $p_\ell[k]$ varied from cycle to cycle, but we assume for simplicity that there is no such variation.

A special case in which the integral in *(13.42)* simplifies is when the input $u(t)$ is piecewise constant, taking the value $p_\ell[k]$ in configuration ℓ of the kth cycle. After some changes of variable to simplify the integral, the model *(13.43)* becomes

$$x(t_k + \tau_{\ell+1}) = e^{A_\ell(\tau_{\ell+1} - \tau_\ell)}x(t_k + \tau_\ell) + \left(\int_0^{\tau_{\ell+1} - \tau_\ell} e^{A_\ell \xi} B_\ell d\xi \right) p_\ell[k] \quad (13.44)$$

Symmetry in the form and operation of a power circuit often permits an expression of the type *(13.43)* or *(13.44)*, derived for one configuration within a cycle, to be used for other "equivalent" configurations within the cycle, with only minor permutations of variables. The resonant converter in Examples 12.3 and 12.6, for example, displays such symmetry.

The transition times $\tau_{\ell+1}$, $\ell = 0, \ldots, N - 1$, are of two types. Some transition times correspond to *directly controlled* changes in the conducting states of switches. Examples of this are transistors turning on and off or (forward biased) thyristors turning on, in response exclusively to external control signals. These directly controlled transition times are determining variables, just as the $p_\ell[k]$ are. The remaining transition times correspond to *indirectly controlled*, state-dependent changes in switch status. Examples of this are diodes turning on when the reverse bias on them falls to zero, diodes and thyristors turning off when the current through them falls to zero, or controlled switches operating under the action of state-dependent feedback. These indirectly controlled transition times fall in the category of auxiliary variables, of the sort referred to in connection with the generalized state-space model *(12.35)*.

We now combine the expressions obtained from *(13.43)* for the configurations from $\ell = 0$ to $\ell = N - 1$, equating the final state of one configuration to the initial state of the next, to obtain the desired cycle-to-cycle description. The result takes the form of the first equation in the generalized state-space description *(12.35)*, if $x(t_k)$ is denoted by $x[k]$:

$$x[k + 1] = \phi\left(x[k], p[k], w[k] \right) \quad (13.45)$$

The auxiliary vector $w[k]$ comprises all the indirectly controlled transition times. We obtain the input vector $p[k]$ by stacking: (1) all the determining vectors $p_\ell[k]$; (2) all the directly controlled transition times; and (3) any remaining determining variables, such as those that specify the external reference signals applied to a feedback controller. (Determining variables in the third category actually do not appear in *(13.45)*. However, they will appear in the constraint equations that determine the indirectly controlled transition times, so we include them in the definition of $p[k]$ to keep the notation manageable.)

We now identify the constraint equations that determine $w[k]$. If $\tau_{\ell+1}$ is an indirectly controlled transition time, some state-dependent signal in configuration

ℓ of the circuit reaches a threshold level at time $t_k + \tau_{\ell+1}$. By equating the signal to the threshold, we obtain a constraint equation that involves $x(t_k + \tau_{\ell+1})$ and the determining variables that govern configuration ℓ. Using (13.43) we can write the constraint in the form:

$$0 = \sigma_{\ell+1}\Big(x(t_k + \tau_\ell), p[k], \tau_\ell, \tau_{\ell+1}\Big) \qquad (13.46)$$

This constraint could depend explicitly on k under certain conditions, but we take (13.46) to be time invariant for simplicity.

Next we combine all such constraint equations for the indirectly controlled transition times, and use (13.43) for $\ell = 0, 1, \ldots$ to relate the constraints back to the state $x[k]$ at time t_k. In this way we obtain a set of constraints of the form given in connection with the generalized state-space description (12.35), namely:

$$0 = \sigma\Big(x[k], p[k], w[k]\Big) \qquad (13.47)$$

We can also associate a set of output equations of the form (12.36) with the generalized state-space model (13.45) and (13.47). The variables of interest may be sampled values of components of the state vector, averages computed over the cycle, or harmonic components over the cycle. We can represent all of these cases, and many others, in the form:

$$y[k] = \gamma\Big(x[k], p[k], w[k]\Big) \qquad (13.48)$$

It is beyond the scope of this book to discuss the issues involved in numerical solution or simulation of piecewise LTI models. We simply note here that such simulation involves two tasks: (1) simulating the LTI model that governs any given switch configuration; and (2) recognizing the event that signals the transition from one configuration to the next.

13.5 LINEARIZING DISCRETE-TIME MODELS

In Section 13.4, as well as Examples 12.5 and 12.6, we showed how nonlinear, time invariant, sampled-data models in generalized state-space form arise when we describe the cycle-to-cycle behavior of piecewise LTI models for power electronic circuits. If the steady-state behavior of the power circuit is periodic, which is almost always the case, the steady state of the corresponding sampled-data model is constant, as should be evident from Fig. 13.4. Once we know the constant steady state of the sampled-data model, we can find the periodic solution of the piecewise LTI model directly, if desired.

In this section we show that linearization of a nonlinear generalized state-space model yields a linear model in ordinary state-space form. Doing the linearization of a time invariant sampled-data model at a constant steady state results in an LTI sampled-data model. Such an LTI discrete-time model is, explicitly or implicitly, the basis for most careful small-signal stability studies in power electronics. We discuss the analysis of stability and transfer function properties for discrete-time LTI models in Section 13.6.

Our starting point is the discrete-time generalized state-space description in *(12.35)* and *(12.36)* or *(13.45)*, *(13.47)*, and *(13.48)*. We can easily specialize the results to ordinary discrete-time state-space descriptions by setting $w[k]$ and $\sigma(\cdot)$ to zero. The linearization procedure parallels that for continuous-time systems in Section 13.2.

As in the continuous-time case, we use capital letters to denote the nominal values of the variables in *(13.45)*, *(13.47)*, and *(13.48)*. Thus in the nominal operating condition, $x[k] = X[k]$, $p[k] = P[k]$, $w[k] = W[k]$, and $y[k] = Y[k]$. The nominal values can be any values that together satisfy the equations, though the most important case for us is when the nominal solutions are constant. We now consider deviations from nominal and mark these quantities with a $\sim$, so that away from the nominal solution we have

$$x = X + \tilde{x} \qquad p = P + \tilde{p} \qquad w = W + \tilde{w} \quad \text{and} \quad y = Y + \tilde{y} \qquad (13.49)$$

If the deviations are small, we deduce from *(13.45)*, *(13.47)*, and *(13.48)* that, to a first-order approximation, the deviations satisfy

$$\tilde{x}[k+1] \approx \frac{\partial \phi}{\partial x} \tilde{x}[k] + \frac{\partial \phi}{\partial p} \tilde{p}[k] + \frac{\partial \phi}{\partial w} \tilde{w}[k]$$

$$0 \approx \frac{\partial \sigma}{\partial x} \tilde{x}[k] + \frac{\partial \sigma}{\partial p} \tilde{p}[k] + \frac{\partial \sigma}{\partial w} \tilde{w}[k] \qquad (13.50)$$

$$\tilde{y}[k] \approx \frac{\partial \gamma}{\partial x} \tilde{x}[k] + \frac{\partial \gamma}{\partial p} \tilde{p}[k] + \frac{\partial \gamma}{\partial w} \tilde{w}[k]$$

These expressions result from substituting *(13.49)* into *(13.45)*, *(13.47)*, and *(13.48)* and then expanding $\phi(\cdot)$, $\sigma(\cdot)$, and $\gamma(\cdot)$ in multivariable Taylor series, up to and including linear terms. The partial derivative symbols denote Jacobian matrices. For example, $\partial \phi / \partial x$ denotes a matrix whose entry in the ith row and jth column is the partial derivative of the ith component of $\phi(\cdot)$ with respect to the jth component of x, namely, $\partial \phi_i / \partial x_j$. These partial derivatives, in general, are all functions of x, p, and w and must all be evaluated at the nominal solution, with $x = X, p = P$, and $w = W$, when they are used in *(13.50)*.

The linearized model in *(13.50)* is itself in generalized state-space form, with the deviations $\tilde{x}$, $\tilde{p}$, $\tilde{w}$, and $\tilde{y}$ now constituting the state variables, inputs, auxiliary variables, and outputs, respectively. The major simplification, though, is that this description is linear, because all the coefficient matrices are known as soon as we specify the nominal solution. The linearity allows us to solve explicitly the constraint expression in *(13.50)* for the auxiliary variables $\tilde{w}[k]$, provided the matrix $\partial \sigma / \partial w$ is invertible (which it will be, if the classification of variables into state variables, input or determining variables, and auxiliary variables has been done correctly):

$$\tilde{w}[k] \approx -\left[\frac{\partial \sigma}{\partial w}\right]^{-1} \left[\frac{\partial \sigma}{\partial x} \tilde{x}[k] + \frac{\partial \sigma}{\partial p} \tilde{p}[k]\right] \qquad (13.51)$$

Using *(13.51)* to eliminate the auxiliary variables in *(13.50)*, we get a linear description in ordinary state-space form:

$$\tilde{x}[k+1] \approx \mathcal{A}[k]\tilde{x}[k] + \mathcal{B}[k]\tilde{p}[k]$$
$$\tilde{y}[k] \approx \mathcal{E}[k]\tilde{x}[k] + \mathcal{F}[k]\tilde{p}[k] \tag{13.52}$$

where

$$\mathcal{A}[k] = \frac{\partial \phi}{\partial x} - \frac{\partial \phi}{\partial w}\left[\frac{\partial \sigma}{\partial w}\right]^{-1}\frac{\partial \sigma}{\partial x}$$

$$\mathcal{B}[k] = \frac{\partial \phi}{\partial p} - \frac{\partial \phi}{\partial w}\left[\frac{\partial \sigma}{\partial w}\right]^{-1}\frac{\partial \sigma}{\partial p}$$

$$\mathcal{E}[k] = \frac{\partial \gamma}{\partial x} - \frac{\partial \gamma}{\partial w}\left[\frac{\partial \sigma}{\partial w}\right]^{-1}\frac{\partial \sigma}{\partial x} \tag{13.53}$$

$$\mathcal{F}[k] = \frac{\partial \gamma}{\partial p} - \frac{\partial \gamma}{\partial w}\left[\frac{\partial \sigma}{\partial w}\right]^{-1}\frac{\partial \sigma}{\partial p}$$

If the original description was in ordinary state-space form rather than generalized form, the auxiliary variables and constraints would be absent, so $\mathcal{A}$, $\mathcal{B}$, $\mathcal{E}$, and $\mathcal{F}$ would simply be given by $\partial \phi/\partial x$, $\partial \phi/\partial p$, $\partial \gamma/\partial x$, and $\partial \gamma/\partial p$, respectively.

If the original nonlinear model is time invariant, so that $\phi(\cdot)$, $\sigma(\cdot)$, and $\gamma(\cdot)$ are not explicitly dependent on the time index k, and if in addition the nominal solution is constant, all the partial derivatives that define the linearized model will be constant. The linearized model will therefore be LTI in this case—the most important case in power electronics.

We obtained the generalized state-space model *(13.45)*, *(13.47)*, and *(13.48)* by piecing together the descriptions *(13.43)* and constraints *(13.46)* for successive configurations of a piecewise LTI model. This underlying structure suggests an alternative route to the linearized model *(13.52)* in the case of piecewise models. We can linearize the description *(13.43)* and constraint *(13.46)* for *each configuration* and then piece together the resulting *linearized* models. We leave it to you to check that this alternative route produces the same linearized model *(13.52)*.

Computing a Constant Nominal Solution   Suppose that the nominal solution of the sampled-data model is constant: $X[k] = X$, $P[k] = P$, and $W[k] = W$. To compute this constant nominal solution, we need to solve a time invariant system of equations of the form *(13.45)* and *(13.47)*, with these constant values substituted into them:

$$X = \phi(X, P, W)$$
$$0 = \sigma(X, P, W) \tag{13.54}$$

Here, P is known, as it is the specified vector of determining variables or inputs for the steady state, but X and W have to be found. The equations sometimes have a structure that permits a simple closed-form solution. For example, sometimes there are no auxiliary variables and the first equation in *(13.54)* takes the form:

$$X = \Phi(P)X + \beta(P) \tag{13.55}$$

where $\Phi(\cdot)$ is an $n \times n$ matrix. Rearranging *(13.55)* to group terms in X on one side and then solving for X, we obtain

$$X = [I - \Phi(P)]^{-1}\beta(P) \tag{13.56}$$

More generally, however, the system *(13.54)* is nonlinear and has to be solved by iterative numerical methods. A standard Newton–Raphson approach assumes we have a good initial approximation of the constant steady-state solution and then uses the corresponding linearized model (specifically, the first two expressions in *(13.50)*, with $\tilde{p} = 0$) to improve the approximation. The partial derivatives at each iteration are evaluated at the current estimate of the steady state. (Without a good initial guess, the iteration process may converge to some spurious solution of the nonlinear equations or may fail to converge at all.)

EXAMPLE 13.4
Linearized Sampled-Data Models for Switching Converters
under Duty-Ratio Control and Current-Mode Control

Let's take as our starting point the approximate sampled-data model obtained in *(12.34)* of Example 12.4:

$$x[k+1] = \phi\left(x[k], v_{\text{in}}[k], d_k\right) \tag{13.57}$$

Linearizing *(13.57)*, we find

$$\tilde{x}[k+1] \approx \frac{\partial \phi}{\partial x}\tilde{x}[k] + \frac{\partial \phi}{\partial v_{\text{in}}}\tilde{v}_{\text{in}}[k] + \frac{\partial \phi}{\partial d}\tilde{d}_k \tag{13.58}$$

The partial derivatives must be evaluated at the nominal operating condition. This nominal solution corresponds to the steady state obtained by fixing $d_k = D$, $v_{\text{in}} = V_{\text{in}}$, and having $x[k+1] = x[k] = X$ for all k.

If we neglect the terms in T^2, the model is

$$x[k+1] \approx (I + TA_{d_k})x[k] + TB_{d_k}v_{\text{in}}[k] \tag{13.59}$$

where

$$A_{d_k} = d_k A_1 + d'_k A_0 \quad \text{and} \quad B_{d_k} = d_k B_1 + d'_k B_0 \tag{13.60}$$

The corresponding steady state is governed by an equation in the form of *(13.55)* and can be computed using *(13.56)*, which in this case gives

$$X = \begin{bmatrix} I_L \\ V_C \end{bmatrix} = -A_D^{-1}B_D V_{\text{in}} \tag{13.61}$$

This happens to be exactly the same equation that we obtained for the steady state of the averaged model in *(13.7)*. Evaluating the derivatives in the linearized model *(13.58)–(13.60)* at the steady state, we find

$$\frac{\partial \phi}{\partial x} = I + TA_D$$

$$\frac{\partial \phi}{\partial v_{\text{in}}} = TB_D \qquad\qquad (13.62)$$

$$\frac{\partial \phi}{\partial d} = T\left[(A_1 - A_0)X + (B_1 - B_0)V_{\text{in}}\right]$$

Substituting these expressions in *(13.58)*, we obtain a linearized sampled-data model for duty-ratio control.

We showed in Example 12.5 that in current-mode control the duty ratio becomes an auxiliary variable, governed by a constraint of the form:

$$0 = \sigma\left(x[k], v_{\text{in}}[k], i_P[k], d_k\right) \qquad\qquad (13.63)$$

For the up/down converter, this constraint is

$$0 \approx i_L[k] + \frac{d_k T v_{\text{in}}[k]}{L} - i_P[k] + S d_k T \qquad\qquad (13.64)$$

As noted in Example 12.5, we can solve for d_k from the constraint *(13.64)* and substitute it into *(13.59)* to obtain an ordinary state-space model. We can then linearize the resulting model if we so desire. However, we can obtain the same result with less messy computations by retaining the generalized state-space form and using *(13.53)*. We have already done part of the work in obtaining *(13.58)*. Linearizing the constraint, we get

$$0 \approx \frac{\partial \sigma}{\partial x}\tilde{x}[k] + \frac{\partial \sigma}{\partial v_{\text{in}}}\tilde{v}_{\text{in}}[k] + \frac{\partial \sigma}{\partial i_P}\tilde{i}_P[k] + \frac{\partial \sigma}{\partial d}\tilde{d}_k \qquad\qquad (13.65)$$

For the up/down converter, we find from *(13.64)* that:

$$\frac{\partial \sigma}{\partial x} = [\, 1 \quad 0\,]$$

$$\frac{\partial \sigma}{\partial v_{\text{in}}} = \frac{DT}{L}$$

$$\frac{\partial \sigma}{\partial i_P} = -1 \qquad\qquad (13.66)$$

$$\frac{\partial \sigma}{\partial d} = T\left(S + \frac{V_{\text{in}}}{L}\right)$$

Using these expressions to solve *(13.65)* for $\tilde{d}_k$ and substituting into *(13.58)–(13.60)* results in the desired linearized sampled-data model for current-mode control. We defer the detailed expression to Example 13.6, where we use the linearized model to examine the stability of the nominal operating condition. Note that we have written the partial derivatives in *(13.62)* and *(13.66)* in terms of D for simplicity, but you can easily rewrite them in terms of the nominal value I_P of the commanded peak current.

A continuous-time parallel of the current-mode control results in Examples 12.5 and 13.4 can also be worked out. To do so, we must combine the averaged model in *(13.6)* of Example 13.1 with a constraint equation of the form:

$$0 = c\Big(x(t), \overline{v}_{\text{in}}(t), i_P(t), d(t)\Big) \qquad (13.67)$$

to obtain a *continuous*-time generalized state-space model. Here, $i_P(t)$ is a slowly varying signal that is the analog of $i_P[k]$ in Example 13.4; in fact, $i_P[k]$ may be thought of as the sample value of $i_P(t)$ at $t = kT$. We leave you to work out the detailed form of this constraint, using Fig. 12.9 for guidance. Linearization of this model proceeds as in the discrete-time case, with the obvious changes to *(13.53)*.

13.6 ANALYSIS OF DISCRETE-TIME LTI MODELS

We have shown in Section 13.5 that discrete-time LTI models commonly arise in power electronics as descriptions of perturbations from cyclic steady state, when we examine samples of circuit variables taken once per cycle. Such models constitute the main route to analyzing the stability of cyclic nominal operation in power circuits. They are also the basis for designing discrete-time or digital control systems.

In this section we develop some of the basic tools and results needed to treat discrete-time LTI systems. The focus of our discussion is a model in the form of *(12.27)* and *(12.28)*, repeated here for convenience:

$$\begin{aligned} x[k+1] &= \mathcal{A}x[k] + \mathcal{B}p[k] \\ y[k] &= \mathcal{E}x[k] + \mathcal{F}p[k] \end{aligned} \qquad (13.68)$$

The results can, of course, be applied to a linearized model in the form of *(13.52)*, when its coefficient matrices are constant.

13.6.1 Time-Domain Solution

Unlike the continuous-time case, the time-domain solution for discrete time is easy to describe. For the initial condition $x[0]$ and input $p[0]$, we can use *(13.68)* to determine $x[1]$ and $y[0]$. Knowing $x[1]$ and $p[1]$, we can similarly determine $x[2]$ and $y[1]$. Clearly, we can continue this iteration indefinitely, yielding

$$x[k] = \mathcal{A}^k x[0] + \sum_{i=0}^{k-1} \mathcal{A}^{k-1-i} \mathcal{B}p[i] \qquad (13.69)$$

with a corresponding expression for the output. Compare this solution to the solution in *(13.39)* for a continuous-time system.

The first term on the right in *(13.69)* constitutes the response to initial conditions alone, in the absence of external inputs, and is the natural response or free response or zero-input response. The second term is the forced response or zero-state response. Asymptotic stability of the natural response is evidently equivalent to having the entries of $\mathcal{A}^k$ decay asymptotically to zero as k goes to infinity. The conditions needed for this decay to happen can be identified most easily if we first obtain a transform-domain solution.

13.6.2 Transform-Domain Solution

The transform that is appropriate for the solution of time invariant difference equations is the (one-sided) $\mathcal{Z}$ transform. The $\mathcal{Z}$ transform of a sequence $x[k]$ is denoted by $x(z)$ or $\mathcal{Z}\{x[k]\}$ and defined by:

$$x(z) = \mathcal{Z}\{x[k]\} = \sum_{k=0}^{\infty} x[k]z^{-k} \qquad (13.70)$$

The definition also applies without modification to vector or matrix sequences.

For instance, the scalar geometric sequence $x[k] = a^k$, with a possibly complex ratio a, has the transform $x(z) = 1+az^{-1}+a^2z^{-2}+\cdots$, which can be condensed to $x(z) = (z - a)^{-1}z$. (More precisely, the condensation can be carried out for $|z| > |a|$, which is the so-called region of convergence for the transform of this sequence. For the one-sided transform, however, we can get by without worrying further about regions of convergence. All the transforms we compute have a common region of convergence for large enough $|z|$, so all the operations we carry out on them will be legitimate.) Scalar geometric sequences play the same role for discrete-time LTI systems as exponentials do for continuous-time systems, because they have the same property of constant fractional change per unit time. We often refer to a geometric sequence of the form a^k as a *discrete-time exponential* of *ratio a*. (The ratio a is also commonly referred to as the *frequency* of the discrete-time exponential or as its *multiplier*.)

There is a simple way to compute the inverse $\mathcal{Z}$ transform of a given $x(z)$: Expand $x(z)$ in a power series in negative powers of z and then identify the coefficient of z^{-k} as the kth element $x[k]$ of the discrete-time sequence. We can easily do this expansion if the transform is rational in z and has only one pole. For example, if the transform of a sequence $x[k]$ is given as $(z - a)^{-1}$, writing the transform as $z^{-1} + az^{-2} + a^2z^{-3} + \cdots$ identifies the sequence as being $x[0] = 0$, $x[1] = 1$, $x[2] = a$, $x[3] = a^2$, and so on.

For more complicated rationals, we first obtain a partial fraction expansion as a sum of simpler, single-pole terms, each of which we can then inverse transform quite easily. (There is also an integral formula for the inverse transform, but we do not use it.) The two examples given so far suggest that a pole at $z = a$ in the transform corresponds to a discrete-time exponential of ratio a in the time response. The partial fraction construction shows that this assertion is indeed true. Note that the exponential decays if and only if the pole has a magnitude less than 1.

What makes the $\mathcal{Z}$ transform valuable in solving state-space equations is the following *shift theorem*, which plays an analogous role to the differentiation theorem of Laplace transforms in *(13.12)*:

$$\mathcal{Z}\{x[k+1]\} = z\mathcal{Z}\{x[k]\} - zx[0] \qquad (13.71)$$

This result is easily obtained from the definition. Applying it to *(13.68)* gives

$$zx(z) - zx[0] = Ax(z) + Bp(z) \qquad (13.72)$$

Gathering terms in $x(z)$ on one side and solving for it yields

$$x(z) = (zI - A)^{-1}zx[0] + (zI - A)^{-1}Bp(z) \qquad (13.73)$$

This expression is the transform-domain version of *(13.69)*. Note that $(zI - A)^{-1}z$ is the transform of the sequence A^k for $k = 0, 1, \ldots$, and $(zI - A)^{-1}$ is the transform of the sequence $0, I, A, A^2, \ldots$; the scalar versions of these results appeared as examples in the discussion following *(13.70)*. Also note that, as with Laplace transforms, the product of transforms corresponds to the (discrete-time) convolution of the associated sequences.

The following alternative way of recovering the time-domain sequence associated with $(zI - A)^{-1}$ gives an insight into the properties of the solution and leads us to the condition for asymptotic stability. Each entry of $(zI - A)^{-1}$ has as its denominator the polynomial $\det(zI - A)$, which we call the characteristic polynomial of the system or of A, as in continuous-time systems. The roots of this polynomial are again called the characteristic roots or the eigenvalues of A. We also refer to them as *characteristic ratios* or *natural ratios*. Now, expanding each entry of $(zI - A)^{-1}$ as a sum of partial fractions, we obtain terms whose poles are the natural ratios of the system. The associated time function is therefore a sum of discrete-time exponentials at the natural ratios of the system.

Asymptotic Stability   With the preceding interpretation of the time-domain sequences corresponding to the entries of $(zI - A)^{-1}$, you can see from the comments following *(13.73)* that the entries of the sequence A^k, as well as the natural response of the system, are weighted sums of discrete-time exponentials at the natural ratios. Hence the natural response will decay asymptotically to zero if and only if all the natural ratios have magnitudes less than 1, that is, lie strictly inside the unit circle in the complex plane. This condition is therefore necessary and sufficient for asymptotic (and actually exponential) stability. If the natural ratios are all 0, the natural response goes to zero in a finite number of steps; we refer to the system as *deadbeat* in this case.

We can also show from *(13.69)* or *(13.73)* that the forced response will display discrete-time exponentials at the natural ratios, in addition to terms that reflect the properties of the input. As with continuous-time systems, exponential stability of the natural response turns out to be sufficient to guarantee a bounded response to bounded inputs.

EXAMPLE 13.5

Stability of a Sampled-Data Model for an Up/Down Converter under Duty-Ratio Control

We start by computing the characteristic polynomial of the LTI sampled-data model that we obtained in (13.58) and (13.62) of Example 13.4 for the case of duty-ratio control of an up/down converter. We need to find the characteristic polynomial of the matrix $\partial\phi/\partial x = I + TA_D$ in (13.62). Using the expression for A_D from (13.6), we obtain for an up/down converter the characteristic polynomial:

$$\det\begin{bmatrix} z - 1 & -TD'/L \\ TD'/C & z - 1 + T/(RC) \end{bmatrix} = (z - 1)\left(z - 1 + \frac{T}{RC}\right) + \frac{T^2 D'^2}{LC} \tag{13.74}$$

This characteristic polynomial has a simple relation to that of the averaged model in Example 13.2, namely, $\det(sI - A_D)$. Note that:

$$\det(zI - I - TA_D) = T^2 \det\left(\frac{z - 1}{T}I - A_D\right) \tag{13.75}$$

so substituting $(z - 1)/T$ for s in the characteristic polynomial of the averaged model yields (apart from the constant factor T^2) the characteristic polynomial of the sampled-data model. Thus λ_1 is a characteristic root of A_D if and only if $1 + T\lambda_1$ is a characteristic root of $I + TA_D$. Whereas the natural response of the sampled-data model contains a geometric series of the form $(1 + T\lambda_1)^k$, that of the averaged model contains a continuous-time exponential of the form $e^{\lambda_1 t}$, whose values at the sampling instants are $(e^{\lambda_1 T})^k$. For small enough T, $e^{\lambda_1 T} \approx 1 + \lambda_1 T$, and the predictions of the two models are comparable. We noted in Example 13.2 that the characteristic roots of the averaged model constitute a lightly damped complex pair. You should verify that the corresponding characteristic roots of the sampled-data model form a complex conjugate pair within, and close to, the unit circle.

We can take our stability analysis one step further, treating the case of current-mode control in an up/down converter.

EXAMPLE 13.6

Stability of a Sampled-Data Model for an Up/Down Converter under Current-Mode Control

Turning to the sampled-data model for current-mode control in Example 13.4, we need to find the characteristic polynomial of the matrix:

$$\mathcal{A} = \frac{\partial\phi}{\partial x} - \frac{\partial\phi}{\partial d}\left[\frac{\partial\sigma}{\partial d}\right]^{-1}\frac{\partial\sigma}{\partial x} \tag{13.76}$$

in (13.53), where the partial derivatives are given by (13.62) and (13.66). Denoting $(LS + V_{\text{in}})^{-1}$ by μ to simplify the notation, we find that:

$$\mathcal{A} = \begin{bmatrix} 1 - \mu(V_{\text{in}} - V_C) & TD'/L \\ -TD'/C - \mu(I_L L/C) & 1 - (T/RC) \end{bmatrix} \tag{13.77}$$

from which finding $\det(zI - \mathcal{A})$ is straightforward.

If there is no stabilizing ramp, then $S = 0$ and $\mu = 1/V_{\text{in}}$. Using the steady-state relationships in *(13.61)*, we find in this special case that $\mathcal{A}$ simplifies to:

$$\mathcal{A} = \begin{bmatrix} -D/D' & TD'/L \\ -(TD'/C) - (DL/RCD'^2) & 1 - (T/RC) \end{bmatrix} \qquad (13.78)$$

The corresponding characteristic polynomial is

$$\det(zI - \mathcal{A}) = \left(z + \frac{D}{D'}\right)\left(z - 1 + \frac{T}{RC}\right) + \frac{TD}{RCD'} + \frac{T^2 D'^2}{LC} \qquad (13.79)$$

For small T, one root of the polynomial is close to (and slightly smaller than) $1 - (T/RC)$, and the other is close to (and slightly greater than) $-D/D'$. This result leads us to predict that the circuit will be unstable for $D > 0.5$, as one of the roots will then fall out of the unit circle, becoming more negative than -1.

We gain a more detailed insight into the dynamics of the circuit by examining $\mathcal{A}$ in *(13.78)* under the assumption that T is negligibly small. Recalling that this $\mathcal{A}$ governs the behavior of perturbations away from nominal in the absence of control, *(13.78)* shows that with $S = 0$ and $T \approx 0$ we have

$$
\begin{aligned}
\tilde{\imath}_L[k+1] &\approx (-D/D')\tilde{\imath}_L[k] \\
\tilde{v}_C[k+1] &\approx \tilde{v}_C[k] - (DL/RCD'^2)\tilde{\imath}_L[k]
\end{aligned}
\qquad (13.80)
$$

The first expression shows that the inductor current perturbations form a discrete-time exponential of ratio $-D/D'$, alternating in sign from one sample to the next, so the perturbations vary at one half the switching frequency. For $D < 0.5$, the magnitude of the inductor current perturbations decays exponentially. For $D > 0.5$, the magnitude of the perturbations grows, soon becoming large enough to fall beyond the purview of the linearized model (but limited in any case by the current threshold). This condition is an example of *ripple instability*, which we referred to at the beginning of Chapter 12. Simple calculations with the approximate inductor current waveforms in Fig. 13.5(a) and (b) confirm these predictions regarding stability of the perturbations.

The second expression in *(13.80)* shows that the capacitor voltage perturbations stay essentially constant, apart from a forcing term involving $\tilde{\imath}_L[k]$. This term varies at one half the switching frequency and, for $D < 0.5$, has an exponentially decaying magnitude.

Evidently, steady-state operation with period T is impossible for $S = 0$ and $D > 0.5$, and the waveforms assume more complicated forms, corresponding either to periodic operation at some multiple of T (subharmonic operation) or to "chaotic" variation from cycle to cycle. These possibilities are represented in Fig. 13.5(c) and (d), respectively. Although the dc/dc conversion function is basically unaffected—the duty ratio on a longer time scale takes the necessary value—such operation is generally undesirable because of problems in filtering the switching ripple.

The stabilizing ramp in current-mode control is introduced precisely in order to overcome the stability limit on the allowable duty ratio. This sawtooth waveform *entrains* the converter waveforms, maintaining them at the switching frequency when they would otherwise display subharmonic or aperiodic behavior. We leave it to you to explore the effect of $S > 0$ on the characteristic roots of $\mathcal{A}$ and to obtain an approximate analysis based on simple calculations with the inductor current waveform in Fig. 12.9, Example 12.5.

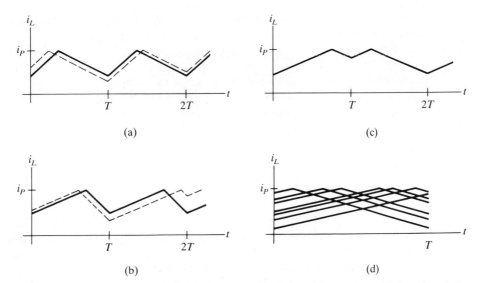

Figure 13.5 Inductor current waveforms with no stabilizing ramp: (a) for $D < 0.5$; (b) for $D > 0.5$; (c) subharmonic operation for $D > 0.5$, period $2T$; (d) several cycles superimposed to show "chaotic" operation for $D > 0.5$.

13.6.3 Transfer Functions and Frequency Response

Transfer functions are computed by setting initial conditions to 0 in the $\mathcal{Z}$ transform relationships. Using (13.73), we see that the transfer function matrix from p to x is $(zI - A)^{-1}B$ and the transfer function matrix from p to y is

$$\mathcal{G}(z) = \mathcal{E}(zI - A)^{-1}B + \mathcal{F} \qquad (13.81)$$

The transfer function is especially important for its role in defining the frequency response of a sampled-data system. The key result here is the following discrete-time version of (13.29)–(13.32). If the input to the system is of the form:

$$p[k] = p_0 a_0^k \qquad (13.82)$$

the output is of the form:

$$y[k] = \mathcal{G}(a_0)p_0 a_0^k + \cdots \qquad (13.83)$$

assuming that a_0 is not a pole of $\mathcal{G}(z)$. (The terms that are not shown correspond to discrete-time exponentials at the natural ratios of the system.) You can verify (13.83) by carrying out a partial fraction expansion of $y(z) = \mathcal{G}(z)p(z)$ and inverse transforming.

If the input sequence is obtained by sampling the continuous-time exponential $p_0 e^{j\omega_0 t}$ every T seconds,

$$p[k] = p_0(e^{j\omega_0 T})^k \qquad (13.84)$$

Comparing this expression with *(13.82)* and using *(13.83)* shows that the corresponding output is

$$y[k] = \mathcal{G}(e^{j\omega_0 T})p_0(e^{j\omega_0 T})^k + \cdots \tag{13.85}$$

From *(13.85)* you can easily see the effect of the system on inputs obtained by sampling a sinusoidal waveform, such as that in *(13.29)*. Specifically, if the input to a single-input, single-output (SISO) discrete-time LTI system is

$$p[k] = p_0 \sin(\omega_0 kT + \theta) \tag{13.86}$$

the output is

$$y[k] = |\mathcal{G}(e^{j\omega_0 T})|\, p_0 \sin\!\left[\omega_0 kT + \theta + \angle \mathcal{G}(e^{j\omega_0 T})\right] + \cdots \tag{13.87}$$

We term $\mathcal{G}(e^{j\omega T})$ the *frequency response* of the sampled-data system. The result *(13.87)* shows its role in defining the response to samples of a sinusoid at any given frequency.

The frequency response of a sampled-data LTI system differs in an important way from that of a continuous-time system: It is a *periodic* function of ω, with period $2\pi/T$, because $e^{j\omega T} = e^{j(\omega + 2\pi/T)T}$ for all ω. This difference reflects the fact that the samples of a sinusoid at the frequency ω, taken regularly at intervals of T, coincide with the samples of a sinusoid at a frequency displaced from ω by any integer multiple of $2\pi/T$; see Fig. 13.6. We refer to this phenomenon as *aliasing* and say that these sinusoids are *aliases* of each other. Hence $y[k]$ in *(13.87)* is unchanged if ω_0 is increased by any integer multiple of $2\pi/T$.

Because $\mathcal{G}(z)$ has real coefficients, $\mathcal{G}(e^{-j\omega T})$ is the complex conjugate of $\mathcal{G}(e^{j\omega T})$. We therefore need to know only the frequency response over the interval $0 \le \omega \le \pi/T$. The response for $-\pi/T \le \omega \le 0$ can be obtained by conjugation, and the response for any other frequency can be obtained by invoking the periodicity of the frequency response. Hence the highest frequency of interest when we describe or probe a sampled-data system is $\omega_{\max} = \pi/T$. The same limit is

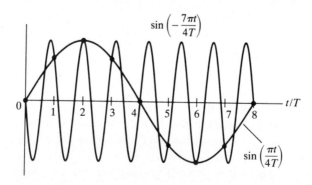

Figure 13.6   Aliasing of sampled sinusoids.

suggested by the fact that the fastest possible variation in the samples occurs when their signs alternate from cycle to cycle, corresponding to a period of $2T$, or a frequency of $2\pi(1/2T) = \pi/T$. This frequency is called the *Nyquist frequency*.

EXAMPLE 13.7

Frequency Response of a Sampled-Data Model for an Up/Down Converter under Duty-Ratio Control

We return to the approximate linearized sampled-data model for an up/down converter under duty-ratio control obtained in *(13.58)* and *(13.62)* of Example 13.4. The transfer function from duty-ratio perturbations $\tilde{d}$ to perturbations $\tilde{v}_C$ in the sampled capacitor voltage can be computed as:

$$\mathcal{G}(z) = \begin{bmatrix} 0 & 1 \end{bmatrix}\left(zI - \frac{\partial\phi}{\partial x}\right)^{-1}\frac{\partial\phi}{\partial d} \qquad (13.88)$$

where the partial derivatives are as given in *(13.62)*. However, there is a shortcut to the answer. In Example 13.5 we related the characteristic polynomials of the linearized sampled-data model here and the linearized averaged model in *(13.8)* of Example 13.2. Exploiting this relationship again, we obtain the desired transfer function simply by replacing s with $(z - 1)/T$ in the transfer function *(13.24)*.

We find the frequency response for this linearized sampled-data model by replacing z by $e^{j\omega T}$ in $\mathcal{G}(z)$. The result is plotted in Fig. 13.7 up to the switching frequency $\omega_S = 2\pi/T = 3.14\times 10^5$ rad/s. The frequency response of the linearized averaged model is plotted for comparison, repeated from Fig. 13.2(b). Because $(e^{j\omega T} - 1)/T \approx j\omega$ for $\omega T \ll 1$, it is not surprising that the two agree for low frequencies. In fact, the agreement happens to be good until about $\omega T = 0.5$ in this case. Beyond this frequency, and below $\omega_S/2$, the phase of the averaged model levels off at $-\pi/2$ rad, whereas the sampled-data model continues to add phase delay, leveling off at $-\pi$ rad. The magnitudes agree closely until they come nearer to $\omega_S/2$.

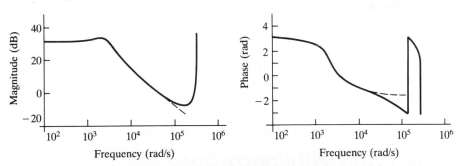

Figure 13.7   Frequency response of the linearized sampled-data model (——) and linearized averaged model (– – –) of the up/down converter.

The response of the sampled-data model beyond $\omega_S/2$ is consistent with the symmetry and periodicity properties mentioned earlier. We could compute the response of the averaged model beyond $\omega_S/2$. However, the response at these frequencies has no relevance to the actual behavior of the converter—for reasons discussed in Example 11.6 and immediately after—so we do not show it.

Recall that the typical control for a switching converter involves a circuit similar to that of Fig. 11.9, Example 11.6, so what is really of interest is the response of the circuit to perturbations $\tilde{m}(t)$ in the modulating signal $m(t)$. We have already argued that for *slow* variations in $m(t)$, $d(t) \approx m(t)/K$, so we can apply the averaged model. The sampled-data model is not restricted to slow variations in $m(t)$.

To see the effect of variations in $m(t)$ on the response of the linearized sampled-data model, let's suppose that:

$$m(t) = K[D + \epsilon \sin(\omega t)] \qquad (13.89)$$

where the amplitude ϵ is a small constant. Then $\tilde{m}(t) = K\epsilon \sin(\omega t)$ and we can easily deduce from Fig. 11.9 that:

$$\tilde{d}_k = \epsilon \sin[\omega(k + D + \tilde{d}_k)T]$$
$$\approx \epsilon \sin(\omega kT + \omega DT) \qquad (13.90)$$

provided $\tilde{d}_k \ll D$. The phase offset ωDT or time offset DT is a result of the duty ratio in each cycle being determined by the value of $m(t)$ at the instant this signal crosses the sawtooth ramp, rather than at the beginning of the cycle. As the frequency response from $\tilde{d}$ to $\tilde{v}_C$ is given by $\mathcal{G}(e^{j\omega T})$, we see that:

$$\tilde{v}_C[k] \approx |\mathcal{G}(e^{j\omega T})| \epsilon \sin[\omega kT + \omega DT + \angle \mathcal{G}(e^{j\omega T})] + \cdots \qquad (13.91)$$

This expression holds even for $\omega > \pi/T$, so long as $\tilde{d}_k$ is sufficiently small. However, if $\omega > \pi/T$, the sequence $\tilde{d}_k$ in *(13.90)* can also be produced by an $\tilde{m}(t)$ that varies at an alias frequency *below* π/T. The corresponding response mimics the response to the low-frequency alias. For example, varying $\tilde{m}(t)$ at the switching frequency produces the same effect as a constant $\tilde{m}(t)$. Hence experimental frequency-response measurements using a network analyzer can produce misleading results above π/T—the network analyzer looks only at the input-frequency component of the output, but the dominant part of the output response usually will be at some lower frequency. We leave you to explore further the relationships among $m(t)$, d_k, $d(t)$, $\bar{v}_C(t)$, and $v_C[k]$ for variations in $m(t)$ at various frequencies.

Example 13.7 suggests how information provided by a sampled-data model can be used to describe the behavior of the underlying continuous-time model. You need to keep in mind the relation between the two classes of models whenever you use a sampled-data model.

Notes and Bibliography

Reference [10] of Chapter 11 discusses state-transition matrices, controllability, observability, transfer functions, stability, and other such topics from Chapters 11 and 13. (It also covers all the control design we do in Chapter 14.) However, we recommend [1] for a deeper study of the theory of LTI state-space systems.

Our description of piecewise LTI models in Section 13.4 is based on [2], which contains many references to related earlier work. Issues in the simulation of piecewise LTI models using differential equation solvers are discussed in [3]. In [4] the switches are modeled as small or large resistors, depending on their conducting state, so the use of stiff-equation solvers becomes critical. The simulation approaches in [5] and [6] exploit heavily the LTI nature of each configuration. They explicitly form equations such as *(13.43)* and *(13.46)*, and use numerically determined time-functions that approximate the matrix exponential. An approach based on averaged models and corrections to them is described in [14], which also provides some additional references on simulating power circuits. Efficient computation of the periodic steady state is treated in [7].

Important early references on current-mode control are [8]–[10]. The idea of a stabilizing ramp to entrain the converter waveforms for duty ratios greater than 0.5 is already in [8]. The continuous-time model that we refer to after Example 13.4 is derived in [11]. (This derivation fixes an error in one of the two dynamic models presented in [10]. We use the other model from [10] in Section 14.2, though our derivation is slightly different.) Careful studies of chaotic behavior in power converters are described in [12].

The value of approximate sampled-data models such as those in Examples 12.4, 12.5, and 13.4–13.7 is emphasized in [11]. The decomposition into slow and fast subsystems that is evident in *(13.80)* bears further study. There may be systematic ways to derive "hybrid" models that, for instance, combine slowly varying averaged models with switched models. Reference [7] of Chapter 11 illustrates a similar idea, with models that interrelate local averages of some variables with local ω-components of other variables.

There are papers in the power electronics literature that show how to evaluate the frequency response of a piecewise LTI model beyond the Nyquist frequency. The predicted responses agree well with experimental measurements using a network analyzer. However, these results are of doubtful value because of the aliasing artifacts referred to at the end of Example 13.7. A good discussion of the frequency response of a sampled-data system can be found in [13], which is an excellent text on digital control.

1. T. Kailath, *Linear Systems* (Englewood Cliffs, New Jersey: Prentice-Hall, 1980).
2. G. C. Verghese, M. E. Elbuluk and J. G. Kassakian, "A General Approach to Sampled-Data Modeling for Power Electronic Circuits," *IEEE Trans. Power Electronics* 1:76–89 (April 1986).
3. R. J. Dirkman, "The Simulation of General Circuits Containing Ideal Switches," in *IEEE Power Electronics Specialists Conference (PESC)* (Blacksburg, June 1987), 185–194.
4. R. Nilssen and O. Mo, "KREAN, A New Simulation Program for Power Electronic Circuits," in *IEEE Power Electronics Specialists Conference (PESC)* (San Antonio, June 1990), 506–511.
5. C.-C. Liu, C. H. K. Chang, T.-T. Hsiao and J. M. Bocek, "A Fast Decoupled Method for Time-Domain Simulation of Power Converters," in *IEEE Power Electronics Specialists Conference (PESC)* (Kyoto, April 1988), 748–755.
6. A. M. Luciano and A. G. M. Strollo, "A Fast Time-Domain Algorithm for the Simulation of Switching Power Converters," *IEEE Trans. Power Electronics* 5:363–370 (July 1990).
7. Y. Kuroe, T. Maruhashi and N. Kanayama, "Computation of Sensitivities with Respect to Conduction Time of Power Semiconductors and Quick Determination of Steady State for Closed-Loop Power Electronic Systems," in *IEEE Power Electronics Specialists Conference (PESC)* (Kyoto, April 1988), 756–764.
8. C. W. Deisch, "Simple Switching Control Method Changes Power Converter Into a

Current Source," in *IEEE Power Electronics Specialists Conference (PESC)* (Syracuse, June 1978), 300–306.

9. A. Capel, G. Ferrante, D. O'Sullivan, and A. Weinberg, "Application of the Injected Current Model for the Dynamic Analysis of Switching Regulators with the New Concept of LC^3 Modulator," in *IEEE Power Electronics Specialists Conference (PESC)* (Syracuse, June 1978), 135–147.

10. S. P. Hsu, A. Brown, L. Resnick, and R. D. Middlebrook, "Modeling and Analysis of Switching DC-to-DC Converters in Constant-Frequency Current-Programmed Mode," in *IEEE Power Electronics Specialists Conference (PESC)* (San Diego, June 1979), 284–301.

11. G. C. Verghese, C. A. Bruzos and K. N. Mahabir, "Averaged and Sampled-Data Models for Current Mode Control: A Reexamination," in *IEEE Power Electronics Specialists Conference (PESC)* (Milwaukee, June 1989), 484–491.

12. J. H. B. Deane and D. C. Hamill, "Analysis, Simulation and Experimental Study of Chaos in the Buck Converter," in *IEEE Power Electronics Specialists Conference (PESC)* (San Antonio, June 1990), 491–498.

13. K. J. Astrom and B. Wittenmark, *Computer Controlled Systems: Theory and Design* (Englewood Cliffs, New Jersey: Prentice-Hall, 1990).

14. P. T. Krein and R. M. Bass, "A New Approach to Fast Simulation of Periodically Switching Power Converters," in *IEEE Industry Applications Society Annual Meeting* (Seattle, October 1990).

Problems

13.1 Verify that, if $f(x, u, t) = M(u, t)x$ for some matrix $M(u, t)$, then $\partial f / \partial x = M(u, t)$.

13.2 Examples 13.1 and 13.2 dealt with the steady state, linearization, and transfer functions of switched and averaged models of the up/down converter. Work out the detailed derivation of the linearized switched model, accounting for the fact that $\tilde{q}$ is restricted to the values 0 and 1. Extend the results in those examples to the case where the capacitor ESR $R_C \neq 0$. Also determine the transfer functions in the averaged model from duty-ratio perturbations and input voltage perturbations to inductor current perturbations. Specialize your results to $R_C = 0$ to check the results in the examples and also compare them to the results obtained directly from the linearized averaged circuits in Chapter 11. How do your analysis and results change if the switches, instead of being modeled as a short circuit when they conduct, are represented by a (small) resistance in series with a (small) dc voltage source?

13.3 Repeat Examples 13.1 and 13.2 as well as Problem 13.2 for other switching converters described in Chapters 6 and 7, tackling at least the following: down (or buck), up (or boost), flyback, and forward converters.

13.4 Verify the responses to initial conditions shown in Fig. 13.3, Example 13.3 for an up/down converter. Also show their relationship to the waveforms in Fig. 11.3, in which we presented the response to step changes in the source, with and without feedforward.

13.5 The general time-domain solution in *(13.40)* for the LTI model in *(13.10)* was derived using transform-domain arguments. For a direct check in the time domain, confirm that the expression in *(13.40)* indeed satisfies the differential equation in *(13.10)* and takes the specified intial value $x(t_0)$ at t_0.

13.6 Show that the "rotation matrix" used to analyze the resonant converter in Example 12.6 is indeed the state-transition matrix of the circuit in each of its topological states.

13.7 Verify the expression in *(13.44)* for the evolution of the state $x(t)$ when the input $u(t)$ is piecewise constant.

13.8 Repeat the analysis of (a) the up/down converter in Example 12.5 and (b) the resonant converter in Example 12.6, but now follow the prescription given in Section 13.4 for analyzing piecewise LTI models.

13.9 We briefly described how the linearized model in *(13.50)* could be used in an iterative solution of *(13.54)* for computation of the steady state in a sampled-data model. Obtain a more detailed algorithm, and test it on the sampled-data model for an up/down converter under current-mode control, referring to Examples 12.5 and 13.4. Take the commanded peak current $i_P[k]$ to be constant at $I_P = 7$ A, assume no stabilizing ramp $(S = 0)$, and let all other parameter values be the same as those in Example 11.1.

13.10 Derive a continuous-time generalized state-space model for the average behavior of an up/down converter under current-mode control, following the route suggested in the paragraph following Example 13.4. Determine its constant steady state. Then linearize this model around the steady state and determine the transfer function from perturbations $\tilde{\imath}_P$ in the commanded peak current to perturbation $\tilde{v}_o$ in the output voltage. As a check, one pole of your transfer function should go to $-\infty$ when there is no stabilizing ramp $(S = 0)$, leaving a transfer function with just one pole. Suppose we decide to allow perturbations $\tilde{s}$ around a nominal stabilizing-ramp slope of S. What is the transfer function from $\tilde{s}$ to $\tilde{v}_o$?

13.11 Repeat Problem 13.10 for the other switching converters listed in Problem 13.3.

13.12 The stability of the sampled-data model for the up/down converter in Example 13.6 was studied by means of the matrix $\mathcal{A}$ in *(13.77)*. Verify that the entries of this matrix are as claimed. Extend the exact and approximate analysis in that example to the case of a nonzero stabilizing ramp, $S \neq 0$, studying the locus of characteristic ratios as S varies. Verify your approximate analysis by appropriate calculations with the inductor current waveform in Fig. 12.9. What would be a good choice for S?

13.13 Guided by the results in Example 13.7 on the frequency response of the sampled-data model of the up/down converter, explore the relationships among $m(t)$, d_k, $d(t)$, $\overline{v}_C(t)$, and $v_C[k]$ for variations in $m(t)$ at various frequencies.

13.14 For the sampled-data model of the up/down converter under current-mode control, as given in Example 13.4, determine the transfer functions from perturbations $\tilde{\imath}_P$ and $\tilde{v}_{in}$ to $\tilde{v}_o$. With the nominal value of i_P fixed at $I_P = 7$ A, and with all other parameter values the same as those in Example 11.1, plot the associated frequency responses for varying values of the stabilizing-ramp slope S.

13.15   Obtain linearized sampled-data models of operation in discontinuous conduction for the switched converters treated in Problems 12.9 and 12.10, for both duty-ratio control and current-mode control. Evaluate the stability of each model, compute relevant transfer functions, and compare your results with the results of Problems 11.3–11.5.

13.16   Find a linearized sampled-data model for the phase-controlled rectifier drive in Problem 12.11, relating perturbations $\tilde{\alpha}$ in the firing angle to perturbations $\tilde{i}_d$ in the armature current. Determine the associated transfer function.

13.17   Find a linearized sampled-data model for the resonant dc/dc converter in Problem 12.7 and determine the associated characteristic polynomial. Using the notation in Example 12.3, let $L = 200\ \mu\mathrm{H}$, $C = 100$ nF, $V_1 = 14$ V, and $f_S = 40$ kHz. Plot the locus of natural ratios as V_2 varies from 0 V to 13 V.

13.18   This problem suggests one of many approaches to finding a state-space *realization* of a scalar, rational transfer function $h(s)$, that is, a state-space model whose transfer function is $h(s)$.

(a) Suppose $h(s) = h_0/(s + a)$. Find a first-order realization.

(b) If $h(s) = (e_1 s + e_0)/(s^2 + a_1 s + a_0)$, verify that a realization of the form *(12.25)* and *(12.26)* is obtained by choosing

$$A = \begin{bmatrix} 0 & 1 \\ -a_0 & -a_1 \end{bmatrix} \quad B = \begin{bmatrix} 0 \\ 1 \end{bmatrix} \quad E = \begin{bmatrix} e_0 & e_1 \end{bmatrix} \quad F = 0$$

(c) An $h(s)$ whose numerator degree does not exceed its denominator degree can be expanded into a partial fraction expansion comprising a constant and terms are as simple as those in (a) and (b). Show how state-space realizations of each of these terms can be combined to yield a state-space realization of $h(s)$.

Chapter 14

Feedback
Control Design

THE subject of control design is vast. We can do no more in this chapter than suggest, through examples, how the dynamic models we have developed for power circuits can be mated with the systematic design methods available in the general control literature.

Most of the literature on control has been developed around general dynamic models specified in transfer-function or state-space form. The detailed development of appropriate models for a particular area is necessarily left to those working in it. This is why we devoted considerable effort in Chapters 11–13 to obtaining dynamic models for power circuits. These models take the form of circuits, transfer functions, or state-space descriptions. With these models in hand, we can connect with and make use of the large and active literature on control design.

We begin this chapter by building on the introduction to classical control presented in Section 11.5. Our concern there was to provide an overview of the issues of stability, performance, and robustness in the context of continuous-time LTI models and to show some of the resulting constraints on the loop and compensator transfer functions. Although we examined good and bad candidate controllers for power circuits in examples, we stopped short of illustrating the classical approach to actually *designing* good controllers. Section 14.1 lays out the basis for a widely used design strategy and applies this strategy to the same up/down converter that we used as a running example in Chapters 11–13.

Classical control also gives us the tools in many cases to deal with systems that have more than one possible feedback loop, as we show in Section 14.2. For example, many multiloop control systems can be satisfactorily designed by sequentially designing single feedback loops, as illustrated by the case of current-mode control. We again use the up/down converter to provide examples of multiloop control design.

Turning to control design with state-space models, we consider what can be accomplished by measuring and feeding back complete state information, rather than information only about the output to be controlled. Section 14.3 outlines some control design possibilities for such *state feedback* methods, which include various nonlinear control approaches. State-space methods in power electronics so far have been used primarily for modeling, simulation, or stability assessment, along the lines described in Chapters 12 and 13. We expect to see greater use of state-space methods and other modern approaches to control design, but to go further than the brief overview in Section 14.3 is beyond the scope of this book.

The preceding frameworks for control design with continuous-time models have natural parallels for discrete-time or sampled-data models, as we mention in Section 14.4. We use the case of a high-power-factor ac/dc converter as a simple example of discrete-time control design.

14.1 CLASSICAL CONTROL DESIGN

The configuration and notation we assume here are those introduced in Section 11.5 and redrawn more simply in Fig. 14.1. We showed in that section that certain important performance objectives could be translated into requirements for the nominal loop gain $\ell(j\omega) = h(j\omega)g(j\omega)$. These in turn impose requirements on the frequency response $h(j\omega)$ of the compensator for a given nominal plant frequency response $g(j\omega)$. We usually want a high loop gain magnitude at low frequencies, where the reference signals and disturbances are significant, and a small loop gain magnitude at high frequencies, where the measurement noise and modeling errors are large. For good transient performance, we want the crossover frequency ω_c (at which the magnitude of the loop gain is 1) to be as high as possible. The control design task in this setting is to choose the compensator so that these requirements are met, while maintaining stability of the closed-loop system.

To proceed further with this formulation, we need to relate the stability of the closed-loop system to the loop gain. The key result here is the Nyquist stability criterion.

14.1.1 The Nyquist Stability Criterion

Recall from Section 11.5.2 that for stability we require the sensitivity function $\mu(s)$ to not have any poles in the right half plane. Equivalently, its denominator $1+\ell(s)$ must not have any zeros in the right half plane. The Nyquist criterion allows us to check the latter condition by testing the loop gain $\ell(j\omega)$. We merely state the

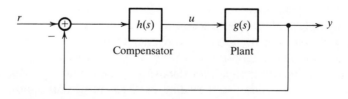

Figure 14.1   Configuration for classical control design.

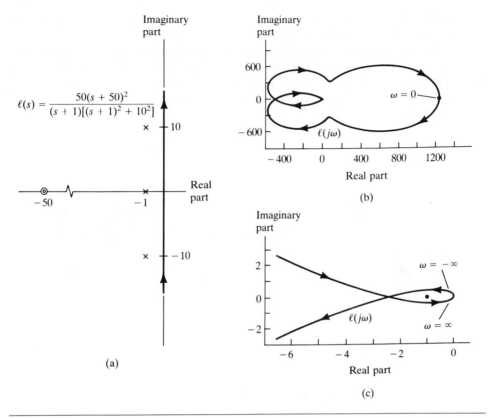

$$\ell(s) = \frac{50(s + 50)^2}{(s + 1)[(s + 1)^2 + 10^2]}$$

(a)

(b)

(c)

Figure 14.2   (a) Poles (**x**) and zeros (**o**) of a loop transfer function $\ell(s)$. (b) The Nyquist plot $\ell(j\omega)$. (c) The region near the origin of the Nyquist plot, on an expanded scale.

Nyquist criterion; you can find the proof in standard undergraduate texts on control. Our main concern is its implications for control design.

To obtain a simple statement of the Nyquist criterion, we restrict ourselves to the case where $\ell(s)$ has no poles on the imaginary axis, so that $\ell(s)$ is well defined for all imaginary values of s. This requirement would seem to rule out some of the idealized models that we routinely use, such as lossless LTI circuits and the ideal integrators introduced by PI controllers. However, if we take account of the small but inevitable parasitic losses associated with such circuits or components, their poles shift slightly into the left half plane. Of course, whenever we invoke such parasitic elements in order to obtain a loop gain that is well defined at all frequencies, we must check to see that any conclusions we eventually obtain are not very sensitive to the actual values assumed for the parasitics.

The Nyquist Plot   Consider the loop transfer function $\ell(s)$ whose pole/zero configuration is shown in Fig. 14.2(a). As s moves upward along the imaginary axis in Fig. 14.2(a), from $-j\infty$ to $j\infty$, $\ell(s)$ takes values along the path labeled $\ell(j\omega)$ in Fig. 14.2(b). This representation of the loop gain is termed its *Nyquist plot*. The region near the origin of the Nyquist plot is shown on an expanded scale in Fig. 14.2(c). The Nyquist plot is evidently well defined for any loop transfer function that does not have poles on the imaginary axis. The plot is symmetrical because $\ell(s)$ has real coefficients, so that $\ell(-j\omega) = \ell^*(j\omega)$, where * denotes the complex conjugate. We could therefore first plot $\ell(j\omega)$ as ω varies from 0 to ∞, and then reflect this *partial* Nyquist plot across the real axis to obtain the full Nyquist plot.

The Nyquist Criterion   Suppose that the Nyquist plot passes through the point -1 at some frequency ω_p, so that $\ell(j\omega_p) = -1$. In this case, $1 + \ell(s)$ has a zero at $s = j\omega_p$, the closed-loop system has a pole at this location on the imaginary axis, and the system is therefore unstable. Physically, the condition $\ell(j\omega_p) = -1$ means that a sinusoidal oscillation of frequency ω_p can be sustained in the closed loop.

   For the more typical situation where the Nyquist plot does not pass through -1, it can be shown that the number of unstable closed-loop poles equals the sum of:

1. the number of unstable poles of the loop transfer function $\ell(s)$; and
2. the net number of times that the Nyquist plot encircles the point -1 clockwise (with counterclockwise encirclements counted as negative clockwise encirclements).

This result is due to Nyquist, and his stability criterion follows from it immediately: The closed-loop system is stable if and only if the net number of counterclockwise encirclements of the point -1 equals the number of unstable poles of the loop transfer function $\ell(s)$.

   For the special but important case where $\ell(s)$ is stable, the Nyquist criterion tells us that the closed-loop system is stable if and only if the Nyquist plot has no net encirclements of the point -1. This is the case illustrated in Fig. 14.2(b) and (c), which show one clockwise encirclement and one counterclockwise encirclement.

   The simplicity and generality of the Nyquist criterion make it an exceptionally powerful tool. An immediate consequence, for example, is the result on robust stability that we quoted in Section 11.5.4: If the loop gain of a stable closed-loop system is perturbed from $\ell(j\omega)$ to $\ell(j\omega) + \tilde{\ell}(j\omega)$, while keeping the number of unstable open-loop poles unchanged, and if $|\tilde{\ell}(j\omega)| < |1 + \ell(j\omega)|$, the system remains closed-loop stable. The inequality ensures that the Nyquist plot is not perturbed enough to change the number of encirclements of -1, so stability is preserved.

14.1.2 A Design Approach

We cannot easily proceed further in extracting useful general guidelines for control design without narrowing our focus a little. Thus we consider the design implications of the Nyquist criterion only for the special but important case of a *stable* loop transfer function $\ell(s)$. The details of the design approach that we outline next need to be modified for cases involving an unstable $g(s)$ and/or unstable $h(s)$, although the Nyquist criterion still plays a key role in any such modification. We add one other restriction: We always choose the compensator to give a large and positive loop gain at low frequencies, so $\ell(0) \gg 1$.

For any physical system, $\ell(s)$ has more poles than zeros, so that as ω increases from 0 to ∞ the magnitude of the loop gain eventually decreases steadily. The partial Nyquist plot thus begins on the positive real axis for $\omega = 0$ and has the general form of a contracting clockwise spiral as ω increases. For ω approaching ∞, the plot approaches the origin asymptotically at an angle of $-\delta\pi/2$ rad, where δ is the number of poles plus the number of right-half-plane zeros minus the number of left-half-plane zeros. There may be local departures from this general form, with counterclockwise movements around frequencies determined by the locations of the left-half-plane zeros and expansion rather than contraction in the vicinity of resonances.

With a stable $\ell(s)$, the Nyquist criterion tells us that for closed-loop stability we should have no net encirclements of the point -1. A convenient and usually very successful strategy for meeting this condition is to ensure that, as ω increases, the (partial) Nyquist plot enters the unit circle without encircling the point -1 and is then confined to the unit circle for higher frequencies. This is the case in Fig. 14.3(a). To obtain stability *robustness*, we should keep -1 *well away* from the region enclosed by the Nyquist plot, so that perturbations of $\ell(j\omega)$ caused by modeling errors do not shift the Nyquist plot so much that it encloses -1.

The magnified view of the desired form of the Nyquist plot shown in Fig. 14.3(a) marks the gain crossover frequency ω_c, namely, the frequency where the

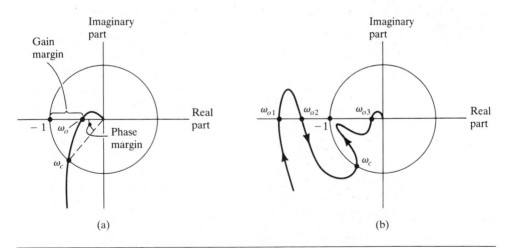

(a)   (b)

Figure 14.3   (a) Magnified view of the desired behavior of the Nyquist plot near the crossover points. (b) More complicated behavior near crossover.

loop gain magnitude is unity, $|\ell(j\omega_c)| = 1$. The amount by which $\angle\ell(j\omega_c)$ exceeds $-\pi$ rad is referred to as the *phase margin*. Also marked in the figure is the phase crossover frequency ω_o, defined by $\angle\ell(j\omega_o) = -\pi$ rad. The factor $1/|\ell(j\omega_o)|$ by which the loop gain magnitude at ω_o is smaller than unity is termed the *gain margin*.

The gain and phase margins serve as simple measures of the distance of the Nyquist plot from the point -1 and hence as measures of stability robustness. For the plot in Fig. 14.3(a), the phase margin is about $\pi/4$ rad (or 45°), and the gain margin is about 2 (or 6 dB). These margins allow us to tolerate an additional phase delay of $\pi/4$ rad in the loop gain at the gain crossover frequency, or a doubling of the loop gain magnitude at the phase crossover frequency, before the Nyquist plot encloses -1 and the closed-loop system goes unstable. We would consider such margins to be safe if we were fairly sure of our models in the vicinity of the crossover points.

However, the situation may be too complicated to convey with just a pair of crossover points. For example, the Nyquist plot shown in Fig. 14.3(b) has multiple crossover frequencies, each of which may be given an interpretation. The phase crossover at ω_{o2}, for instance, indicates that *reducing* the loop gain magnitude by the factor $|\ell(j\omega_{o2})|$ causes the system to go unstable, which shows that we have a *conditionally stable* system. (Transients that produce saturating signals often lead to behavior similar to that encountered by reducing the loop gain in a linear model. This leads us to avoid conditionally stable designs.) In this particular instance, however, even knowing the characteristics of all the crossover frequencies is not enough to indicate that the Nyquist plot goes disastrously close to -1 for a frequency between ω_c and ω_{o3}. Although we can often deduce the main features of a Nyquist plot by examining a few points on the plot, we cannot always do so.

14.1.3 Using Bode Plots

The information in the (partial) Nyquist plot can be displayed equivalently by a pair of plots showing the magnitude and phase of the loop gain as functions of frequency. As mentioned in Section 11.5.3, these plots are called *Bode plots* when we use log–log scales to display the magnitude and linear–log scales to display the phase. Figure 14.4(a) shows the Bode plots corresponding to the pole/zero configuration and Nyquist plot in Fig. 14.2. The log-magnitude is traditionally measured in decibels (dB), with the loop gain in dB being $20 \log_{10}|\ell(j\omega)|$. The frequency axis is marked off in *decades*, with each decade representing a factor of ten in frequency.

Note that the Bode plots of a *product* of two frequency responses are obtained by *adding* the respective Bode plots of the two factors, because the log-magnitudes and phases of the factors add. For example, as:

$$\log|\ell(j\omega)| = \log|h(j\omega)| + \log|g(j\omega)| \tag{14.1}$$

and

$$\angle\ell(j\omega) = \angle h(j\omega) + \angle g(j\omega) \tag{14.2}$$

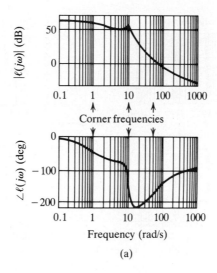

Figure 14.4 (a) Bode plots for the frequency response associated with the pole/zero configuration and Nyquist plot in Fig. 14.2.

we can easily obtain the Bode plots of the loop gain $\ell(j\omega)$ by adding the plots of the frequency response $h(j\omega)$ of the compensator to those of the frequency response $g(j\omega)$ of the power circuit. The Bode plots of $h(j\omega)$ or $g(j\omega)$ in turn are the sums of the plots associated with their individual poles and zeros, which happen to be easy to sketch. These characteristics and the connection with the Nyquist plot make Bode plots very useful tools for design exploration.

Figure 14.4(b) shows the Bode plots for some factors associated with real and complex poles and zeros. The Bode plots of the *reciprocals* of these factors are simply derived from the given plots: They are just the *negatives* of the plots in Fig. 14.4(b).

Note that the "corner" frequency in each log-magnitude plot in Fig. 14.4(b) is directly determined by the associated pole or zero location. Except for a small range of frequencies around the corner frequency, the slope at frequencies below the corner frequency is essentially 0, and above the corner frequency is essentially an integer multiple of 20 dB/decade. This integer multiple is -1 for each pole and $+1$ for each zero. Hence, for a complex pole or zero pair, the multiple is -2 or $+2$, respectively. The detailed behavior in the vicinity of the corner frequency for a complex pair is determined by the relative damping. The lighter the damping, the sharper is the resonance at the corner frequency (as noted in Section 9.1.2).

The phase plots also have a simple structure. The phase a low frequencies is essentially 0 rad for a real pole or zero in the left half plane and π rad for one in the right half plane. Complex pole or zero pairs have a phase of essentially 0 rad at low frequencies. The phase changes by an integer multiple of $\pi/2$ rad as ω increases past the corner frequency. This integer multiple is -1 for each

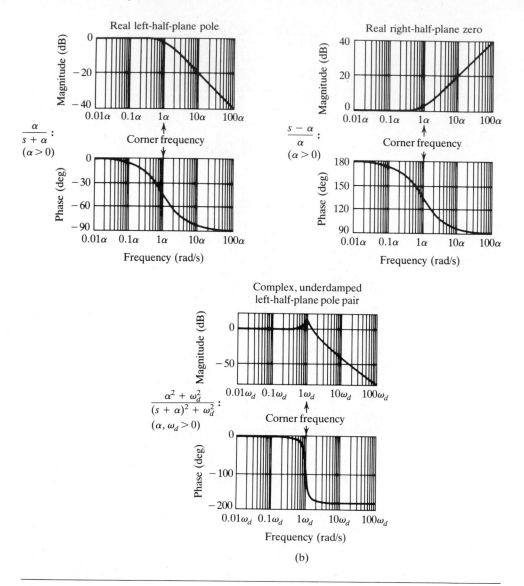

Figure 14.4 (cont.) (b) Bode plots for frequency responses of some pole and zero factors.

left-half-plane pole or right-half-plane zero and +1 for each left-half-plane zero or right-half-plane pole. The change in phase happens mainly in the decade below and above the corner frequency. For a real pole or zero, the change is approximately uniform over these two decades. For a complex pair, the lighter the damping, the sharper the phase change is around the corner frequency.

The Bode plots for a more complicated frequency response, such as that in

Fig. 14.4(a), can be understood as the sum of the plots for its constituent poles and zeros. The log-magnitude plot has slopes that are essentially integer multiples of 20 dB/decade between the corner frequencies of these poles and zeros, and the phase plot is approximately an integer multiple of $\pi/2$ rad, except for transition regions around the corner frequencies. Of course, when two corner frequencies are less than two decades apart, the Bode plots in the frequency interval between them are dominated by features of the transition.

14.1.4 Designing the Bode Plots of the Loop Gain

The preceding discussion and the study of simple cases such as those in Fig. 14.4 show that, so long as we are away from the corner frequencies, we can actually approximate well the phase of the loop gain at a frequency ω from the slope of its log-magnitude Bode plot and some auxiliary information about right-half-plane zeros. For the case of stable $\ell(s)$ with $\ell(0) > 0$, which is what we are restricting ourselves to, the result is as follows.

Suppose the slope of the log-magnitude plot at ω is $20n(\omega)$ dB/decade, so $n(\omega)$ is approximately an integer in the regions between corner frequencies. Then, if $\ell(s)$ has no right-half-plane zeros with corner frequencies below ω, $\angle \ell(j\omega) \approx n(\omega)\pi/2$ rad; if the slope is approximately -20 dB/decade, the phase is approximately $-\pi/2$ rad, and so on. For the more general case where there are $r(\omega)$ right-half-plane zeros whose corner frequencies are below ω, the result is

$$\angle \ell(j\omega) \approx [n(\omega) - 2r(\omega)]\pi/2 \text{ rad} \qquad (14.3)$$

The approximate result in (14.3) has the virtue of being simple enough to yield simple design guidelines. A candidate compensator obtained by such approximations can be subjected to a more refined analysis on a second pass through the design.

Our design approach in Section 14.1.2 has two immediate consequences. As indicated in Fig. 14.3(a), we require the phase at the gain crossover frequency ω_c to be greater than $-\pi$ rad by an amount equal to the phase margin. Hence (14.3) implies the constraint $n(\omega_c) - 2r(\omega_c) > -2$. At crossover $n(\omega_c)$ must be negative, and if we limit ourselves to crossing over away from corner frequencies, it must be approximately an integer. We must therefore have $r(\omega_c) = 0$ and $n(\omega_c) = -1$. Our conclusions are then the following.

1. We must obtain crossover below the corner frequencies of all right-half-plane zeros. Thus right-half-plane zeros impose a limit on the attainable closed-loop bandwidth. It turns out that the limitations imposed by right-half-plane zeros are fundamental, not a consequence of the particular design approach that we have followed.

2. The slope of the log-magnitude of the loop gain at crossover must be approximately -20 dB/decade. This sets a limit on how fast we can roll off the loop gain magnitude from the desired large values at low frequencies to the required low values at high frequencies.

Before proceeding, we simplify our terminology. For the rest of this chapter, whenever we refer to the slope of a log-magnitude plot at ω, we mean $n(\omega)$ itself rather than the slope in dB/decade.

EXAMPLE 14.1

Compensator Design for an Up/Down Converter under Duty-Ratio Control

We showed in Examples 11.1, 11.3, and 11.11 and in the discussion preceding Example 11.12 that some simple approaches to controlling an up/down converter led to unsatisfactory results. We tried open-loop control and proportional feedback control, with and without feedforward. We now follow the design approach just outlined.

Nominal Model We use the linearized model obtained in Examples 11.9 and 13.2. In those examples we derived the nominal transfer function $g(s)$ between duty ratio perturbations and output voltage perturbations. The pole/zero diagram and Bode plots of the associated frequency response were shown in Fig. 13.2, using the parameter values given in Example 11.1. The Bode plots are repeated here in Fig. 14.5(a).

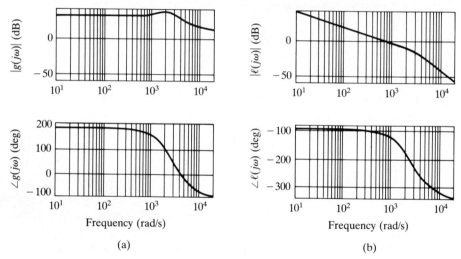

Figure 14.5   (a) Bode plots of the nominal frequency response of the up/down converter. (b) Bode plots of the loop gain with integral control.

The resonant peak in the log-magnitude plot marks the corner frequency associated with the complex pole pair. Beyond this, the log-magnitude starts to roll off with a slope of -2, but changes to a slope of -1 after the corner frequency associated with the right-half-plane zero. The phase angle correspondingly decreases by π radians in the vicinity of the resonance and a further $\pi/2$ in the vicinity of the corner frequency of the right-half-plane zero. Note that $\angle g(0) = \pi$ rad, that is, $g(0)$ is negative because of our choice of polarity for the definition of the output voltage v_o. Hence we require that the dc gain $h(0)$ of the compensator be negative in order to get $\ell(0) > 0$.

Compensator Design For the particular parameter values in our example, there is little separation between the corner frequencies of the pole pair and the zero, which makes designing for crossover in this region difficult. Moreover, such a design would be quite sensitive to the detailed features of the Bode plots in this region, and these features (such as the shape of the resonant peak) depend strongly on the component values and duty ratio. Let's therefore be less ambitious and aim for a loop crossover frequency that is below the resonant frequency.

Because the slope of $|g(j\omega)|$ below resonance is 0, the compensator's log-magnitude must roll off with slope -1 at crossover. A simple way to obtain a high loop gain at low frequencies and to roll off with a slope of -1 is to introduce a pole at (or near) the origin into the loop transfer function, so as to obtain (approximate) integral control action. Let's therefore pick $h(s) = -\beta/s$ (or $-\beta/(s + \epsilon)$ for some small $\epsilon > 0$), where β is a positive constant that we choose so as to obtain the desired crossover frequency. The negative sign in $h(s)$ is needed to obtain $h(0) < 0$.

The Bode plots of the resulting loop gain are shown in Fig. 14.5(b). Note that beyond crossover the loop gain magnitude continues to roll off, and its slope settles down to -2 beyond the corner frequency of the right-half-plane zero. To obtain a good gain margin and to limit the effects of modeling errors and switching ripple, we must sustain the roll-off beyond crossover. (A more refined converter model that included an equivalent series resistance or ESR of value R_C in series with the capacitor C would show the presence of a further zero at $-1/R_C C$. To counteract the effect of the ESR zero and maintain the roll-off beyond crossover, we could include an additional pole at approximately $-1/R_C C$ in the compensator.)

The integrator gain β fixes the crossover frequency ω_c. We need to pick ω_c small enough that the resonant peak in the loop gain magnitude is sufficiently below unity gain to give an adequate gain margin. The magnitude of the peak without the integrator is around 31 dB at 2000 rad/s. To obtain a gain margin of 9 dB, for example, we must pick β so that the integrator introduces an attenuation of 40 dB at 2000 rad/s. This requirement leads to the choice $\beta = 20$. The corresponding crossover frequency can be computed to be around 740 rad/s.

Evaluation We need to analyze and simulate several aspects of the performance of the preceding design in order to understand the trade-offs involved and to develop a basis for modifying or accepting the design. We describe some of the ingredients of such an evaluation here.

The block diagram in Fig. 14.6(a) represents the linearized model of the closed-loop converter under integral control. It shows the integral feedback from output voltage perturbations to duty-ratio perturbations and also the dependence of the output perturbations on perturbations in the supply voltage. We have already derived the various transfer functions shown; see *(11.8)* and *(11.18)* in particular. An easy computation shows the *audio susceptibility* transfer function to be

$$\frac{\tilde{v}_o(s)}{\tilde{v}_{\text{in}}(s)} = \frac{-(D'D/LC)s}{s[s^2 + (1/RC)s + (D'^2/LC)] - \beta(I_L/C)[s - (V_{\text{in}}/LI_L)]} \qquad (14.4)$$

This expression may be used to study the response of the closed-loop converter to small variations in the supply voltage. It generally also provides a reasonable approximation of the response to large variations. Note that—as expected with integral control—the dc value of this transfer function is 0, so the output is insensitive to constant offsets in the input voltage. The frequency response associated with this transfer function is shown in Fig. 14.6(b), and indicates poor rejection of input ripple at frequencies near 1700 rad/s. For example, the attenuation of input ripple at 120 Hz (740 rad/s) is less than 5 dB.

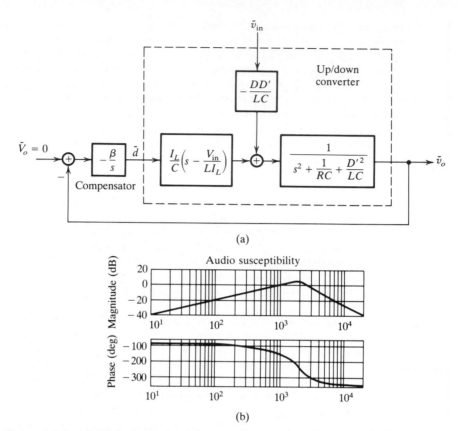

(a)

(b)

Figure 14.6 (a) Block diagram of the linearized model of the closed-loop converter with integral control. (b) Bode plots of the audio susceptibility frequency response, from input voltage perturbations to output voltage perturbations.

The denominator of the transfer function in *(14.4)* is the characteristic polynomial of the closed-loop system, and its roots yield the system poles. The locus of the closed-loop poles in our integral control approach, as the compensator gain β varies from 0 to its design value of 20, is shown in Fig. 14.7(a). The complex poles of the closed-loop system are slightly closer to the imaginary axis than the poles of the open-loop converter alone. We therefore expect the time-domain response in closed loop to settle somewhat more slowly after a disturbance than in the open-loop case.

Figure 14.7(b) shows simulations of the response of an open-loop design with feedforward of the supply voltage, as in Fig. 11.3(b), Example 11.1, and the response of our closed-loop design here, without feedforward, when the source voltage steps from 12 V to 8 V. These simulations use a nonlinear averaged model that is less idealized than the model used for the simulations in Fig. 11.3. We model the diode and transistor as nonideal switches, representing them by a 10 mΩ resistor in series with a 0.4 V dc source when they conduct. We also assume a capacitor ESR of 100 mΩ and an inductor ESR of 40 mΩ. As a result, the open-loop scheme, despite its feedforward, does not perform as well as it did in Example 11.1 and has significant steady-state error. Offsets in the nominal duty ratio,

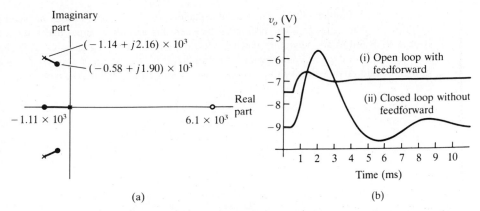

(a)   (b)

Figure 14.7 (a) Pole–zero diagram showing the open-loop poles (**x**) and zero (**o**) of the loop transfer function and the closed-loop poles (•). (b) Response of the nonideal up/down converter model to a step change in source voltage from 12 V to 8 V: (i) open loop with feedforward; (ii) closed loop without feedforward.

due to uncertainty about the exact value of V_{in} and implementation constraints, would cause additional error.

The initial response of the closed-loop system without feedforward resembles that of the open-loop system without feedforward, which we presented in Fig. 11.3(a). The output voltage magnitude changes by over 3 V before the feedback through the integral control becomes strong enough to pull it back toward nominal. Note that we manage to attain zero steady-state error despite the nonidealities in the circuit. The transient response of the closed-loop system would be far better if we incorporated feedforward.

The integral control approach here is *not* robust to decreases in the load. For instance, if the load were dropped a factor of 10 by increasing the load resistor R from 2 Ω to 20 Ω, the resonant peak of the open-loop system would increase by about 20 dB, the distance of the open-loop poles to the imaginary axis would drop by a factor of 10, and the right-half-plane zero would move farther out by a factor of 10. If we retained the integral control designed for a load of 2 Ω, the closed-loop system would be unstable. We evidently need a different controller if such load variations are anticipated. The current-mode control design described later in this chapter is better behaved in this respect.

Implementation Figure 14.8 is a schematic representation of a possible implementation of our integral compensation, and you should compare it with Fig. 11.9, Example 11.6,

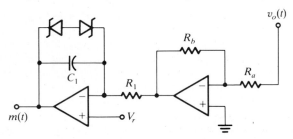

Figure 14.8 Schematic representation of a possible integral control implementation for the up/down converter.

and Fig. 11.20, Problem 11.11. The output voltage $v_o(t)$ is first fed through an inverting amplifier of gain $-R_b/R_a$. The difference between this amplifier output and a reference voltage V_r is fed to an integrator, which produces the modulating signal $m(t)$. This signal is compared with the sawtooth waveform of amplitude K to generate the duty ratio. Some straightforward computations show that the relationships among the parameters are

$$-\frac{R_a}{R_b}V_r = V_{\text{ref}}$$

$$\frac{V_r}{K} = D \qquad\qquad (14.5)$$

$$\frac{1}{KR_1C_1}\frac{R_b}{R_a} = \beta$$

Here V_{ref} is the reference value for the output voltage, which we specified to be -9 V in Example 11.1, D is the nominal duty ratio, 0.43, and $\beta = 20$, according to the computation we have just made.

The Zener diodes across the integrator capacitor C_1 in Fig. 14.8 limit $m(t)$ to an acceptable range between 0 and K, preventing *integrator windup*. Without these diodes, C_1 could charge up during extended transients to values well outside the range in which its voltage affects the duty ratio—and might then take a long time to come back within the range where the duty ratio responds to the feedback control signal. Usually, V_r is raised gradually to its nominal value when the circuit is first powered up. This so-called *soft start* limits the rate at which the duty ratio rises initially, and thereby reduces stresses in the circuit during start-up.

Alternative Compensators   A simple integral control is not the only possibility for up/down converters in this classical single-loop feedback configuration. By using a more elaborate and less conservative controller, we may be able to obtain a higher crossover frequency and closed-loop bandwidth, while maintaining an adequate stability margin. For example, the compensator shown in Fig. 14.9(a) can accomplish the task if there is sufficient sep-

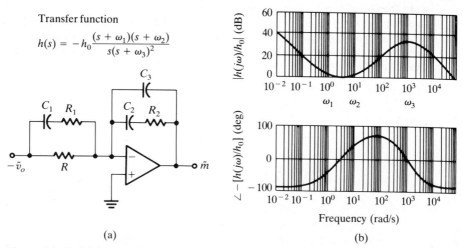

Transfer function

$$h(s) = -h_0\frac{(s + \omega_1)(s + \omega_2)}{s(s + \omega_3)^2}$$

(a)

(b)

Figure 14.9   (a) Alternative compensator. (b) Bode plots of its frequency response.

aration between the resonant frequency of the converter transfer function and the corner frequency associated with its right-half-plane zero. (With the particular parameter values in our example, the separation is less than a decade, which is too small.) The Bode plots of this compensator's frequency response are given in Fig. 14.9(b). If we pick ω_1 and ω_2 to lie below but near the resonant frequency of the up/down converter, set ω_3 to lie below the corner frequency of its right-half-plane zero, and choose the gain so that crossover occurs between ω_2 and ω_3, we may be able to obtain a faster closed-loop system while maintaining an adequate stability margin.

14.2 MULTILOOP CONTROL

The concepts and tools of classical control are aimed at designing feedback controllers for single feedback loops, where one measured signal is fed back to the controller, and one control signal is set by it. More commonly, however, several signals are available for feedback and several control signals need to be set. The usual way in which we use classical control to deal with this multiloop control task is to *sequentially* design single-input, single-output loops that are obtained by pairing the outputs and inputs in some fashion. The specific features of the system and control problem often suggest a natural pairing of signals and a sequence for the control design.

EXAMPLE 14.2

Speed Control of a dc Machine

Consider using a phase-controlled rectifier to control the rotor speed of a dc machine in the face of speed-dependent load torques. Our sole control signal is the firing angle α of the rectifier. The signals that can be fed back include the voltage v_d applied to the armature, the armature current i_d, and the rotor speed ω_r.

A block diagram of a typical multiloop "cascade" design for speed control is shown in Fig. 14.10. The outer loop (or major loop or speed loop) compares the actual rotor speed with the commanded speed. The error indicates the electromagnetic torque that needs to be applied to the motor. As the average armature current $\bar{i}_d$ (averaged over the source period)

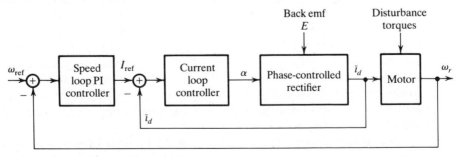

Figure 14.10 Multiloop configuration for speed control of a dc motor using a phase-controlled rectifier.

determines the average electromagnetic torque, the speed loop controller specifies a desired average armature current I_{ref}.

The inner loop (or minor loop or current loop) is designed as outlined in Examples 11.2, 11.10, and 11.12, to make $\bar{i}_d$ follow I_{ref} independently of the speed or, equivalently, the back emf. The purpose of this loop is thus to convert the phase-controlled rectifier into an approximate current source. So long as the variations in I_{ref} are considerably slower than the bandwidth of the current loop, we can assume that $\bar{i}_d \approx I_{ref}$. A more refined approximation would be

$$\bar{i}_d(s) \approx \frac{1}{sT_i + 1} I_{ref}(s) \qquad (14.6)$$

where T_i is the dominant time constant of the current loop. These approximations of the inner current loop greatly simplify the analysis of the outer speed loop. However, the approximations are reasonable only if we take care to make the bandwidth of the outer loop significantly smaller than that of the inner loop.

With the preceding approximations, we can easily design a PI controller for the speed loop, using the design strategy described earlier. If the closed-loop system is stable and driven only by constant signals, the output of the PI controller is constant in the steady state. Its input must then be 0; that is, the speed error must be 0. The output of the PI controller in steady state is thus automatically set to the value of I_{ref} needed to generate the torque required by the load. We omit further consideration of the details. Once an initial design has been obtained, a more detailed and accurate model for the current loop can be used to assess overall system performance more accurately, using simulations and analysis. The initial design can then be refined.

The same cascade approach can also be extended to provide position control. For this, a position control loop is constructed outside the speed loop, with the position error being the input to a position controller. A proportional control in this loop ensures that when the position error is 0, the speed command is 0.

14.2.1 Current-Mode Control

A similar multiloop approach in the case of high-frequency switching converters leads to the idea of current-mode control, which we mentioned in Examples 11.6, 12.5, 13.4, and 13.6. Rather than having the controller set the duty ratio in each cycle, we cause it to set the peak inductor current or peak switch current in each cycle. The most commonly used method involves turning on the switch every T seconds and turning it off when the inductor current i_L or the switch current reaches the specified peak value i_P—or a threshold related to i_P. The duty ratio thus becomes an auxiliary variable, implicitly determined by i_L and the control variable i_P. In this section we discuss the case of converters operating in continuous conduction. The discontinuous conduction mode is actually easier to analyze, and we leave it to you to determine the requisite modifications, with Examples 11.5 and 11.8 as a starting point.

Figure 12.9 showed the key features of the inductor current waveform of a switched converter under current-mode control. The threshold signal for the inductor current in each cycle is obtained by subtracting a stabilizing ramp of slope S

from the control signal i_P. As we noted in Example 13.6, the stabilizing ramp entrains the converter waveforms, maintaining them at the switching frequency. The ripple in the inductor current waveform in Fig. 12.9 is somewhat exaggerated. In normal operation the ripple is usually smaller, although it needs to be large enough to permit robust operation of the threshold detection that underlies current-mode control. If the ripple is too small, sensing the current variations reliably in the presence of switching and other noise is difficult.

Ideally, current-mode control results in an inner loop that regulates the inductor current i_L and its average $\bar{i}_L$ at approximately i_P. The controller in the outer ("current programming" or voltage regulation) loop specifies the value of i_P, and thereby $\bar{i}_L$, needed to regulate the average output voltage $\bar{v}_o$ at a desired value. We now examine the conditions under which we come close to this ideal operation.

If ST and $V_{in}T/L$ are small, the offset between i_P and $\bar{i}_L$ is small in the steady state. For the offset to be small during transients as well, we must ensure that the *slew rate* V_{in}/L of the inductor current is significantly higher than the rate of change of i_P. We can interpret this second condition as requiring the outer loop to have a much smaller bandwidth than the inner loop.

Under the preceding conditions, we can neglect the offset and dynamics of the inner loop—the dynamics attributable to the inductor—in the initial design of the outer loop controller. This approach allows us to obtain a simplified model for the outer loop and makes the design of its controller much easier. We can then assess and refine the controller through simulation and analysis with more detailed and accurate models that take into account the offset and dynamics of the inner loop. Of course, if the conditions for approximately ideal operation are not satisfied, we are forced to work with the detailed models from the beginning.

We have already shown how to obtain more detailed dynamic models for current-mode control. Examples 12.5 and 13.4 showed how to obtain a sampled-data model for the dynamics of the outer loop, with the inner loop closed. We used this model in Example 13.6 to analyze stability of the inner loop and bring out the role of the stabilizing ramp. It could further be used to design sampled-data controllers for the outer loop. We have also suggested (following Example 13.4 and in Problem 13.10) that continuous-time averaged models for current-mode control can be developed analogously to sampled-data models.

The fact that current-mode control approximately transforms the inductor into a controlled current source not only facilitates multiloop control design but also underlies other properties and applications of current-mode control. For example, down (or buck or buck-derived) converters, which have no switch interposed between the inductor and the load, become relatively insensitive at the output to variations in the input voltage. Also, we can easily enforce power sharing when several paralleled converters under current-mode control feed a common load. Designing a compensator that can handle both continuous and discontinuous operation is usually simpler with current-mode control, because the inductor current dynamics does not affect either mode significantly. In up (or boost) converters, there is no switch between the inductor and the supply, so we can use current-mode control to draw a specified current waveform from the supply. This feature is commonly used in high-power-factor converters, as we shall see in Example 14.5.

Another advantage of current-mode control is that we obtain direct control of the peak switch current in each cycle, and hence an automatic current limiting capability. This capability allows us to protect the converter from the effects of overloads or transformer saturation. (In implementing such a capability, we must guard against premature turnoff caused by current spikes generated at the switching instants by nonideal behavior of the switches.)

EXAMPLE 14.3

Compensator Design for an Up/Down Converter under Current-Mode Control

We discussed duty-ratio control of an up/down converter in Example 14.1. Current-mode control provides an illuminating contrast.

Nominal Model   We begin by deriving a simplified averaged model for an up/down converter under current-mode control, using the approximation $\bar{i}_L \approx i_P$. Our starting point is the averaged model in Fig. 11.11, Example 11.7, repeated here in Fig. 14.11, but with $\bar{i}_L$ replaced by i_P.

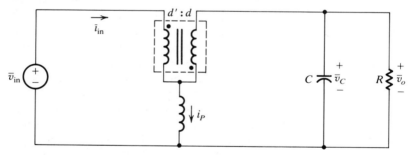

Figure 14.11   Averaged circuit of an up/down converter under current-mode control, assuming that $\bar{i}_L \approx i_P$.

We need a description of the circuit that does not involve the duty ratio d, as d is only implicitly determined under current-mode control. Because d just specifies the ideal transformer, a direct way to obtain a description that does not involve d is through a dynamic power balance on the averaged circuit. The transformer is a lossless element and so does not enter the power balance equation.

The fact that a "power" balance holds for the averaged quantities follows directly from *Tellegen's theorem*. This remarkable theorem states that $\sum_k v_k i_k = 0$, where $\{v_k\}$ and $\{i_k\}$ are *any* sets of variables that satisfy Kirchhoff's voltage and current law equations, respectively, on a given circuit topology. The sum is taken over all the branches in the circuit (with a consistent sign convention for the voltages and currents on all the branches). The voltages and currents in our averaged circuits satisfy KVL and KCL, so we can apply this equation—which we are calling a power balance—to them.

For the circuit in Fig. 14.11, a power balance yields

$$\bar{v}_C C \frac{d\bar{v}_C}{dt} + \bar{v}_C \frac{\bar{v}_C}{R} + L \frac{di_P}{dt} i_P = \bar{v}_{\text{in}} \bar{i}_{\text{in}} = \bar{v}_{\text{in}} \left(i_P + C \frac{d\bar{v}_C}{dt} + \frac{\bar{v}_C}{R} \right) \qquad (14.7)$$

We can now linearize this around a constant nominal operating condition and express the resulting LTI model in the transform domain. Denoting nominal values by capitals and perturbations by $\sim$ and replacing $\bar{v}_C$ by the average output voltage $\bar{v}_o$, we find

$$\tilde{v}_C(s) = \tilde{v}_o(s) = \frac{(V_{\text{in}} - sLI_P)R}{sRC(V_o - V_{\text{in}}) + (2V_o - V_{\text{in}})} \tilde{i}_P(s)$$

$$+ \frac{V_o + RI_P}{sRC(V_o - V_{\text{in}}) + (2V_o - V_{\text{in}})} \tilde{v}_{\text{in}}(s) \qquad (14.8)$$

Using the steady-state relationships $V_o = -(D/D')V_{\text{in}}$ and $I_P D' = -V_o/R$ to simplify these expressions, we obtain

$$\tilde{v}_o(s) = a_0 \frac{1 - sT_z}{1 + sT_c} \tilde{i}_P(s) + b_0 \frac{1}{1 + sT_c} \tilde{v}_{\text{in}}(s) \qquad (14.9)$$

where

$$T_c = \frac{1}{1 + D} RC \qquad T_z = \frac{LI_P}{V_{\text{in}}} = \frac{D}{D'^2} \frac{L}{R}$$

$$a_0 = -\frac{D'}{1 + D} R \qquad b_0 = -\frac{D^2}{D'(1 + D)} \qquad (14.10)$$

For the parameter values in Example 11.1, the control-to-output transfer function $g(s)$ in (14.9) has a pole at $-(1/T_c) = -3247$ rad/s, and a right-half-plane zero at $1/T_z = 6095$ rad/s. Compare these pole and zero locations with those in Fig. 13.2 for the case of duty-ratio control. The zero locations are identical. However, there is only one pole for current-mode control because we have neglected the dynamics of the inner loop. This pole is better damped than the poles that govern duty-ratio control.

The negative sign of $g(0)$ is an artifact of our sign convention for the output voltage. We cancel it by choosing our compensator transfer function $h(s)$ to also have a negative dc value, so that the dc loop gain $g(0)h(0)$ will be positive. To avoid having to account for these signs in the discussion that follows, we work in terms of $-g(s)$ and $-h(s)$. The Bode plots of the frequency response $-g(j\omega)$ are shown in Fig. 14.12(a). For our parameter values, the corner frequencies of the pole and zero are separated by less than a factor of 2, so the corners are not really visible.

Compensator Design   In order to have a good closed-loop bandwidth, we want a compensator that results in a crossover frequency not far below the frequency of the zero. Note that the frequency response of $-g(s)$ in Fig. 14.12(a) has a magnitude slope of about -1 in the region between the pole and zero frequencies, with an associated phase of $-\pi/2$ rad at a frequency of around 4000 rad/s. We therefore aim to have crossover occur at this frequency, with $-h(s)$ contributing little gain or phase at this frequency. However, we want the compensator to provide significant gain at lower frequencies and significant attenuation at higher frequencies. The Bode plots for the frequency response of a $-h(s)$ that has the

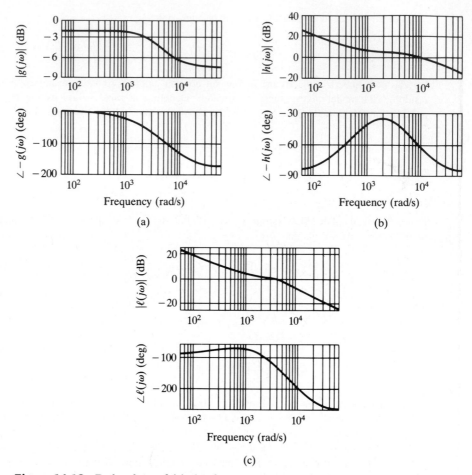

Figure 14.12   Bode plots of (a) the frequency response of $-g(s)$, where $g(s)$ is the control-to-output transfer function; (b) the frequency response of $-h(s)$, where $h(s)$ is the transfer function of the proposed compensator; and (c) the loop gain with this compensator.

required features are shown in Fig. 14.12(b), corresponding to the transfer function:

$$h(s) = -h_0 \frac{1 + (s/\omega_1)}{s[1 + (s/\omega_2)]} \qquad (14.11)$$

with $h_0 > 0$. This transfer function is easily realized with an op-amp circuit like that in Fig. 14.13.

For an initial attempt at picking compensator parameters, let's choose the corner frequencies of the compensator to be $\omega_1 = 600$ rad/s and $\omega_2 = 6000$ rad/s. Then $|h(j\omega)|$ at $\omega = 4000$ rad/s is $1.402 \times 10^{-3} h_0$, while $|g(j\omega)| = 0.603$. To obtain crossover at this frequency we must have $h_0 = 1/(1.402 \times 10^{-3} \times 0.603) = 1183$. The loop gain corresponding to this choice is shown in Fig. 14.12(c). The crossover is at 4000 rad/s, as expected, and the phase margin is around 50°.

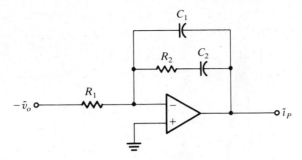

Figure 14.13 An op-amp realization of the transfer function $h(s)$.

Evaluation We mention only a few of the many tests that should be made in evaluating the candidate design. It is important first to check that the conditions we used to justify setting $\tilde{\imath}_L \approx i_P$ are indeed satisfied. We leave it to you to make the checks.

The response of the closed-loop converter to small input voltage variations can be studied using the linearized model shown in Fig. 14.14(a). To determine the response of the converter to an input step as large as -4 V, corresponding to the input voltage dropping from 12 V to 8 V, we may wish to use a more detailed nonlinear model. Figure 14.14(b) shows the response of both the linearized averaged model and a nonlinear switched model to a -4 V step in the input voltage. The nonlinear simulation assumes an ideal converter and uses $S = -V_o/L$ (the "optimum" slope, in the sense that it causes inductor current perturbations to disappear in essentially one cycle). Note the improvement over the case of duty-ratio control. Another difference from duty-ratio control is that increasing the load resistance from 2 Ω to 20 Ω actually improves the response to the input step. Again, we leave it to you to check this.

Further Issues Numerous other issues need to be dealt with before the design can be considered anywhere near complete. We may want to explore what happens with other compensator choices. For instance, a simple integral control can be used if we are satisfied with a crossover frequency sufficiently below the converter's pole frequency of 3247 rad/s.

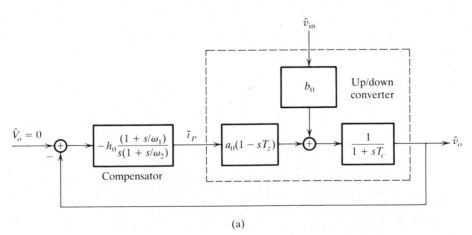

(a)

Figure 14.14 (a) Block diagram of the linearized model of the closed-loop converter.

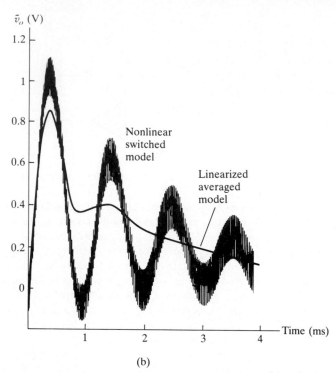

(b)

Figure 14.14 (cont.)   (b) Response of the closed-loop converter to a step change in the input voltage from 12 V to 8 V, estimated using the linearized averaged model and a switched nonlinear model.

Also, either compensator will probably need additional roll-off beyond crossover, to cancel the effect of the zero contributed by capacitor ESR.

Various implementation issues need to be dealt with. We need to decide whether to sense the switch current or the inductor current and whether to do this with a sense resistor or a current transformer. The slope S of the stabilizing ramp has to be fixed on the basis of the largest duty ratio that we expect to encounter. Note that when the supply drops to 8 V in our example, we have $D > 0.5$, so without the stabilizing ramp the converter would not operate at the switching frequency. The control circuitry has to be arranged so that signal strengths are appropriate at each point, to permit reliable operation in the presence of various sources of noise, such as current spikes at the switching instants. More detailed consideration of such issues is beyond our scope here.

Sequential design of multiloop control systems is not always possible or useful. In many cases, we have to consider all possible feedback paths simultaneously. This leads to the study of matrix generalizations of the classical LTI control results we have been using so far. We then have to work with plant, compensator, and loop transfer *matrices* that act on *vectors* of signals.

The requisite matrix extensions of classical control theory to deal with issues of stability, performance, and robustness in multi-input, multi-output systems have been under development over the last three decades. Although there are natural extensions of most classical LTI results, there are also unavoidable complications and subtleties. Nevertheless, matrix extensions of most classical results are now available and well supported by various computer tools. Their study has also opened up new approaches to control design, which are active areas of research. However, it is well beyond the scope of this book to consider the applications of these extensions to power electronics.

In the rest of this chapter, we illustrate only how feedback may be used in a multiloop system to alter the stability properties of a nominal model and modify the rate at which an operating condition is attained. We do little more than mention other aspects of performance or issues of robustness.

EXAMPLE 14.4

Multiloop Control of an Up/Down Converter

In Example 14.1 we examined duty-ratio control of switching converters, in which the duty ratio is perturbed as a dynamically varying function of perturbations in the average output voltage. We now consider a modification of this approach, taking a cue from the current-mode control scheme discussed in Example 14.3. Let's make the duty-ratio perturbations a function of perturbations in the average output voltage *and* the average inductor current (not the instantaneous inductor current, which is what current-mode control is based on). The particular control law that we have in mind sets

$$\tilde{d}(t) = -h_I \tilde{\imath}_L(t) - h_V \tilde{v}_o(t) - h_N \int^t \tilde{v}_o(\xi) d\xi \tag{14.12}$$

where h_I, h_V and h_N are *constant* feedback gains, which we choose so that the closed-loop system has specified natural frequencies.

The last term in *(14.12)* represents integral control action. If the closed-loop system is stable and driven only by constant signals, all variables must settle to constant values in the steady state. The integrand $\tilde{v}_o$ in *(14.12)* must therefore settle to zero to prevent the integral from contributing a time-varying term to the equation. Hence, so long as we can find feedback gains that stabilize the system, we can guarantee zero steady-state error in the output voltage.

With *(14.12)* serving as a model for the feedback path, we need to introduce the transfer functions that constitute our model of the forward path. In Examples 11.7, 11.9, 13.1, and 13.2, we have already shown how to derive the transfer functions in the following transform domain relationships:

$$\tilde{\imath}_L(s) = \frac{1}{a(s)}[b_1(s)\tilde{d}(s) + b_2(s)\tilde{v}_{\text{in}}(s)]$$

$$\tilde{v}_o(s) = \frac{1}{a(s)}[c_1(s)\tilde{d}(s) + c_2(s)\tilde{v}_{\text{in}}(s)] \tag{14.13}$$

Our earlier results show that for the up/down converter (with no capacitor ESR):

$$a(s) = s^2 + \frac{1}{RC}s + \frac{D'^2}{LC}$$

$$c_1(s) = \frac{I_L}{C}\left(s - \frac{V_{in}}{LI_L}\right) \qquad (14.14)$$

$$c_2(s) = \frac{-D'D}{LC}$$

Some straightforward computations similarly show that:

$$b_1(s) = \frac{V_{in}}{LD'}\left(s + \frac{1+D}{RC}\right)$$

$$ \qquad (14.15)$$

$$b_2(s) = \frac{D}{L}\left(s + \frac{1}{RC}\right)$$

To get an equation that describes the closed loop, we now substitute *(14.13)* in the Laplace transformed version of *(14.12)*. Some rearrangement of the resulting equation then yields

$$\left(s[a(s)+h_I b_1(s) + h_V c_1(s)] + h_N c_1(s)\right)\tilde{d}(s)$$

$$= -\left(s[h_I b_2(s) + h_V c_2(s)] + h_N c_2(s)\right)\tilde{v}_{in}(s) \qquad (14.16)$$

In *(14.16)*, $\tilde{v}_{in}$ represents an external drive, and $\tilde{d}$ is now an internal variable of the closed feedback loop. This equation allows us to identify the characteristic polynomial of the closed-loop system as:

$$a_H(s) = s[a(s) + h_I b_1(s) + h_V c_1(s)] + h_N c_1(s) \qquad (14.17)$$

For the up/down converter, *(14.17)* is a third-order polynomial with three gains, so it turns out that we can actually obtain an *arbitrary* characteristic polynomial by picking the gains appropriately.

To illustrate, we can use the parameter values of the up/down converter in Example 11.1 and pick the feedback gains to place the closed-loop eigenvalues at the locations shown in Fig. 14.15(a), namely, $-2477 \pm j4170$ and -4000. (We mention the basis for this choice at the end of Section 14.3.1.) The corresponding closed-loop characteristic polynomial is $[(s + 2477)^2 + 4170^2](s + 4000)$. Equating it to the polynomial in *(14.17)* yields the gains $h_I = 0.119$, $h_V = -0.093$, and $h_N = -430$. Figure 14.15(b) shows the response of $\tilde{v}_o$, $\tilde{i}_L$ and the duty-ratio perturbations $\tilde{d}$ in the resulting LTI model for a -4 V step in $\tilde{v}_{in}$. This response provides a reasonable approximation to the response of the converter to a drop in supply from 12 V to 8 V. The response compares favorably with what we obtained using duty-ratio control and current-mode control.

Note that, even though it is possible in principle to obtain an arbitrary closed-loop characteristic polynomial, we need to keep in mind the various constraints and assumptions that underlie our model. Most importantly, the closed-loop time constants should be signif-

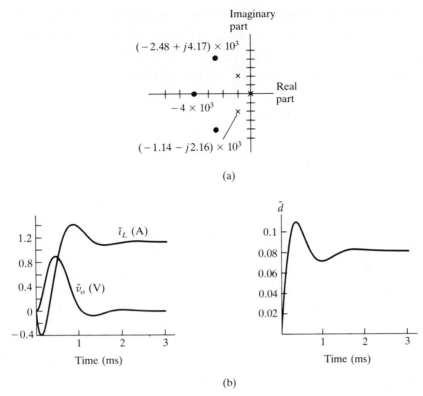

Figure 14.15 (a) Open-loop poles (**x**) and chosen closed-loop poles (•). (b) Closed-loop response of the linearized model to a −4 V step change in the input voltage.

icantly larger than twice the switching period, and the gains should be small enough for the duty-ratio perturbations to not leave the allowable range of $-D$ to D'. These considerations keep us from becoming too ambitious in our choice of closed-loop natural frequencies.

In the following section we show how to recognize the preceding simultaneous multiloop design as an instance of state feedback in an LTI model.

14.3 STATE FEEDBACK

The central idea of feedback control is to make the control inputs responsive to the state of the system. This immediately suggests that the variables that should ideally be measured and fed back to the controller are the state variables associated with some model of the system. If these variables cannot be measured directly, we can attempt to estimate them from available measurements. These simple ideas form the basis for a variety of linear and nonlinear control designs.

14.3.1 Pole Placement by LTI State Feedback

Consider the LTI state-space model *(12.25)* or *(13.10)*, repeated here for convenience:

$$\frac{dx}{dt} = Ax(t) + Bu(t) \qquad\qquad (14.18)$$

where the state vector x has n components and the control input u has m components. Suppose that we choose the control vector to be

$$u(t) = Hx(t) + w(t) \qquad\qquad (14.19)$$

where H is a constant state-feedback gain matrix of dimension $n \times m$, and $w(t)$ is some external input. The resulting closed-loop system is then governed by the LTI state-space model:

$$\frac{dx}{dt} = (A + BH)x(t) + Bw(t) \qquad\qquad (14.20)$$

Its stability is determined by its natural frequencies, which are the eigenvalues of $A + BH$ or, equivalently, the roots of its characteristic polynomial $a_H(s) = \det(sI - A - BH)$.

The multiloop control law in Example 14.4 is an example of LTI state feedback. Without integral control, that is, when $h_N = 0$ in *(14.12)*, the control is an LTI function of the state variables $\tilde{i}_L$ and $\tilde{v}_C$ of the converter and corresponds to LTI state feedback in the linearized averaged model *(13.8)*. When integral control is present, the control law corresponds to LTI state feedback in an *augmented* LTI state-space model whose state variables are $\tilde{i}_L$, $\tilde{v}_C$, and the output of the integrator used for integral control.

To what extent can the stability properties of the closed-loop system be modified from those of the open-loop system by choice of the state-feedback gain matrix H? In Example 14.4 we were able to obtain an arbitrary closed-loop polynomial by state feedback. This turns out to be a general property of LTI state feedback, provided the open-loop system satisfies a *controllability* condition. Controllability refers to the ability of the control inputs to independently affect all the state variables and can be checked if desired by certain tests on the matrix pair A, B. For a controllable LTI system, the closed-loop natural frequencies under LTI state feedback can be chosen arbitrarily (except for the constraint that complex natural frequencies should appear in conjugate pairs). The control literature usually refers to this as the *pole placement* property of LTI state feedback. The stability properties of a controllable LTI system can therefore be modified arbitrarily by LTI state feedback.

If the system is not controllable, some of the natural frequencies of the open-loop system remain natural frequencies of the closed-loop system, no matter how the control is chosen. This situation may not be of concern if these uncontrollable natural frequencies are sufficiently damped. If not, we must consider finding additional control inputs.

In low-order examples, we can proceed with control design for pole placement as in Example 14.4. First we compute the closed-loop characteristic polynomial $a_H(s)$ as a function of the entries of H. Then we equate it to the desired characteristic polynomial, that is, equate its n coefficients to those of the desired polynomial. The result is n equations involving the entries of H. If the system is controllable, these equations are guaranteed to be solvable for H. Note, however, that the n equations involve nm unknowns, so the solution will be unique only if $m = 1$, that is, only if there is just a single control variable. When $m \geq 2$, we can try to use the extra degrees of freedom in choosing H to satisfy other objectives, such as keeping the magnitudes of the entries of H within acceptable limits.

Let's now turn to the choice of closed-loop natural frequencies. Typically, the desirable region in the complex plane corresponds to an area similar to the shaded area in Fig. 14.16. This represents a compromise among several objectives. To get a sufficiently fast decay of the transients, we want the real parts of all the natural frequencies to be more negative than some negative number, say, $-\alpha$. However, we should not attempt to place these natural frequencies too far into the left half plane. For one thing, large feedback gains and correspondingly large control signals are needed to do so. Even if we can generate large control signals, they may move us out of the domain in which linear models are reasonable. Also, linear models such as our averaged models are often derived under the assumption that transients vary slowly. These are the reasons that the desirable region in Fig. 14.16 has a boundary at $-\beta$. Finally, we want the imaginary parts to be not much greater than the real parts in order to avoid unduly oscillatory behavior.

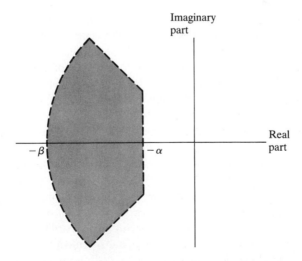

Figure 14.16 Desirable locations for the closed-loop natural frequencies.

Although the closed-loop pole locations determine important features of the transient response, they reveal little about other critical aspects of performance or about robustness. For example, pole locations alone do not tell us the peak values of circuit variables, or whether stability will be preserved when there are deviations from the nominal model. Hence pole placement is only an entry into the design process and has to be followed up by additional analysis, simulation, and modifications of the initial design.

LTI state feedback in discrete-time models is treated in exactly the same way, with the obvious changes in *(14.18)–(14.20)*. Again, if the open-loop system is controllable, the natural ratios of the closed-loop system can be placed arbitrarily by LTI state feedback. Placing all the natural ratios at 0 causes any transients to vanish in a finite number of time steps, resulting in what is called a *deadbeat* system (as mentioned in Section 13.6.2). Although it represents an attractive theoretical possibility, a deadbeat system generally requires large gains and can also be quite sensitive to variations in the gains or other parameters.

Linear–Quadratic Regulation   Pole placement is not the only route to choosing an appropriate state-feedback gain. An alternative approach has been extensively developed in the control literature and is well supported by computer tools. It is based (in the continuous-time case) on choosing the control u in *(14.18)* to minimize a quadratic cost of the form:

$$\int_0^\infty \left(\|x(\xi)\|^2 + r\|u(\xi)\|^2 \right) d\xi \qquad (14.21)$$

Here $\|x\|$ denotes the length of the vector x, and similarly for $\|u\|$. The positive weighting parameter r reflects the relative costs of, or penalties on, control and state excursions. More general quadratic costs—for example, costs that differentially penalize the different components of the state and control vectors—can be handled with very little additional difficulty. Note that the criterion does not directly penalize *peak* excursions of state or control variables. Large peak excursions of short duration could still yield a small integrated squared error.

The optimal control with this criterion is given by an LTI state feedback, $u(t) = H^*x(t)$. The optimal gain H^* can be computed from the parameters of the problem. The optimal control is guaranteed to stabilize the system, provided the system is controllable. This control law also is robust to significant gain and phase variations on the control inputs. For the case of a single control input, the optimal control law ensures that $|1 + \ell(j\omega)| \geq 1$, where $\ell(j\omega)$ is the loop gain. It follows from this that we have at least a 60° phase margin and can maintain stability when the gain drops by as much as a factor of 2 (6 dB), or when it increases indefinitely. Significant time varying nonlinearities can also be tolerated. Of course, all these properties are for the idealized model *(14.18)*. You must keep in mind the limitations of the model.

The preceding approach is referred to in the control literature as *linear–quadratic* or *LQ regulation*, because it involves a linear model and a quadratic cost criterion. In practice, we would simulate the closed-loop system with the optimal LQ gain for one choice of r (or of other weighting parameters, in more elaborate formulations). If the simulations show that state or control excursions will be excessive under anticipated operating conditions, we adjust r appropriately and iterate through the design. (The control law in Example 14.4 was actually designed using this LQ approach, although we presented it in the example as a pole-placement controller.) There is also a complete discrete-time parallel to the continuous-time LQ framework.

Observers   State-feedback controllers require measurements of the state variables. If these variables are not measured, they can be estimated from available measurements using an *observer*.

The core of an observer for the model *(14.18)* is a real-time simulator of the model, implemented as in Fig. 12.1. To the input of each integrator in the simulator we add a correction term. This term is proportional to the error between the available measurements and our estimates of these measurements, which we derive from the simulator. By choosing the proportional gains appropriately, we can ensure that the state variables of the simulator converge exponentially to those of the model *(14.18)*, provided the model satisfies a certain *observability* condition. We can adjust the convergence rates arbitrarily—in the ideal case—by adjusting the gains.

Further consideration of observers is beyond our scope here. We simply note that using an observer instead of direct state feedback usually leads to some degradation in performance and robustness. In power electronics, it often pays to measure state variables directly.

14.3.2 Nonlinear State Feedback

Discussing nonlinear state-feedback schemes in any generality is difficult, because they are designed to match or exploit the specific nonlinearities of a given model. However, the piecewise LTI nature of typical power converter models makes them amenable to certain design strategies. We mention only one general approach, based on choosing switching surfaces. Specific instances of this approach can be found in all categories of power converters.

Switching Surfaces   Our control in power electronics is essentially limited to deciding when to turn switches on or off. In the intervals between switch operations, the state trajectories are usually well described by LTI models. As we noted in Sections 12.4.3 and 13.4, the end point of the state trajectory for one switch configuration or topological state becomes the initial point for the state trajectory in the succeeding switch configuration. This allows us to describe entire state trajectories, including the effects of switching. A direct approach to control design is therefore to divide the state space into regions that are separated by *switching surfaces* and to specify which switches turn on or off when a state trajectory arrives at each of these surfaces.

Switching surfaces may be fixed or varied as a function of external commands and feedback signals. The design of switching surfaces can be trivial for first-order systems and relatively simple for second-order systems, because the state trajectories and switching surfaces—or curves—are easy to visualize in the state plane. Higher dimensional systems can sometimes be dealt with satisfactorily.

Three types of switching surfaces are encountered. The first, which we call a *refracting* surface, introduces a discontinuity in trajectories that arrive at the surface but allows them to continue on past the surface. A switching surface is refracting when the components of the state velocity vectors normal to the surface point in the same direction on both sides of the surface. This situation is shown in Fig. 14.17(a).

The second type, which we call a *coasting* surface, is composed of trajectories of the system formed after the switching. This is illustrated in Fig. 14.17(b). When a trajectory hits a coasting surface, its normal velocity drops to zero and its subsequent motion lies in the surface. Figure 14.17(b) suggests how coasting surfaces for two configurations of a second-order system may be joined to produce cyclic steady-state behavior.

The third type of switching surface, which the control literature calls a *sliding* surface, introduces a more severe discontinuity. The normal components of the state velocity vectors in this case both point toward the switching surface, as indicated in Fig. 14.17(c). Consequently, a trajectory that arrives at the surface remains on the surface, and the system is then said to be in a *sliding mode*. In practice, we impose a switching discipline that prevents the switching frequency (and associated switching losses) from becoming excessive, so the trajectory will actually "chatter" back and forth across the surface. However, we can imagine that in the limit of

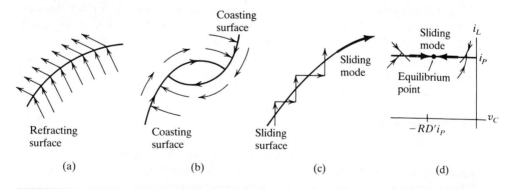

Figure 14.17   Switching surfaces: (a) refracting surface; (b) coasting surface; (c) sliding surface; and (d) sliding surface in current-mode control.

infinitely fast switching, the motion is more like a sliding than a chattering. The duty ratio during this fast switching must be such that the resulting average of the velocity vectors on the two sides lies along the surface. This average velocity along the surface determines the state equation during sliding. The sliding surface must be designed to attract trajectories and to induce an acceptable sliding motion once a trajectory hits the surface.

Several switching control laws in power electronics approach sliding-mode control in the limit of infinite-frequency switching. It is usually best to get an initial idea of the behavior of such systems by assuming ideal sliding—refining the analysis later by building on the insights obtained from the idealized analysis. We have already shown such an approach in the current-mode control design of Example 14.3. The switching surface there corresponded to the line $i_L = i_P$ in the state space, as shown in Fig. 14.17(d). We ignored offset and dynamics of the inner loop in our analysis, which corresponds to assuming ideal sliding. In that example, the switching line was actually varied slowly by the outer control loop, as a function of the deviations of the output voltage from the desired value. Hysteretic controllers provide another class of examples that are often best understood if we use sliding-mode ideas prior to more detailed analysis. (We mentioned a hysteretic controller for current-mode control in Example 12.5.)

14.4 DIGITAL CONTROL

We noted in Chapters 12 and 13 that discrete-time models are well suited to describing power electronic circuits, because these circuits are usually operated cyclically, with controls that are specified by a finite number of variables in each cycle. The state variables in these models comprise circuit quantities sampled each cycle. We have shown in Sections 13.5 and 13.6 that discrete-time models are important in assessing the stability of cyclically operated power converters. These models also form a natural starting point for the design of *discrete-time* or *digital controllers*.

Digital controllers process sequences of numbers (in contrast to analog controllers, which process continuous-time waveforms). In our applications, the numbers are generally obtained by sampling continuous waveforms using analog-to-digital (A/D) converters. The discrete-time processing may involve only simple operations with latches, adders, registers, and so on, or may involve digital signal processing (DSP) chips or microprocessors capable of more elaborate computations and equipped with significant memory. The numbers produced by the digital controller are translated back into continuous-time signals, using digital-to-analog (D/A) converters, and these signals are then applied to the system. A typical D/A converter interpolates the samples in some fashion (usually a simple *zero-order hold*, which just remembers or holds the most recent value). A multiplying D/A converter uses the sample values to set the gain of an analog multiplier in the system.

Digital controllers—and microprocessor-based controllers in particular—have historically been seen as more appropriate for relatively large and slow applications. In such applications their flexibility, ability to carry out complicated computations and table look-ups, and insensitivity to noise and aging more than offset the cost. The required sampling and processing speeds also are manageable in these cases. In power electronics, the major application of digital controllers has been to motor drives. With advances in the capabilities of digital processors, and with more demanding requirements on power converters, we expect to see increasing attention paid to digital control possibilities.

We have already mentioned that the results on continuous-time state feedback in Section 14.3 all have natural discrete-time parallels. Similarly, the classical control results in Sections 14.1 and 14.2 have their discrete-time parallels. Developing these parallels is beyond the scope of this book, however. We shall be content with one representative example.

EXAMPLE 14.5

Digital Control of a High-Power-Factor ac/dc Switching Preregulator

Switching regulators are often operated off an ac supply by first rectifying the sinusoidal ac voltage and then filtering it with a large capacitor. A major difficulty with this approach, however, is that the source current, far from being sinusoidal, consists of a tall, narrow charging pulse in each cycle. This distorted current waveform has high harmonic content and leads to poor power factors, in the range of 0.5 to 0.7. Ideally, we want the source current to be sinusoidal, so as to minimize harmonic interference with other loads operating off the same source. We also want this sinusoid to be in phase with the source voltage, so as to maximize the power available for given constraints on the peak source current. (See Section 3.4 for detailed discussion of these requirements.)

An increasingly popular solution to this problem is to insert a high-frequency switching preregulator, such as that in Fig. 14.18, between the rectified ac supply and the switching regulator load. Let T_L denote the period of the (rectified) input voltage $v_{in} = V |\sin(\pi t / T_L)|$ and T_S the switching period of the preregulator, so that $T_L \gg T_S$. The circuit is basically an up (or boost) converter, and the input current equals the inductor current. We operate the circuit under current-mode control but make the reference i_P for the inductor current proportional to v_{in} so:

$$i_P(t) = \mu v_{in}(t) \qquad (14.22)$$

This approach leads to an input current that approximates a rectified sinusoid in phase with the input voltage. The proportionality factor μ in (14.22) is varied from one input cycle to the next, that is, every T_L seconds, so as to regulate the output voltage v_o around a desired nominal value V_o. With proper control, this configuration yields power factors in the range of 0.95 to 0.99, reduces the total harmonic distortion of the input current to as low as 3 percent, accommodates large variations in the input-voltage amplitude, permits a smaller capacitor to be used, and provides a better output voltage to drive the downstream regulators. We now develop and analyze one possible digital control scheme for this converter.

We assume, as in Example 14.3, that the inner loop of the current-mode controller performs ideally and we therefore make the approximation $i_L \approx i_P$. This approximation

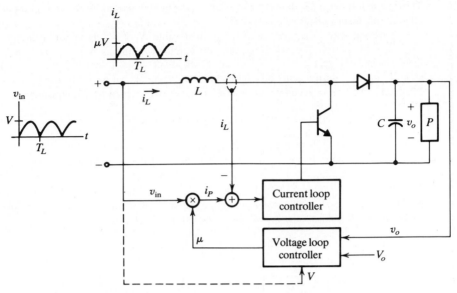

Figure 14.18 High-power-factor ac/dc switching preregulator.

turns out to yield tractable models and satisfactory controllers in many cases. We also model the load on the preregulator as a constant-power load that draws a power P. This is a reasonable model for a load that comprises high-efficiency regulated switching converters. Other loads, such as resistive or current-source loads, can be incorporated into the following development if desired, with a slight increase in complexity.

Assuming lossless switching in the preregulator and ignoring the switching ripple, we can now write the following power balance equation in the kth cycle, with μ in this cycle denoted by $\mu[k]$:

$$\frac{1}{2}C\frac{dv_o^2}{dt} = v_{in}i_{in} - \frac{1}{2}L\frac{di_L^2}{dt} - P$$

$$= \mu[k]v_{in}^2 - \mu^2[k]\frac{1}{2}L\frac{dv_{in}^2}{dt} - P$$

(14.23)

where we have substituted i_P for both i_L and i_{in} and then made use of (14.22). This equation is useful in obtaining efficient simulations that ignore details at time scales smaller than T_S.

We simplify (14.23) still further to obtain the desired discrete-time model. Integrating the equation over the kth cycle, and denoting v_o^2 at the beginning of the kth cycle by $x[k]$, we find that:

$$\frac{1}{2}C(x[k+1] - x[k]) = T_L\left(\mu[k]\frac{V^2}{2} - P\right)$$

or

$$x[k+1] = x[k] + \frac{T_L V^2}{C}\mu[k] - \frac{2T_L}{C}P$$

(14.24)

This is a first-order, LTI, discrete-time, state-space model, driven by the control input $\mu[k]$ and the (unknown) disturbance input P.

A natural way to regulate x to the desired value $X = V_o^2$ is to use a discrete-time version of PI control. The block diagram corresponding to this is shown in Fig. 14.19. We have isolated the factor $C/T_L V^2$ in the block diagram as a separate gain, in order to simplify the final design equations. Instead of the integrator that would be used in an analog PI scheme, we use an accumulator, whose output σ is governed by the equation:

$$\sigma[k+1] = \sigma[k] + (x[k] - X) \tag{14.25}$$

If the closed-loop system in Fig. 14.19 is stable and P is constant, the system reaches a constant steady state. In this steady state, the output of the accumulator is constant, so its input will be 0. In other words, if we can find gains h_1 and h_2 that stabilize the system, in the steady state $v_o^2 = x = X = V_o^2$, or $v_o = V_o$, as desired. (Actually, the value of v_o at the beginning of each input cycle is not quite the same as the average value of v_o over the cycle; the former, not the latter, will be regulated to V_o.)

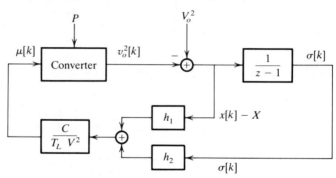

Figure 14.19 A discrete-time PI control design for the high-power-factor converter.

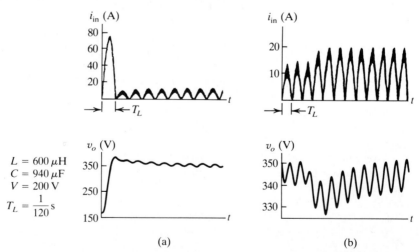

Figure 14.20 Response of the closed-loop system to (a) a large perturbation in the output voltage and (b) a step in load power.

A state-space model for the closed-loop system is

$$\begin{pmatrix} \sigma[k+1] \\ x[k+1] \end{pmatrix} = \begin{pmatrix} 1 & 1 \\ h_2 & 1+h_1 \end{pmatrix} \begin{pmatrix} \sigma[k] \\ x[k] \end{pmatrix} - \begin{pmatrix} 1 \\ h_1 \end{pmatrix} X - \begin{pmatrix} 0 \\ \frac{2T_L}{C} \end{pmatrix} P \qquad (14.26)$$

for which the characteristic polynomial is $z^2 - (h_1 + 2)z + 1 + h_1 - h_2$.

The simulation in Fig. 14.20(a) shows the response of the circuit to a large perturbation in the output voltage, with the gains h_1 and h_2 picked so as to place both natural ratios at 0.5, corresponding to the characteristic polynomial $(z - 0.5)^2$. The simulation in Fig. 14.20(b) displays the response to a step increase in load power from 1100 W to 1650 W.

Notes and Bibliography

Reference [10] of Chapter 11 is a very good text for all the LTI control approaches—classical and modern—that we discuss here. References [12]–[14] in Chapter 11 represent the state of the art at the interface between control theory and computer-aided application to control design. Reference [13] in Chapter 13 is an excellent text on digital control, though somewhat advanced. The special focus of [1] is linear–quadratic control.

Compensator design for high-frequency switching converters is studied in [2], with comparisons among different approaches and different converters. Classical compensator design using averaged models is clearly illustrated by means of design examples in [3], in the context of current-mode control of switching converters. (The correction in reference [11] of Chapter 13 will induce a few changes in the models used in [3], but the effects of the changes are not significant here.) A useful comparison of duty-ratio control and current-mode control for continuous and discontinuous conduction is presented in [4].

Various practical issues and possibilities in compensator design and implementation can be gleaned by studying application notes such as [5]. The guided tour in [6] of a detailed two-page schematic of a computer power supply makes evident the various control challenges beyond compensator design, and also shows some standard ways to handle these challenges.

Tellegen's theorem, which we invoked to justify a "power" balance for the averaged circuit in Fig. 14.11, is the subject of [7]. Some applications of the theorem to power electronics are also mentioned in [7].

Examples of power electronic control systems that switch on coasting surfaces are presented in [8], which deals with high-frequency switching converters, and in [9] and [10], which are devoted to series resonant converters. In [10], an analog controller designed using a linearized sampled-data model is used to regulate the resonant converter trajectory at the nominal. Control of power electronic circuits using sliding surfaces is discussed in [11], [12], and [13].

The use of energy-related variables in power electronics can often lead to elegant analyses and control schemes. This is illustrated by the results in [9], [10], [14], [15] and Example 14.5, for instance. The latter example is explored further in [16].

1. B. D. O. Anderson and J. B. Moore, *Optimal Control—Linear Quadratic Methods* (Englewood Cliffs, New Jersey: Prentice-Hall, 1990).
2. D. M. Mitchell, *Switching Regulator Analysis* (New York: McGraw-Hill, 1988).
3. R. D. Middlebrook, "Modeling Current-Programmed Buck and Boost Regulators," *IEEE Trans. Power Electronics*, 4: 36–52 (January 1989).

4. D. M. Sable, R. B. Ridley and B. H. Cho, "Comparison of Performance of Single-Loop and Current-Injection-Control for PWM Converters which Operate in Both Continuous and Discontinuous Modes of Operation," in *IEEE Power Electronics Specialists Conference (PESC)* (San Antonio, June 1990), 74–79.

5. *Unitrode Applications Handbook 1987–88* (Unitrode Corporation, Lexington, Massachusetts).

6. P. Horowitz and W. Hill, *The Art of Electronics* (Cambridge, England: Cambridge University Press, 1989).

7. P. Penfield, Jr., R. Spence, and S. Duinker, *Tellegen's Theorem and Electrical Networks* (Cambridge, Massachusetts: MIT Press, 1970).

8. W. W. Burns, III and T. G. Wilson, "Analytic Derivation and Evaluation of a State Trajectory Control Law for DC-DC Converters," in *IEEE Power Electronics Specialists Conference (PESC)* (Palo Alto, June 1977), 70-85.

9. R. Oruganti, J. J. Yang and F. C. Lee, "Implementation of Optimal Trajectory Control of Series Resonant Converter," *IEEE Trans. Power Electronics* 3: 318–327 (July 1988).

10. M.-G. Kim and M.-J. Youn, "An Energy Feedback Control of Series Resonant Converter," in *IEEE Power Electronics Specialists Conference (PESC)* (San Antonio, June 1990), 59–66.

11. R Venkataramanan, A. Sabanovic, and S. Ćuk, "Sliding Mode Control of DC-to-DC Converters," in *IEEE International Conf. on Industrial Electronics and Control Instrumentation* (1985), 251–258.

12. H. Sira-Ramirez, "Sliding Motions in Bilinear Switched Networks," *IEEE Trans. Circuits and Systems* 34: 919–933 (August 1987).

13. S. R. Sanders, G. C. Verghese and D. E. Cameron, "Nonlinear Control of Switching Power Converters," *Control—Theory and Advanced Technology* 5: 601–627 (December 1989).

14. D. Gouttenegre and B. Velaerts, "Modeling and Analysis of DC-DC Converters Control by Power Equalization," in *IEEE Power Electronics Specialists Conferences (PESC)* (Kyoto, April 1988), 960–967.

15. S. R. Sanders and G. C. Verghese, "Lyapunov-Based Control for Switched Power Converters," in *IEEE Power Electronics Specialists Conference (PESC)* (San Antonio, June 1990), 51–58.

16. V. J. Thottuvelil, D. Chin, and G. C. Verghese, "Hierarchical Approaches to Modeling High-Power-Factor AD-DC Converters," in *Second IEEE Workshop on Computers in Power Electronics* (Bucknell University, August 1990).

Problems

14.1   This problem refers to the Nyquist plot in Fig. 14.2.

(a) Verify that the Nyquist plot is correct and that there are no net encirclements of -1, so that the closed-loop system is stable. Check by directly computing the characteristic polynomial of the closed-loop system and verifying that its roots are in the open left half plane. (Rather than evaluating the roots of this third-order polynomial, you will find it simpler to use the fact that the roots of the polynomial $s^3 + \alpha s^2 + \beta s + \gamma$ are in the open left half plane if and only if $\alpha > 0$, $\gamma > 0$, and $\alpha\beta > \gamma$. This is a particular instance of the Routh–Hurwitz test referred to in Section 11.5.2.)

(b) If the loop gain in this case is decreased sufficiently from its nominal value, the closed-loop system becomes unstable. Use the Nyquist plot to estimate the factor by which the nominal gain must be decreased before the onset of this instability and check, using the characteristic polynomial. How many unstable poles are there when this instability sets in?

(c) If the loop gain is decreased still further than in (b), the closed-loop system becomes stable again. Use the Nyquist plot to estimate the factor by which the nominal gain must be decreased before this happens and check, using the characteristic polynomial.

14.2   Check that the Bode plot in Fig. 14.4(a) is consistent with the Nyquist plot in Fig. 14.2. Verify the Bode plots in Fig. 14.4(b), and use them to justify *(14.3)*, which approximately relates the phase of the loop gain to the slope of its magnitude and the number of right-half-plane zeros. Also verify that the Nyquist plot for a stable $\ell(s)$ approaches the origin at an angle of $-\delta\pi/2$, where δ is defined in Section 14.1.2.

14.3   Check (and elaborate on) the results and claims regarding duty-ratio control of the up/down converter in Example 14.1. How does the closed-loop simulation in Fig. 14.7(b) compare with the response you would predict using the linearized model in Fig. 14.6(a)? Also compare with a simulation that combines the controller in this example with the averaged model in *(12.13)*. For the alternative compensator in Fig. 14.9, determine h_0, ω_1, ω_2, and ω_3 in terms of the resistor and capacitor values. Also find a state-space description of it.

14.4   Check (and elaborate on) the results and claims regarding current-mode control of the up/down converter in Example 14.3. Are the conditions that justify neglecting the offset and dynamics of the inner loop satisfied for the parameter values in the example? Compute the locus of closed-loop poles for $R = 2\ \Omega$ and $R = 20\ \Omega$ as the gain h_0 in *(14.11)* is varied from 0 to 2000.

14.5   Check (and elaborate on) the results and claims regarding multiloop control of the up/down converter in Example 14.4. With the voltage gain h_V and integral gain h_N fixed at the values in the example, determine and sketch the locus of closed-loop natural frequencies as the current gain h_I varies upward from 0. Obtain a simulation of the closed-loop system obtained by combining the control law in Example 14.4 with the averaged model in *(12.13)*. How does its response to a step in the input voltage compare with the results in Fig. 14.15(b)?

14.6   Using classical control ideas and results from Problems 11.3, 11.5, and 12.5, generate candidate controllers for an up/down converter operating in discontinuous conduction under (a) duty-ratio control and (b) current-mode control. Discuss the potential advantages and disadvantages of your controllers. How well do they handle transitions to continuous conduction?

14.7   Using classical control ideas, generate candidate controllers for other high-frequency switching converters operating under (a) duty-ratio control and (b) current-mode control in (i) continuous conduction and (ii) discontinuous conduction. Examine down (buck) and up (boost) converters in particular.

14.8   Develop multiloop control with integral control action for the case of continuous conduction in down and up converters.

14.9 Explicitly show that the multiloop control design in Example 14.4 corresponds to state feedback in an augmented state-space model that comprises the up/down converter along with the integrator used for integral control.

14.10 Consider the multiloop configuration for motor speed control in Example 14.2.

(a) Suppose the transfer function that relates the motor speed w_r to the average armature current $\bar{i}_d$ is $K/(sT_r + 1)$. Using the model in *(14.6)* for the current loop, show how to design the PI controller in the speed loop so as to obtain a crossover between $1/T_r$ and $1/T_i$. Also explore how the phase margin depends on the PI controller parameters.

(b) Draw a block diagram to show how the speed-control approach can be extended to obtain position control. Verify that a proportional controller in this outermost loop will result in zero steady-state error, provided the closed-loop system is stable and is driven by constant signals.

14.11 Consider the digital controller for the high-power-factor ac/dc switching preregulator in Example 14.5. What gains h_1 and h_2 are needed to place both natural ratios at 0.5? What gains are needed to make the system deadbeat?

14.12 Suppose that you want to implement state feedback in the discrete-time LTI system $x[k + 1] = \mathcal{A}x[k] + \mathcal{B}p[k]$ but that the time required at each sampling instant to compute the control is not negligible. You may then be forced to make the present control depend on the state at the *previous* instant. Suppose that you use the control law $p[k] = \mathcal{H}_X x[k - 1] + \mathcal{H}_P p[k - 1]$.

(a) Write a state-space description for the augmented closed-loop system whose state vector at time k comprises $x[k]$ and $p[k]$ (the variables in $p[k]$ represent the memory of the controller and are therefore natural state variables). If the original system is controllable, you can place the natural ratios of the augmented system at arbitrary locations by properly choosing $\mathcal{H}_X$ and $\mathcal{H}_P$.

(b) Use the preceding results to modify the digital controller for the high-power-factor ac/dc switching preregulator in Example 14.5, assuming now that the controller requires one time step for its calculations.

14.13 Obtain a continuous-time averaged model for the high-power-factor ac/dc switching preregulator in Example 14.5 by taking the local average of *(14.23)* over an interval T_L. Show that under appropriate assumptions the result is a continuous-time LTI model for $\bar{v}_o^2$. How is the model modified if the load is a resistor R in parallel with the constant-power load P?

Components
and Devices

Chapter 15

Components:
An Overview

A power circuit is composed of only a few kinds of components: switches, capacitors, and inductors (or transformers). In its ideal form, each of these components is lossless and capable of operating at any frequency. In the ideal switch, for instance, the voltage across it is zero when it is on, the current flowing through it is zero when it is off, and the transition between these two states occurs infinitely fast. Although we did not explicitly discuss the means by which the actual switches are turned on and off (a topic covered in Chapter 22), an ideal switch responds infinitely fast to its drive signal and requires zero drive power. And, of course, there is no limit to the off-state voltage and on-state current of the ideal switch. Similarly, the ideal capacitor and the ideal inductor are purely energy storage elements, and they each store energy in only one field type. The ideal capacitor contains no magnetic energy, and the ideal inductor contains no electric energy. Furthermore, neither element dissipates any energy.

In reality, the components we use fall short of this ideal description, and so our power circuits are neither 100% efficient nor capable of operating at an infinitely high switching frequency. Improvements in power circuit behavior depend on development of components with characteristics that are closer to ideal. Thus you need to understand the fabrication and material technologies on which power electronic components depend. Only through such an understanding can you properly apply power circuit components and predict developments in the field.

As an example of the impact of advances in components consider how the power supply for low dc voltage electronic loads has changed over the last 40 years. In the early 1950s, semiconductor power diodes became a practical replacement for vacuum tube rectifiers. As the on-state voltage of a semiconductor diode (about 1 V) was much lower than that of a vacuum rectifier (about 30 V), the efficiency of the power supply increased significantly. By the mid-1960s, the switching speeds of bipolar power transistors had improved enough to make high-frequency dc/dc converters operating in the 10–20 kHz range possible. This improvement allowed replacement of linear-regulator technology with the lighter, smaller, and

405

more efficient switching regulator. By the late 1970s, the power MOSFET became economically practical and permitted efficient operation in the 200 kHz to 1 MHz range, which further reduced the size of the required switching power supply. In each case, a component improvement permitted the "breakthrough" in power electronic technology.

15.1 PRACTICAL SEMICONDUCTOR SWITCHES

Typical current and voltage waveforms for a semiconductor switching device are shown in Fig. 15.1. When the switch is on, it exhibits a forward voltage drop V_{on}, which gives a conduction loss $p = V_{on}I$ as current I flows through it. Similarly, when the switch is off, it carries a small *leakage current* that produces an off-state loss. The leakage current is usually small enough that we can neglect this off-state loss. But when the switch is a Schottky diode, or when the power circuit tries to impose a voltage across a switch that is greater than the switch's *breakdown voltage*, the leakage current and consequent off-state loss can be significant.

A semiconductor switch also needs time to make the transition from one state to the other. During this *switching time*, $v_{sw}i_{sw} \neq 0$ and the device dissipates energy. The total energy dissipated in the turn-on and turn-off transitions, multiplied by the switching frequency, f_{sw}, gives the net *switching loss* for the device.

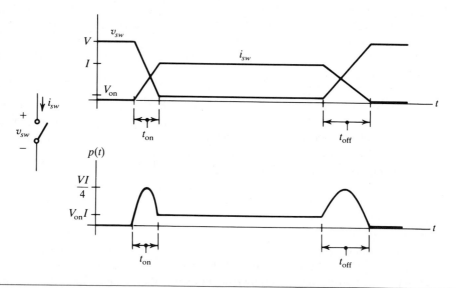

Figure 15.1   Typical voltage, current, and dissipation waveforms for a power semiconductor switch.

15.2 PRACTICAL ENERGY STORAGE ELEMENTS

We can explain some of the dissipation that occurs in real energy storage elements by the resistance of the capacitor's leads and plates and of the inductor's winding. However, calculating the loss is not simple because we must take into account the *skin effect*, which dictates where current flows in a conductor, and the *proximity effect*, which dictates how much circulating, or *eddy*, currents are induced in nearby wires or other conductive material. In general, to reduce this I^2R dissipation, we must make the energy storage element physically larger to incorporate more conductor area. Then, when the conductors are large enough for the skin and proximity effects to dominate the total loss, conductor fabrication becomes more complicated.

In addition to this I^2R dissipation, the *change* of stored energy in their magnetic or dielectric materials also causes losses in inductors and capacitors. In general, these losses are a nonlinear function of ac field strength, frequency, dc bias, and temperature. Both inductors and capacitors can be made from a variety of materials, which implies tradeoffs. One material might have less loss at a given frequency than another, but have a lower permeability (or permittivity), a lower maximum field strength, or a higher cost.

Dissipation is not the only troublesome imperfection of inductors and capacitors. Because these elements can be used as filters, we also care about whether their impedances deviate from the anticipated $j\omega L$ or $1/j\omega C$. Capacitors have a parasitic series inductance resulting from the loop formed by their plates, their leads, and their connection to other components. Inductors have a parasitic parallel capacitance caused by the proximity of two windings at different potentials. We can neglect these parasitic components for both elements if the frequency is low enough. However, at very high frequencies the "capacitor" acts like an inductor and the "inductor" acts like a capacitor. As a result, a filter having these components does not behave as expected above a certain frequency.

15.3 POWER SEMICONDUCTOR DEVICES

In this section our purpose is to outline the main power semiconductor devices available today—diodes, transistors, and thyristors—and to discuss their terminal characteristics. In Chapters 16–19 we explain in more detail the physical behavior of these devices.

At this point, however, you should know that current in a semiconductor is carried by both positive charges (holes) and negative charges (electrons). When one of these charge carriers dominates the total concentration of carriers, we call it the *majority carrier*. We call the other type of carrier the *minority carrier*. Devices whose behavior is dominated by the distribution of majority carriers (*majority carrier devices)*, the field-effect transistor, for example, can be switched much more quickly than can those devices that rely on the change of minority carrier distributions *(minority carrier devices)*, such as the bipolar junction transistor.

15.3.1 The Power Diode

The diode is a switch whose state cannot be explicitly controlled, but is instead determined by the behavior of the circuit in which it is used. If the circuit tries to impose a positive voltage across the diode, it will turn on, and if the circuit tries to impose a negative current through the diode, it will turn off. There are two types of diodes: the bipolar diode (based on a pn semiconductor junction) and the Schottky diode (based on a metal–semiconductor junction). The schematic symbols and the i–v characteristics for these devices are shown in Fig. 15.2.

At the rated current I_F, the forward voltage V_F is typically 1 V for a bipolar diode, and 0.6 V or less for a Schottky diode. Note that above I_F the forward voltage increases at a faster rate with current than it does below I_F. The reason is that the resistive drop in the diode, which increases linearly with current, becomes comparable to, and then exceeds, the junction voltage, which increases logarithmically with current. The reverse voltage rating of a power diode V_{BR}, sometimes called the *peak inverse voltage* (PIV), is determined by reverse leakage current. For a bipolar diode, V_{BR}, which can be as high as 5 kV, is usually specified at a leakage current I_R between 1 μA and 1 mA, depending on the I_F rating of the diode. For reverse voltages greater than V_{BR}, the leakage current increases rapidly. For the same die size and voltage, the leakage current of a Schottky diode is much higher than that of a bipolar diode. For both diodes, leakage current increases exponentially with temperature.

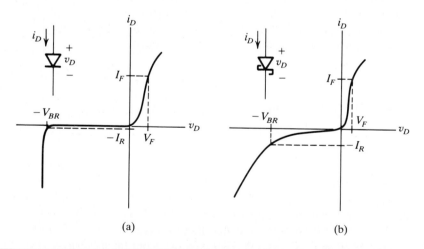

(a)   (b)

Figure 15.2   Symbols and i–v characteristics of power semiconductor diodes:
(a) the bipolar junction diode;   (b) the Schottky diode.

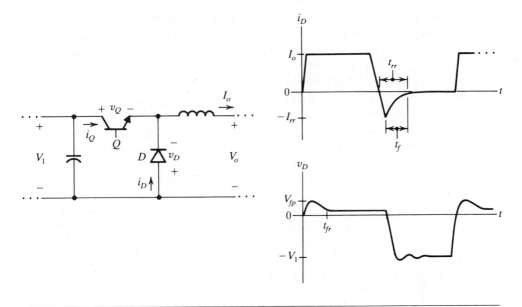

Figure 15.3 Switching waveforms for a bipolar diode used in a down converter.

The Bipolar Diode In addition to the presence of a nonzero forward drop and leakage current, the switching behavior of a bipolar diode departs from that of an ideal diode. Figure 15.3 shows typical voltage and current waveforms during turn on and turn off of a bipolar diode in a down converter. Note that when the device first begins to carry current, its forward voltage rises to V_{fp}, known as the *forward recovery voltage*, a value higher than the static curve of Fig. 15.2(a) predicts. Typical values for V_{fp} range between 5 V and 20 V. This phenomenon occurs because the excess charge requires time to build up to its final value in the device, and before it does, the resistance of the diode is very high. During the *forward recovery time* t_{fr}, the power dissipation in the diode is much higher than predicted from the diode's static characteristics. Also, this transient voltage is impressed across other devices in the power circuit (the transistor in this converter, for example), and this additional voltage stress must be considered when rating these other devices.

As the waveforms in Fig. 15.3 also show, when the bipolar diode turns off, its current first goes negative; then its reverse voltage begins to rise and the current returns to zero. This phenomenon is called *reverse recovery*, and it lasts for a time called the *reverse recovery time* t_{rr}. It occurs because the excess carriers in the device must be removed before the junction can withstand a reverse voltage. The peak value of the reverse current, $-I_{rr}$, depends on how fast the power circuit allows the diode current to fall and how much impedance is in series with the diode. For example, if Q in the down converter of Fig. 15.3 turned on instantaneously,

I_{rr} would be limited only by the parasitic inductance of the diode, connections, and leads. Under some conditions, especially when the parasitic circuit inductance is reduced, I_{rr} can even exceed I_F. When I_{rr} is too large, a turn-on snubber (discussed in Chapter 23) must be placed in series with the transistor to limit it. Another practical consequence of reverse recovery is that until the reverse current peaks at $-I_{rr}$ and starts to decrease to zero, the diode is still on. Therefore reverse recovery limits the maximum frequency at which we can use a diode.

Large values of $-I_{rr}$ increase the current stress on other semiconductor devices in the power circuit. During time t_f, which can be very short compared to t_{rr}, di/dt is high, and the parasitic inductance through which the reverse current flows imposes a high Ldi/dt voltage across these other devices. For this reason, we sometimes refer to diodes in terms of their "snappiness." A snappy diode is one whose current quickly drops to zero after I_{rr} is reached; in other words, the di/dt of these diodes is very high. In a nonsnappy, or *soft recovery*, diode, the recovery current takes a relatively long time to return to zero from $-I_{rr}$. That is, it exhibits a relatively small recovery di/dt.

In general, the higher the reverse voltage rating of a bipolar diode, the longer are its forward and reverse recovery times. Reducing these recovery times (by reducing minority carrier lifetimes) increases the forward voltage of the device. For this reason, diodes designed for low-frequency applications, such as 50- or 60-Hz rectifiers, have lower forward drops than diodes designed for 20–100-kHz high-frequency dc/dc or PWM converters. For instance, a typical 200-V *rectifier grade* diode has a reverse recovery time of 50 μs and a forward drop of 1.2 V at the rated current. A typical 200-V, 30-A, *fast recovery* diode has a reverse recovery time as low as 50 ns, but a forward drop of 1.6 V.

The Schottky Diode   A Schottky diode does not generate significant excess carrier concentrations when conducting. Instead it carries current primarily by the drift of majority carriers. For this reason it does not have the forward or reverse recovery problems of a bipolar diode, although its junction space-charge layer capacitance must still be charged and discharged. As the reverse voltage rating of the diode is raised above 100 V, however, an important fraction of its total current is carried by minority carriers. As this happens, the device begins to exhibit recovery characteristics similar to those of a bipolar diode. When applied conservatively, Schottky diodes exhibit a forward drop of about 0.4 V less than that of comparably rated silicon bipolar junction diodes.

15.3.2 The Power Transistor

There are two types of power transistors: the bipolar junction transistor (BJT) and the metal-oxide-semiconductor field-effect transistor (MOSFET). The schematic symbols and the i–v characteristics for these two devices are shown in Fig. 15.4. The BJT is a minority carrier device, so like the bipolar diode, excess charge

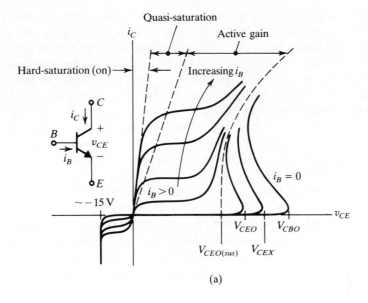

(a)

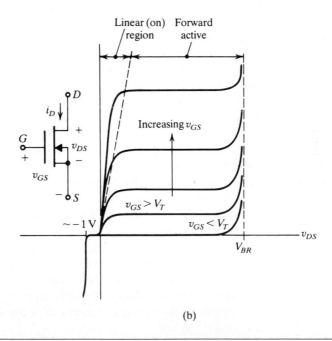

(b)

Figure 15.4 Schematic symbols and static i–v characteristics for the npn BJT and the n-channel MOSFET: (a) BJT; (b) MOSFET.

must be supplied or removed to turn it on or off. As a majority carrier device the MOSFET does not have stored minority carriers, but it does have junction and oxide capacitances that must be charged or discharged to turn the device on or off. Therefore neither the BJT nor the MOSFET can be switched instantaneously.

Note that both the BJT and MOSFET have a maximum permissible off-state voltage. They both also have a nonzero on-state voltage that depends on how much current is flowing and how hard the transistor is driven. Between the on and off-states is a region of operation referred to as the *forward active region*. For fixed-base current (or gate voltage), the collector (or drain) current of both transistors remains roughly constant over the full range of the device voltage in this region. Because power transistors are used as switches that are either on or off, the forward active region of operation is of interest only during switching, when it must be traversed.

The Power BJT   To be in the on state, the BJT requires the presence of base current i_B. The ratio of on-state collector current i_C to the base current necessary to support this value of i_C is an important parameter called the *dc current gain*, h_{FE} (or β). A typical curve of h_{FE} as a function of collector current is shown in Fig. 15.5. Note that, although h_{FE} can be quite high for low collector currents, it falls off rapidly with high currents. The rated current of a power BJT, $I_{C(\max)}$, is usually well above the point at which h_{FE} starts to fall. Thus, unlike signal transistors for which $h_{FE} > 200$ is common, values for h_{FE} of 5 to 20 at rated current are typical for power transistors.

Base current also affects the off-state characteristics of a BJT, as shown in Fig. 15.4(a). If the base current is sufficiently negative (resulting from reverse biasing the base-emitter junction during turn off), the collector voltage can go as high as the collector-base junction breakdown voltage V_{CBO}.[†] The device rating, V_{CEX}, gives the voltage limit of the device with a specified reverse voltage (X) between base and emitter. If $i_B = 0$ in the off state, the collector voltage can go only as high as V_{CEO}, which is lower than V_{CBO} because transistor action is amplifying the collector leakage current. In either case, as the collector leakage current grows, the device voltage folds back to the *sustaining* voltage $V_{CEO(\text{sus})}$, the voltage at which the leakage current amplified by the transistor action is in balance with, or sustains, the collector current. We explain the breakdown phenomenon of a BJT in more detail in Chapter 18.

Figure 15.6 shows typical current and voltage waveforms for a high-voltage bipolar junction transistor as it is turned on and off with a clamped inductive load. Note that during turn on, the collector voltage falls quickly at first but then "tails off" for the last few volts, reaching its final value only after the *voltage tail time* t_{vt}. This behavior is similar to the forward recovery of a bipolar diode, and is

---

[†]This subscript notation defines the terminal conditions for the parameter. For example, V_{CBO} is the collector-base voltage with the emitter open.

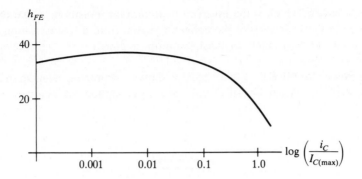

Figure 15.5 A typical relationship between current gain and collector current for a power BJT. The current rating of the transistor is $I_{C(max)}$.

caused by the gradual drop in the resistance of the transistor collector region as the carrier concentrations in this region increase.

At the turn-off transition, i_B may be negative for a period of time before the collector voltage begins to rise. During this interval, carriers in excess of those needed to keep the transistor on are being removed. The charge associated with these excess carriers is called the *storage charge*; the time required to remove the change is called the *storage delay time* t_s. Note that after the storage charge is removed, the collector voltage rises slowly at first but then finishes its rise quickly.

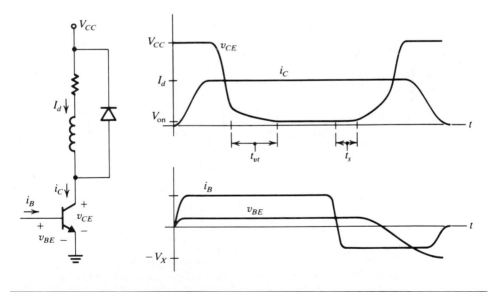

Figure 15.6 Typical switching waveforms for a BJT.

This behavior, which is the inverse of the turn-on transition, is due to the gradual change in the resistance of the collector region from a low to a high value as the transistor passes through its quasi-saturation region.

The Power MOSFET   Figure 15.7(a) shows a simple representation of the MOS-FET structure. When a positive voltage v_{GS} is applied from gate to source, a con-

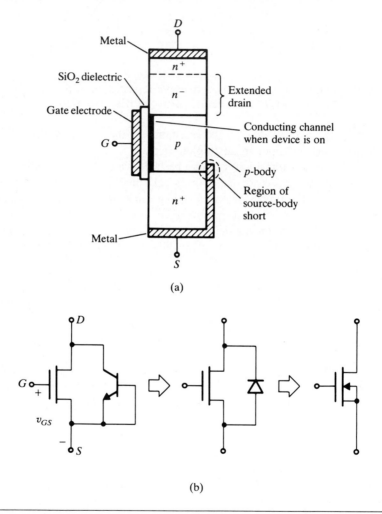

(a)

(b)

Figure 15.7   The power MOSFET: (a) simplified physical structure;   (b) the parasitic npn transistor shown with its base shorted to its emitter by the source contact, the equivalent parasitic body diode, and the power MOSFET symbol that represents the presence of the source-body short.

ducting n-type channel is formed beneath the gate electrode in the p-region, which is known as the *body* of the device. This channel connects the drain to the source, turning the device on.

The MOSFET turns on when v_{GS} exceeds the *threshold voltage* V_T, which is typically 3–5 V. The gate acts like a capacitor, so when the MOSFET is either on or off, the gate power is zero, making possible the use of gate drive circuits that are simple and efficient compared to those necessary to drive a BJT. Energy is required to switch the MOSFET, however, as the process is equivalent to charging or discharging a capacitor. For this reason the gate drive circuit is required to supply energy at an appreciable average power if the MOSFET is operated at high frequencies.

When on, the MOSFET acts like a resistor of value $R_{DS(on)}$, which is composed of two parts. The first is the *channel resistance* arising from the conducting channel created by the electric field beneath the gate electrode. The second is the resistance of the relatively thick silicon region comprising the *extended drain* region unique to power MOSFETs. The extended drain is a lightly doped region into which the drain-body space charge layer grows when the MOSFET is off. When the MOSFET is fully on, the drain resistance dominates $R_{DS(on)}$ unless the device is rated for voltages of less than about 100 V. In that case the extended drain is short, and the channel resistance becomes a significant fraction of the total on-state resistance.

Unlike the channel resistance, whose value varies with v_{GS}, the drain resistance is of constant value. Therefore $R_{DS(on)}$ for most MOSFETs is independent of v_{GS} for v_{GS} higher than a few volts above the threshold voltage V_T.

The most significant part of $R_{DS(on)}$ for devices rated above about 100 V results from the lightly doped drain region, so its value for these devices is strongly related to the transistor's breakdown voltage. Thus we can show that in an ideal device,

$$R_{DS(on)} \propto V_{BR}^{2.5}/\text{Area} \qquad (15.1)$$

For example, a device designed to withstand 400 V would have $4\sqrt{2}$ times the resistance of a device designed to withstand only 200 V, if they both have the same die area. Actual devices obey this relationship only at the higher voltage levels (above approximately 100 V). For devices rated below 100 V, $R_{DS(on)}$ is higher than *(15.1)* predicts because the channel resistance becomes significant and because less of the total die area is actually used.

The three silicon layers comprising the MOSFET of Fig. 15.7(a) create a parasitic npn transistor structure. To prevent this parasitic device from affecting the behavior of the MOSFET, the p-body and n^+ source regions are shorted by the source contact, as shown. The result is the parasitic bipolar diode across the MOSFET shown in Fig. 15.6(b). This *body diode* is sometimes useful, but in general, it has relatively slow reverse recovery characteristics. The MOSFET symbol indicates the presence of the source-body short, and the arrow indicates that the body is p-type and the channel is n-type.

Figure 15.8 shows typical voltage and current waveforms for the MOSFET as it turns on and off with an inductive load. Note that the device does not begin to change state until the gate voltage equals V_T (typically 5 V). During each transition, as the drain voltage either falls or rises to its final value, the gate voltage stays constant at approximately the threshold voltage V_T. This plateau in the gate voltage waveform is caused by the negative feedback effect of the drain-gate capacitance, which we explain in more detail in Chapter 22. The speed at which a MOSFET turns on or off is determined by the rate at which its parasitic capacitances can be charged or discharged. The more current the gate drive circuit can deliver or sink, the faster the device switches.

Safe Operating Area Except for voltage breakdown, the i–v characteristics shown in Fig. 15.4(a) and (b) do not indicate the operational limits of the device. Figure 15.9 shows a transistor's i–v characteristics with several other limits imposed. The area within these limits is called the *safe operating area*, or SOA, of the device.

One boundary of the SOA is the maximum rated device current $I_{(max)}$. The current-carrying capabilities of the wire bonds connecting the die to its package and of the device itself are responsible for this limit. The limit is a function of how long the current lasts. A BJT's upper current limit for very short pulses is typically less than a factor of two above its dc limit, while a MOSFET's upper current limit is typically three to four times higher.

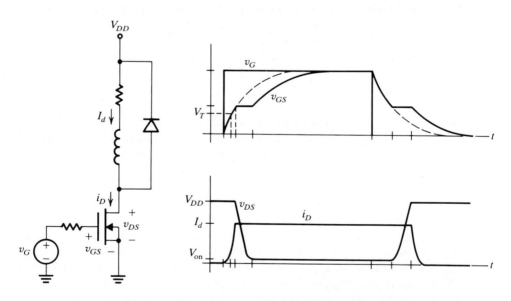

Figure 15.8 Typical switching waveforms for a power MOSFET.

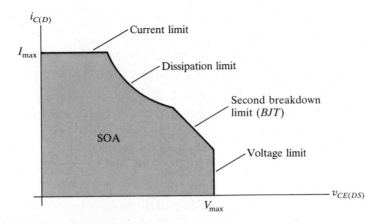

Figure 15.9 The safe operating area (SOA) of a transistor.

The power rating of the device forms another boundary of the SOA. This maximum power increases as the pulse duration decreases for constant repetition rate.

The final boundary of the SOA depends on a phenomenon called *second break-down*. This is a problem that occurs only in BJTs when both the voltage and current are high. It is a destructive breakdown that causes a self-sustaining collapse in the voltage in a localized region of the collector-base junction. All the current flowing through the device focuses on this spot and creates such a large dissipation in such a small area that the device fails. An ideal MOSFET does not experience this problem, but its parasitic bipolar transistor can. But because the base of the parasitic bipolar is shorted to its emitter, second breakdown is seldom a problem in practical MOSFETs.

The Darlington Connection   The Darlington connection of two bipolar transistors, shown in Fig. 15.10(a), results in a composite device having an effective current gain that is approximately the product of the individual gains, that is, $h_{FE} = I_C/I_B \approx h_{FE_1} h_{FE_2}$. It is a particularly useful connection for power transistors because the base current necessary to drive the pair is considerably less than that needed to drive Q_2 alone. The smaller base current allows a reduction in the current rating of the base drive circuit, simplifying its design. The Darlington connection can be made using discrete devices, but because of its utility, transistor manufacturers have made it available as a monolithically integrated device.

The Darlington connection has two disadvantages. The first is that the on-state drop of the pair is higher than the on-state voltage of Q_2 alone could be. The reason is that the lowest on-state voltage of the pair is composed of the hard-saturation voltage of Q_1 plus the base-emitter voltage of Q_2, whereas Q_2, if used

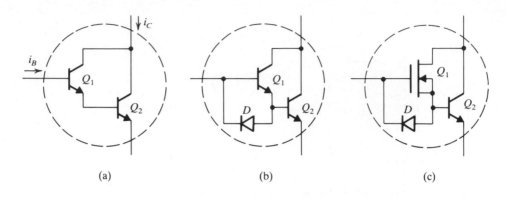

Figure 15.10 The Darlington connection: (a) the basic connection of two bipolar transistors; (b) the connection of (a) with the addition of a diode to improve its turn-off speed; (c) a Darlington connection using a MOSFET for the drive transistor Q_1.

alone, has a hard-saturation voltage equal to its base-emitter voltage *minus* its base-collector voltage. This condition is easy to see for signal-processing transistors, in which the saturation voltage of Q_2 alone could be on the order of 0.2 V, whereas the Darlington pair can have an on-state voltage of no less than 0.8 V, the saturated collector-emitter voltage of Q_1 plus the base-emitter voltage of Q_2. The saturation voltage of a power device usually has an appreciable contribution from the resistance of the collector region, however, making the difference between a single device and a Darlington pair less dramatic. A Darlington pair exhibits an on-state voltage of between 2 and 5 V, depending on its current and voltage ratings.

The second disadvantage of the Darlington connection is that when the drive circuit tries to turn the pair off, it can pull current from the base of Q_1 only. The base current of Q_2 goes to zero when Q_1 turns off, and therefore the charge stored in the base of Q_2 is not forcibly removed. Instead it decays to zero by the process of recombination, resulting in a long turn-off time. An effective solution to this second problem is to add a diode between the bases of Q_1 and Q_2, as shown in Fig. 15.10(b), so that the drive circuit has access to the base of Q_2 during turn off when $I_B < 0$ and the diode is on. In a monolithic Darlington connection, the diode is usually a separate device mounted in the same package and wire-bonded to the die containing the Darlington pair.

We can also use a MOSFET for Q_1, as shown in Fig. 15.10(c). The result is a composite device having the MOSFET's ease of drive and the BJT's on-state characteristics. The larger the die area of the MOSFET, the smaller is its on-state resistance, and the lower is the on-state voltage of the pair.

The Insulated Gate Bipolar Transistor   The insulated gate bipolar transistor (IGBT) is a monolithically integrated Darlington-like connection of a MOSFET driving a BJT, as shown in Fig. 15.11(a). It has the benefits of the drive simplicity of a MOSFET and the low forward drop per unit area of a BJT. Because of the way the two transistors are integrated, we can protect the MOSFET from the full off-state voltage of the IGBT by the space-charge layer at the collector junction of the BJT. The result is the need for dedication of a much smaller device area to the MOSFET. Note that because the two transistors are of opposite polarity (an n-channel MOSFET and a pnp BJT), the gate is driven with respect to the collector of the BJT.

The turn-on transition time of the IGBT is that of its BJT. The turn off is much slower than would be possible with the BJT alone, however, because there is no way to aid the turn-off process by pulling negative current from the base of the BJT. Figure 15.11(b) shows how the collector current makes a quick change

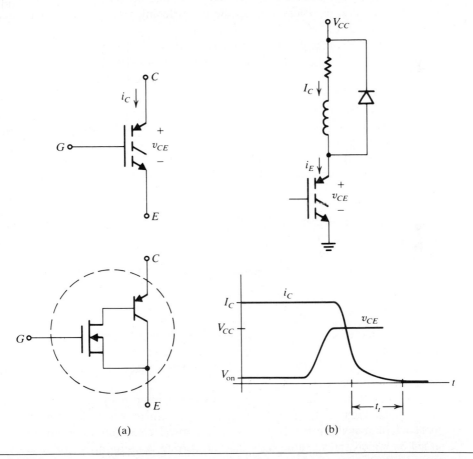

(a)   (b)

Figure 15.11   (a) The IGBT and its equivalent connection of MOSFET and BJT. (b) Waveforms at turn off showing the slow "tailing" effect during t_t.

when the MOSFET first turns off, but then slowly tails to zero. The initial step corresponds to that fraction of the device current that had been flowing through the MOSFET. The tail is the characteristic of a BJT turning off with an open circuit base. Reducing the minority carrier lifetime in the base of the BJT will give a faster turn-off time—but at the expense of a higher on-state voltage.

The IGBT has a parasitic thyristor in parallel with it that can be turned on if the device current gets too high or if the rate of rise of the off-state voltage is too great. When this happens, the device latches on, and it cannot be turned off from the gate.

15.3.3 The Thyristor

The name *thyristor* is generic for devices composed of four silicon layers (p-n-p-n) and exhibiting a regenerative internal mechanism that latches the device in the on state. As a family, thyristors are capable of switching voltages up to 6 kV and currents up to 5 kA.

Four members of the thyristor family are used as switches in power circuits: the silicon controlled rectifier (SCR), the triode ac switch (TRIAC), the gate turn-off thyristor (GTO), and the MOS controlled thyristor (MCT). As you have already seen in Chapter 5, the SCR is only semicontrollable; it can be turned on at a specified time, but it must be turned off by the action of the circuit in which it is connected. The GTO and MCT, on the other hand, can be turned both on and off from their control (gate) terminal. But they can only block forward voltage. The TRIAC is similar to the SCR, but can conduct and withstand voltage in either direction. In many respects it can be thought of as a pair of SCRs connected in anti-parallel.

The Silicon Controlled Rectifier   The most common form of the thyristor is the silicon controlled rectifier (SCR). Figure 15.12(a) shows its schematic symbol and i–v characteristic. If $v_{AK} > 0$, a current applied to the gate turns the SCR on. This action initiates an internal regenerative mechanism that latches the device in the on state, even if the gate current is then removed. The gate power required to switch the SCR is therefore very small.

The voltage V_{BO} in the i–v characteristic is the *forward break over voltage*. When subjected to a forward voltage above this value, the leakage current in the device is sufficient to initiate the internal regenerative mechanism in the absence of gate current, and the SCR turns on. The on-state voltage of an SCR is typically between 1 and 3 V.

As an SCR turns on, it first begins to carry current in the regions of the die near the gate lead. Only after some time (the length of which depends on how finely interdigitated are the gate and cathode metal contacts on the die) has the turned-on region spread sufficiently that the current is uniformly distributed over the entire device. During this *current-spreading* process, in which only a fraction of the SCR is conducting, it is important to keep the total anode current low enough

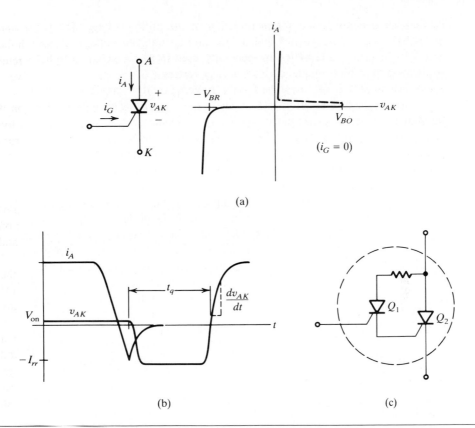

(a)

(b)   (c)

Figure 15.12   The SCR: (a) schematic symbol and its static i–v characteristic;   (b) reverse recovery, showing minimum off-state time t_q;   (c) Darlington-like connection of two SCRs.

to avoid a localized region of excessively high current density, which can destroy the device. For this reason, a maximum rate-of-rise of anode current is specified for all SCRs. Typically, di_A/dt must be kept less than 1000 A/μs.

When the SCR turns off, it displays the same reverse recovery phenomenon as a bipolar diode, and there is a period in the turn-off process during which $i_A < 0$. Even after $i_A = 0$, however, the SCR is incapable of blocking a forward voltage until the charge remaining in the device has decayed below a critical value. The time required for this process is called the *circuit commutated turn-off time*, t_q, and ranges from about 5 to 100 μs. Reapplying a forward voltage to the SCR before t_q risks retriggering the regenerative process and turning the device on again.

There are ways to reduce t_q by adjusting the SCR's design, although an additional consequence is a higher on-state voltage. For this reason there are two standard grades of SCRs: *rectifier grade* and *inverter grade*. Rectifier grade SCRs have a t_q in the 100 μs range and an on-state voltage between 1 and 2 V. These devices are generally used for rectification of the utility voltage. Inverter grade SCRs have a t_q of between 5 and 30 μs, and an on-state voltage of between 2 and 3 V. Inverter grade SCRs are generally used in motor drives or other PWM applications and high-frequency resonant converters.

When an SCR is off, it can be falsely triggered on if its anode–cathode voltage v_{AK} has too high a positive rate of change, dv_{AK}/dt. The current that flows through the junction capacitance when v_{AK} is changing has the same effect as gate current, and if it is large enough, it will turn the SCR on. For this reason, all SCRs have a maximum dv_{AK}/dt rating, typically between 200 and 2000 V/μs. The time t_q is usually specified under the assumption that the subsequent rise of the anode voltage is at this maximum rate, which is a worst-case condition.

We can connect two SCRs in a Darlington-like way, as shown in Fig. 15.12(c), in order to reduce the gate current requirement. In large devices, Q_1 is integrated on the same die as the main SCR, Q_2, and is known as an *amplifying gate*. A special version of this device uses a photosensitive SCR for Q_1. A photo-diode (LED) is then mounted in the same package and optically coupled to Q_1. The combination provides a means of electrically isolating the gate drive circuit from the SCR. However, these *opto-SCRs* often have a very low dv_{AK}/dt rating.

There is another special version of the SCR, called the *asymmetrical SCR* (ASCR), in which the reverse blocking capability of the device has been sacrificed for a much shorter t_q and a higher dv_{AK}/dt rating. The ASCR is useful in power circuits that have a diode in antiparallel with each controllable switch, and therefore require no reverse blocking capability.

The Gate Turn-Off Thyristor   The gate turn-off thyristor (GTO) is an SCR in which special provisions made during fabrication enable a large negative gate current to stop the internal regenerative process. The device then turns off. Two parameters of the SCR are compromised in the GTO: forward drop, which is typically 1 V more than a comparably rated SCR, and reverse blocking capability, which is typically less than 50 V.

Figure 15.13(a) shows the schematic symbol for the GTO. Like the SCR, it has maximum dv_{AK}/dt and di_A/dt ratings. It also has a maximum anode current that can be turned off from the gate terminal. The *turn-off gain* of the GTO is defined as the ratio of the anode current to the peak negative gate current required to turn it off. Gains on the order of 3–4 are typical.

The mechanism employed to turn off a GTO is to force the anode current into regions of the device area having low regenerative gain. Therefore, as this current-focussing process is taking place, the anode current must decrease in order to avoid localized regions of high current density and device failure. Any parasitic or load inductance in series with the GTO will prevent i_A from decreasing at the necessary rate. For this reason, GTOs are always employed with a snubber circuit, as shown in Fig. 15.13(b), to provide an alternative path for the current in the circuit inductance L. This snubber also determines the rate-of-rise of v_{AK}.

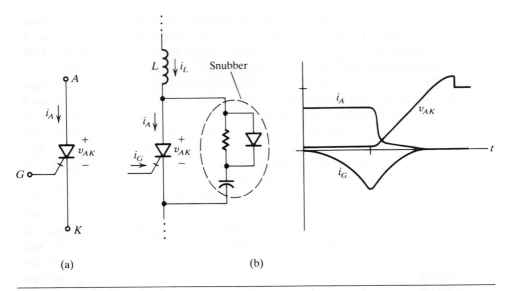

(a)   (b)

Figure 15.13   (a) The GTO symbol.   (b) GTO switching waveforms when a snubber circuit is used to provide an alternate path for i_L.

The MOS Controlled Thyristor   The MOS controlled thyristor (MCT) is a new thyristor device that offers the advantages of the GTO without its high turn-off gate drive-current requirements. An equivalent circuit for the device is shown in Fig. 15.14. Two MOSFETs are integrated with the SCR. The p-channel MOSFET,

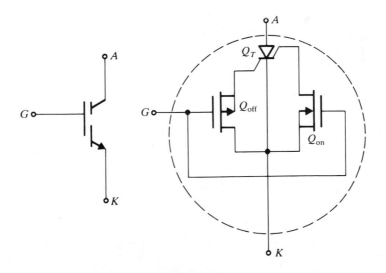

Figure 15.14   The MCT and its equivalent circuit.

labeled Q_{off}, is connected between the gate and cathode of the SCR. The n-channel MOSFET used to turn the device on, Q_{on}, is connected between the anode of the SCR and one of its internal layers. We have taken the liberty of identifying this internal connection point by the upside-down gate lead on the SCR in the figure.

The MCT is turned on by driving its gate terminal approximately 5 V positive with respect to the cathode, which turns on Q_{on} and supplies "gate" current to Q_T. This action initiates the normal SCR regenerative mechanism. Turn off is accomplished by lowering the gate voltage approximately 10 V below the cathode voltage, which turns on Q_{off} and shunts current away from the region of the SCR supporting regeneration, thus killing the regenerative process and turning Q_T off.

At this time the MCT is not commercially available. Even the symbol shown in Fig. 15.14 is tentative. It is however, a device with considerable potential.

The Triode ac Switch   The triode ac switch (TRIAC), whose symbol and i–v characteristic are shown in Fig. 15.15, is a thyristor capable of bilateral conduction. Because it is a slow device, having a dv/dt limit on the order of 100 V/μs, it is used almost exclusively for electric utility frequency applications such as light dimming and speed control of appliances and hand held power tools. The TRIAC is bilateral, so we do not use the terms "anode" and "cathode" to identify its terminals. Instead, we identify the switched terminals as main terminals 1 and 2 (MT$_1$ and MT$_2$). We use MT$_1$ as the reference terminal for the gate and MT$_2$ voltages.

We can turn the TRIAC on with positive or negative gate current—and with MT$_2$ either positive or negative. The TRIAC conducts in quadrant I of its i–v plane if MT$_2$ is positive when gate current is applied, and in quadrant III if MT$_2$ is negative. The four possible triggering modes are therefore I+, I−, III+, and III−, with the "+" and "−" referring to the direction of gate current. The triggering sensitivity of the TRIAC is different in each of the triggering modes, with I+

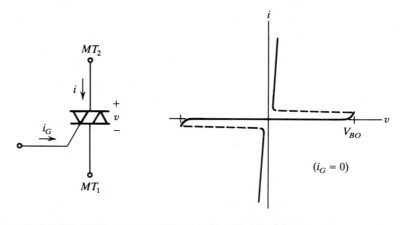

Figure 15.15   Symbol and i–v characteristic for the TRIAC.

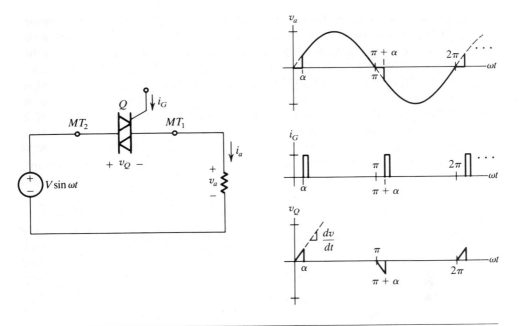

Figure 15.16 The TRIAC used as an ac controller to illustrate the commutating dv/dt to which it is subjected.

and III— being the most sensitive (requiring the least gate current) and therefore the most frequently used. In any of the modes, the gate characteristic, measured between G and MT_1, is similar to a forward biased diode.

In practically all applications the TRIAC must block voltage immediately after turning off, so we replace the SCR parameter t_q with a parameter called the *commutating dv/dt* for the TRIAC. The meaning of this parameter is illustrated in Fig. 15.16, which shows a TRIAC, Q, being used as an ac controller. When Q turns off at the zero-crossing of i_a, its voltage v_Q rises at an initial rate ωV. This rate of rise is the commutating dv/dt to which the TRIAC is subjected. It must be less than the maximum value specified by the manufacturer or the TRIAC will turn on again before the next gate current pulse. This limitation makes TRIACs unsuitable for use at frequencies above approximately 400 Hz.

15.4 CAPACITORS

Several different types of capacitors are used in power circuits. They differ in the dielectric employed and in physical form. The choice of capacitor type depends on the particular application. In addition to its capacitance value, a capacitor designed for power circuit applications usually has a maximum rms current specification because of internal dielectric and I^2R heating.

The dielectric material supporting the electric field in a capacitor is character-ized by two parameters: its *dielectric strength*, which is the value of the electric field at which the dielectric breaks down; and its *dielectric constant*, K. The capacitance value achievable with a material is proportional to K, which is the permittivity of the material normalized to ϵ_o, the permittivity of free space.

An important parameter characterizing capacitors is the *dissipation factor D*, which is the ratio of the real part of the capacitor's admittance to its imaginary part. The admittance is frequency dependent, as is D, and therefore is usually presented in graphic form. A nonzero dissipation factor means that the capacitor exhibits losses. These losses are a function of construction method and the physical properties of the dielectric material.

15.4.1 Film Capacitors

For applications that require large currents but relatively little capacitance, such as in snubbers or resonant tanks, a *film capacitor* is a common choice. It is simply two layers of conductive foil separated by insulating film. This long dielectric–conductor sandwich is usually rolled into a cylinder. The two foil electrodes are offset so that only one sticks out beyond the dielectric at either end of the cylinder. Terminals are then connected to the edges of these electrodes.

The dielectric film is a thin, insulating sheet of plastic. Although many plas-tics can be used, the one most commonly used in power electronic capacitors is polypropylene, with polycarbonate a distant second. Polypropylene has more than twice the dielectric strength of polycarbonate (about 16 kV/mil for polypropylene versus 7.5 kV/mil for polycarbonate). The electrode can be vacuum deposited on the dielectric film to produce *metallized film*, yielding an electrode thickness of 0.01–0.05 μm, or a separate sheet of metal foil, 5–15 μm thick, can be used. Capacitors using the *metal-foil* construction are capable of carrying considerably higher currents than those made with metallized film.

The dielectric properties of polypropylene are very stable when subjected to changes in temperature. The material's dielectric constant ($K \approx 2.3$) changes only a few percent when the temperature changes from 25°C to 105°C. Polycarbonate has a slightly higher dielectric constant ($K \approx 3.0$) and can be used at temperatures as high as 125°C. The dielectric loss in both materials is very low and is stable under temperature variations (a D of approximately 0.02% for polypropylene and 0.1% for polycarbonate over a frequency range of 1–100 kHz is typical). The film capacitor's lead and electrode resistance are the dominant sources of loss.

The parasitic inductance of a film capacitor depends on the capacitor package's geometry and size. Typical resonant frequencies (where $\omega L = 1/\omega C$) range from 100 kHz for a 50-μF capacitor to 10 MHz for a 0.01-μF capacitor.

15.4.2 Electrolytic Capacitors

Electrolytic capacitors offer far greater capacitance per unit volume than do film ca-pacitors. The tradeoff is much higher series resistance and an inability to withstand a reverse voltage of significant value. Because of their unipolar voltage limitation, they are most often used as dc filter capacitors, in which a large capacitance is needed but the current (per unit of capacitance) is relatively small. There are two major types of electrolytic capacitors: aluminum and tantalum. Tantalum capacitors have more capacitance per volume, but they are more expensive because of the material's higher cost.

Aluminum Electrolytic Capacitors   The first step in producing aluminum electrolytic capacitors is to chemically etch an aluminum foil, making its surface very porous. This process increases the surface area of the foil (which is connected to the positive terminal of the capacitor) by a factor of as much as 100. A very thin layer of aluminum oxide, Al_2O_3, is then formed on the foil's surface. This oxide, which has a dielectric constant of $K = 8.4$, is the dielectric in the capacitor. A liquid electrolyte forms the other "plate" of the capacitor as it comes in contact with the oxide. A second foil of aluminum makes electrical contact to this electrolyte and is the negative terminal of the capacitor. As the electrolyte is not a very good conductor compared to a metal, the resistance of an electrolytic capacitor is relatively high.

The behavior of the electrolyte is the reason that the life of electrolytic capacitors is short compared to that of other components used in power circuits. The electrolyte slowly escapes through the seal of the package until enough has left to change dramatically the electric characteristics of the capacitor. The rate at which this occurs is temperature dependent—the higher the temperature, the shorter the life. You should always remember that the loss of electrolyte occurs during storage as well as during operation of the power circuit, although usually at a much slower rate because the capacitor is cooler. Before the electrolyte runs out, the most common cause of failure of an aluminum electrolytic capacitor is a short circuit through the oxide.

Tantalum Electrolytic Capacitors   Tantalum oxide, Ta_2O_5, has a dielectric constant approximately three times that of Al_2O_3, giving tantalum capacitors a significant size advantage over aluminum capacitors. Three different methods are used to construct tantalum electrolytics. The *tantalum foil* construction is identical to that used for making aluminum electrolytics. The *wet slug* and *solid* constructions each start with a porous tantalum pellet made by sintering powdered tantalum. A wire, which forms the positive terminal, is attached to this pellet. In the wet slug construction, the negative electrode is a liquid electrolyte, which is absorbed by the porous pellet. The negative electrode of the solid tantalum electrolytic is manganese dioxide, created on the Ta_2O_5 surface by another chemical reaction.

15.4.3 Multilayer Ceramic Capacitors

Multilayer ceramic capacitors are made by stacking many layers of thin, unfired ("green") ceramic sheets on which a conductor has been screen printed using conductive inks. This sandwich is then sintered to form a solid, monolithic capacitor.

Table 15.1   Class II Ceramic Capacitor Nomenclature

LOW T (°C)		HIGH T (°C)		CHANGE FROM 25°C VALUE (%)	
X	−55	4	+65	A	±1.0
Y	−30	5	+85	B	±1.5
Z	+10	6	+105	C	±2.2
		7	+125	D	±3.3
		8	+150	E	±4.7
				F	±7.5
				P	±10
				R	±15
				S	±22
				T	+22, −33
				U	+22, −56
				V	+22, −82

There are two types of ceramic dielectrics, designated Class I and Class II. Class I ceramics have a dielectric constant of between 10 and 500 that is very stable (typically ±3%) over wide ranges of voltage, time, and temperature. The dissipation factor of these materials is also low, typically 0.1% at 1 MHz. Capacitors designated NPO (or COG) are made of class I ceramic.

Class II ceramics have much higher dielectric constants than class I materials and come in two subgroups. The first has a dielectric constant that ranges from 500 to 3000 and is relatively stable over voltage, time, and temperature. The second has a relatively unstable dielectric constant ranging from 3000 to 20,000. A three-symbol code is used to specify the performance of these Class II ceramics. The first and second symbols specify the low and high limits of an operating temperature range. The third symbol specifies by how much the capacitance value at the temperature extremes will change from its value at 25°C. Table 15.1 defines this nomenclature. Two common types are X7R (±15% from −55°C to +125°C) and Z5U (+22%, −56% from +10°C to +85°C).

Multilayer ceramic capacitors can have very low series resistance and inductance and so are good capacitors to use at high frequencies. However, they are quite expensive compared to electrolytic capacitors, particularly for low-voltage ratings. In those cases, making the ceramic layers only as thick as required to achieve the necessary dielectric strength is mechanically difficult.

15.5 INDUCTORS AND TRANSFORMERS

Almost without exception, inductors and transformers used in power electronic circuits are custom-designed for the specific application. Seldom can these components be obtained from a manufacturer as a standard product. As a result, you must become familiar with the materials and manufacturing techniques used in producing magnetic components, subjects covered in Chapter 20. As bases for that presentation, in Chapter 2 we discussed the behavior of transformers and their circuit models, and here we provide an overview of loss mechanisms, materials, and the common physical forms of inductors and transformers.

15.5.1 Magnetic Materials

Most magnetic components used in power circuits contain a core of high-permeability material. Use of a core results in a smaller structure than otherwise, and the core prevents magnetic fields from extending great distances from the component. If they existed, these stray fields could cause interference or induction heating of other components or structures. Two broad classes of materials are used to make cores: The most common are ferrous alloys; the others are magnetic ceramics known as *ferrites*.

The great variety of ferrous alloys are available in several forms. These alloys differ in permeabilities, loss characteristics, and mechanical workability. For low-frequency applications (below 1 kHz), iron with a low silicon content, often known as *transformer steel* is generally used. At higher frequencies other elements are added to the iron to increase its permeability and lower its loss. These more exotic magnetic steels—containing various amounts of cobalt, nickel, molybdenum and chromium—can be used to frequencies of about 20 kHz. They are very expensive relative to the silicon–iron materials. The ferrous alloys are available in either sheet or powdered form. When powdered, they can be used at frequencies higher than those cited. However, both their permeability and saturation flux density are considerably less than those of the bulk material.

Ferrites are ceramics made of ferrous oxide, zinc, and either manganese or nickel. Many proprietary types exist, and each is identified by a unique designation—for example, 4C4 or 3C6. Their maximum flux densities are about one third that of the ferrous alloys, and at temperatures above about 150°C, their magnetic characteristics deteriorate markedly. Cores made of this material are limited in size because of their brittleness. Ferrites exhibit much lower losses at high frequencies than do ferrous alloys and thus can be used at much higher frequencies. The Mn materials are useful to approximately 1 MHz, while the Ni materials are acceptable to about 10 MHz.

15.5.2 Magnetic Core Geometries

Figure 15.17 shows some of the commonly used core shapes. The 'C,' 'E,' and 'pot' cores are the easiest to utilize, because their windings can be made on an open bobbin. The bobbin is then slipped over the legs of the 'C' or 'E' cores or the center post of the 'pot' core. A toroidal core, however, must be wound by threading the wire through the core, a process that can be automated, but which is generally more expensive than winding on a bobbin.

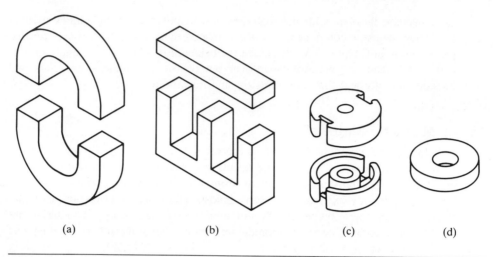

(a)   (b)   (c)   (d)

Figure 15.17   Various core geometries used in power circuits: (a) the double 'C'
core;   (b) the 'E' core with an 'I' closing piece;   (c) the 'pot' core.
(d) the toroidal core.

15.5.3 Losses in Magnetic Components

Loss in a magnetic component arises from two sources: the resistance of the copper
windings; and hysteresis and eddy currents in the core. At high frequencies, the
copper loss is aggravated by skin and proximity effects, but in general we can calculate it in a relatively straightforward manner. Core loss, however, is a nonlinear
function of both frequency and flux level. For a particular material, core manu-

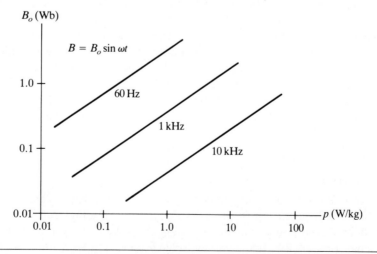

Figure 15.18   A core-loss graph for ordinary transformer steel.

facturers generally provide data giving core loss in W/kg as a graph parametric in frequency and flux level, as shown in Fig. 15.18 for ordinary transformer steel. For ferrites, the loss data is normally provided in W/cm^3.

Problems

15.1   Sketch the transistor voltage and current waveforms v_Q and i_Q for the down converter of Fig. 15.3. In what ways does the diode affect the voltage and current ratings of the transistor?

15.2   The down converter of Fig. 15.19(a) is operating at a frequency of 20 kHz and delivering 500 W to the load. The diode is ideal, but the transistor has an on-state drop of 1.5 V, a leakage current of 2 μA, and nonzero switching times whose values are shown in Fig. 15.19(b). What is the efficiency of the converter? What is the efficiency if the switching frequency is increased to 50 kHz?

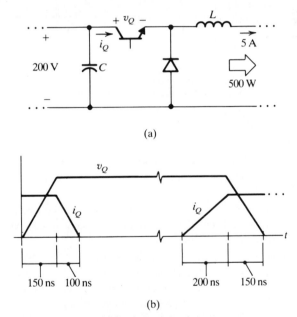

(a)

(b)

Figure 15.19   (a) The down converter whose switching dissipation is determined in Problem 15.2.   (b) Transistor voltage and current waveforms during switching for the converter of (a).

15.3   The diode in the direct down converter of Fig. 15.20 is in series with 500 nH of parasitic lead inductance, and the diode current i_D is as shown. Calculate and plot the diode and transistor voltages v_D and v_Q and the transistor current i_Q. If the converter is operating at 200 kHz, what are the switching losses in the diode and transistor?

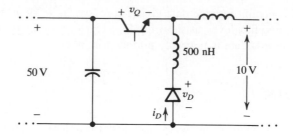

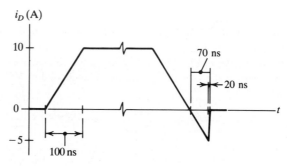

Figure 15.20   The direct down converter, whose switching losses are analyzed in Problem 15.3, and its diode current waveform.

15.4   The converter of Fig. 15.19(a) is operating at 100 kHz, and the base drive signal to the transistor has a duty ratio of 50%. If the transistor voltage and current waveforms during switching are as shown in Fig. 15.19(b), what is the output voltage of the converter?

15.5   What is the maximum effective duty ratio that we can achieve in the 100-kHz converter of Problem 15.4 if we require the diode to turn on for an arbitrarily short time before the transistor turns on again?

15.6   A simple light dimmer circuit is shown in Fig. 15.21. When the TRIAC, Q_T, is off, C charges through R_1 and the lamp. The device Q_S is a *silicon bilateral switch* (SBS), one of a family of *trigger devices* designed to be used in thyristor gate drive circuits. Think of it as a gateless TRIAC having symmetrical and low breakover voltages, $\pm V_{BO}$. If the magnitude of the voltage across Q_S exceeds V_{BO}, the SBS will turn on and thereafter behave as a TRIAC.

In this circuit, the dimming level is controlled by varying R_1. Note that the dimmer is connected between the hot (H) and dimmed hot (DH) nodes and does not require access to the neutral wire (N), which is often not available at a wall switch being retrofitted with a dimmer. The SBS has specified breakover voltages of $V_{BO} = \pm 27$ V. Determine and sketch the lamp voltage v_ℓ. Assume that you can model the lamp as a 100-Ω resistor.

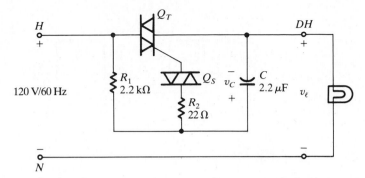

Figure 15.21 The light dimmer circuit of problem 15.6. The TRIAC is fired by the SBS, Q_S.

15.7 An inverter grade SCR, Q, is used to excite a resonant tank circuit, as shown in Fig. 15.22. The SCR has a reverse recovery time of $t_{rr} = 10$ μs and a recovery characteristic so snappy that you can assume $t_f \approx 0$. The diode D and the 20-Ω resistor form a snubber to maintain continuity of energy in L when Q turns off. The SCR is turned on at $t = 0$. Calculate and sketch i_a and v_a.

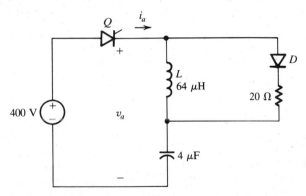

Figure 15.22 The circuit of Problem 15.7, in which v_a and i_a are calculated for Q having a recovery time of $t_{rr} = 15$ μs.

15.8 The circuit of Fig. 15.23 is a PWM inverter that generates a sinusoidal short-term average voltage $\bar{v}_a$, so that $i_a = 200 \sin w_a t \approx \bar{v}_a / R$; that is, at the output frequency w_a, the LR load appears to be resistive. The PWM frequency is 5 kHz, and the switching sequence for $\bar{v}_a > 0$ is as shown. The diodes in the high-frequency bridge switches D_1 and D_3 have reverse recovery times of $t_{rr} = 0.5$ μs, and t_f is a negligible part of this time. Because of this nonzero recovery time, an inductor, $L_s = 1$ μH, is placed in series with the source V_d to limit the reverse recovery current when S_1 and S_3 toggle. The diode D_s and resistor R_s are necessary to dissipate the excess energy trapped in L_s at the end of the D_1 or D_3 recovery period. What is the required power rating of R_s?

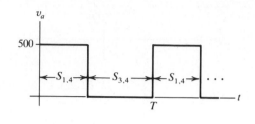

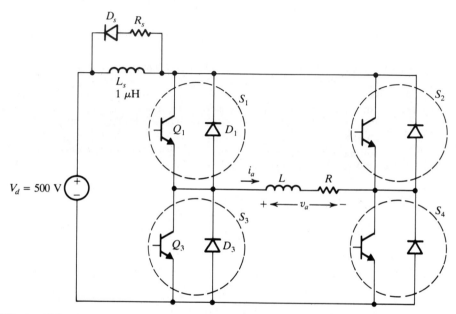

Figure 15.23   A PWM inverter with source inductance introduced to limit the peak reverse recovery currents of D_1 and D_3. The power rating of R_s is determined in Problem 15.8.

15.9   In discussing the SCR, we said that the opto-SCR was unsuitable for use as a bridge switch because of its low dV_{AK}/dt rating. Explain why.

15.10   We stated that the current gain of a Darlington pair is *approximately* $h_{FE_1} h_{FE_2}$. If $h_{FE_1} = h_{FE_2} = 20$, what is the error in h_{FE} that results from using this approximation?

Chapter 16

Review of Semiconductor Devices

IN this chapter we review the fundamentals of semiconductor device behavior: the existence of holes and electrons, the methods by which charge transport occurs, and the physical details of a pn junction in thermal equilibrium. We then derive the electric behavior of the bipolar diode, the bipolar junction transistor (BJT), and the metal-oxide-semiconductor field-effect transistor (MOSFET). We do not consider the effects of high voltage or high-current density in these analyses. In Chapters 17–19 we describe how these high stresses affect the design and operation of power semiconductor devices.

16.1 ELEMENTARY PHYSICS OF SEMICONDUCTORS

In this section our purpose is to review the basic physics of semiconductor devices. We first discuss the creation of holes and electrons, which are the positive and negative charge carriers in a semiconductor. We then describe the mechanisms of charge transport and analyze a pn junction in thermal equilibrium.

16.1.1 Charge Carriers in a Semiconductor

The physical characteristic of semiconductors that distinguishes them from metals is the presence of both negative *and* positive mobile charge carriers. The negative charge carrier is the electron (as it is for a metal); the positive charge carrier is called a *hole*.

The Intrinsic Semiconductor   Figure 16.1(a) shows a two-dimensional representation of a pure silicon lattice. Such a lattice is called *intrinsic* to distinguish it from an *extrinsic* lattice, or one that has a small percentage of atoms of an impurity present in the lattice. Each silicon atom forms a covalent bond with its four neighboring atoms, sharing its four outer shell electrons so that all the atoms form completed shells of eight electrons.

435

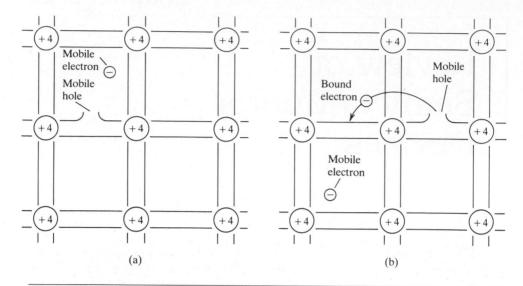

(a)   (b)

Figure 16.1   An intrinsic silicon lattice: (a) creation of a mobile electron–hole
pair by a broken covalent bond;   (b) hole movement caused by the
"sliding" of bound electrons adjacent to the hole.

Although the silicon lattice is stable at room temperature, some electrons have
sufficient thermal energy to break their bonds, creating a sea of *mobile (free) electrons*. Besides creating a mobile electron, each broken bond becomes a positively
charged location, called a *hole* in the lattice. In a metal, this positive charge remains tied to the atom and is therefore immobile. But in a semiconductor, very
little energy is required for a valence electron forming a bond with an adjacent
atom to replace the electron whose escape caused this first broken bond, as shown
in Fig. 16.1(b). In essence the hole (which is the absence of a bound electron) has
moved. Although the details of this movement are complicated, the hole in general
acts like a mobile, positively charged particle with a positive mass.

In a sample of intrinsic semiconductor, one hole exists for every mobile electron (from here on we use the term "electron" to mean "mobile electron"). Owing
to the dielectric relaxation phenomenon, which drives any net space charge to zero
in a conductive medium, the holes and electrons reside in equal concentrations
throughout the sample. We label these concentrations n and p (in units of #/cm^3)
for electrons and holes, respectively. In intrinsic material, $n = p$ because the freeing
of an electron always leaves behind a broken bond, creating a hole.

Because both electrons and holes can move, their concentrations can vary
spatially, while at the same time maintaining spatial charge neutrality. This is the
significant difference between a metal and a semiconductor. The absence of mobile
positive charge carriers in a metal prevents the creation of nonuniform electron
distributions. If a nonuniform distribution were present, it could not be neutralized

by a corresponding distribution of positive charges, and the resulting electric field would force a uniform redistribution of electrons (dielectric relaxation).

Electrons and holes may *recombine* if they encounter one another. For low concentrations of mobile carriers, the recombination rate is approximately proportional to the product of their concentrations, or pn. In intrinsic material, with only the equilibrium thermal energy causing the creation (*generation*) of hole–electron pairs, the concentrations of holes and electrons are those necessary for the recombination and generation rates to be equal. This state is called *thermal equilibrium*, in which:

$$n_o = p_o = n_i \qquad (16.1)$$

where the subscript "o" indicates thermal equilibrium, and n_i is the *intrinsic concentration*.

The generation rate increases with temperature, so the value of n_i changes with temperature. Its specific temperature dependence is

$$n_i = CT^{3/2}e^{-E_g/2kT} \qquad (16.2)$$

where E_g is the energy band gap in the semiconductor (1.1 eV for silicon), k is Boltzmann's constant, T is temperature in K, and C is a constant. For silicon, $n_i \approx 1.4 \times 10^{10}/\text{cm}^3$ at 25°C.

The Extrinsic Semiconductor   To change the ratio of the concentration of holes to electrons in thermal equilibrium, we can dope the silicon with elements from either column V or column III of the periodic table. Atoms of elements from column V, such as phosphorous or arsenic, have five valence electrons. When such an atom forms a covalent bond with four neighboring silicon atoms, as shown in Fig. 16.2(a), its fifth outer shell electron is only loosely bound to the atom. Very little energy is required to free it, thus creating a free electron. Conversely, elements from column III, such as boron, have three valence electrons. When such an atom forms a bond with silicon atoms, as shown in Fig. 16.2(b), the shell is incomplete (less than eight electrons). A nearby valance electron can move to this shell with very little change in energy, thus creating a hole.

At room temperature the thermal energy is high enough to free the fifth valence electron from most of the atoms of the column V impurity. These electrons become part of the sea of mobile electrons. For this reason, we call column V elements *donor* impurities. The positively ionized atom left behind does not behave as a hole, because it takes too much energy for an adjacent valance bond electron to replace the lost electron. Instead, the positive charge remains fixed at the site of the donor atom. The concentration of these fixed, ionized donor sites depends on the way in which the impurity was introduced into the lattice. Its value can vary spatially and is designated as $N_D^+(x)$.

Similarly, at room temperature most of the column III impurities have filled their valence shells by accepting an electron from a nearby silicon atom, creating a

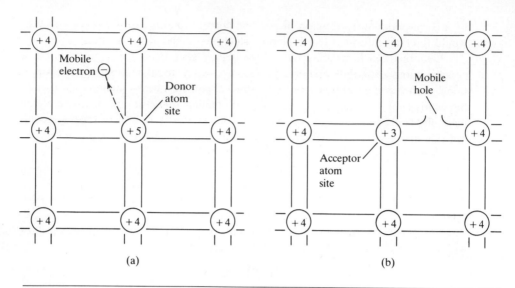

Figure 16.2 Extrinsic semiconductor lattices showing the creation of mobile holes and electrons: (a) a lattice doped with a column V (donor) impurity; (b) a lattice doped with a column III (acceptor) impurity.

hole in the process. For this reason, we call column V elements *acceptor* impurities. The acceptor atom with a completed outer shell now has a net negative charge and remains fixed in the lattice with a spatially dependent concentration, designated $N_A^-(x)$.

Some of the electrons and holes in an extrinsic semiconductor also are created in pairs by thermal generation. However, the concentration of the species contributed by the impurity is dominated by the ionization process. For example, the concentration of electrons in material doped with donor impurities is approximately equal to the concentration of the donors, or N_D. Holes, however, are still created only by the breaking of silicon atom bonds, and therefore the concentrations of holes and electrons are no longer equal. To determine the carrier concentrations we must now combine the constraint of charge neutrality with the balance of thermal generation and recombination.

The charge neutrality constraint that results from the dielectric relaxation phenomenon, which forces any space charge density to zero throughout the sample, requires that:

$$q(p_o - n_o + N_D^+ - N_A^-) = 0 \qquad (16.3)$$

At room temperature and above, the concentrations of ionized impurities are very close to the total impurity concentrations N_D and N_A. That is,

$$N_D^+ \approx N_D$$
$$N_A^- \approx N_A \qquad (16.4)$$

Because the impurity concentration is always very small compared to the concentration of silicon atoms, the thermal generation rate in extrinsic silicon is nearly the same as it is in intrinsic silicon. This generation is balanced by recombination in thermal equilibrium. The $n_o p_o$ product in thermal equilibrium, which is proportional to the recombination rate, is thus the same as for intrinsic material, or:

$$n_o p_o = n_i^2 \qquad (16.5)$$

but now $n_o \neq p_o$.

We can find the thermal equilibrium concentrations n_o and p_o for an extrinsic semiconductor from *(16.3)* and *(16.5)*. The donor and acceptor concentrations usually differ by several orders of magnitude in an extrinsic semiconductor material. Therefore the total impurity concentration is dominated by the concentration of one type of impurity. This dominant impurity must have a concentration much greater than n_i, or the material will behave no differently from the intrinsic material. For example, if $N_D \gg N_A$,

$$n_o \approx N_D$$
$$p_o \approx \frac{n_i^2}{N_D} \qquad (16.6)$$

In this case there are many more electrons than holes, and the material is called *n type*. Similarly, when $N_A \gg N_D$,

$$p_o \approx N_A$$
$$n_o \approx \frac{n_i^2}{N_A} \qquad (16.7)$$

and the material is called *p type*.

16.1.2 Charge Transport in Semiconductors

Charge transport (current flow) in a semiconductor is created by two mechanisms: *drift*, which is charge flow forced by an electric field; and *diffusion*, which is flow produced by a concentration gradient.

Drift Current   The presence of an electric field exerts a $q\vec{E}$ force on each charged particle, which accelerates the particle until it encounters an obstruction. The result is an average velocity proportional to the electric field and a *drift current*:

$$\vec{J}^{\text{drift}} = q\mu\vec{E} \qquad (16.8)$$

where the constant of proportionality μ is the *mobility*, whose value varies with both material and carrier type. The electric field $\vec{E}$ is called the *drift field*. Conduction in metals is by drift.

Diffusion Current   Carriers "piling up" and forming gradients cause a net *diffusion* of carriers from regions of high concentration to regions of low concentration, resulting in a *diffusion current*. The average velocity of a carrier is proportional to the size of the gradient and its direction is opposite to that of the gradient (down the gradient). The diffusion current density for positively charged particles is

$$\vec{J}^{\text{diff}} = -qD\vec{\nabla}\mathcal{K}(x, y, z) \qquad (16.9)$$

where $\mathcal{K}(x, y, z)$ is the spatially dependent concentration, and the constant of proportionality D, between gradient and average carrier velocity, is the *diffusion constant*. Its value, like that for mobility, depends on carrier type and material. For negatively charged particles the diffusion current flows in the opposite direction.

Diffusion is a transport mechanism that does not occur in metals because dielectric relaxation prevents maintenance of a nonuniform charge distribution. But as semiconductors have two types of mobile charge carriers—holes and electrons— an equal nonuniform distribution of both can exist without creating a net charge. The existence of diffusion as a charge transport mechanism in semiconductors is the essential difference between them and metals. This mechanism allows us to make electronic devices from semiconductor materials.

In one dimension, we can express the drift and diffusion current densities in terms of the physical and electric properties of the material:

$$J^{\text{drift}} = J_e^{\text{drift}} + J_h^{\text{drift}} = q\mu_e nE + q\mu_h pE \qquad (16.10)$$

$$J^{\text{diff}} = J_e^{\text{diff}} + J_h^{\text{diff}} = qD_e\frac{\partial n}{\partial x} - qD_h\frac{\partial p}{\partial x} \qquad (16.11)$$

The variables n, p, and E are functions of position x. Therefore, even if the total current density $J = J^{\text{drift}} + J^{\text{diff}}$ is constant throughout the sample, the individual drift and diffusion currents, as well as the individual hole and electron currents, are generally functions of position.

The parameters μ_e and μ_h are the mobilities of the electrons and holes, respectively. They are complicated functions of temperature, carrier concentration, and electric field. However, for the purpose of first-order device analysis, you can consider them to be constants that relate the average velocity of a charged particle to the strength of the electric field. Their values at room temperature for silicon are listed in Table 16.1.

The parameters D_e and D_h are the diffusion constants of the electrons and holes, respectively. They relate the average velocity of a charged particle to the spacial gradient of its concentration. The diffusion constant for each carrier type is related to the mobility of that carrier by the *Einstein relation:*

$$\frac{D_e}{\mu_e} = \frac{D_h}{\mu_h} = \frac{kT}{q} \qquad (16.12)$$

For a material in which the current density is proportional to the electric field, we define the constant of proportionality as the *conductivity* of the material σ. From *(16.10)* you can see that, for a semiconductor,

$$\sigma = q\mu_e n + q\mu_h p \qquad (16.13)$$

Table 16.1 Parameters of Various Semiconductors at 25°C

PARAMETER	SILICON	GERMANIUM	GALLIUM–ARSENIDE	UNITS
E_g	1.12	0.66	1.42	eV
n_i	1.4×10^{10}	3×10^{13}	1.8×10^6	/cm^3
μ_e	1360	3900	8500	cm^2/V-s
μ_h	490	1900	400	cm^2/V-s
D_e	34	100	213	cm^2/s
D_h	12	50	10	cm^2/s
τ_e	$\approx 10^{-6}$	$\approx 10^{-3}$	$\approx 10^{-8}$	s
τ_h	$\approx 10^{-6}$	$\approx 10^{-3}$	$\approx 10^{-8}$	s
ϵ	11.8	16	13.1	ϵ_o(F/m)

Note that σ is a function of the carrier concentrations. In extrinsic material, the concentration of one carrier type is usually much higher than that of the other, so the conductivity is dominated by one of the terms in *(16.13)*.

EXAMPLE 16.1

A Conductivity Calculation

The main terminals of power devices are most often on opposite surfaces of the silicon die in order to maximize ohmic contact area. The die is usually mounted so that these surfaces are in the horizontal plane, and current flowing between the terminals is then flowing vertically. Hence, these power devices are called *vertical* devices. One consequence of this design is that the device current must flow through the substrate of the die (the piece of silicon on which the device is made), and this substrate has resistance. Figure 16.3 shows a square die of n-type silicon with a size and thickness representative of the substrate of a 100-A device. What is the voltage drop across the substrate when the device is carrying a rated current of 100 A?

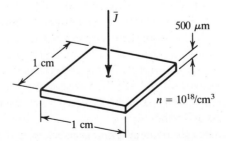

Figure 16.3 Silicon chip representing a power-device substrate.

We first calculate the resistance between the top and bottom surfaces of the die:

$$R = \frac{\ell}{A\sigma} \approx \frac{\ell}{A(q\mu_e n)} = \frac{0.5 \times 10^{-1}}{1(1.6 \times 10^{-19} \times 200 \times 10^{18})} = 1.56 \text{ m}\Omega \qquad (16.14)$$

Note that the mobility value used in this calculation is 200 cm^2/V-s, rather than the value 1360 cm^2/V-s of Table 16.1. The reason is that the value given in Table 16.1 is for doping concentrations low enough so as not to cause the lattice structure to deviate much from that of intrinsic silicon. However, at a concentration of 10^{18}/cm^3, the impurity atoms in the lattice begin to "get in the way" of the mobile carriers, reducing their mobility.

With the device carrying 100 A, the voltage drop across the substrate is $1.56 \times 10^{-3} \times 100 = 156$ mV. Compared to a power diode forward voltage of 1 V, this substrate drop is small but not negligible.

16.1.3 The pn Junction in Thermal Equilibrium

When n-type and p-type materials are "brought together" to form a junction, infinite gradients of both hole and electron concentrations initially appear at the point of contact (known as the metallurgical junction). The resulting diffusion currents sweep holes away from the p side, and electrons away from the n side, leaving behind the immobile charge of the ionized impurity atoms. This immobile *space charge*, which is positive on the n side and negative on the p side, creates an electric field pointing from the n side toward the p side of the junction.

The presence of this electric field now causes hole and electron drift currents to flow in directions opposite to their diffusion currents; holes drift back toward the p side and electrons drift back toward the n side. We are assuming that there is no external stimulation of the junction—and that it is in thermal balance with its surroundings—so a steady state is reached in which the net electron and hole currents are independently zero. This condition is called *thermal equilibrium*. Note that although the total current of each carrier type is zero, each has large, but equal and oppositely directed, drift and diffusion currents.

The net charge density $\rho(x)$, electric field $E(x)$, and electrostatic potential $\psi(x)$ within the pn junction in thermal equilibrium are shown in Fig. 16.4. The region of nonzero charge density in the vicinity of the junction is called the *space-charge layer* (SCL). The rest of the device is known as the *neutral region*.

The variation of the space-charge density is difficult to determine analytically, but near the pn junction, where the material is almost completely depleted of mobile carriers, $\rho(x)$ is a constant proportional to the density of ionic impurity. Near the edge of the SCL, the charge density tapers back to zero over a distance of a characteristic length known as a *Debye length*. As this distance is typically short compared to the length of the SCL, the charge distribution is often approximated as stepping to zero at the edge of the SCL, as shown by the solid lines in Fig. 16.4. This is the *abrupt SCL model*.

We can find the electric field from Gauss's law. It is zero in the neutral regions: if it were not, a drift current, for which there would be no equal and opposite diffusion current, would flow. A net current would result, but one cannot exist if

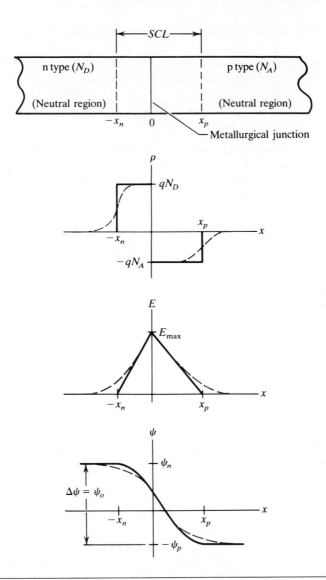

Figure 16.4 The space-charge $\rho(x)$, electric field $E(x)$, and electrostatic potential $\psi(x)$ across a pn junction in thermal equilibrium. The dashed lines show actual distributions; the solid lines are used for the abrupt SCL model.

the device is in thermal equilibrium. Within the SCL the electric field is not zero, but triangular in shape. Using the abrupt SCL model, we obtain the peak electric field:

$$E_{max} = \frac{qN_D x_n}{\epsilon_{Si}} = \frac{qN_A x_p}{\epsilon_{Si}} \qquad (16.15)$$

where x_n and x_p are the SCL lengths in the n-type and p-type materials, respectively.

We can obtain the potential difference $\Delta\psi$ between the two neutral regions by integrating the expression for the electric field from x_n to x_p:

$$\Delta\psi = E_{max}\left(\frac{x_n + x_p}{2}\right) \qquad (16.16)$$

This potential difference is known as the *built-in potential* and is represented by the symbol ψ_o. It is a function of the doping concentrations of the n and p regions.

In thermal equilibrium, a good model for the distribution of holes and electrons in a semiconductor at temperatures of interest to us ($-60°C$ to $250°C$) is that of a gas of noninteractive massive particles under the influence of a gravitational potential. Boltzmann statistics then apply, and we can express the concentration of free holes and electrons in terms of the electrostatic potential $\psi(x)$ as:

$$n_o(x) = n_i e^{q\psi(x)/kT} \quad \text{and} \quad p_o(x) = n_i e^{-q\psi(x)/kT} \qquad (16.17)$$

where $\psi(x)$ is referenced to a point in the semiconductor at which $n_o = p_o = n_i$.

Using (16.17), we can express the ratio of carrier concentrations at two different points, x_1 and x_2, in the semiconductor as the exponential of the potential difference between the two. That is,

$$\frac{n_o(x_1)}{n_o(x_2)} = e^{q(\psi(x_1)-\psi(x_2))/kT} \quad \text{and} \quad \frac{p_o(x_1)}{p_o(x_2)} = e^{-q(\psi(x_1)-\psi(x_2))/kT} \qquad (16.18)$$

We know the concentrations of the carriers in the two neutral regions ($n_n = N_D$ and $p_n = n_i^2/N_D$ in the n region and $p_p = N_A$ and $n_p = n_i^2/N_A$ in the p region). Hence, we can determine ψ_o as a function of doping by using (16.18) with x_1 located in the neutral n region, and x_2 located in the neutral p region:

$$\psi_o = \psi(x_1) - \psi(x_2) = \frac{kT}{q}\ln\left[\frac{N_A N_D}{n_i^2}\right] \qquad (16.19)$$

If $N_A = N_D = 10^{16}/cm^3$, then at $25°C$, $\psi_o = 0.7$ V.

Knowing the potential difference, we can use (16.16) and (16.19) to find the peak electric field and the length of the SCL:

$$E_{max} = \sqrt{\frac{2qN_D\psi_o}{\epsilon_{Si}\left(1 + \frac{N_D}{N_A}\right)}} \qquad (16.20)$$

and

$$x_n = \left(\frac{N_A}{N_D}\right)x_p = \sqrt{\frac{2\epsilon_{Si}\psi_o}{qN_D\left(1 + \frac{N_D}{N_A}\right)}} \qquad (16.21)$$

EXAMPLE 16.2

SCL Width in an Asymmetrically Doped pn Junction

Most pn junctions have significantly different impurity concentrations on either side of the junction. Such a junction is called an *asymmetrical junction*. In this example we show that the SCL extends primarily into the more lightly doped side of the junction.

We assume that in the pn diode of Fig. 16.5(a), $N_A = 10^{18}$ cm^{-3} on the p side and $N_D = 10^{15}$ cm^{-3} on the n side. The resulting built-in potential ψ_o is

$$\psi_o = \frac{kT}{q} \ln\left[\frac{N_A N_D}{n_i^2}\right] = 0.73 \text{ V} \qquad (16.22)$$

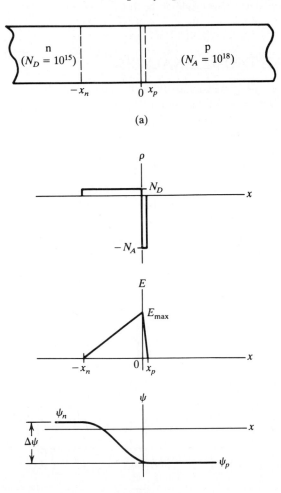

(a)

(b)

Figure 16.5 (a) An asymmetrically doped abrupt junction pn diode. (b) Space-charge, electric field, and potential distributions for the diode of (a).

The distances that the SCL extends into the n and p regions in thermal equilibrium are therefore:

$$x_n = \sqrt{\frac{2\epsilon_{Si}\psi_o}{qN_D\left(1 + \frac{N_D}{N_A}\right)}} = 0.96 \ \mu m \qquad (16.23)$$

$$x_p = \frac{N_D}{N_A}x_n = 0.96 \text{ nm} \qquad (16.24)$$

Figure 16.5(b) shows the space-charge density, electric field, and electrostatic potential for this device. Note that, as x_p is so much smaller than x_n, nearly all the voltage drop appears across the n region side of the SCL.

16.2 SIMPLE DIODE ANALYSIS

The simple one-dimensional diode that we analyze in this section is shown in Fig. 16.6. We assumed the abrupt SCL model and indicated the edges of the SCL with dashed lines. Voltage v_A is applied to the terminals of the diode as shown.

16.2.1 Dependence of the SCL Width on v_A

The voltage v_A must appear across the neutral regions of the diode and/or the SCL. Although we do not go through the derivation here, under *low-level injection* most of v_A appears across the SCL. By "low-level injection" we mean that throughout the diode the minority carrier concentrations remain small compared to the majority carrier concentrations, even though the diode may not be in thermal equilibrium. Under this condition, the voltage across the SCL is

$$\Delta\psi = \psi_o - v_A \qquad (16.25)$$

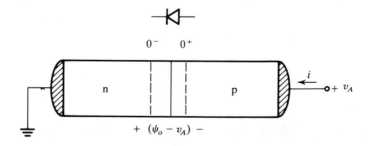

Figure 16.6 Simple one-dimensional diode with voltage v_A applied to its terminals.

We find the peak electric field and the n and p side widths of the SCL by substituting $\psi_o - v_A$ for ψ_o in (16.20) and (16.21):

$$E_{max} = \sqrt{\frac{2qN_D(\psi_o - v_A)}{\epsilon_{Si}\left(1 + \frac{N_D}{N_A}\right)}} \tag{16.26}$$

$$x_n = \sqrt{\frac{2\epsilon_{Si}(\psi_o - v_A)}{qN_D\left(1 + \frac{N_D}{N_A}\right)}} \tag{16.27}$$

$$x_p = \sqrt{\frac{2\epsilon_{Si}(\psi_o - v_A)}{qN_A\left(1 + \frac{N_A}{N_D}\right)}} \tag{16.28}$$

Note that a positive v_A reduces the SCL width, whereas a negative v_A increases the width. Under forward bias, the applied voltage never gets so big that the SCL width reaches zero. Under reverse bias, the SCL can grow substantially if $|v_A| \gg \psi_o$.

16.2.2 Incremental Capacitance of the SCL

Changing the voltage across the SCL of a pn diode also changes $\rho(x)$. This change of charge causes a current to flow through the diode, allowing us to model the SCL as a capacitor. Unlike the common parallel plate capacitor, however, the charge in this junction capacitor is not directly proportional to the voltage across it. The reason is that an incremental change in the physical distance separating the "plates," $x_n + x_p$, creates an incremental change in charge. The relationship between the stored SCL charge, Q, and v_A is

$$Q = qN_D x_n A = A\sqrt{2q\epsilon_s N_D(\psi_o - v_A)} \tag{16.29}$$

where A is the cross-sectional area of the device.

We can make this nonlinear relationship between charge and voltage linear at any applied voltage to find the incremental capacitance C_{inc} of the SCL. In reverse bias, where $|v_A| \gg \psi_o$, this capacitance is

$$C_{inc} = \frac{dQ}{d|v_A|} \approx A\frac{\sqrt{q\epsilon_s N_D/2}}{\sqrt{|v_A|}} \propto \frac{1}{\sqrt{|v_A|}} \tag{16.30}$$

16.2.3 Carrier Concentrations at the Edge of the SCL

Recall that (16.18) gives the relationship between the carrier concentrations at two points and the potential difference between them. But (16.18) applies only in thermal equilibrium. Strictly speaking, therefore, we cannot use it to determine concentrations throughout the diode when an applied voltage, v_A, is present. However, because for reasonable values of v_A the SCL is disturbed only slightly from thermal equilibrium, we can use *(16.18)* for points within the SCL. Specifically, we use *(16.18)* to relate the carrier concentrations at the n- and p-region edges of the SCL:

$$\frac{n_p(0^-)}{n_n(0^+)} = e^{-q(\psi_o - v_A)/kT}$$

$$\frac{p_n(0^+)}{p_p(0^-)} = e^{-q(\psi_o - v_A)/kT}$$

$$(16.31)$$

where $n_p(0^-)$ is the concentration of electrons at the p-region edge of the SCL, and so on. We can rewrite these equations as:

$$\frac{n_p(0^-)}{n_n(0^+)} = \frac{n_{po}}{n_{no}} e^{qv_A/kT}$$

$$\frac{p_n(0^+)}{p_p(0^-)} = \frac{p_{no}}{p_{po}} e^{qv_A/kT}$$

$$(16.32)$$

where n_{po} is the concentration of electrons in the neutral n region *at thermal equilibrium*, and so on. We also know that $n_{no} = N_D$, $p_{po} = N_A$, $n_{po} = n_i^2/N_A$, and $p_{no} = n_i^2/N_D$.

Although *(16.32)* tells us the ratio of carrier concentrations under any applied voltage, we still need one more relationship to determine the actual values. This relationship comes from our previous assumption that the diode is in low-level injection, which means that $n_n(0^+) \approx n_{no}$ and $p_n(0^-) \approx p_{po}$. Substituting these approximations into *(16.32)* yields

$$n_p(0^-) = n_{po} e^{qv_A/kT}$$

$$p_n(0^+) = p_{no} e^{qv_A/kT}$$

$$(16.33)$$

Thus *(16.33)* relates the carrier concentrations at the SCL edges to the applied voltage v_A. We can obtain the carrier concentration profiles $n(x)$ and $p(x)$ in the two neutral regions by applying these boundary conditions to the solution of one-dimensional *diffusion equations* for holes and electrons. In doing so we focus on the *excess* carrier concentration, which is the difference between the total and the equilibrium concentrations. We denote this excess concentration with a " ′ ". For example, $n_p'(0^-) = n_p(0^-) - n_{po}$.

16.2.4 Determination of Diode Current

The total diode current density J, which is independent of x, is

$$J = J_e(x_i) + J_h(x_i)$$

$$(16.34)$$

where

$$J_e(x_i) = q\mu_e n(x_i)E(x_i) + qD_e \frac{\partial n(x)}{\partial x}\bigg|_{x_i}$$

$$J_h(x_i) = q\mu_h p(x_i)E(x_i) - qD_h \frac{\partial p(x)}{\partial x}\bigg|_{x_i}$$

(16.35)

We now posit that in extrinsic material under low-level injection—if minority carrier current is a significant part of the total current—then the drift component of minority carrier current is negligible; that is, diffusion dominates the minority carrier current flow. Our argument goes as follows.

If the drift of majority carriers (holes in the p region, for example) is important, the drift of minority carriers at the same place will be unimportant. The reason is that the minority concentration is many orders of magnitude less than the majority concentration because of our assumption of low-level injection. Moreover, both concentrations are subjected to the same electric field, so the drift components of their currents are orders of magnitude different. Therefore, if the minority carrier current is important (that is, if it is a significant part of the total current), it must be so because of its diffusion component.

The realization of this important result makes bipolar device analysis simpler. For the minority current density, we can now avoid the nonlinearity caused by the product of carrier concentration and electric field in (16.35), because this term is, by our assumption, negligible. As a result, the continuity equation for the minority carriers, which in general is

$$\frac{1}{q}\frac{\partial J_e}{\partial x} + G - R = \frac{\partial n}{\partial t}$$

(16.36)

for the p region, now simplifies to:

$$D_e \frac{\partial^2 n}{\partial x^2} + G - R = \frac{\partial n}{\partial t}$$

(16.37)

The term R in (16.37) is the recombination rate of carrier pairs per unit volume and time, and the term G is the generation rate of carrier pairs. (The latter includes not only those carriers generated thermally, but also any carriers generated by other means, such as light.) The net recombination rate $R - G$ is zero in thermal equilibrium but is not zero when the device is out of thermal equilibrium.

For our analysis we assume a very simple form for $R - G$, one in which the net recombination rate is related linearly to the *excess* concentration of minority carriers. The constant of proportionality is $1/\tau_e$ or $1/\tau_h$, where τ_e and τ_h are known as the *minority carrier lifetime* for electrons or holes, respectively.

As we are assuming that the regions of the diode outside the SCL are neutral, $n' = p'$ and

$$\frac{\partial n'}{\partial x} = \frac{\partial n}{\partial x} = \frac{\partial p'}{\partial x} = \frac{\partial p}{\partial x}$$

(16.38)

If we combine these relationships with our assumed recombination model, and assume dc steady state conditions, *(16.37)* becomes

$$D_e \frac{\partial^2 n'(x)}{\partial x^2} - \frac{n'(x)}{\tau_e} = 0 \qquad (16.39)$$

or the diffusion equation for electrons when they are in the minority. We can write a similar equation for holes. Knowing the boundary conditions at the edges of the SCL from *(16.33)*, we can solve these diffusion equations for the excess carrier concentrations in the two neutral regions of the diode.

The solution to *(16.39)* is a pair of real exponentials (one growing, the other decaying) with characteristic lengths $\sqrt{D_e \tau_e}$. If we assume the bar to be infinitely long, we can discard the growing term for physical reasons, leaving

$$n'_p(x) = n'_p(0)e^{-x/L_e} \qquad (16.40)$$

where

$$L_e = \sqrt{D_e \tau_e} \qquad (16.41)$$

is the *diffusion length* for electrons. It is the mean distance that an electron can be expected to diffuse before it is lost to recombination. We now see that "infinitely long" means long compared to a diffusion length.

The coefficient $n'_p(0)$ in *(16.40)* is related to the applied voltage through *(16.33)*, so we can rewrite *(16.40)* in terms of v_A as:

$$n'_p(x) = n_{po}\left(e^{qv_A/kT} - 1\right)e^{-x/L_e} \qquad (16.42)$$

A similar equation describes the profile of excess holes $p'_n(x)$ in the n region.

We assume that, because the SCL is relatively narrow, no appreciable generation or recombination occurs within it, even though the carrier concentrations are not at their equilibrium values. Because of this assumption, if we know the electron current density at any point within the SCL, we know it everywhere within the SCL. Similarly, if we know the hole current density at one point, we know it everywhere. Therefore, because we know the electron current density at the p-region edge of the SCL and the hole current density at the n-region edge of the SCL, we know these current densities everywhere within the SCL. Thus:

$$J_e(0^+) = J_e^{\text{diff}}(0^+) = \frac{qD_e n_{po}}{L_e}\left(e^{qv_A/kT} - 1\right)$$

$$J_h(0^-) = J_h^{\text{diff}}(0^-) = \frac{qD_h p_{no}}{L_h}\left(e^{qv_A/kT} - 1\right) \qquad (16.43)$$

We can now obtain the total diode current density by summing these two components:

$$J = \left[\frac{qD_h p_{no}}{L_h} + \frac{qD_e n_{po}}{L_e}\right]\left(e^{qv_A/kT} - 1\right)$$

$$= \left[\frac{qD_h n_i^2}{N_D L_h} + \frac{qD_e n_i^2}{N_A L_e}\right]\left(e^{qv_A/kT} - 1\right) \qquad (16.44)$$

which is the ideal semiconductor diode equation. The term in brackets is the *saturation current* density J_s. It is the magnitude of the *reverse leakage* current density of a diode.

EXAMPLE 16.3

Concentration Profiles and Currents in a Short Diode

In our diode analysis we assumed that the terminal contacts of the diode were infinitely far from the junction. As we found, infinite in this case means many diffusion lengths. In some diodes, however, the terminals are much closer together. The purpose of this example is to determine the carrier profiles and current densities in such diodes.

Figure 16.7 shows a pn diode with p and n regions that are ℓ_p and ℓ_n long, respectively. We assume that these lengths are very short compared to the corresponding minority carrier diffusion lengths, that is, L_e on the p side and L_h on the n side.

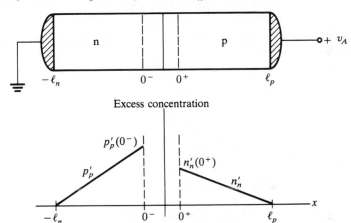

Excess concentration

Figure 16.7 A short diode and its excess carrier profiles when it is forward biased.

We must use both the growing and decaying solutions to the diffusion equations, *(16.39)*, in order for the short diode to meet the boundary conditions at the SCL edge and at the ohmic contact (where the excess carrier concentrations are zero). The resulting solutions are of the form $\sinh(x/L_{e,h})$ and $\cosh(x/L_{e,h})$. The argument in these functions is always small in our diode, so the solutions are very close to straight lines. In other words, we can expect most of the excess carriers to diffuse from the junction to the ohmic contact before they are lost to recombination. Therefore, if we ignore the recombination term in *(16.37)*, we see that the solutions to the diffusion equations are straight lines. In essence, we define a short diode as one in which recombination in the neutral regions can be ignored.

Figure 16.7 shows the excess carrier profiles based on our assumptions. The electron and hole current densities in the SCL are

$$J_e(0^+) = J_e^{\text{diff}}(0^+) = \frac{qD_e n_{po}}{\ell_p}\left(e^{qv_A/kT} - 1\right)$$

$$J_h(0^-) = J_h^{\text{diff}}(0^-) = \frac{qD_h p_{no}}{\ell_n}\left(e^{qv_A/kT} - 1\right)$$

(16.45)

and the total diode current density is

$$J = \left[\frac{qD_h p_{no}}{\ell_n} + \frac{qD_e n_{po}}{\ell_p} \right] \left(e^{qv_A/kT} - 1 \right) \tag{16.46}$$

We have assumed that the SCL width is small compared to ℓ_n and ℓ_p, so the lengths of the neutral regions are not significantly reduced by the SCL. This condition is generally true for forward bias conditions. Under reverse bias, however, the SCL width becomes much wider. As it does, the actual lengths that should be used in *(16.46)* get smaller and the reverse leakage current therefore gets larger. This increase in leakage current is not important until the SCL starts to reach the edge of the device, in which case the current begins to increase very rapidly. This condition is called *punch-through.* It is one breakdown mechanism that could limit the maximum reverse voltage V_{BR} that can be placed across a device. If the SCL punches through on the n side first, then:

$$V_{BR} \approx \frac{qN_D \ell_n^2}{2\epsilon_s} \left(1 + \frac{N_D}{N_A} \right) \tag{16.47}$$

As most devices are designed to avoid this condition, the maximum reverse voltage is instead limited by *avalanche*, a phenomenon that we discuss in Chapter 17.

16.3 THE BIPOLAR TRANSISTOR

The bipolar junction transistor (BJT) consists of three layers of doped semiconductor forming two junctions. The layers can be arranged as npn or pnp. Most power transistors are of the npn variety because of the superior performance of this arrangement. We first examine the behavior of the BJT using a very simple one-dimensional model.

16.3.1 Physical Model and Terminal Currents

Figure 16.8(a) is a one-dimensional model of an npn bipolar transistor. The doping concentrations in the emitter, base, and collector are N_{DE}, N_{AB}, and N_{DC}, respectively. The width of the base region is W, and the widths of the emitter and collector regions are long compared to the diffusion lengths in these regions.

You can apply everything you learned in the analysis of the pn diode directly to the analysis of this transistor. We assume low-level injection throughout the device, so that applied terminal voltages appear only across the SCLs, and minority carriers flow only by diffusion. We therefore know the excess carrier concentrations at the edge of each SCL as a function of the voltage applied between the neutral regions on either side of the junction. Hence, we can solve the diffusion equations for the carrier concentrations in the neutral regions.

Figure 16.8(b) shows the minority carrier concentrations for forward active operation, the operating condition in which the base-emitter junction is forward biased and the base-collector junction is reverse biased. Also, we assume that the lifetime of minority carriers in the base is long or, alternatively, that the diffusion length of minority carriers in the base L_B is long compared to the base width W.

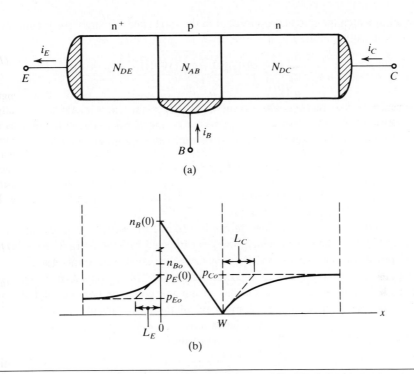

Figure 16.8 (a) One-dimensional model of an npn bipolar transistor.   (b) Minority carrier concentrations for operation in the forward active gain region.

From the minority carrier distributions of Fig. 16.8(b), we can calculate the minority currents at the edges of an SCL and then sum them to obtain the total current density at the junction. The total current flowing right to left at $x = 0$ is the emitter current, and that at $x = W$ is the collector current:

$$J_E = J(0) = \frac{qD_E}{L_E}\left(p_E(0) - p_{Eo}\right) + \frac{qD_B n_B(0)}{W} \qquad (16.48)$$

$$J_C = J(W) = \frac{qD_C}{L_C}p_{Co} + \frac{qD_B n_B(0)}{W} \qquad (16.49)$$

where the directions of J_E and J_C are as defined in Fig. 16.8(a).

However, in (16.48) and (16.49) we ignore the difference between the emitter and collector currents resulting from the recombination of charge in the base. We assume that an electron injected at the emitter reaches the collector. Although these lost carriers generally are insignificant compared to the total number of electrons flowing from emitter to collector, the holes required to recombine with them may be an important part of the base current. In calculating the base current we must therefore account for recombination in the base. If we again use a recombination model in which the rate is proportional to the total excess charge, the recombination current density supplied to the base J_{BR} is

$$J_{BR} = \frac{Q_B}{\tau_B} = \frac{q[n'_B(0) - n'_B(W)]W}{2\tau_B} \approx \frac{qn_B(0)W}{2\tau_B} \qquad (16.50)$$

where the approximations are a result of $n'_B(0) \gg n_{Bo}$ and $n'_B(W) = -n_{Bo}$. We can now specify the base current in terms of the emitter and collector currents, *(16.48)* and *(16.49)*, and the additonal recombination component, *(16.50)*:

$$J_B = (J_E - J_C) - J_{BR}$$

$$= \frac{qD_E}{L_E}\left(p_E(0) - p_{Eo}\right) - \frac{qD_C}{L_C}p_{Co} + \frac{qn_B(0)W}{2\tau_B} \qquad (16.51)$$

We can relate each of the three terms in *(16.51)* to a different physical process. The first term is the rate at which holes are injected into the emitter from the base. This process is known as *reverse injection* and usually accounts for the largest component of base current in a power transistor. The second term is the rate at which holes are injected into the base from the collector because of the reverse biased collector junction. This component of the base current is usually neglible. The third term is the rate at which holes must be supplied to support recombination in the base. Figure 16.9 shows this relationship.

16.3.2 Device Parameters

Two related parameters characterize the current gain of a bipolar transistor. The ratio of collector current to base current is called the *short-circuit common emitter-current gain β_F*, or:

$$\beta_F = \frac{i_C}{i_B} \qquad (16.52)$$

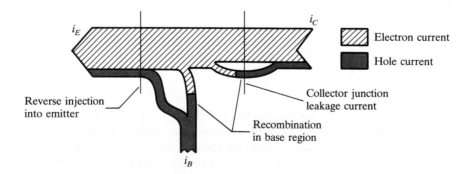

Figure 16.9   Current flow in a one-dimensional npn transistor, showing the origin of base current.

The ratio of collector current to emitter current is called α_F, the *short-circuit common base-current gain*, or:

$$\alpha_F = \frac{i_C}{i_E} \qquad (16.53)$$

The subscript F denotes forward operation of the device, uppercase implying a ratio of total quantities. As $i_B = (i_E - i_C)$, the relationship between these two current gains is

$$\beta_F = \frac{\alpha_F}{1 - \alpha_F} \qquad (16.54)$$

Because an ideal transistor requires no base current and $\beta_F = \infty$, we may view the two mechanisms giving rise to the base current shown in Fig. 16.9 as *defects* of two types. The first is the *recombination defect* δ_R and is the result of the impossibility of achieving a zero base width W or an infinite base lifetime τ_B. The second is the *emitter defect* δ_E, which reflects the impossiblity of creating an emitter junction which when forward biased does not cause reverse injection. That is, it does not inject majority carriers (holes in this case) from the base into the emitter.

These defects are defined mathematically as the ratio of the corresponding component of the base current to the minority current (electrons in this case) injected into the base from the emitter. In terms of the physical parameters of the transistor, these ratios are

$$\delta_R = \frac{i_{BR}}{i_e(0)} = \frac{qn_B(0)W/2\tau_B}{qD_Bn_B(0)/W} = \frac{W^2}{2L_B^2} \qquad (16.55)$$

$$\delta_E = \frac{i_h(0)}{i_e(0)} = \frac{qD_E}{L_E} \frac{(p_E(0) - p_{Eo})}{qD_Bn_B(0)/W} \approx \frac{D_EWN_{AB}}{D_BL_EN_{DE}} \qquad (16.56)$$

where i_{BR} is the recombination component of base current, and $i_h(0)$, the hole current at $x = 0$, is the component of the base current supplying the holes for reverse injection.

From *(16.55)* you can see the desirability of keeping the base very short compared to L_B in order to minimize δ_R. From *(16.56)* you can see that the emitter should be doped as heavily as possible—compared to the base—in order to keep δ_E small.

Two other parameters that are useful in relating the behavior of a transistor to its physical construction are the *base transport factor* and the *emitter efficiency*. The base transport factor α_T is the ratio of the electron current at the collector junction and the electron current injected at the emitter. It is related to the recombination defect by:

$$\alpha_T = 1 - \delta_R \qquad (16.57)$$

The emitter efficiency γ_E is the ratio between the electron current injected into the base from the emitter and the total emitter current. It is related to the emitter defect by:

$$\gamma_E = \frac{1}{(1 + \delta_E)} \qquad (16.58)$$

From these definitions, we can show that:

$$\frac{i_C}{i_B} = \beta_F = \frac{1 - \delta_R}{\delta_E + \delta_R} = \frac{\alpha_T \gamma_E}{1 - \alpha_T \gamma_E} \qquad (16.59)$$

and

$$\frac{i_C}{i_E} = \alpha_F = \frac{1 - \delta_R}{1 + \delta_E} = \alpha_T \gamma_E \qquad (16.60)$$

EXAMPLE 16.4

Calculating the Gain of a Transistor

For the transistor in Fig. 16.8(a), assume that $N_{DE} = 3 \times 10^{18}$ cm^{-3}, $N_{AB} = 10^{17}$ cm^{-3}, and $N_{DC} = 10^{15}$ cm^{-3}. The minority carrier lifetime in both the base and collector regions is 1 μs, but as the emitter is so heavily doped, its lifetime is 10 ns. The width of the base region is $W = 5$ μm, and the emitter and collector regions are long compared to their diffusion lengths.

We first calculate the diffusion lengths in the three regions. At the low levels of doping in the base and collector, the diffusion constant for holes in the collector is $D_C = 15$ cm^2/s, and for electrons in the base it is $D_B = 40$ cm^2/s. The diffusion lengths in these two regions are therefore $L_B = 63$ μm and $L_C = 39$ μm. In the emitter, however, the doping concentration is so high that the mobilities, and therefore the diffusion constants, are reduced from their values for intrinsic material. We assume that $D_E = 5$ cm^2/s for holes in the emitter, which gives $L_E = 2.2$ μm.

With these numbers, we can now calculate the defects for this transistor:

$$\delta_R = \frac{W^2}{2L_B^2} = 5 \times 10^{-4} \qquad (16.61)$$

and

$$\delta_E = \frac{D_E W N_{AB}}{D_B L_E N_{DE}} = 3.8 \times 10^{-3} \qquad (16.62)$$

These defect values result in a forward current gain of:

$$\beta_F = \frac{1 - \delta_R}{\delta_E + \delta_R} = 79 \qquad (16.63)$$

Note that in this example the emitter defect is the dominant cause of base current. This condition is typical for power transistors.

16.3.3 The Ebers–Moll Model of an npn Transistor

Figure 16.10 is the *Ebers–Moll* (large-signal) *model* of an npn transistor. It is the combination of a forward model, consisting of the base-emitter diode and the collector-current source, and a reverse model, consisting of the base-collector diode and the emitter-current source. The forward model describes the transistor when its base-emitter junction is forward biased, and the reverse model describes the transistor when its base-collector junction is forward biased. The Ebers–Moll model describes the large-signal behavior of the transistor for any set of terminal conditions.

Our analysis of the npn transistor in its forward active region in Section 16.3.1 yielded an analytic representation of the forward portion of the Ebers–Moll model, except that we neglected reverse leakage at the collector junction, I_{CS}. This neglect was equivalent to saying that the leakage current in the reverse model's diode was very small compared to that in the forward model's current source.

We can determine the defects and gains for the reverse operation by simply replacing the emitter parameters with those of the collector in *(16.55)* and *(16.56)*. The recombination defect δ_R remains the same, but the emitter defect does not, because the emitter in the reverse region of operation is the collector in forward operation. Most transistors have heavily doped emitters to improve the forward model's gain and lightly doped collectors to withstand high collector voltages. Thus the emitter defect in reverse operation is much larger than that in the forward operating region. For example, using the parameters in Example 16.4, we find the reverse emitter defect to be

$$\delta_{Er} = \frac{D_C W N_{AB}}{D_B L_C N_{CE}} = 1.9 \qquad (16.64)$$

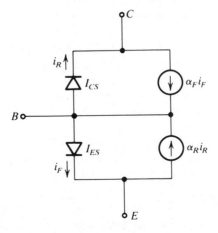

Figure 16.10   The Ebers–Moll model of an npn transistor.

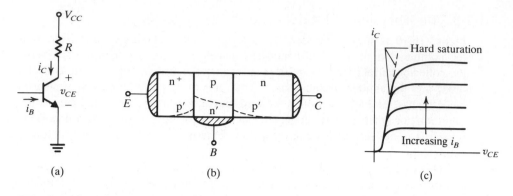

Figure 16.11 The saturation region of bipolar transistor operation: (a) common emitter amplifier circuit; (b) excess carrier profiles in a saturated npn transistor; (c) the collector characteristics for an npn transistor.

This value for the emitter defect yields reverse current gains of $\beta_R = 0.52$ and $\alpha_R = 0.34$.

16.3.4 Saturation

When we operate a transistor in its forward active region, the more base current we deliver, the higher the collector current becomes. If the load on the transistor causes the collector-emitter voltage to drop as the collector current rises—as it would in the common-emitter amplifier circuit of Fig. 16.11(a)—a point can be reached at which the base-collector junction becomes forward biased. This condition, in which both the collector and emitter junctions are forward biased, is called *saturation*. Figure 16.11(b) shows the excess carrier profiles in the transistor under this condition.

As a transistor enters saturation from the forward active region, its base current must be increased to support the additonal requirements of increased base charge and minority carrier injection into the collector (the reverse injection of reverse operation). The gain of the transistor therefore appears to be decreasing. If the base current is increased further, a point is reached where for a given collector current, the collector-emitter voltage cannot be reduced further because both diodes are strongly forward biased. This point is called *hard saturation*, and is identified in the typical collector i–v characteristics of an npn transistor of Fig. 16.11(c).

16.4 THE MOSFET

Figure 16.12(a) shows a lateral MOSFET geometry. The term lateral (as opposed to vertical) refers to the direction of current flow between drain D and source S when the MOSFET is conducting. Both the drain and the source are made of heavily doped n-type silicon and the *body*, or *substrate*, is made of relatively lightly doped p-type silicon. The gate G is a conductive plate (usually made of polysilicon, but sometimes made of metal) that is electrically isolated from the rest of the device by an oxide insulator called the *gate oxide*.

Looking just at the silicon structure of the device, we would conclude that no current could flow between the drain and source terminals because they are isolated by a pair of back-to-back pn junctions (n^+pn^+). This, in fact, is the case when the MOSFET is off. It will withstand a voltage in either direction, and the SCL supporting this voltage will grow mostly into the p region. One end of the gate oxide is therefore exposed to this off-state voltage, as well.

If we assume that the drain and source are both grounded and apply a positive voltage to the gate, an electric field perpendicular to the plane of the gate appears in the oxide layer. This field originates on positive charge residing on the gate electrode and terminates on negative charge within the p region of the silicon. At low gate voltages, this negative charge comes from the immobile acceptor ions that are left unneutralized when holes flow away from the oxide–silicon interface. In other words, an SCL grows into the silicon from the interface, as shown in Fig. 16.12(b).

Because of the electric field in the SCL, the potential in the silicon just under the oxide is higher than it is in the neutral p region. The silicon is in thermal

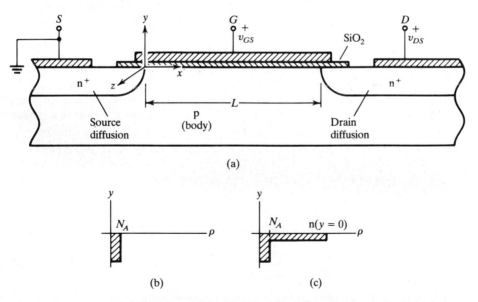

(a)

(b)   (c)

Figure 16.12 The lateral MOSFET: (a) transistor structure; (b) field terminating charge distribution in the p region at low values of v_{GS};
(c) field terminating charge distribution in the p region for $v_{GS} >$
V_T, where V_T is the gate threshold voltage.

equilibrium (no current is flowing), so the relationships between carrier densities and the potential given by *(16.17)* apply here. Therefore, although the concentration of holes decreases as the potential at the surface increases, the concentration of electrons is growing. When the potential becomes great enough, there actually are more electrons than holes, and a thin layer of effectively n-type silicon, called a *channel*, is created just beneath the oxide. When this condition occurs, there is no longer a pn junction between the drain and source. Instead, there is an n^+nn^+ structure that acts like a resistor and permits current to flow. It is a very large resistor, however, because the concentration of electrons under the oxide is very low.

As we continue to increase the gate voltage, the concentration of electrons in the n-type layer grows. Eventually the electron concentration equals that of the original hole concentration N_A, a condition called the *onset of inversion*. At this gate-source voltage, known as the the *threshold voltage*, V_T, the resistance of the n-type channel begins to fall rapidly as the gate voltage is raised. For $v_{GS} > V_T$, the concentration of electrons becomes very large, as shown in Fig. 16.12(c), and the resistance of the channel becomes a strong function of the gate voltage.

The MOSFET of Fig. 16.12 is an *n-channel enhancement mode* MOSFET. The term enhancement refers to the fact that the channel conductance is enhanced by the application of gate voltage. The device is off if $v_{GS} = 0$. It is also possible to make p-channel MOSFETs by reversing the dopant types, in which case the gate voltage must be negative with respect to the source in order for the device to be turned on.

16.4.1 MOSFET Characteristics

In the preceding discussion we assumed a zero drain-source voltage, or at least one that was negligible compared to V_T. Such an assumption results in a negative charge distribution in the channel that is independent of x. Actually, if we assume that a current is flowing from drain to source, the voltage at the drain edge of the channel will be higher than it is at the source end. The electric field in the oxide will therefore be weaker at the drain edge, and the number of electrons on which it terminates will be correspondingly less. As a result, the conductivity of the channel varies with x.

For $v_{DS} < (v_{GS} - V_T)$, the drain current as a function of v_{GS} and v_{DS} is

$$i_D = \frac{\mu_e Z C_o}{L}\left[(v_{GS} - V_T)v_{DS} - \frac{v_{DS}^2}{2}\right] \qquad (16.65)$$

where C_o is the oxide capacitance per unit area, L is the channel length, and Z is the z-axis dimension of the channel.

If v_{DS} is small compared to $(v_{GS} - V_T)$, the quadratic term in *(16.65)* may be ignored and the drain current is directly proportional to the drain-source voltage.

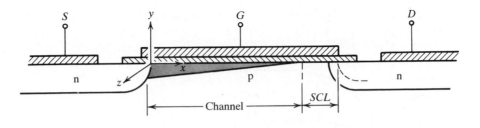

Figure 16.13 Inverted channel and SCL in a MOSFET with $v_{DS} > v_{GS} - V_T$.

The ratio i_D/v_{DS} is a conductance of value:

$$\frac{i_D}{v_{DS}} = g_{ch} = \frac{\mu_e Z C_o}{L}(v_{GS} - V_T) \qquad (16.66)$$

Because the relationship between drain voltage and drain current is linear for any gate voltage, this region of operation is commonly called the *linear* region. Note that this name corresponds to the fully on condition of the device, or the saturation condition in bipolar transistor terminology.

When v_{DS} reaches a value such that the voltage across the oxide at the drain end of the channel is reduced to V_T, that is, when $v_{DS} = v_{GS} - V_T$, the electron density at that end of the channel falls to zero. In other words, the region under the gate near the drain is no longer inverted. Further increases in v_{DS} are absorbed by a space-charge layer that grows at the junction between the drain and this noninverted region as shown in Fig. 16.13.

The length of the SCL is typically very small compared to the distance between the drain and source diffusions. Therefore the length of the channel region in Fig. 16.13 is approximately L, independent of v_{DS}. Similarly, the concentration of electrons in the channel and the voltage from one end to the other ($v_{GS} - V_T$) are independent of v_{DS}. For this reason, when v_{DS} exceeds ($v_{GS} - V_T$), the current flowing through the channel no longer depends on the drain-source voltage, and the following relationship exists between the drain current and the gate voltage:

$$i_{D(\text{sat})} = \frac{\mu_e Z C_o}{2L}(v_{GS} - V_T)^2 \qquad (16.67)$$

We call this region of operation *saturation*, although it corresponds to a condition we call forward active in bipolar transistor terminology.

Figure 16.14 shows the complete common-source i–v characteristics for an n-channel MOSFET. Note that, although these characteristics appear similar to those of the bipolar transistor, the value of drain current depends on the gate voltage, rather than the base current.

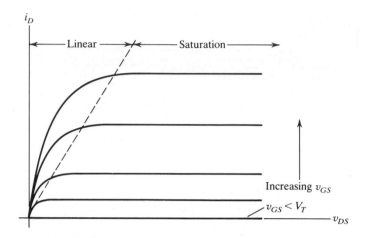

Figure 16.14 Common-source i–v characteristics for an n-channel MOSFET.

16.4.2 The Dependence of V_T on Substrate Potential

The value of v_{GS} at which the channel becomes inverted ($v_{GS} = V_T$) depends on the potential of the neutral p-type substrate, or body. Once the inverted channel is formed, it has (at its source end) the same potential as the source. The lower the potential in the neutral p region is, compared to the source potential, the larger the voltage drop across the depleted SCL region is and the larger the electric field at the oxide interface is. We can multiply this electric field, adjusted for the different dielectric constants of silicon and oxide, by the thickness of the oxide to find V_T.

Although we do not develop explicit relationships here, note that the lower the substrate potential is, the higher the threshold voltage is, and vice versa. For this reason, connecting the substrate to a known potential is important, so that the threshold voltage will have a fixed value. Because the substrate forms a pn diode with both the drain and the source, connecting it to the lower of the drain and source potentials is necessary. Normally, the source is the lower potential, so a typical MOSFET has its substrate connected to its source, forming what is known as the *source-body short*.

Notes and Bibliography

There are many textbooks and reference books on the fundamentals of semiconductor devices that take the student from the basic physics of carrier concentrations and charge transport through to the i–v characteristics of modern electronic devices. Volumes 1 and 2 of the SEEC series [1,2] are some of the oldest, yet still very informative and useful, presentations on semiconductor physics and bipolar devices. Volume 1 provides the physical background needed to understand all semiconductor devices, and Volume 2 applies this understanding to the bipolar diode and transistor.

The Modular Series on Solid State Devices [3,4,5,6] is a more recent effort to explain the fundamentals of semiconductor devices. Volume 1 presents the fundamentals of semiconductor physics, and then Volumes 2, 3, and 4 give extensive coverage of the pn diode, the bipolar junction transistor, and field-effect devices, respectively. The material is very accessible, although detailed, the figures are informative, and there are many problems at the end of each chapter. It is a excellent set of texts to read as a review and an advancement of the material presented in an introductory course on semiconductor devices.

1. R. B. Adler, A. C. Smith, and R. L. Longini, *Introduction to Semiconductor Physics*, Semiconductor Electronics Education Committee, Volume 1 (New York: John Wiley & Sons, 1964).
2. P. E. Gray, D. DeWitt, A. R. Boothroyd, and J. F. Gibbons, *Physical Electronics and Circuit Models of Transistors*, Semiconductor Electronics Eductaion Committee, Volume 2 (New York: John Wiley & Sons, 1964).
3. R. F. Pierret, *Semiconductor Fundamentals*, Modular Series on Solid State Devices, Volume 1 (Massachusetts: Addison-Wesley, 1983).
4. G. W. Neudeck, *The pn Junction Diode*, Modular Series on Solid State Devices, Volume 2 (Massachusetts: Addison-Wesley, 1983).
5. G. W. Neudeck, *The Bipolar Junction Transistor*, Modular Series on Solid State Devices, Volume 3 (Massachusetts: Addison-Wesley, 1983).
6. R. F. Pierret, *Field-Effect Devices*, Modular Series on Solid State Devices, Volume 4 (Massachusetts: Addison-Wesley, 1983).

Problems

16.1   The intrinsic carrier concentration in a semiconductor is a strong function of temperature. In silicon, it is

$$n_i^2 = C_1 T^3 e^{-12800/T}$$

where T is in Kelvins. The electron and hole mobilities also vary with temperature. We can approximate their relationship to T as:

$$\mu_e = C_2 T^{-2.2} \quad \text{and} \quad \mu_h = C_3 T^{-2.2}$$

From the values given in Table 16.1 for n_i, μ_e, and μ_h at 300K, find the values of these parameters at $-50°C$ and $+150°C$.

16.2   An n-type silicon sample has a conductivity of $5 \times 10^{-2} \ \Omega^{-1}\text{cm}^{-1}$ at 300K. Based on your results in Problem 16.1, find the conductivity of this sample at 200K and 400K. (*Note:* The sample may not be extrinsic at all three temperatures.)

16.3   Gallium–arsenide (GaAs) is a compound semiconductor with a structure similar to silicon except that it consists of alternating Ga and As atoms. (That is, each Ga atom has four As atoms as nearest neighbors, and each As atom has four Ga atoms as nearest neighbors.) The ion core of Ga has a charge of +3q and the ion core of As has a charge of +5q (q is the electronic charge).

(a) Draw a two-dimensional representation of what you expect to happen when Mg, Al, or Si is substituted for a Ga atom or when Si, P, or S is substituted for an As atom (a total of six cases). For each case, state whether the impurity is a donor, an acceptor, or neither.

(b) Assume that Si atoms are added to GaAs with a concentration of 10^{15} cm^{-3} and that Si will substitute for As four times as often as it substitutes for Ga. Calculate the conductivity of the resulting sample at room temperature using the following parameters for GaAs:

$$n_i^2 = 80 \times 10^{12} \text{ cm}^{-6} \quad \mu_e = 8800 \text{ cm}^2/\text{V-s} \quad \text{and} \quad \mu_h = 400 \text{ cm}^2/\text{V-s}$$

16.4   Assume that you have a chamber that extends to $\pm\infty$ in the x-direction and has a unit cross-sectional area in the yz-plane. Within this chamber particles are distributed uniformly in the yz-plane but have the following distribution in the x-direction at $t = 0$:

$$C(x) = (1 + \cos kx) \text{ cm}^{-3}$$

These particles obey the following diffusion equation:

$$\frac{\delta C(x,t)}{\delta t} = D \frac{\delta^2 C(x,t)}{\delta x^2}$$

Find an expression that describes $C(x,t)$ for $t > 0$. (*Hint:* Use complex notation and assume a solution of the form:

$$C(x,t) = C_o + Re\left\{C_1 e^{at} e^{bx}\right\}$$

The resulting solution provides an estimate of the time required for an excess minority carrier profile in a bipolar device to flatten out after the excitation has been turned off.)

16.5   Calculate the built-in potential ψ_o for a pn junction when both sides of the junction are doped at (a) 10^{14} cm^{-3}, (b) 10^{16} cm^{-3}, and (c) 10^{19} cm^{-3}. For each case, calculate the reverse voltage at which the peak electric field is 3×10^5 V/cm. (This is approximately the voltage at which avalanche, a breakdown mechanism, occurs.) For each voltage, find the distance that the SCL extends into either the n or p regions.

16.6   For each of the cases in Problem 16.5, find the applied forward bias voltages at which high-level injection is reached. High-level injection is defined as the point at which the maximum excess minority carrier concentration equals the dopant level.

16.7   The diode shown in Fig. 16.15 has a p region that is doped much more heavily than its n region. Assume that ℓ, the width of both regions, is very small compared to the characteristic diffusion length of the minority carrier.

(a) Find an expression for the total current flowing through the device when a voltage v_A is applied with the polarity shown. Assume that the SCL width is small compared to the length of the device. Evaluate this current when $v_A = 0.6$ V.

(b) Find an expression for x_n, the distance that the SCL extends into the n region, as a function of the applied voltage v_A. Calculate this distance for $v_A = +0.5$, 0, and -100 V.

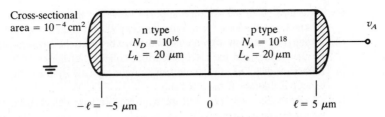

Figure 16.15 The diode structure analyzed in Problem 16.7.

    (c) When $v_A < 0$, x_n can become a significant fraction of the length of the n region. Derive an expression for the leakage current that flows under reverse bias that accounts for this extension of the SCL. Assume no recombination or generation in the SCL. Graph this current as a function of the applied reverse voltage. (You should observe a condition, called punch-through, where the leakage current gets very large.)

16.8 A forward biased voltage, v_A, is applied across a diode with neutral regions that are the same length as the corresponding minority carrier diffusion length, as shown in Fig. 16.16. (Ignore the width of the SCL.)

    (a) Find, sketch, and label p', n', J_e (drift and diffusion), and J_h (drift and diffusion) in the two neutral regions.

    (b) Find, sketch, and label the electric field in the neutral regions. Integrate the expression for this field to determine the total voltage drop across the neutral regions.

    (c) Assuming that all the applied voltage appears across the SCL, find the voltage at which the maximum excess carrier concentration is only a factor of 5 away from the dopant concentration. At this voltage, what is the total voltage drop across the neutral regions?

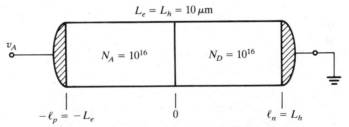

Figure 16.16 A diode whose neutral regions have lengths equal to their minority carrier diffusion lengths. This diode is analyzed in Problem 16.8.

16.9 An npn transistor's base has width W that is small compared to its diffusion length L_e. Assume that the applied voltages are such that the excess electron concentration at the base side of the base-emitter SCL edge is $n'(0^+)$ and that the excess electron concentration at base side of the base-collector SCL edge is zero.

    (a) Assuming that the minority carrier lifetime in the base is infinite, solve for the excess electron concentration in the base, and approximate the net recombination current density in the base as a function of the real lifetime, τ_e, and $n'(0^+)$.

    (b) Find the ratio of this recombination current to the electron current injected by the emitter. (This is the approximate base defect of the transistor.)

(c) Instead of using the assumption in (a), solve for the exact excess electron concentration in the base for a finite minority carrier lifetime τ_e.

(d) Using your result in (c), find the difference between the electron current injected by the emitter and the electron current collected at the collector. Express this difference as a ratio of the electron current injected by the emitter. (This is the exact base defect of the transistor.)

(e) At what value of W/L_e will the approximate base defect differ by a factor of two from the exact base defect?

16.10 The npn transistor with doping levels and dimensions shown in Fig. 16.17 has an infinite minority carrier lifetime throughout.

(a) Find α_F, α_R, β_F, β_R, I_{ES}, and I_{CS} for this transistor.

(b) If the applied base-emitter voltage is $v_{BE} = 0.6$ V and $v_{BC} = 0$ V, find the current that flows into the base terminal.

(c) With $v_{BE} = 0.6$ V, at what base collector voltage v_{BC} will the base terminal current be twice that found in (b)? How much has the collector-terminal current changed from its value under the conditions of (b)? (*Note:* This value could be considered the onset of saturation.)

(d) For the base-collector voltage found in (c), sketch the excess carrier profiles in the transistor.

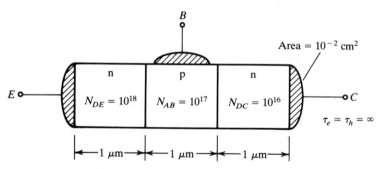

Figure 16.17 Dimensions and doping concentrations for the transistor analyzed in Problem 16.10.

16.11 The npn transistor shown Fig. 16.18 has a base whose width is small compared to the minority carrier diffusion length L_e, but emitter and collector lengths that are very long compared to their (equal) diffusion lengths L_h.

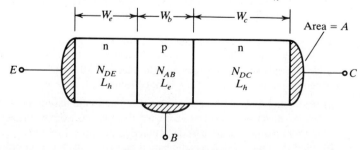

Figure 16.18 The transistor analyzed in Problem 16.11.

(a) For the forward mode condition ($v_{BE} > 0$ and $v_{BC} \approx 0$), sketch and label the excess minority carrier concentrations and the hole and electron currents (both drift and diffusion) throughout the transistor.

(b) Repeat (a) for the reverse mode condition ($v_{BC} > 0$ and $v_{BE} \approx 0$).

(c) What is the emitter defect of this device?

16.12 Sketch and label the SCL charge density as a function of v_A for $-1 \text{ V} \leq v_A \leq \psi_o$ for a pn junction with doping concentrations $N_D = N_A = 10^{16} \text{ cm}^{-3}$ and $A = 1 \text{ cm}^2$. Using your sketch, illustrate the distinction between the large-signal junction capacitance $C_o = Q/v_A$ and the incremental capacitance $C_i = dQ/d \mid v_A \mid$.

16.13 The following questions relate to the asymmetrical pn diode shown in Fig. 16.19.

(a) Find the charge in the SCL as a function of the applied voltage v_A.

(b) Find an expression for the large signal junction capacitance $C_o = Q(v_A)/|v_A|$. Graph and label it as a function of v_A for $-10 \text{ V} \geq v_A \geq -200 \text{ V}$. (*Note:* For this and the following parts, assume that $|v_A|$ is very much larger than ψ_o, so that $(\psi_o - v_A) \approx -v_A$.)

(c) Find an expression for $C_i = dQ(v_A)/d \mid v_A \mid$. Graph and label it as a function of $-v_A$ for $-10 \text{ V} \geq v_A \geq -200 \text{ V}$.

(d) Find an expression for $C_{\text{energy}} = E_c(v_A)/(0.5v_A^2)$, the ratio between the stored electrostatic energy and one half the voltage squared. Graph and label it as a function of $-v_A$ for $-10 \text{ V} \geq v_A \geq -200 \text{ V}$. (*Note:* The proper way to calculate energy in a capacitor is

$$E_c(v_A) = \int_0^{Q(v_A)} v \, dQ = \int_0^{v_A} v \frac{dQ}{dv} \, dv = \int_0^{v_A} v C_i \, dv$$

ignoring the effect of ψ_o near $v = 0$.)

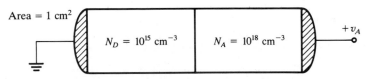

Area = 1 cm^2

$N_D = 10^{15} \text{ cm}^{-3}$ $N_A = 10^{18} \text{ cm}^{-3}$ $+ v_A$

Figure 16.19 The pn diode that is the subject of Problem 16.13.

16.14 An abrupt junction bipolar diode with ohmic contacts has a voltage, v_A, applied across its terminals, as shown in Fig. 16.20. Use the abrupt depletion region model for the SCL.

(a) If $N_A = N_D = 10^{16} \text{ cm}^{-3}$, what is the junction's built-in potential ψ_o?

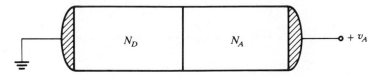

N_D N_A $+ v_A$

Figure 16.20 The abrupt junction bipolar diode analyzed in Problem 16.14.

(b) For the doping concentrations of (a), at what applied voltage will the space-charge layer be (i) half as long as it is when $v_A = 0$ V? (ii) ten times as long as it is when $v_A = 0$ V?

(c) If you want the maximum electric field strength in the SCL to be $E_{max} = 3 \times 10^5$ V/cm when $v_A = -200$ V, what value of doping concentration should you use? (Assume a symmetrical junction where $N_A = N_D$.)

Chapter 17

Power Diodes

POWER semiconductor devices are distinguished by the high current densities at which they operate while on and the high voltages they must withstand when off. In this chapter we address the effects of these requirements on the design and performance of power diodes. Because of its relatively simple structure and physical operation, we use the diode to introduce the principles governing the behavior of power semiconductor devices. We then apply these principles to the transistor and thyristor in Chapters 18 and 19.

In this chapter we also describe the operation and behavior of the power Schottky diode, a device that uses a metal semiconductor junction and has a lower on-state voltage than the bipolar junction diode. It also has the advantage of turning on and off much faster. However, Schottky diodes lose these advantages when they are designed to support reverse voltages in excess of about 100 V, and at all voltage ratings they exhibit much higher leakage currents than do bipolar diodes.

17.1 THE BIPOLAR DIODE UNDER REVERSE BIAS

The characteristics limiting the reverse voltage rating of a bipolar diode are leakage current and avalanche breakdown. Although we discussed both phenomena in Chapter 16, a number of simplifying assumptions were not consistent with the behavior of a device designed specifically for power applications.

17.1.1 Leakage Current

In Chapter 16 we showed that a forward biased diode whose neutral regions are in low-level injection—and are long compared to the minority carrier diffusion lengths—have the excess carrier profile shown in Fig. 17.1 and an i–v relationship of:

$$J = \left[\frac{qD_h n_i^2}{N_D L_h} + \frac{qD_e n_i^2}{N_A L_e}\right]\left(e^{qv_A/kT} - 1\right) = J_s\left(e^{qv_A/kT} - 1\right) \qquad (17.1)$$

In reverse bias, that is, $v_A \ll 0$, $J = J_s$, where J_s is the leakage-current density.

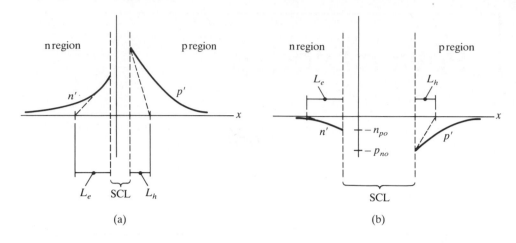

Figure 17.1 Excess minority carrier profiles in a long diode: (a) forward bias:
(b) reverse bias.

At room temperature and under forward bias, J_s is typically very small compared to J. For instance, if we assume that $v_A = 0.6$ V, J is 2.6×10^{10} times greater than J_s. Such a small leakage current results in negligible reverse bias power dissipation, even when the diode is blocking several thousand volts.

However, J_s is proportional to n_i^2, which is a strong function of temperature. That is,

$$n_i^2 \propto T^3 e^{-E_g/kT} \tag{17.2}$$

At 25°C, $n_i^2 \approx 2 \times 10^{20}/\mathrm{cm}^6$, but at 175°C (a typical maximum operating temperature for a diode) $n_i^2 = 3.8 \times 10^{27}/\mathrm{cm}^6$. This seven order of magnitude increase gives a leakage current whose effects are no longer negligible.

Carrier Generation in the SCL   We know that *(17.1)* does not account for all the current in a reverse biased diode. As discussed in Chapter 16, electron–hole pairs are thermally generated throughout the device at a rate that depends on the temperature. In thermal equilibrium, this generation is exactly canceled by recombination, which occurs at a rate proportional to the product of the carrier concentrations. The result is that $n_o p_o = n_i^2$ *everywhere* in the device. When the diode is reverse biased, however, it is no longer in thermal equilibrium. The excess carrier concentrations within the space-charge layer and the adjacent diffusion regions are substantially below their equilibrium values n_o and p_o. In these regions, then, the thermal generation rate, which is independent of n and p, exceeds the recombination rate. The carrier concentrations do not build up because the thermally generated carriers are swept out of the SCL by the intense electric field that exists there. Holes flow to the p region's contact and electrons flow to the n region's contact. These carrier flows give a component of the measured reverse bias leakage current not accounted for by *(17.1)*.

The generation rate in the depleted region is proportional to n_i, which is a function only of material and temperature. If we define a characteristic lifetime τ_{SCL} so that the generation rate is $G = n_i/\tau_{SCL}$, the total current density caused by thermal generation in the SCL is

$$J_{SCL} \approx q(L_e + L_h + \ell_p + \ell_n)G = \frac{q(L_e + L_h + \ell_p + \ell_n)n_i}{\tau_{SCL}} \qquad (17.3)$$

where $(L_e + L_h + \ell_p + \ell_n)$ is the width of the SCL and adjacent diffusion regions.

Because the width of the SCL and the diffusion regions are so long in a power device, we make τ_{SCL} as large as possible. This lifetime is related to, but is usually longer than, the low-injection–level minority carrier lifetimes τ_e and τ_h. In general, J_{SCL} is much larger than J_s at room temperature, but at high operating temperatures, J_s is larger.

EXAMPLE 17.1

Calculation of Leakage-Current Components

For the diode described by (17.1), assume that $\tau_e = \tau_h = 0.2$ μs, $\tau_{SCL} = 10$ μs, and $N_A = N_D = 3 \times 10^{15}/cm^3$. These doping concentrations are such that for a reverse voltage of $v_A = -100$ V, the SCL extends approximately 4.6 μm into both the n and p regions, that is, $\ell_p + \ell_n = 9.2$ μm.

We calculate J_s at 25°C, using (17.1) and parameters from Table 16.1, obtaining $2.4 \times 10^{-10}A/cm^2$. For this value of J_s, a forward biased current density of 3 A/cm² requires $v_A = 0.58$ V. This current density is very low compared to the typical current density for a power diode but is about the level at which this diode will no longer be in low-level injection. Although we do not yet have an accurate model for the diode in high-level injection, we can expect it to carry a forward current density of 50 A/cm², resulting in $v_A \approx 1$ V. Operating at this current density, the diode's dissipation is 50 W/cm².

The minority carrier diffusion lengths in this diode are $L_e = 28$ μm and $L_h = 17$ μm. The total width of the SCL and diffusion regions is therefore approximately 55 μm at -100 V. At room temperature, this width yields a thermally generated leakage current $J_{SCL} = 1.25 \times 10^{-6}A/cm^2$. Although this leakage current is much larger than J_s, it causes a reverse bias power dissipation of only 0.125 mW/cm², which is negligible compared to the forward bias power dissipation.

At 175°C, $J_s = 4.6 \times 10^{-3}A/cm^2$ and $J_{SCL} = 5.4 \times 10^{-3}A/cm^2$. With these currents, the reverse bias dissipation is 1 W/cm², which is still small but is quickly becoming important. At 200°C, $J_s = 28 \times 10^{-3}A/cm^2$, $J_{SCL} = 13 \times 10^{-3}A/cm^2$, and the reverse dissipation is 30 W/cm².

A precise quantitative prediction of leakage current requires consideration of effects in addition to those discussed here. For instance, temperature affects lifetimes, and generation within the SCL is not uniform. However, we can draw some important conclusions from the preceding analysis:

1. Generation in the wide SCL and diffusion regions of a power device will yield substantially more leakage current at 25°C than is predicted by the classic diffusion model, (17.1).

2. To keep the leakage current acceptably small, the minority carrier lifetime in the SCL should be as long as possible.

3. The component of leakage current resulting from thermal generation in the SCL, J_{SCL}, grows as n_i, which approximately doubles for every $11°C$ increase in T between $-50°C$ and $200°C$. But J_s grows as n_i^2, so it eventually dominates at high temperatures. This increase in reverse leakage current limits the maximum operating junction temperature of a power device.

17.1.2 Avalanche Breakdown

We assumed in Chapter 16 that applying an electric field to a semiconductor causes the holes and electrons to flow with a velocity proportional to the electric field. Actually, the particles frequently collide with the lattice imperfections (impurity atom sites, physical defects, or phonons). What we really meant was that the *average* velocity of the particles is proportional to the electric field. If the electric field is strong enough, the velocity of a particle becomes large enough that, during a collision, energy is imparted to the lattice to create an electron–hole pair. This process is called *impact ionization*. It is a multiplicative process: The newly generated particles are accelerated by the field, collide with the lattice, and create additional electron–hole pairs, and so on.

As the reverse bias voltage across a diode increases, so does the peak electric field in its SCL. When the field reaches a magnitude of about 3×10^5 V/cm, the impact ionization process begins. The number of electron–hole pairs generated in the SCL by impact ionization grows very quickly with increasing electric field above this value. Because these generated carriers add to the thermally generated ones to give more reverse leakage current, the diode's reverse bias characteristic looks like that shown in Fig. 17.2. When impact ionization reaches an infinite rate, we say that the diode is in *avalanche breakdown*.

We can model the impact ionization process by defining *impact ionization*

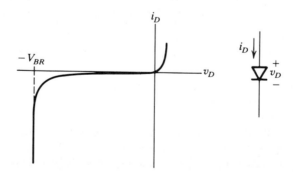

Figure 17.2   The i–v characteristic of a power diode, showing the fast rise in reverse leakage current when the avalanche breakdown voltage $-V_{BR}$ is reached.

coefficients for electrons, α_n, and holes, α_p. These coefficents give the number of electron–hole pairs that a single electron (or hole) would create at a given field strength if it moved 1 cm in the direction of the field. Actual measurements suggest that these coefficients are proportional to the exponential of the negative inverse of the electric field:

$$\alpha_n = a_n e^{-b_n/E}$$
$$\alpha_p = a_p e^{-b_p/E} \tag{17.4}$$

For silicon, $a_n = 7 \times 10^5$/cm, $a_p = 1.6 \times 10^5$/cm, $b_n = 1.23 \times 10^6$ V/cm, and $b_p = 2 \times 10^5$ V/cm.

Using these definitions, we can compute $M(x)$, the total number of electron–hole pairs generated in a one-dimensional SCL of width $\ell = \ell_n + \ell_p$ by a single electron–hole pair at a distance x from the n region edge of the SCL:

$$M(x) = 1 + \int_0^x \alpha_n M(x')\, dx' + \int_x^\ell \alpha_p M(x')\, dx' \tag{17.5}$$

By differentiating *(17.5)* and solving for $M(x)$, we get

$$M(x) = M(0) \exp\left[\int_0^x (\alpha_n - \alpha_p)\, dx'\right] \tag{17.6}$$

Combining *(17.5)* and *(17.6)* gives

$$M(x) = \frac{\exp\left[\int_0^x (\alpha_n - \alpha_p)\, dx'\right]}{1 - \int_0^\ell \alpha_p \exp\left[\int_0^x (\alpha_n - \alpha_p)\, dx'\right] dx} \tag{17.7}$$

Avalanche breakdown occurs when $M(x)$ reaches infinity, or when

$$\int_0^\ell \alpha_p \exp\left[\int_0^x (\alpha_n - \alpha_p)\, dx'\right] dx = 1 \tag{17.8}$$

For any pn junction, we can use the doping levels (and profiles, as no junction is truly abrupt) to find the electric field throughout the SCL for a specific reverse voltage. We can then use this electric field in *(17.8)* to find the avalanche breakdown voltage.

The onset of avalanche breakdown occurs approximately when the peak electric field in the SCL reaches a critical value, E_c. This statement is not strictly accurate, because the avalanche process requires that impact ionization occur over some minimum distance so the multiplication process can sustain itself. Because the gradient of the electric field is much greater in the SCL of a heavily doped diode than in the SCL of a lightly doped diode, the distance over which ionization occurs is shorter. The peak electrical field can therefore be greater in a heavily doped diode before avalanche begins. For a doping level of 10^{16}/cm^3 (typical of a 40–50 V device), $E_c \approx 3.7 \times 10^5$ V/cm, but for a doping level of 10^{14}/cm^3 (typical of a 800–1000 V device), $E_c \approx 2.1 \times 10^5$ V/cm.

EXAMPLE 17.2

Designing an Asymmetric Junction to Withstand 200 V

The asymmetrically doped p^+n diode shown in Fig. 17.3 is to be designed to support a reverse voltage of $v_A = -200$ V. We will use $E_c = 3 \times 10^5$ V/cm for the critical electric field at which avalanche occurs, a magnitude appropriate for the doping level we will calculate.

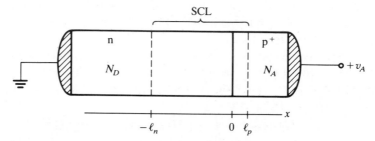

Figure 17.3 The SCL in an asymmetrical abrupt junction diode.

If the peak electric field is E_c when $v_A = -200$ V, we can easily find the corresponding length of the SCL:

$$v_A = \frac{1}{2}(\ell_p + \ell_n)E_c \qquad (17.9)$$

so

$$\ell_p + \ell_n \approx \ell_n = \frac{2 \times 200}{E_c} = 13.3 \ \mu m \qquad (17.10)$$

This is the minimum length of the n region. From this value of ℓ_n, which occurs when the peak electric field is E_c (that is, when $v_A = -200$ V), we can determine the maximum possible n-region doping concentration N_D:

$$N_D = \frac{\epsilon_{Si}E_c}{q\ell_n} = 1.4 \times 10^{15}/cm^3 \qquad (17.11)$$

Junction Termination   The pn junctions of practical devices must eventually reach the surface of the silicon die. For instance, Fig. 17.4 shows a typical diode structure in which p-type impurities are deposited on a lightly doped n-type substrate—in an area left open by an oxide mask—and then diffused into the substrate by placing the wafer in a diffusion furnace. Note that a lateral diffusion accompanies the vertical diffusion. The resulting junction curves to the surface with a radius approximately equal to the vertical diffusion depth.

Two phenomena cause this junction to break down at a lower voltage than the one-dimensional analysis leading to (17.8) predicts. First, the electric field is stronger near the curvature than it is in the middle of the diode, where the SCL edges are essentially parallel planes. Second, the maximum electric field at which

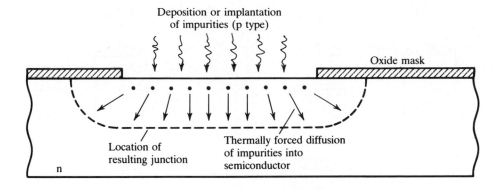

Figure 17.4 Impurity deposition and diffusion through an oxide mask resulting in a junction that curves to the surface.

breakdown occurs is lower at the surface of a semiconductor than it is in the bulk material because of lattice damage at the surface.

A number of techniques have been developed to reduce the electric field intensity at the surface of a power semiconductor device. These are *field plates*, *floating field rings*, and *beveled edge* junction termination. These techniques create boundary conditions that reduce the field intensity where the junction exhibits curvature and where it meets the surface of the device.

17.2 THE PIN DIODE

A reverse biased, asymmetrically doped diode supports nearly all the voltage across the SCL in the lightly doped n^- region. Making a vertical power diode by simply diffusing the p^+ region into an n^- substrate, as shown in Fig. 17.5(a), would result in a large n^- substrate resistance and a large resistive component of the forward drop. The reason is that the substrate must be about 500 μm thick to maintain mechanical integrity, which introduces a large series resistance. However, because the SCL has a width of between 10 and 200 μm, sandwiching a relatively thin n^- layer between p^+ and n^+ layers, as shown in Fig. 17.5(b), can lower the resistive component of the forward drop.

Doping the n^- region so lightly that it is nearly intrinsic further improves performance. The resulting electric field magnitude is shown in Fig. 17.5(c). Because the field distribution is almost rectangular, the width of the i region is half that of the n^- region of Fig. 17.5(b) for the same voltage rating. The diode of Fig. 17.5(c) is called a *pin diode*. In a practical device, the intrinsic region is really very slightly n type (ν) or p type (π), and the device is known as a pνn or pπn diode. Almost all power diodes have this type of structure, which we refer to generically as a *pin structure*.

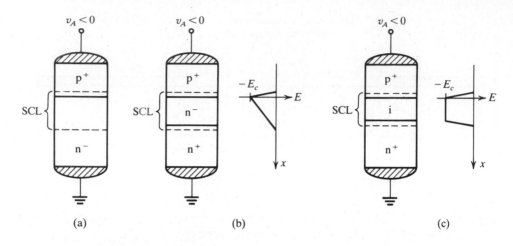

Figure 17.5 Evolution of the pin diode structure: (a) a conventional asymmetric diode with a large n^- substrate resistance; (b) the diode of (a) with the neutral part of the n^- region replaced by a low resistance n^+ region; (c) the diode of (b) with the n^- region replaced by an intrinsic region. The resulting field distributions for (b) and (c) are also shown.

EXAMPLE 17.3

Breakdown Voltage of a pin Diode

If the doping concentration in the center layer of the diode shown in Fig. 17.5(b) is $N_D = 10^{14}/cm^3$, what is the relationship between the breakdown voltage and the width of the center layer W_ν? We assume that all the voltage is dropped across the ν region. From Gauss's law we can determine $E_\nu(x)$, the field in this region:

$$\frac{dE_\nu}{dx} = +\frac{qN_D}{\epsilon_{Si}} = +1.6 \times 10^7 \text{ V/cm}^2 \tag{17.12}$$

$$E_\nu(x) = E_{p\nu} + 1.6 \times 10^7 x \tag{17.13}$$

When the device just reaches avalanche breakdown, the field at the $p\nu$ junction is $-E_c$, and the breakdown voltage V_{BR} is

$$V_{BR} = \int_0^{W_\nu} -E_\nu(x)\,dx = E_c W_\nu - 8 \times 10^6 W_\nu^2 \tag{17.14}$$

This result is graphed in Fig. 17.6 for $E_c = 2 \times 10^5$ V/cm, which is a value of critical field typical of this doping level.

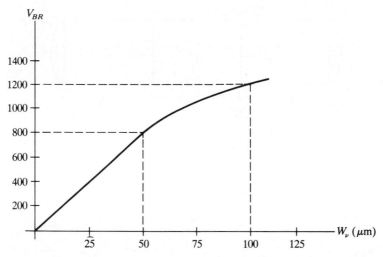

Figure 17.6 Breakdown voltage as a function of the length of ν region for a pνn diode with a doping concentration of $N_D = 10^{14}/\text{cm}^3$.

17.2.1 Forward Bias

We based the analysis leading to the pn diode i–v characteristic, *(17.1)*, on the assumption that both sides of the junction were in low-level injection. This assumption is not valid for the intrinsic region of a pin diode because in this region $n \approx n' \gg n_o$ and $p \approx p' \gg p_o$ at any reasonable injection level. When the total carrier concentration in a region is dominated by the excess concentration, as in this case, we say that the region is in *high-level injection*. One consequence of high-level injection is that $n \approx p$ in the neutral regions. High-level injection is also called *conductivity modulation*, because the conductivity of the material is no longer determined by the majority doping level; it is now a function of injection level.

The diode we chose for analysis is shown in Fig. 17.7, along with the minority carrier concentrations. We assume that the width of the intrinsic region W_i is comparable to the characteristic diffusion length. The intrinsic region is in high-level injection everywhere, but the heavily doped p$^+$ and n$^+$ regions remain in low-level injection. Because the end regions are so heavily doped, the diffusion of minority carriers in these regions is negligible, as it was in the heavily doped side of the asymmetric diode analyzed in Chapter 16. Only hole current flowing by drift is significant in the p$^+$ region, and only electron current flowing by drift is significant in the n$^+$ region. These holes and electrons flow into the intrinsic region where they recombine. It is this recombination that determines the terminal current.

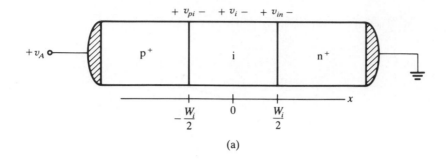

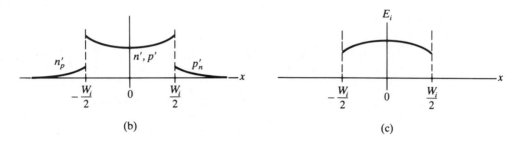

Figure 17.7 (a) A one-dimensional pin diode structure. (b) Carrier concentrations with the intrinsic region in high-level injection. (c) The electric field in the intrinsic region.

Because $n = p'$, in high-level injection, the recombination rate for holes and electrons is the same and is characterized by the *ambipolar lifetime*, τ_a. Furthermore, as the majority and minority concentrations are the same, we can no longer make the approximation that the minority current flows only by diffusion. The continuity equations in the intrinsic region must now contain the drift term that we eliminated in our analysis of the diode in low-level injection. Including the drift term, the continuity equations are

$$\frac{\partial n}{\partial t} = \frac{1}{q}\frac{\partial J_c}{\partial x} - \frac{n'}{\tau_a}$$
$$= \mu_e\left(n\frac{\partial E}{\partial x} + E\frac{\partial n}{\partial x}\right) + D_e\frac{\partial^2 n}{\partial x^2} - \frac{n'}{\tau_a} \tag{17.15}$$

$$\frac{\partial p}{\partial t} = -\mu_h\left(p\frac{\partial E}{\partial x} + E\frac{\partial p}{\partial x}\right) + D_h\frac{\partial^2 p}{\partial x^2} - \frac{p'}{\tau_a} \tag{17.16}$$

The concentrations are equal to each other and dominated by their excess in high-level injection, so we can eliminate E from *(17.15)* and *(17.16)* to give

$$\frac{\partial n'}{\partial t} \approx D_a \frac{\partial^2 n'}{\partial x^2} - \frac{n'}{\tau_a} \qquad (17.17)$$

$$\frac{\partial p'}{\partial t} \approx D_a \frac{\partial^2 p'}{\partial x^2} - \frac{p'}{\tau_a} \qquad (17.18)$$

where D_a is the *ambipolar diffusion constant*,

$$D_a = \frac{2D_e D_h}{D_e + D_h} \qquad (17.19)$$

Because $n' = p'$, *(17.17)* and *(17.18)* are identical. We solve them to determine the spatial dependencies of the carrier concentrations and the diffusion components of the currents in the intrinsic region. As in our previous analyses, the concentrations vary exponentially with x. But the characteristic diffusion length, called the *ambipolar diffusion length*, $L_a = \sqrt{D_a \tau_a}$, is now the same for both holes and electrons.

There is also a drift component of current in the intrinsic region, and the drift field gives rise to a voltage drop, v_i, which adds to the on-state voltage of the diode. Using a straightforward calculation, we can obtain the value of this drift field, E_i:

$$J_e = q\mu_e \left(\frac{kT}{q} \frac{dn'}{dx} + n' E_i \right) \qquad (17.20)$$

$$J_h = q\mu_h \left(-\frac{kT}{q} \frac{dp'}{dx} + p' E_i \right) \qquad (17.21)$$

Because $J_e + J_h = J$ (the total diode current), which is constant throughout the device, and $n' = p'$,

$$E_i = -\frac{kT}{q} \left(\frac{\mu_e - \mu_h}{\mu_e + \mu_h} \right) \frac{1}{n'} \frac{dn'}{dx} + \frac{J}{q(\mu_e + \mu_h)n'} \qquad (17.22)$$

The first term in *(17.22)* is the field necessary to ensure quasi-neutrality when $\mu_e \neq \mu_h$ (and, therefore, $D_e \neq D_h$). The second term is simply the ohmic drop in the conductivity modulated intrinsic region. The conductivity of this region now depends on injection level, and is $\sigma_i = q(\mu_e + \mu_h)n'$.

An interesting consequence of high-level injection is that the electric field E_i is independent of injection level and therefore of current. We can show this condition as follows. The solution to the diffusion equation, *(17.17)*, is a pair of exponentials with characteristic length L_a. Thus the first term in *(17.22)*, in which the spatial derivative of n' is divided by n', does not depend on n'. And as we assumed that all the current flowing through the diode results from recombination within the intrinsic region—and as this recombination is proportional to the excess carrier concentration—we know that $J \propto n'$. The second term in *(17.22)* is therefore also independent of n'. Thus the electric field E_i and therefore the voltage v_i are dependent only on physical parameters and the length of the intrinsic region.

We now determine the i–v characteristic of the pin diode of Fig. 17.7(a) by assuming that $\mu_e = \mu_h$ and $D_e = D_h$ in the intrinsic region. This is a reasonably accurate assumption in regions of very high-level injection. From *(17.19)*, you can see that if $D_e = D_h$, then $D_e = D_h = D_a$. Although there is no ambipolar mobility parameter, we will specify the mobility of holes and electrons as μ_a for consistency.

As $J_e \approx 0$ at $x = -W_i/2$, we can solve *(17.20)* for $E_i(-W_i/2)$ and then solve for the total diode current J:

$$E_i\left(\frac{-W_i}{2}\right) = -\left(\frac{kT}{q}\right)\frac{1}{n'}\frac{dn'}{dx} \tag{17.23}$$

$$J_h\left(\frac{-W_i}{2}\right) = J = q\mu_a\left(p'E - \frac{kT}{q}\frac{dp'}{dx}\right) \tag{17.24}$$

$$= -2qD_a\left.\frac{dp'}{dx}\right|_{(-W_i/2)} \tag{17.25}$$

Similarly, at $x = +W_i/2$, $J_h \approx 0$ and:

$$J_e\left(+\frac{W_i}{2}\right) = J = -2qD_a\left.\frac{dn'}{dx}\right|_{(W_i/2)} \tag{17.26}$$

Note that *(17.23)* and *(17.26)* are boundary conditions on the concentrations p' and n', which are equal. Using these conditions to solve *(17.17)* and *(17.18)* in the steady state gives

$$n' = p' = \left(\frac{J\tau_a}{2qL_a}\right)\frac{\cosh(x/L_a)}{\sinh(W_i/2L_a)} \tag{17.27}$$

We can now determine E_i by substituting *(17.27)* into *(17.22)* and setting $\mu_e = \mu_h = \mu_a$:

$$E_i = \frac{J}{2q\mu_a n'} = \left(\frac{kT}{q}\right)\frac{1}{L_a}\frac{\sinh(W_i/2L_a)}{\cosh(x/L_a)} \tag{17.28}$$

Fig. 17.7(c) shows this field.

We can find the voltage v_i by integrating E_i from $-W_i/2$ to $W_i/2$:

$$v_i = \int_{-W_i/2}^{W_i/2} E_i\, dx = \frac{2kT}{q}\sinh\left(\frac{W_i}{2L_a}\right)\tan^{-1}\left[\sinh\left(\frac{W_i}{2L_a}\right)\right] \tag{17.29}$$

For $W_i/2 \le L_a$,

$$v_i \approx \frac{2kT}{q}\left(\frac{W_i}{2L_a}\right)^2 \tag{17.30}$$

and for $W_i/2 \gg L_a$,

$$v_i \approx \frac{\pi kT}{2q} e^{(W_i/2L_a)} \qquad (17.31)$$

Note that v_i is independent of J, as we expected. Also note that if $W_i/2 \leq L_a$, $v_i \leq 50$ mV at room temperature, which is very small compared to a typical value for the diode's total applied voltage, v_A. If $W_i/2 \gg L_a$, however, v_i can be very large.

The excess carrier concentrations at the intrinsic-region edges of the two SCLs are proportional to the exponential of the two junction voltages v_{pi} and v_{in}.

$$n'(-W_i/2) = n_i \left(e^{qv_{pi}/kT} - 1 \right) \qquad (17.32)$$

$$n'(W_i/2) = n_i \left(e^{qv_{in}/kT} - 1 \right) \qquad (17.33)$$

where v_{pi} and v_{in} are the voltages across the p$^+$–i and n$^+$–i junctions, respectively. In our example $n'(-W_i/2) = n'(W_i/2)$, so, $v_{pi} = v_{in}$.

If we ignore v_i (compared to the junction voltages), then $v_{pi} = v_{in} = v_A/2$, where v_A is the applied terminal voltage. The diode i–v relationship is

$$J = \frac{2qL_a n_i}{\tau_a} \tanh\left(\frac{W_i}{2L_a} \right) \left(e^{qv_A/2kT} - 1 \right) \qquad (17.34)$$

17.2.2 Minority Carrier Diffusion Currents in the End Regions

In the preceding analysis we assumed that the gradient of the excess carrier concentrations in the p$^+$ and n$^+$ end regions of the pin diode were so small that we could neglect the minority carrier diffusion currents in these regions. However, this assumption is not a good one for very high current densities.

As end-region diffusion currents become a more important fraction of the total current, the voltage drop across the intrinsic region, v_i, increases. The reason is that when we ignored the minority currents in the n$^+$ and p$^+$ regions, the total diode current was determined by the concentration gradients in the intrinsic region and given by (17.25) or (17.26). However, when the diffusion currents of minority carriers in the end regions are significant, $J_e(-W_i/2)$ and $J_h(W_i/2)$ are less than the total diode current, and the excess concentrations in the intrinsic region are correspondingly lower. Thus the modulated conductivity of the intrinsic region is lower and the ohmic contribution to v_i is higher.

High-Level Injection Effects At high current densities, the end-region diffusion currents become important because the intrinsic edges of the SCL are in high-level injection. The threshold of high-level injection is usually defined as the injection level at which the excess minority carrier concentration at the junction edge is equal to the equilibrium majority concentration. As $n' = p'$, high-level injection also implies that $n \approx p$. The consequence of high-level injection in the intrinsic region is that the electron concentration at the p$^+$ edge of the p$^+$–i SCL is higher than predicted by the low-level injection approximation. Because the n$^+$ region is still in low-level injection, the minority current still is principally diffusive, that is,

$$J_e\left(\frac{-W_i}{2}\right) = \frac{qD_e n'_p(-W_i/2)}{L_e} \tag{17.35}$$

where L_e is the diffusion length of electrons in the p$^+$ region. If the diode were in low-level injection on both sides of the junction, $n'_p(-W_i/2)$ would be a very small number because the p$^+$ region is so heavily doped. However, if the intrinsic side of this junction is deep into high-level injection, $n'_p(-W_i/2)$ is much larger than predicted by (16.33), for which we assumed that $n_n \approx n_{no}$. Therefore, for a given junction voltage, the current $J_e(-W_i/2)$ is higher than predicted by our previous analysis based on an assumption of low-level injection. Let's now determine the minority concentration at an SCL edge when the other edge of the SCL is in high-level injection.

The ratio of the carrier concentrations at the two SCL edges of a diode are related to the voltage across the SCL, v_j, by equations (16.32), repeated here for convenience:

$$\frac{n_p}{n_n} = \frac{n_{po}}{n_{no}} e^{qv_j/kT}$$
$$\frac{p_n}{p_p} = \frac{p_{no}}{p_{po}} e^{qv_j/kT} \tag{17.36}$$

Note that in the pin diode the junction voltage does not equal the applied voltage, v_a. When one side of the junction (the n side for example), is in high-level injection by a factor of γ (that is, $p_n = n_n = \gamma n_{no}$, where $\gamma \gg 1$, the minority carrier concentration on the other side of the junction is γ times higher than predicted by (16.33):

$$n_p = \gamma n_{po} e^{qv_j/kT} \tag{17.37}$$

Because the middle region of a pin diode is very lightly doped, γ can be very large. For instance, if the magnitude of v_{pi} causes an excess carrier concentration in the intrinsic region at the p$^+$–i SCL edge of $10^{18}/\text{cm}^3$, then $\gamma \approx 10^8$. If we assume that the p$^+$ region is doped at $N_{Ap} = 10^{18}/\text{cm}^3$, its equilibrium minority carrier concentration is $n_{po} \approx 10^2/\text{cm}^3$. From (17.37) then, $n_p \approx 10^{18}/\text{cm}^3$. This concentration is high enough to create a minority diffusion current in the p$^+$ region that is significant compared to the recombination current in the intrinsic region. However, we need values for the diffusion constants and lifetimes in the two regions to make an exact comparison. The point is that, as the diode current increases, the end-region diffusion currents become a larger fraction of the total current, reducing injection into the i region. The conductivity modulation of the i region is thus reduced, so v_i—and, consequently, the forward drop of the diode—increases.

Heavy Doping Effects   The p^+ and n^+ end regions of the pin diode are heavily doped to reduce their resistance and to keep the minority carrier diffusion currents as small as possible. When the doping level is approximately $10^{17}/cm^3$ or larger, which is typical for a power diode, the large number of impurity atoms reduces the energy gap in these regions by an amount ΔE_g. This is known as *bandgap narrowing*. Although we do not discuss this phenomenon in detail, its effect is to raise the equilibrium carrier concentrations so that:

$$n_o p_o = n_{iE}^2 \qquad (17.38)$$

where n_{iE} is known as the *effective intrinsic carrier concentration*. The intrinsic concentration, n_i is related to n_{iE} by:

$$n_{iE}^2 = n_i^2 e^{\Delta E_g / kT} \qquad (17.39)$$

The result of this increase in the $n_o p_o$ product is that both the minority carrier concentration and the minority carrier diffusion current are higher than the values yielded by calculations based on $n_o p_o = n_i^2$. If the end region is doped at $10^{19}/cm^3$, a typical value for ΔE_g is 0.08 eV. At room temperature, this ΔE_g increases $n_o p_o$ to $n_{iE}^2 \approx 25 n_i^2 = 5 \times 10^{21}$.

17.2.3 Reduction of L_a at High Current Levels

Two phenomena reduce the ambipolar diffusion length L_a at the high carrier concentrations reached when the diode's current density gets very large. One is *carrier–carrier scattering*, which reduces the mobility of the carriers and therefore the ambipolar diffusion constant D_a. The other is *Auger recombination*, which reduces the ambipolar lifetime τ_a. As the diode current increases, the corresponding reduction in L_a caused by these two effects makes the voltage across the intrinsic region v_i grow quickly, as expressed by (17.30) or (17.31).

The injection-level dependence of L_a, the increased significance of end-region diffusion currents, and the increase of the $n_o p_o$ product cause the i–v characteristic of a power diode, (17.34), to bend at high currents. Figure 17.8 shows the curve that results from these effects.

17.2.4 Temperature Dependence of J

We show in (17.34) that the diode current is proportional to n_i if the injection level is not too high. The temperature dependence of n_i is

$$n_i \propto T^{3/2} e^{-E_g/2kT} \qquad (17.40)$$

Because n_i increases more rapidly than $e^{qv_j/kT}$ decreases with increasing temperature, J increases with temperature for a given L_a and τ_a. Conversely, the forward drop decreases with temperature if the forward current is fixed. Although we do not discuss here how L_a and τ_a vary with temperature, the effect of these additional

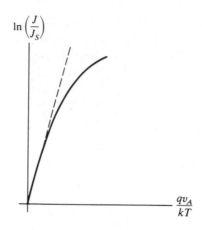

Figure 17.8 The i–v characteristic of a typical power diode, showing the curve bending at high currents because of high-level injection effects.

temperature dependencies is to increase the current flowing through a pin power diode with temperature for a given applied voltage.

17.3 SWITCH TRANSITIONS IN A PIN DIODE

Let's now turn to the transition of a diode between its static on and static off states. For simplicity, we ignore the capacitance of the SCL and focus on the dynamics of the excess stored charge in the neutral regions outside the SCL. We can easily add the effects of the SCL to those we describe here. We also ignore the change in the lifetime in the intrinsic region as the device goes from no excess carriers to high-level injection and the effect of the doping concentration in the intrinsic region on the nature of the transitions.

17.3.1 Forward Recovery

We consider first the turn-on transtion, in which the diode current steps instantly from zero to I_F. As Fig. 17.9(a) shows, the terminal voltage v_A first rises to a value V_{fp} that is much higher than the static forward drop. It then decays to the steady-state value v_F, determined by the current I_F. This process is called *forward recovery*, and V_{fp} is known as the *forward recovery voltage*. The duration of the forward recovery process is called the forward recovery time t_{fr}.

   We can explain the overshoot in the diode voltage by considering how the carrier concentrations in the intrinsic region change during the turn-on transition. Initially, there are no excess carriers. However, as the forward current flows, holes drift through the p$^+$ region and are injected into the intrinsic region. Similarly, the electrons drift through the n$^+$ region and are injected into the intrinsic region.

Over time, these injected carriers accumulate, as indicated by the profiles shown in Fig. 17.9(b).

Consider the carrier profiles at t_1, a short time after the forward current starts flowing. As Fig. 17.9(b) shows, the carriers build up near the junctions but not in the center of the intrinsic region. Some of the holes injected into the intrinsic region at the p^+–i junction add to the accumulation of carriers at this junction, and the rest flow across the intrinsic region to add to the accumulation of carriers at the i–n^+ junction. Similarly, some of the electrons injected at the i–n^+ junction flow across the intrinsic region to add to the accumulation at the p^+–i junction.

How do these injected carriers flow across the intrinsic region? Not by diffusion, because at this time there is no gradient to the carrier profiles in most of the i region. Therefore flow in the middle of the i region must be by drift. The voltage that results from this drift is very large. The reason is that the i region is not in high-level injection, so its conductivity, $\sigma_i = q\mu_e n + q\mu_h p$, is very low and yet the region is carrying the full current I_F. Even if the region is not intrinsic, but is a lightly doped ν or π region, its resistance is high compared to its value upon reaching high-level injection. This resistance is the source of the peak transient voltage V_{fp}, of the waveforms of Fig. 17.9(a).

As time passes, the growing carrier concentrations in the middle region modulate the region's conductivity and reduce its resistance. As a result, the voltage across the middle region drops. When the carrier concentrations reach their final

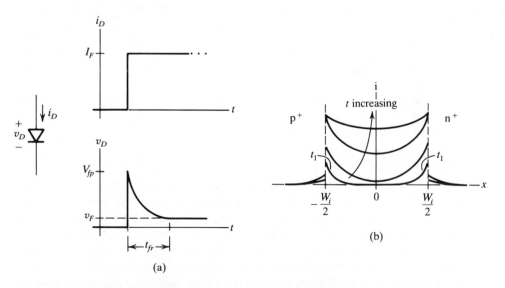

Figure 17.9   Forward recovery of a pin diode: (a) typical diode voltage and current waveforms during the turn-on transition;   (b) the time evolution of excess carrier concentrations in the neutral regions of the device.

value, so too does the diode's voltage. The charge in the middle region at steady state is proportional to the forward current ($I_F \propto Q/\tau_a$). Hence, the time required to supply this charge—in other words, the duration of the forward recovery transition—is approximately τ_a.

EXAMPLE 17.4

Magnitude of the Forward Recovery Voltage

Assume that we have a pνn diode with the center region doped at $N_D = 10^{14}/\text{cm}^3$ and designed to withstand 400 V. This voltage requires a ν-region width of $W_\nu \approx 20 \ \mu\text{m}$. The cross-sectional area of the device is $A = 1 \ \text{cm}^2$, and its forward current steps from zero to 50 A at $t = 0$. What is v_{fp}?

Initially, the resistance of the ν region is

$$R_\nu = \frac{W_\nu}{q(\mu_e n + \mu_h p)A} = 91 \ \text{m}\Omega \qquad (17.41)$$

With $I_F = 50$ A, the drop across this region is 4.6 V.

17.3.2 Reverse Recovery

Consider the diode in the circuit of Fig. 17.10(a), where for $t < 0$ it is carrying a forward current of $I_F = V_F/R_F$. At $t = 0$, the switch connects the diode to the negative voltage V_R through a second resistor, R_R. Because the excess charge in the i region and diffusion regions of the diode cannot change instantaneously, the p$^+$–i and i–n$^+$ junctions remain forward biased for some time after $t = 0$. As the diode voltage is approximately zero during this time, the diode current is negative and equals $-V_R/R_R$. This reverse current aids the removal of excess charge, until the concentrations at the SCL edges become negative and the junction can begin to support a reverse voltage. This process is called *reverse recovery*.

Figure 17.10(b) shows how the carrier profiles in the intrinsic region change as the reverse recovery current flows. Note that, just after $t = 0$, the excess carrier concentrations at the junction edges are still positive, and therefore the junction voltages are also positive. To support the negative current, the excess carrier distributions develop a negative slope near the junction edges. Holes diffuse out of the intrinsic region, are injected into the p$^+$ region, and drift to the left ohmic contact. Similarly, electrons leave the intrinsic region at the i–n$^+$ junction and drift to the right contact. So long as the excess concentrations at the SCL edges are greater than zero, and $|V_R| \gg v_F$, the total diode current equals $-V_R/R_R$. Continuing recovery current flow eliminates the excess carrier concentrations in the intrinsic region, as the profiles in Fig. 17.10(c) show.

After the excess carrier concentrations at the junction edges reach zero, the concentration gradients decrease and the diode current can no longer be maintained at $i_D = -V_R/R_R$. As $|i_D|$ decreases, the drop across R_R also decreases. The

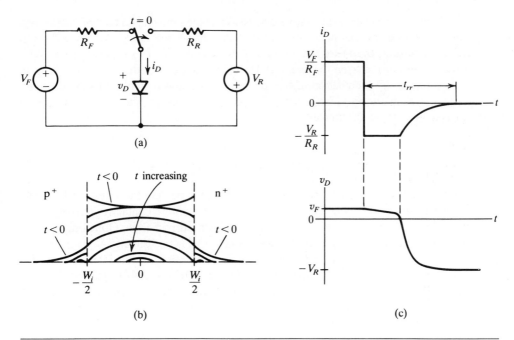

Figure 17.10   (a) A hypothetical circuit used to illustrate reverse recovery of a pin diode.   (b) Carrier profiles in the intrinsic region during the turn-off transition.   (c) Waveforms showing reverse diode current during the reverse recovery time t_{rr}.

difference between V_R and the voltage across R_R appears across the diode SCLs, which begin to widen. The dynamics of this process produce an approximately exponential rise of i_D to zero and fall of v_D to $-V_R$, as Fig. 17.10(c) shows. As Fig. 17.10(c) shows, during the initial phase of the reverse recovery process, the diode voltage changes only slightly from its static forward voltage v_F (the small decrease results from the change in sign of v_i that occurs when the current changes direction).

The duration of reverse current flow in the diode is t_{rr}, the reverse recovery time. Based on our discussion so far, t_{rr} is directly proportional to the reverse current I_R and the charge Q, which is initially stored in the intrinsic region. However, reverse current does not remove all of this stored charge; some of it recombines. How much recombines depends on the magnitude of the negative current compared to that of the forward current. The initial stored charge is $Q = I_F \tau_a$. If $I_R \gg I_F$, almost all the stored charge is swept from the intrinsic region in a time much less than τ_a, and only a small amount will be removed by recombination. But if $I_R \leq I_F$, a substantial fraction of the charge will recombine. In the limit of $I_R = 0$, when all the charge is removed through recombination, the reverse recovery time is $t_{rr} \approx \tau_a$.

From the preceding qualitative discussion of the turn-on and turn-off transitions of a pin diode, we can conclude that the less charge the diode stores for a given forward current, the faster the diode will switch. As $Q = I_F \tau_a$, the way to store less charge is to shorten the lifetime. However, doing so reduces L_a, resulting in less conductivity modulation of the intrinsic region and a higher on-state drop, v_i, across it as given by *(17.30)* and *(17.31)*. For this reason, the forward drop of a fast diode is typically greater than that of a slower diode.

17.4 THE SCHOTTKY BARRIER DIODE

When metal leads are attached to a semiconductor device, the interface between the metal and semiconductor is usually *ohmic*. An ohmic contact is one in which the resistances of the contacting materials govern current flow. Obtaining a good ohmic contact between a semiconductor and a metal is difficult. A great deal of research has been directed at developing an understanding of this interface and the metallurgy necessary to produce ohmic contacts. When we fail to pay attention to these processes, the result is a contact with a poor rectifying characteristic. This rectifying characteristic was the basis for early "cat whisker" crystal receivers and point-contact transistors. However, making a good, large area, rectifying contact is even more difficult than making a good ohmic contact. Diodes made with a metal–semiconductor junction are called *Schottky barrier diodes*. They are an important component in low-voltage (< 50-V) power electronic circuits, because they have lower forward drops than do silicon bipolar diodes. The symbol for a Schottky diode is shown in Fig. 17.11.

In principle, either n-type or p-type semiconductor material can be used to make a Schottky diode, although n-type material is used in power Schottkys because of the higher mobility of electrons. As we demonstrate in this section, Schottky diodes offer two advantages over bipolar diodes. The first is that, for the same current density, the forward voltage drop of a Schottky is several tenths of a volt less than that of a bipolar diode. The second is that the current in a Schottky flows only by drift, so there is no need to accumulate or remove excess carriers, thereby eliminating forward and reverse recovery phenomena. The disadvantage of Schottky diodes is that they have much more leakage current when reverse biased than do bipolar diodes.

17.4.1 The Schottky Barrier

Current flow across a rectifying metal–semiconductor junction is the result of *thermionic emission* of electrons from the metal into the semiconductor and from the semiconductor into the metal. In thermal equilibrium, these two flows are equal and there is no net current. When a voltage is applied to the junction, one direction of electron emission dominates and a net electron current results. We first determine how much energy an electron in a metal must have in order to be emitted

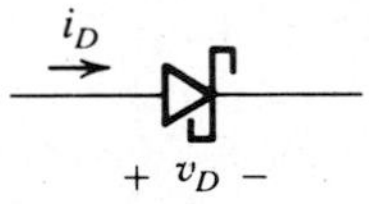

i_D

$+ \ v_D \ -$

Figure 17.11   The circuit symbol for a Schottky barrier diode.

into a vacuum. We then consider the effect of replacing the vacuum with n-type semiconductor material.

When an electron is emitted from a metal's surface, it must contain sufficient energy to overcome the force, F, tending to hold the electron in the metal. A major component of this force is that of the electric field created by the net positive charge left at the metal's surface. The escaping electron with charge $-q$ and its field are shown in Fig. 17.12(a). This field is identical to that which would result if the metal were replaced by an image charge $+q$, located at a distance $-x$ from the surface. A "total escape" is made when the electron is far from the metal's surface, and the total energy required of the electron to get far away is the *work*

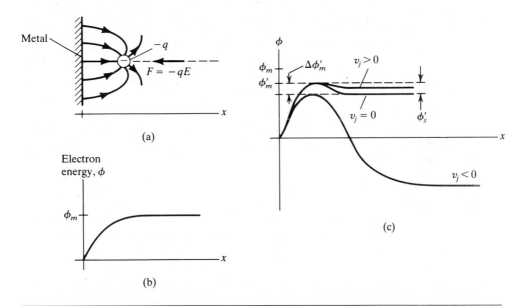

(a)

(b)

(c)

Figure 17.12   (a) An electron escaping from a metal surface and the resulting electric field.   (b) Energy as a function of distance from the metal for the escaping electron of (a).   (c) The energy distribution in the vicinity of a metal–semiconductor junction under conditions of thermal equilibrium ($v_j = 0$), forward bias ($v_J > 0$), and reverse bias ($v_j < 0$).

function ϕ_m of the metal:

$$\phi_m = \int_0^\infty F\,dx \qquad (17.42)$$

This energy, as a function of distance from the metal's surface x, is plotted in Fig. 17.12(b). It shows that electrons must overcome an energy barrier of height ϕ_m to escape into the vacuum.

When an n-type semiconductor instead of a vacuum is brought into contact with metal, electron emission occurs from both the metal into the semiconductor and the semiconductor into the metal. Because the free electrons in the semiconductor are more energetic than those in the appropriately chosen metal, more electrons initially flow into the metal. They leave behind a positive space charge of donor atoms and create a compensating negative charge on the metal's surface. The field created by this space charge opposes the field of Fig. 17.12(a) created by an electron leaving the metal, so the minimum energy required for an electron to escape from the metal, *(17.42)*, is reduced to ϕ'_m.[†] Considering the effect of the SCL field, the electron energy curve in Fig. 17.12(b) changes to that shown for $v_j = 0$ (thermal equilibrium) in Fig. 17.12(c). The asymptote of this energy for large x is the minimum energy of a mobile electron in silicon. Therefore Fig. 17.12(c) also illustrates the potential barrier ϕ'_s that must be overcome by an electron escaping from the semiconductor into the metal when $v_j = 0$. The height of this barrier is principally a function of the electric field created by the SCL. As thermal equilibrium is approached, the SCL grows, reducing ϕ'_m by a negligible amount but increasing ϕ'_s significantly, until the two emission rates are equal and the net electron current across the junction is zero.

The fact that $\phi'_m > \phi'_s$ means that the probability of electron emission from the metal is much lower than it is from the semiconductor. But the metal contains so many more electrons than the semiconductor that the two emission rates can be equal while the barriers are substantially different.

17.4.2 Forward Bias

We can now determine the effect of an applied junction voltage, v_j, on the flow of electrons across the Schottky barrier junction. As for a bipolar diode, v_j will change the SCL width. The junction is highly asymmetric, so nearly all of v_j appears across the semiconductor side of the SCL. If $v_j > 0$ (that is, the diode is forward biased), the electric field of SCL origin is reduced, and the potential distribution becomes that of the curve in Fig. 17.12(c) for $v_j > 0$. The potential barrier to electron emission from the semiconductor has been reduced substantially, but the barrier to emission from the metal has been changed only slightly. The result is a forward

---

[†]The lowering of the potential barrier to ϕ'_m because of an electric field (in this case from the SCL in the semiconductor) is known as the *Schottky effect*.

current that is a strong function of v_j and consists of electrons flowing from the semiconductor to the metal.

Although we do not show the derivation here, the i–v characteristic of a *forward biased* Schottky diode is

$$J_F = J_o\left(e^{qv_j/kT} - 1\right) \qquad (17.43)$$

which is identical in form to the characteristic of a bipolar diode. However, the current coefficient J_o is of a substantially different form from J_s, the leakage current as defined in *(17.1)* for a bipolar diode. The current J_o is given by:

$$J_o = R^* T^2 e^{-q\phi'_m/kT} \qquad (17.44)$$

The parameter R^* is the effective *Richardson constant* for the semiconductor. The current coefficient J_o has a value different from *(17.44)* in reverse bias, as we show later.

The current coefficient J_o for a forward biased silicon Schottky diode is typically four to six orders of magnitude larger than J_s for a silicon bipolar diode. For this reason, the forward drop of a Schottky diode is 0.23–0.34 V lower than that of a bipolar diode operating at the same current density. This lower on-state voltage is a major advantage of using a Schottky diode.

EXAMPLE 17.5

Analysis of a Silicon Schottky Diode

The value of R^* for silicon is approximately 120 A/cm^2-K^2, and a typical value for ϕ'_m for an aluminum to n-type silicon contact is 0.7 V at 25°C. Using these parameters and $T = 25$°C in *(17.44)* yields a forward current coefficient of:

$$J_o = 7.5 \times 10^{-6} \text{ A/cm}^2$$

In comparison, the saturation current of a short-base bipolar diode in low-level injection, with doping concentrations of $N_D = N_A = 10^{16}/$cm^3 and lengths $W_n = W_p = 2.5$ μm is

$$J_s = \frac{qD_e n_i^2}{N_A W_p} + \frac{qD_h n_i^2}{N_D W_n} = 7 \times 10^{-10} \text{ A/cm}^2 \qquad (17.45)$$

At a current density of 100 A/cm^2, the forward drop of the Schottky diode is 0.41 V, and that of the bipolar diode is 0.64 V.

Forward Voltage Drop   The forward drop of a Schottky diode contains a component in addition to v_j that is equal to the voltage across the semiconductor. The resistance of this region has two parts: (1) the lightly doped material into which the SCL grows in reverse bias; and (2) the heavily doped substrate. For low-voltage Schottkys (<40 V), these two resistances are comparable because, even though the lightly doped region has a higher resistivity, the heavily doped region is longer (typically 500 μm versus 5 μm). The total resistance of a 1-cm^2 silicon die might be 0.5–1 mΩ, giving a drift region drop of 50–100 mV at a current of 100 A.

The resistance of the lightly doped region quickly dominates the resistance of the drift region as the Schottky voltage rating increases from approximately 40 V. The reason is that a higher voltage rating requires a longer lightly doped region and a lower doping concentration, both of which increase the resistance of this region. Unlike in the bipolar diode, conductivity modulation does not reduce this resistance when the junction is forward biased. The large resistance is one of the reasons that silicon Schottky diodes are not generally available at voltage ratings above approximately 100 V.

Minority Carrier Flow   So far we have focused on the flow of electrons in a forward biased n-type Schottky. There is also a minority carrier (hole) component of the total current, although it is much smaller than the majority carrier current.

When we apply a voltage across the Schottky barrier junction, the excess minority carrier concentration at the semiconductor edge of the SCL is raised by the exponential of the voltage, as it is in a bipolar junction. Thus some excess carriers are stored in the drift region, and some current will flow by diffusion. However, because the junction voltage is so small relative to that of a bipolar junction, so are the excess concentrations at the SCL edge. Therefore the minority carrier diffusion current is much smaller than the majority carrier drift current, and we generally ignore it.

In addition to its low forward drop, a Schottky with a low reverse voltage rating also has the advantage of very little minority charge storage—and therefore no significant reverse recovery current. But because some minority storage is present in the lightly doped region, and the amount of this stored charge increases with reverse voltage rating, high-voltage Schottkys do exhibit reverse recovery behavior.

EXAMPLE 17.6

Minority Current in a Forward Biased Schottky

What is the minority carrier current density in the Schottky diode of Example 17.5? We can answer this question by assuming that the minority carrier lifetime τ_h in the lightly doped drift region is infinite and in the heavily doped substrate is zero. The distribution of excess holes is, therefore, a straight line between the concentrations at the edge of the SCL and at the interface between the lightly doped region and the substrate, as shown in Fig. 17.13.

We can now determine the diffusion current of holes from the distribution of Fig. 17.13. The excess hole concentration at the SCL edge is

$$p'_n = \frac{n_i^2}{N_D}\left(e^{qv_j/kT} - 1\right) \tag{17.46}$$

and minority hole current is

$$J_h = \frac{qD_h n_i^2}{N_D W_n}\left(e^{qv_j/kT} - 1\right) = J_{oh}\left(e^{qv_j/kT} - 1\right) \tag{17.47}$$

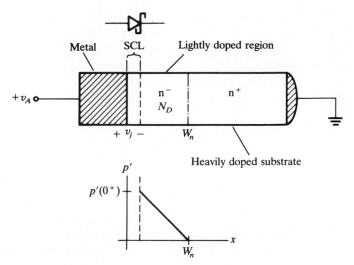

Figure 17.13 Excess minority carrier distribution in a lightly doped region of a Schottky diode.

Using the parameters of Example 17.5, we get $J_{oh} = 9.6 \times 10^{-11}$ A/cm^2. This result is approximately five orders of magnitude smaller than the value of J_o calculated in Example 17.5 and is therefore negligible.

Temperature Dependence of the Forward Drop   The forward voltage drop of a Schottky depends on temperature for two reasons: (1) the i–v characteristic, (17.43), has an explicit temperature dependence; and (2) the resistance of the drift region in the semiconductor increases with temperature. The first mechanism dominates the temperature dependence of the forward drop in almost all practical Schottky diodes.

From (17.43) and (17.44), the relationship between J_F and v_j has a functional dependence on temperature:

$$J_F \propto T^2 e^{-q(\phi'_m - v_j)/kT} \tag{17.48}$$

Because $\phi'_m > v_j$ in forward bias, J_F is a strongly increasing function of temperature if v_j is constant. Conversely, if the current is fixed, v_j will drop as the diode gets hotter.

The resistance of the drift region increases with temperature because electron mobility decreases as temperature increases. The mechanism responsible is the increase in the amplitude of the lattice vibrations. For silicon, the temperature dependency of electron mobility is

$$\mu_e \propto T^{-2.2} \tag{17.49}$$

over the temperature range 25–125°C.

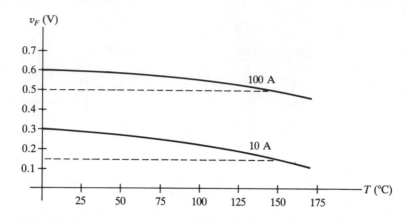

Figure 17.14 The forward voltage drop v_F as a function of temperature for a 50 V, 100 A Schottky diode operating at two different currents.

The forward voltage as a function of T is shown in Fig. 17.14 for a 50 V, 100 A Schottky diode. The influence of the drift region resistance at high currents is apparent from the different slopes of the two curves in the figure.

17.4.3 Reverse Bias

The potential distribution for reverse bias, $v_j < 0$, is shown in Fig. 17.12(c). Under this condition the barrier to the flow of electrons from the semiconductor has been greatly increased, virtually preventing electron flow from the semiconductor to the metal. An important feature of this distribution is that the barrier to electrons leaving the metal has been *reduced* from ϕ'_m, because the field produced at the metal's surface by the SCL has been increased substantially. Reverse current consists primarily of electrons flowing from the metal to the semiconductor. Because of the voltage dependence of the metal barrier height, this current is a strong function of reverse voltage, even though it is still much smaller than the forward current. This reverse voltage dependency makes the leakage current much greater than $-J_o$. The dissipation resulting from this leakage current is the principle determinant of the reverse voltage rating of a Schottky diode.

Leakage Current   Even though the barrier height ϕ'_m has been reduced only a small amount when the Schottky diode is reverse biased, the consequence is significant because the leakage current depends exponentially on this barrier height. The SCL-created field at the metal's surface is enhanced by the reverse voltage, causing ϕ'_m to decrease by an amount $\Delta\phi'_m$. The relationship between this decrease and the applied voltage is

$$\Delta\phi'_m \approx \sqrt{\frac{q}{4\pi\epsilon_{Si}}} \left(\frac{2qN_D|v_j|}{\epsilon_{Si}} \right)^{(1/4)} \tag{17.50}$$

We can illustrate the effect of lowering the barrier height by replacing ϕ'_m in *(17.44)* with $\phi'_m - \Delta\phi'_m$. As the saturation current depends exponentially on this potential, a small $\Delta\phi'_m$ can have a large impact on the reverse leakage current. For example, if $v_j = -50$ V and $N_D = 10^{15}/cm^3$, $\Delta\phi'_m \approx 0.050$ V, which at room temperature will increase the leakage current by a factor of 7.5 over J_o.

Because $\Delta\phi'_m \propto |v_j|^{0.25}$, most of the effect of lowering the barrier height is achieved by the time v_j reaches about -10 V. Figure 17.15 shows a typical reverse bias i–v characteristic for a Schottky diode.

Avalanche in a Schottky Diode As the reverse voltage on a Schottky diode increases, the peak electric field at the metal–semiconductor interface eventually reaches a value high enough for avalanche to occur. This process is identical to avalanche in a bipolar diode, but the avalanching current in a Schottky diode can damage the metal–semiconductor interface. For this reason, all power Schottky diodes have field, or guard, rings surrounding the metal contact, as shown in Fig. 17.16. These rings are additional p^+ diffusions that form p^+n junctions that shield the metal–semiconductor interface from fields high enough to cause the interface to avalanche. The diode is typically designed so that the p^+n junction at the guard ring avalanches before the Schottky junction does.

17.4.4 Switch Transitions in a Schottky Diode

The ideal Schottky diode is a majority carrier device, meaning that, generally, negligible excess charge is stored during any operating condition. Thus its switch transitions do not include forward and reverse recovery phenomena. However, the Schottky diode contains SCL capacitance, which must be charged and discharged during switching. And, as we mentioned at the end of Section 17.4.2, recovery phenomena may be visible in high-voltage devices, where minority current flow is no longer negligible.

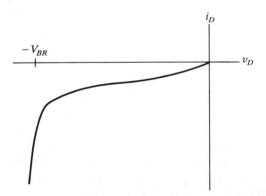

Figure 17.15 Reverse bias i–v characteristic of a Schottky diode, showing the effects of lowering the barrier height and avalanche.

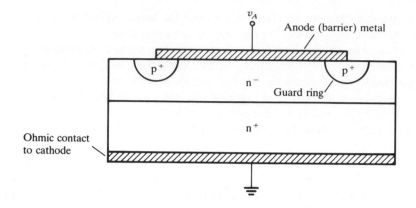

Figure 17.16 Guard ring in a Schottky diode, which prevents the metal–semiconductor interface from avalanching.

17.4.5 Selection of Barrier Metal

Because the potential of the Schottky barrier depends on the work function of the metal contact, we can change the forward and reverse characteristics of the diode by using different metals. Chrome, molybdenum, platinum, and tungsten are typical choices. Chrome gives the lowest junction drop but the highest reverse leakage current, so it is usually reserved for low-temperature applications. Tungsten is commonly used because, although it has the highest forward drop of the metals mentioned, it also exhibits the lowest leakage current and so can be used at high temperatures.

Notes and Bibliography

The best general references on the material presented in this chapter are Ghandi [1], Baliga [2], and Sze [3]. Ghandi's book is devoted to bipolar power devices: the pin diode, the BJT, and the thyristor. It provides a full analysis of the pin diode in forward and reverse bias. It also thoroughly covers the application of junction-termination techniques such as guard rings, field plates, and edge beveling for enhancing a device's breakdown voltage.

Baliga's book, while devoted more to field-effect power devices, does give a thorough discussion of the pin diode and the Schottky diode. It also presents important information on alternative rectifier approaches, such as the synchronous rectifier, the JBS rectifier, and the Gallium Arsenide rectifier. In addition, the chapter on transport physics provides a very good understanding for how mobility and lifetime depend on temperature, dopant concentration, lattice defects, and electric field. Many graphs with either theoretically or empirically determined values are provided. Chapter 3 gives a very extensive coverage of junction breakdown and junction termination techniques.

Sze, while not devoted to power devices, is another source of how and why device parameters vary with temperature, dopant concentration, etc. A comprehensive treatment of the Schottky diode and the physics of thermionic emission is also provided.

1. S. K. Ghandi, *Semiconductor Power Devices: Physics of Operation and Fabrication Technology* (New York: John Wiley & Sons, 1977).

2. B. J. Baliga, *Modern Power Devices* (New York: John Wiley & Sons, 1987).

3. S. M. Sze, *Physics of Semiconductor Devices*, Second Edition (New York: John Wiley & Sons, 1981)

Problems

17.1   Using *(17.2)*, calculate dn_i^2/dT, the change in n_i^2 for a 1 K change in temperature, at $T = 225$ K, $T = 300$ K, and $T = 450$ K. At each temperature, determine the incremental change in temperature that will produce a factor of 10 change in n_i^2.

17.2   Justify the form of *(17.5)* and fill in the missing steps between *(17.5)* and *(17.8)*.

17.3   For the diode we designed in Example 17.2, determine the resistance R_n of the n region. If we assume that the critical electric field remains at $E_c = 3 \times 10^5$ V/cm, how do N_D, ℓ_n, and R_n vary with the breakdown voltage? If the critical field changes with breakdown voltage, in what direction, relative to the change in E_c, would the parameters just calculated change?

17.4   A 77-mm–diameter fast recovery rectifier is rated at 3000 A_{rms}. The vertical profile of this device is shown in Fig. 17.17(a).

    (a) Determine the value of current density at which the device enters high-level injection on the p side of the SCL.

    (b) Assuming no *bulk resistive drop*, that is, no drop outside the SCL, determine the terminal voltage of this diode at rated current. Use the values for diffusion constants and lifetimes given in Fig. 17.17(a).

    (c) What is the terminal voltage at rated current if you do not ignore the drop across the p region outside the SCL? For purposes of this calculation, assume that you can characterize the p region by an infinite lifetime within a diffusion length, L_a, of the junction and by a lifetime of zero throughout the rest of the p-region, as shown in Fig. 17.17(b).

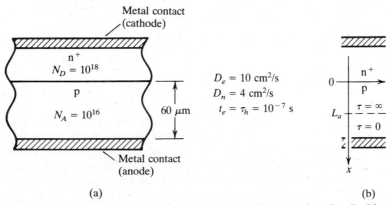

(a)   (b)

Figure 17.17   (a) The 77-mm–diameter fast recovery diode analyzed in Problem 17.4. (b) The lifetime profile used to solve Problem 17.4(c).

17.5   If $\tau_e = 5 \times 10^{-7}$ s in the diode of Problem 17.4, it would no longer be considered a fast-recovery diode. Repeat Problem 17.4(b) and (c) for this case.

17.6   A pνn diode breaks down at a voltage $-V_{BR}$, corresponding to a critical field, E_c, at its pν junction. In terms of E_c, and when $v_A = -V_{BR}$, what is the field at the νn junction if the diode is designed to minimize the ν-region resistance?

17.7   Derive *(17.17)* and *(17.18)*. (*Hint*: multiply *(17.15)* by $\mu_h p'$ and *(17.16)* by $\mu_e n'$ and add the two. Comment on the appropriateness of the approximation imbedded in *(17.17)* and *(17.18)*.)

17.8   Assume, as we did in Section 17.2.1, that the hole and electron mobilities in the middle region of a pin diode are equal but that now the middle region is actually doped slightly n type with $N_\nu = 10^{14}/\text{cm}^3$. Ignore the minority diffusion currents in the end regions, compared to the recombination current in the ν region, and ignore v_ν. For a diode current, J, find $v_{p\nu}$ and $v_{\nu n}$, the voltages across the two junctions, and express J as a function of $v_A = v_{p\nu} + v_{\nu n}$.

17.9   Find the incremental capacitance of the diode described in Problem 17.8 if it is reverse biased by more than a few volts. How does this capacitance vary with voltage?

17.10   An asymmetrically doped p$^+$n diode has an n-region width equal to $W_n \ll L_h$ (this structure is known as a *short-base diode*. What is the rate at which the excess charge in the diffusion regions would disappear if the diode current were abruptly brought to zero from some positive value?

17.11   Assuming a peak electric field of 2×10^5 V/cm, find the doping level and length of the drift region that will permit a silicon Schottky diode to withstand 200 V. What is the resistance of a square centimeter die of such a design? What is the voltage drop produced by this resistance when the forward current is 100 A?

17.12   For the Schottky diode discussed in Problem 17.11, find the ratio between the electron current and the hole-diffusion current when the device is forward biased.

17.13   For the Schottky diode in Example 17.5, find the junction voltage and drift resistance voltage when the diode current density is 100 A/cm^2 at $T = 400$ K. Compare the total voltage drop to that at $T = 300$ K.

17.14   For reverse voltages of values $|v_j| \gg \phi'_m$, graph $\Delta\phi'_m$ and $e^{q\Delta\phi'_m/kT}$ as functions of v_j over the range of 5–50 V. Assume that $N_D = 10^{16}/\text{cm}^3$.

17.15   Starting from the the the continuity equations, *(17.15)* and *(17.16)*, show that for high-level injection the ambipolar diffusion constant is

$$D_a = \frac{2 D_e D_h}{D_e + D_h}$$

as given by *(17.19)*.

17.16   In Section 17.2.4 we stated without a supporting calculation that the positive temperature dependence of n_i was larger than the negative temperature dependence of the Boltzmann factor qv_j/kT. For high-level injection, n_i is replaced by n_{iE} in *(17.34)*. Show that the diode current still increases as temperature increases.

Chapter 18

Power Transistors

IN this chapter we show how the high current and voltage requirements of power transistors affects their design. Our goals are to demonstrate why power transistors display the terminal characteristics they do and the tradeoffs required in designing a transistor. We begin by looking at the power bipolar junction transistor (BJT) and discussing its structure, its off-state and on-state behaviors, and its switching characteristics. We then consider the same features for the power metal-oxide-semiconductor field-effect transistor (MOSFET), a majority carrier device. Finally, we discuss the relatively new insulated gate bipolar transistor (IGBT).

18.1 STRUCTURE OF THE POWER BJT

Figure 18.1(a) shows a cutaway view of a typical npn bipolar power transistor. It is a vertical device, with the collector connection on the bottom (substrate) of the die and the base and emitter connections on the top. Note from the top view in Fig. 18.1(a) that the base and emitter metal traces are *interdigitated*. This pattern is used to keep the distance between the base contact and the centerline of the emitter (line A–A') short, the reasons for which we discuss later.

Looking at the profile of the doping concentration along line A–A', you can see the one-dimensional representation of the transistor shown in Fig. 18.1(b). Note that the collector of this transistor has two regions: a lightly doped ν region and a heavily doped substrate. The purpose of the ν collector region, which is more lightly doped than the p-type base, is to make the space charge layer of the collector–base junction to grow into it, rather than into the base. If this ν region is doped lightly enough, the SCL can then be allowed to "reach through" to the substrate region, as shown in the plot of ρ in Fig. 18.1(b). As you learned when studying the similar pin diode structure, we can nearly double the breakdown voltage for a given ν-region width by allowing this reach through to occur.

When this transistor is in its on state, the entire ν region is in high-level injection and current flows by drift. For this reason, we sometimes refer to the ν region as the *drift region*.

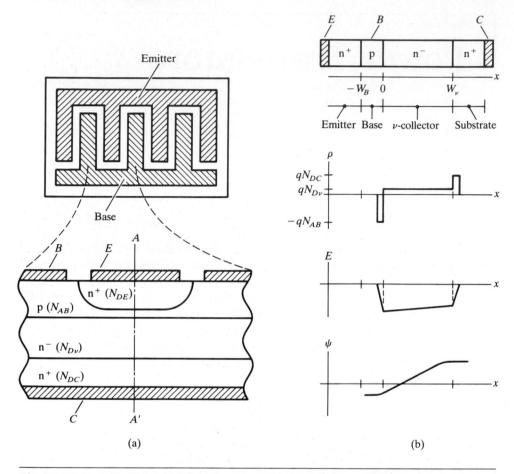

Figure 18.1   The npn power bipolar junction transistor: (a) top and side-section views of the transistor;   (b) one-dimensional representation of the power transistor doping profile along line A–A' bisecting the emitter, and the charge, electric field, and potential profiles of the reverse biased collector–base junction.

18.2 OPERATING REGIONS OF THE POWER BJT

The power BJT structure shown in Fig. 18.1 contains a high-resistance ν region, which is not present in a conventional (signal-level) transistor. When the transistor is in the on state (saturated), this ν region must be conductivity modulated so that its contribution to the on-state drop is small. This condition leads to the presence of four, rather than the usual three, distinct regions of operation.

The first is the off-state region, also known as *cutoff*. In this region both the collector–base and emitter–base junctions are reverse biased, the ν region is depleted, and the collector current is approximately zero. The second is the *forward-active* region, where the collector–base junction is strongly reverse biased, while the emitter–base junction is forward biased. Operation in this region resembles that of a conventional transistor, except at low values of v_{CB}, when the resistance of the ν region becomes important. The third region, called *quasi-saturation*, corresponds to a forward biased p–ν junction that injects excess carriers into the ν region, so that *part* of this region is in high-level injection (conductivity modulated). The fourth region of operation, *hard-saturation*, occurs when the entire ν region is conductivity modulated, and v_{CE} assumes its lowest on-state value for a given current.

The transistor characteristics in each of the four operating regions in the i_C–v_{CE} plane are shown in Fig. 18.2. We now determine the boundaries for each of these regions.

18.2.1 The Cutoff Region

A bipolar transistor in cutoff drops almost all its collector–emitter voltage across the collector–base junction. However, the maximum collector–emitter voltage rating of the transistor is usually less than the avalanche breakdown voltage of this junction. The reason for this difference is the subject of this section.

Collector–Base Voltage Rating   Consider a transistor connection in which the emitter is open circuited and the collector–base junction is reverse biased. The reverse characteristics of this junction are identical to those of a pin diode in Chapter 17: a leakage current exists, with holes flowing into the base region and electrons flowing into the collector substrate. The avalanche breakdown voltage of

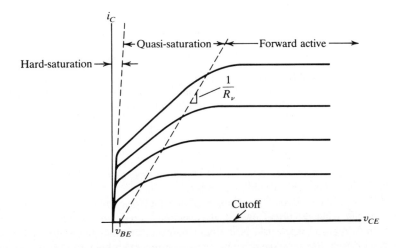

Figure 18.2   The i_C–v_{CE} characteristic of an npn power transistor with a ν collector region.

this collector–base junction is defined as BV_{CBO}, or simply V_{CBO}. The subscripts identify the voltage as that between the collector (C) and base (B) terminals, with the third terminal (emitter) open (O). Given the length W_ν of the ν region and its doping concentration $N_{D\nu}$, we can find V_{CBO} from *(17.13)*.

Collector–Emitter Voltage Rating   The maximum collector–emitter voltage rating of a BJT is a function of the conditions existing at the base terminal. If the base terminal is left open, the only source of majority carriers (holes) to support excess charge in the base is the leakage current of the collector junction. But this leakage current has the same effect as the base current in determining excess base charge and therefore the emitter current. Because of transistor action, this emitter current is larger than the "base" (collector leakage) current by a factor β_F, and as the base is open, this emitter current will increase the collector current by the same amount.

Although this unwanted transistor action can significantly increase the transistor's off-state current, its most important consequence is the effect it has on the maximum voltage that can be supported by the collector junction. The parameter V_{CBO} is the voltage at which avalanche occurs at the collector–base junction. At voltages below this value, impact ionization still occurs, but at a rate given by a value of M, the impact ionization factor, that is relatively small. Thus any current flowing through the collector SCL, from whatever source, is multiplied by M. This additional SCL current also delivers holes to the base region and causes the collector current to increase further by a factor of β_F. What we have is a system with positive feedback. At some point, as the collector voltage is increased (which increases M), the collector current increases without bound, even though the collector junction is not in avalanche breakdown. This voltage level is known as the *sustaining voltage* $V_{CEO(\text{sus})}$.

EXAMPLE 18.1

A Calculation of $V_{CEO(\text{sus})}$

The rate at which the ionization multiplication factor M approaches infinity depends on the reverse collector–base voltage v_{CB} in the following manner [1]:

$$M(v_{CB}) = \frac{1}{1 - \left(v_{CB}/V_{CBO}\right)^m} \qquad (18.1)$$

For an npn transistor, $m = 4$, and for a pnp transistor, $m = 6$. This result comes from a detailed analysis of the ionization integral, *(17.8)*[2].

If i_{eE} is the electron current injected into the base from the emitter in response to the collector leakage current (holes), the electron current that reaches the collector–base junction is $\alpha_T i_{eE}$, where α_T is the base transport factor defined by *(16.57)*. The total collector current, ignoring the original leakage current but taking account the multiplication factor M, is

$$i_C = M \alpha_T i_{eE}$$

To support this electron current, a hole current of:

$$i_{hB} = i_{eE}\left[\frac{1-\gamma_E}{\gamma_E} + (1 - \alpha_T)\right] = i_{eE}\left(\frac{1}{\gamma_E} - \alpha_T\right) \qquad (18.2)$$

must be delivered to the base region, where γ_E is the emitter efficiency, (16.58). The hole current flowing into the base from the collector–base junction is

$$i_{hC} = (M - 1)\alpha_T i_{eE} \qquad (18.3)$$

When these two currents are equal, that is, when

$$M = \frac{1}{\gamma_E \alpha_T} \qquad (18.4)$$

the transistor action is self-sustaining, and an arbitrarily large collector current can flow even though there is no base-terminal current. The value of v_{CE} at which (18.4) is satisfied is the sustaining voltage $V_{CEO(\text{sus})}$. Note that avalanche breakdown of a diode occurs when its voltage causes $M(v_D) = \infty$, but that the voltage limit of the collector–base junction of a transistor occurs at a finite value of $M(v_{CB})$. The voltage at which the collector junction actually breaks down because of avalanche is V_{CBO}, as we have already discussed.

At what collector–base voltage will $M(v_{CB})$ reach the value given by (18.4)? By combining (18.1) and (18.3), we find that:

$$V_{CEO(\text{sus})} = V_{CBO}\left(1 - \gamma_E \alpha_T\right)^{1/m} = \frac{V_{CBO}}{(1 + \beta_F)^{1/m}} \qquad (18.5)$$

As typical values for β in an npn power transistor are between 5 and 20, $V_{CEO(\text{sus})}$ ranges from 64% to 47% of V_{CBO}.

From the result of Example 18.1, we expect $V_{CEO(\text{sus})}$ to be approximately 60% of V_{CBO} in a typical power transistor. However, the specification sheets for actual devices usually show $V_{CEO(\text{sus})}$ to be closer to V_{CBO}. The reason is not that $V_{CEO(\text{sus})}$ is higher than predicted by (18.5), but rather that, owing to the curvature of collector-junction termination, the actual collector–base junction avalanche voltage V_{CBO} is lower than that predicted by the one-dimensional model on which we have been basing our analyses. The sustaining voltage is not similarly reduced, because in the vicinity of the junction's curvature to the surface the effective width of the base is large and the gain of the transistor is therefore very low.

Effect of Base Current on $V_{CEO(\text{sus})}$ Figure 18.3 shows a typical set of i_C–v_{CE} curves (parametric in base current) for a power transistor. Note that for the curve marked $i_B = 0$, which is the condition analyzed in Example 18.1, the collector current is approximately zero until the collector voltage is well above $V_{CEO(\text{sus})}$. As the collector current rises, however, the voltage drops to $V_{CEO(\text{sus})}$ and stays there until the current becomes very large, at which point the voltage again increases.

The reason for this variability in the sustained breakdown voltage is that β_F changes as a function of current. At very low currents, recombination in the base–

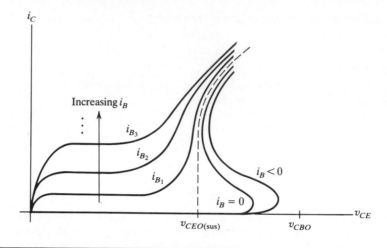

Figure 18.3 Typical i_C-v_{CE} characteristic of an npn power transistor showing the sustaining breakdown voltage.

emitter SCL reduces β_F, and at very high currents, high-level injection effects reduce β_F. These reductions in β_F, which we discuss more completely in Section 18.5, increase the breakdown voltage in accordance with *(18.5)*.

Figure 18.3 also shows that breakdown occurs at a voltage higher than $V_{CEO(\text{sus})}$ if the base current i_B is negative. The reason is that the net hole current delivered to the base is the sum of the current created by impact ionization and the base-terminal current. If the base-terminal current is negative, then M and hence v_{CB}, must be larger to support sustained breakdown. With enough negative base current, we can even achieve the full V_{CBO} breakdown voltage. When breakdown does occur, however, and as the collector current increases, the negative base current becomes a smaller part of the total holes delivered to the base and the collector voltage drops.

18.2.2 Forward-Active Region

From its off state, the transistor enters the forward-active region when the emitter junction is forward biased and minority carriers are injected into the base region. In most of the off-state region, the collector is strongly reverse biased and the entire ν collector is depleted. At small values of reverse bias, however, the ν collector is *not* depleted but presents a region of high resistance to the flow of collector current. The value of this resistance, R_ν, is

$$R_\nu = \frac{W_\nu}{q\mu_e N_{D\nu} A_\nu} \tag{18.6}$$

where A_ν is the cross-sectional area of the ν region.

Figure 18.4(a) shows a simple model for the transistor when it is operating in the forward-active region but at low collector–base voltages. Figure 18.4(b) shows the corresponding minority carrier distributions. The transistor will continue to operate in this region so long as the collector–base junction (ν–p) remains reverse biased. If we assume that $v_{BE} \approx 0.8$ V, this junction is reverse biased if $v_{CE} > i_C R_\nu + 0.8$ V. On the v_{CE}–i_C plane, this boundary between the forward-active and quasi-saturation regions is a straight line with a slope $1/R_\nu$, intersecting the v_{CE} axis at 0.8 V, as shown in Fig. 18.2. The region between this boundary and the $+v_{CE}$ axis defines forward-active operation.

18.2.3 Quasi-Saturation Region

If i_B is fixed and v_{CE} is decreased from its value at the forward-active/quasi-saturation boundary, the base–collector junction becomes forward biased. Because its equilibrium majority concentration is so low, the ν region in the vicinity of the collector junction quickly goes into high-level injection and becomes conductivity modulated. The value of R_ν is therefore reduced in proportion to the fraction of the ν region that is in high-level injection. If the width of this conductivity modulated region is x', then the reduced value of R_ν, R_ν', is

$$R_\nu' = \frac{W_\nu - x'}{W_\nu} R_\nu \qquad (18.7)$$

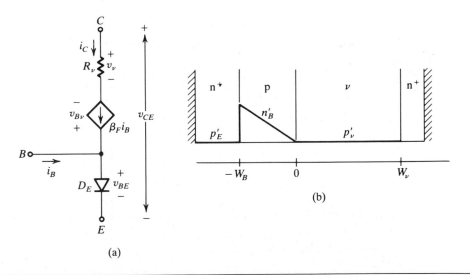

(a)

(b)

Figure 18.4   (a) A simple model for the power BJT operating in the forward-active region.   (b) Excess minority carrier distributions in the power BJT operating in the forward-active region.

The excess carrier distributions corresponding to two values of v_{CE} at constant i_B are shown in Fig. 18.5, for a transistor in the quasi-saturation region. The apparently linear distribution in the ν region is *not* a consequence of $x' \ll L_a$, but results from the conditions imposed by high-level injection.

As we continue to decrease v_{CE} through the quasi-saturation region, v_ν must decrease because it is approximately equal to v_{CE} when both the base–emitter and base–collector junctions are forward biased. The decrease of v_ν is accomplished by decreasing both R'_ν and i_C. If we assume that $i_E \approx i_C$, then i_C is determined by the slope of the minority concentrations at the n^+ and p edges of the emitter-junction SCL. (The emitter current is the sum of the minority diffusion currents, because the emitter and base are in low-level injection.) Thus, if i_C is to decrease, the slope of the excess base concentrations must decrease, resulting in the dashed profiles in Fig. 18.5. Note that the high-level injection extends farther into the ν region, which gives a smaller value of R_ν.

The behavior of the constant i_B curves in quasi-saturation is shown in Fig. 18.2. Eventually $x' = W_\nu$ and the entire ν region is conductivity modulated, which is the boundary between quasi-saturation and hard-saturation.

Because i_C is dropping while i_B stays constant, the current gain β in quasi-saturation is decreasing as v_{CE} decreases. At the boundary between quasi-saturation and hard-saturation, β is significantly smaller than it is at the forward-active edge of quasi-saturation. This difference is apparent from the spacing between the i_B curves in Fig. 18.2.

Carrier Profiles in the Base and ν Collector Regions   The transistor operating in quasi-saturation has its collector current constrained by the circuit. The degree of quasi-saturation (how much of the ν region is conductivity modulated) is therefore controlled by the base current. We are interested in the operating region in which the base current is large enough to conductivity modulate part or all of the ν region. Thus we assume that the ν side of the collector junction is in high-level injection.

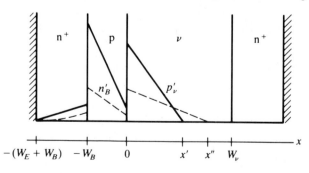

Figure 18.5   Excess carrier distributions corresponding to two values of v_{CE} at constant i_B for a power BJT in quasi-saturation.

If a BJT has a reasonable gain, the collector-region current is dominated by majority flow, electrons in the ν and n^+ regions in this case. Otherwise, the base current would be on the order of the collector current, because the base current is the only source of holes in the collector. The hole current in the collector region is thus approximately zero, and the electron current is equal to the collector-terminal current, which is constrained by the external circuit. The electric field in the ν region is determined by the zero-hole current condition, and the carrier profile is determined by the electron-current condition.

Proceeding as in Section 17.2, we know that the electric field and electron current density in the ν region of Fig. 18.6 are

$$E_\nu(x) = \frac{kT}{q} \frac{1}{n_\nu(x)} \frac{\partial n_\nu(x)}{\partial x} \qquad (18.8)$$

$$J_{e\nu} \approx J_C = q\mu_e n_\nu(x) E_\nu(x) + qD_e \frac{\partial n_\nu(x)}{\partial x} \qquad (18.9)$$

Substituting *(18.8)* into *(18.9)*, and using the Einstein relation, *(16.12)*, we find the collector-current density to be

$$J_C = 2qD_e \frac{\partial n_\nu(x)}{\partial x} \qquad (18.10)$$

which we can solve for the electron distribution to give

$$n_\nu(x) = n_\nu(0) - \frac{J_C}{2qD_e} x \qquad (18.11)$$

However, *(18.11)* is true only for that part of the ν region that is conductivity modulated. Elsewhere, the concentration is approximately equal to its equilibrium value.

EXAMPLE 18.2

A Calculation of Operating Region Boundaries

We want to find the relationship between J_{CQ} and J_{CS}, the values of collector-current density for which the transistor of Fig. 18.6 just reaches the onsets of quasi-saturation and hard-saturation, respectively, as we approach these regions from the forward-active region. We do our analysis for the condition of constant v_{BE}. In a power BJT with a ν region and a narrow base width, W_B, almost all the base current goes to supply reverse hole injection, $J_h(-W_B)$, into the emitter. If v_{BE} is constant, then $J_h(-W_B)$ is also constant. Therefore our condition of constant v_{BE} is approximately the same as constant base current.

The fact that the majority of the base current supplies reverse emitter injection means that little recombination occurs in the p and ν regions. Therefore the collector current is

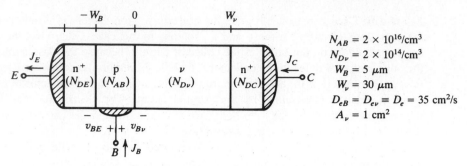

Figure 18.6   The power BJT for which we calculate the boundaries of the quasi-saturation region J_{CQ} and J_{CS}.

approximately equal to the diffusion current of electrons through the base:

$$J_C = \frac{qD_e\left[n_B(-W_B) - n_B(0)\right]}{W_B}$$

$$= \frac{qD_e}{W_B}\left[\frac{n_i^2}{N_{AB}}e^{qv_{BE}/kT} - \frac{n_i^2}{N_{AB}}\frac{n_\nu(0)}{N_{D\nu}}e^{qv_{B\nu}/kT}\right] \qquad (18.12)$$

In writing *(18.12)* we assumed that the transistor is in quasi-saturation; that is, $n_\nu(0) > n_{\nu o} = N_{D\nu}$. Recognizing that $n_\nu(0)/N_{D\nu} = \gamma$, where γ is the high-level injection factor, *(17.37)*, we can rewrite *(18.12)* as:

$$J_C = \frac{qD_e}{W_B}\frac{n_i^2}{N_{AB}}\left[e^{qv_{BE}/kT} - \gamma e^{qv_{B\nu}/kT}\right] \qquad (18.13)$$

The value of $v_{B\nu}$ at which the ν region enters high-level injection is defined as v_α, for which:

$$p_\nu(0) = N_{D\nu} \approx \frac{n_i^2}{N_{D\nu}}e^{qv_\alpha/kT} \qquad (18.14)$$

$$e^{qv_\alpha/kT} = \left(\frac{N_{D\nu}}{n_i}\right)^2 \qquad (18.15)$$

In quasi-saturation, $v_{B\nu} = v_\alpha + v_\gamma$, where v_γ is defined as the junction voltage in excess of v_α. We now rewrite *(18.13)* in terms of v_γ and $N_{D\nu}$, using *(18.15)* to eliminate v_α:

$$J_C = \frac{qD_e}{W_B}\left(\frac{n_i^2}{N_{AB}}\right)\left[e^{qv_{BE}/kT} - \gamma\left(\frac{N_{D\nu}}{n_i}\right)^2 e^{qv_\gamma/kT}\right] \qquad (18.16)$$

The relationship between γ and v_γ (the derivation of which is left to an end-of-chapter problem) is

$$\gamma = e^{qv_\gamma/kT} \qquad (18.17)$$

Substituting *(18.17)* into *(18.16)* yields

$$J_C = \frac{qD_e}{W_B} \frac{n_i^2}{N_{AB}} \left[e^{qv_{BE}/kT} - \left(\frac{N_{D\nu}}{n_i} \right)^2 e^{2qv_\gamma/kT} \right] \qquad (18.18)$$

At the boundary between the forward-active and quasi-saturation regions, $n_B(0) \ll n_B(-W_B)$ and:

$$J_{CQ} = \frac{qD_e}{W_B} \frac{n_i^2}{N_{AB}} e^{qv_{BE}/kT} \qquad (18.19)$$

At the edge of hard-saturation, $n_\nu(W_\nu)$ is just beginning to increase, but still $n_\nu(0) \gg n_\nu(W_\nu)$, so J_C (J_{CS} at this point) as given by *(18.10)* becomes

$$J_{CS} = \frac{2qD_e n_\nu(0)}{W_\nu} = \frac{2qD_e}{W_\nu} \frac{n_i^2}{N_{D\nu}} e^{qv_{B\nu}/kT} \qquad (18.20)$$

Substituting $v_{B\nu} = v_\alpha + v_\gamma$ into *(18.20)* and using *(18.15)* gives

$$J_{CS} = \frac{2qD_e N_{D\nu}}{W_\nu} e^{qv_\gamma/kT} \qquad (18.21)$$

The problem now is to determine v_γ. If we divide *(18.16)* by *(18.19)* and let $J_C = J_{CS}$, we get another equation relating J_{CS}, v_γ, and J_{CQ}:

$$\frac{J_{CS}}{J_{CQ}} = 1 - \frac{(N_{D\nu}/n_i)^2 e^{2qv_\gamma/kT}}{e^{qv_{BE}/kT}} \qquad (18.22)$$

$$e^{qv_\gamma/kT} = \frac{n_i}{N_{D\nu}} e^{qv_{BE}/2kT} \sqrt{1 - \frac{J_{CS}}{J_{CQ}}} \qquad (18.23)$$

Now if we substitute *(18.23)* into *(18.21)*, square both sides, and solve for J_{CS}, we get

$$J_{CS} = \frac{K}{2} \left[\frac{-1}{J_{CQ}} + \sqrt{\frac{1}{J_{CQ}^2} + \frac{4}{K}} \right] \qquad (18.24)$$

where

$$K = \left[\frac{2qD_e n_i}{W_\nu} \right]^2 e^{qv_{BE}/kT} \qquad (18.25)$$

We picked the positive square root in *(18.24)* because, physically, $J_{CS} > 0$.

A numerical calculation will be helpful at this point. If the transistor has the parameters given in Fig. 18.6 and we consider a case in which $J_{CQ} = 100$ A/cm^2, then from *(18.19)* we determine that $v_{BE} = 0.69$ V, and from *(18.25)*, $K = 2.44 \times 10^3$ A^2. Therefore *(18.24)* gives $J_{CS} = 38.7$ A/cm^2.

Because we calculated J_{CS} and J_{CQ} for the same i_B, their ratio is also equal to the ratio of the current gain at the edge of hard-saturation β_S to the gain at the forward-active edge of quasi-saturation β_Q. In this example, β_S is approximately one-third β_Q, illustrating the price paid to drive the transistor all the way through the quasi-saturation region into hard-saturation.

18.3 QUASI-SATURATION VOLTAGE

If the transistor shown in Fig. 18.7 is in quasi-saturation, its collector–emitter voltage is

$$v_{CE} = v_{BE} - v_{B\nu} + v_m + v'_\nu \qquad (18.26)$$

The voltage v'_ν is

$$v'_\nu = R'_\nu i_C \qquad (18.27)$$

where R'_ν is given by (18.7). The voltage v_m is obtained by integrating $E_\nu(x)$, (18.8), from 0 to x':

$$v_m = -\int_0^{x'} E_\nu(x)\,dx = -\frac{kT}{q}\ln\frac{n_\nu(x')}{n_\nu(0)} \qquad (18.28)$$

The point x' is defined by the condition $n_\nu(x') \approx n_{\nu o} = N_{D\nu}$, and $n_\nu(0)$ is

$$n_\nu(0) = \frac{n_i^2}{N_{D\nu}}e^{qv_{B\nu}/kT} \qquad (18.29)$$

Substituting these concentrations into (18.28) gives

$$v_m = v_{B\nu} - \frac{2kT}{q}\ln\frac{N_{D\nu}}{n_i} \qquad (18.30)$$

The first term in (18.30) excludes the voltage across the B–ν junction at the threshold of high-level injection in the ν region, v_α. We can now determine v_{CE} in quasi-saturation by substituting (18.27) and (18.30) into (18.26):

$$v_{CE} = v_{BE} - \frac{2kT}{q}\ln\frac{N_{D\nu}}{n_i} + R'_\nu i_C \qquad (18.31)$$

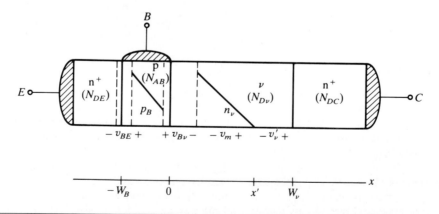

Figure 18.7 A one-dimensional transistor structure, illustrating the different voltage drops between collector and emitter.

EXAMPLE 18.3

A Calculation of v_{CE} at the Boundaries of Quasi-Saturation

The transistor in Fig. 18.8 has the same parameters as the transistor in Fig. 18.6 at $T = 300$ K. What are the values of v_{CE} for this transistor at the boundaries of the quasi-saturation region?

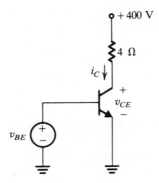

Figure 18.8   A transistor being driven through the quasi-saturation region.

When the transistor is in quasi-saturation, v_{CE} is small compared to 400 V and $i_C \approx 100$ A. Because $A_{\nu} = 1$ cm^2, $J_C = 100$ A/cm^2 and $v_{BE} = 0.69$ V, as in Example 18.2. From *(18.6)* we calculate R_{ν}:

$$R_{\nu} = \frac{W_{\nu}}{q\mu_e N_{D\nu} A_{\nu}} = 67 \text{ m}\Omega$$

The second term in *(18.31)* is

$$\frac{2kT}{q} \ln \frac{N_{D\nu}}{n_i} = 2(0.025) \ln \frac{2 \times 10^{14}}{1.4 \times 10^{10}} = 0.48 \text{ V}$$

At the forward-active region edge of quasi-saturation, $R'_{\nu} = R_{\nu}$ and:

$$v_{CE(Q)} = 0.69 - 0.48 + 0.067(100) = 6.9 \text{ V} \tag{18.32}$$

At the hard-saturation edge of quasi-saturation, $R'_{\nu} \approx 0$ and the collector–emitter voltage is

$$v_{CE(S)} = 0.69 - 0.48 = 0.21 \text{ V} \tag{18.33}$$

In practice, $v_{CE(S)}$ is seldom achieved because the gain of the transistor is so low in the saturation region that it becomes difficult to provide the necessary base current.

18.4 HARD-SATURATION REGION

If, for a given collector current, we drive the transistor with a high enough base current i_B, we will push it into the saturation region. Figure 18.9 shows the carrier profiles for a transistor in this condition. The slopes of the profiles in the base and

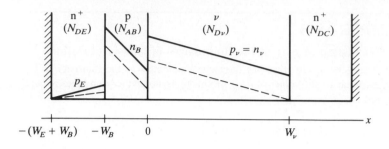

Figure 18.9 Carrier distributions for a transistor in hard-saturation. The dashed profiles are those at the edge of hard-saturation.

ν regions are the same as they were at the onset of saturation, but the levels have shifted upwards.

Although the collector current has remained constant, the hole current injected into the emitter from the base increases by the same factor that the excess electron concentration at $x = -W_B$ has been increased. Furthermore, because the ν–n$^+$ junction is now forward biased, holes originating as base current will be injected into the n$^+$ collector (similar to reverse injection of holes into the emitter). For these reasons, the required base current quickly increases as we try to push the transistor further into the saturation region, and the gain falls rapidly.

18.5 COMMON EMITTER-CURRENT GAIN

The common emitter-current gain $\beta = i_C/i_B$ is a very important parameter for users of power transistors (this gain is often labeled h_{FE} in specification sheets). But unlike users of signal processing transistors, who only want to know β in the forward-active region, we want to know β at the onset of saturation in order to determine the minimum base current needed to ensure saturation. As we mentioned in Example 18.2, the current gain in hard-saturation is not very high in typical power transistors, so we may choose not to push the transistor into this region. Therefore we would also like to know β in the quasi-saturation region.

In modern transistors, base current supplying recombination in the p and ν regions is negligible. The majority of the base current supplies holes injected into the emitter. So long as the base remains in low-level injection at the edge of the emitter–base junction,

$$J_B = \frac{qD_{hE}n_i^2 e^{\Delta E_g/kT}}{N_{DE}W_E}\left(e^{qv_{BE}/kT}\right) \qquad (18.34)$$

where ΔE_g is the amount that the bandgap has narrowed in the emitter as a result of its very heavy donor concentration N_{DE}. (We discussed this effect in Section 17.2.2.)

In the forward-active region, the transistor's collector current is

$$J_{CF} = \frac{q D_{eB} n_i^2}{W_B N_{AB}} \left(e^{q v_{BE}/kT} \right) \tag{18.35}$$

This expression is the same as J_{CQ} in *(18.19)* so we can express J_B in terms of J_{CQ}:

$$J_B = \frac{D_{hE} N_{AB} W_B}{D_{eB} N_{DE} W_E} e^{\Delta E_g/kT} J_{CQ} = \delta_E' J_{CQ} \tag{18.36}$$

where δ_E' is the emitter defect, *(16.56)*, corrected for bandgap narrowing.

We can now express β in the quasi-saturation region in terms of J_{CQ} for the same J_B (or v_{BE}):

$$\frac{J_C}{J_B} = \beta = \frac{1}{\delta_E'} \left(\frac{J_C}{J_{CQ}} \right) \tag{18.37}$$

EXAMPLE 18.4

A Calculation of β at the Boundaries of Quasi-Saturation

We now calculate the gain of the transistor that we considered in Examples 18.2 and 18.3. The emitter parameters are $N_{DE} = 10^{19}/cm^3$, $W_E = 3~\mu m$, and $\Delta E_g = 0.08$ eV.

From *(18.36)* we calculate the emitter defect:

$$\delta_E' = \left(\frac{35}{35} \right) \left(\frac{2 \times 10^{16}}{1 \times 10^{19}} \right) \left(\frac{5}{3} \right) (24.5) = 0.081 \tag{18.38}$$

From *(18.37)* we can calculate the gain of the transistor at the forward-active region edge of the quasi-saturation region:

$$\beta_Q \approx \beta_F = \frac{1}{\delta_E'} = 12.3 \tag{18.39}$$

In Example 18.2 we found that for $v_{BE} = 0.69$ V, the transistor reached hard-saturation at $J_{CS} = 38.7$ A/cm^2, or at approximately $J_{CQ}/2.6$. Therefore the gain at the onset of saturation is $\beta_S = \beta_Q/2.6 \approx 4.7$.

Note from the results of Example 18.4 that the gain of a power BJT is considerably less than the 100–200 exhibited by a typical signal processing transistor. This is a result of the higher value of N_{AB} used in a power transistor. A higher level of base doping gives a lower base resistance, which improves the switching performance of the transistor. The high N_{AB} results in a higher emitter defect, δ_E', and therefore a lower β.

18.5.1 About the Current-Gain Parameter

The power transistor of Examples 18.2–18.4 has a $V_{CEO(sus)}$ of approximately 300 V. Compared to this voltage, the 0.21 V saturation voltage that we calculated in *(18.33)* is much lower than necessary for acceptably small on-state losses. We might, for instance, be satisfied with an on-state voltage of 2 V. This voltage would reduce the base current requirement for a given collector current, because the transistor will not be driven into hard-saturation where β is much smaller than it is in quasi-saturation.

Most specification sheets for power transistors give nominal values for β (or h_{FE}) as a function of collector current while holding v_{CE} constant at a specified value. This value is typically 2 V and is the value we use in the following discussion. A typical relationship between β and i_C is shown in Fig. 18.10. Note that the gain is approximately constant for small collector current. Its slight decrease as the current goes to zero results from the growing importance of recombination within the emitter–base SCL. More interestingly, at high collector currents, the gain is proportional to $1/i_C$. Our purpose in this section is to explain why this loss of gain happens.

Note that v_ν' is the dominant term in *(18.26)* if $v_{CE} = 2$ V. The sum of the first three terms in *(18.26)* is only 0.21 V for the parameters of Example 18.3 and changes little for other practical parameter values. Although v_{BE} varies as we vary the collector current to hold v_{CE} constant, it will change only slightly because of its exponential influence on the collector current. We can therefore assume that $v_\nu' \approx 1.8$ V for all values of collector current. We refer to this voltage as V_o in the following analysis.

Next, we define the current density J_o to be the current that produces $v_\nu = V_o$ at the onset of quasi-saturation:

$$J_o = \frac{V_o}{A_\nu R_\nu} = \frac{V_o q \mu_{e\nu} N_{D\nu}}{W_\nu} \tag{18.40}$$

For current densities J_C greater than J_o, part of the ν region must become conductivity modulated so that v_ν' can remain constant (as constrained by our condition

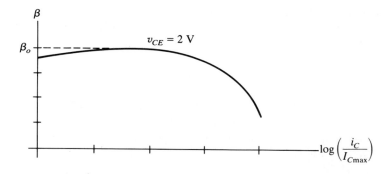

Figure 18.10   A typical specification of common emitter-current gain as a function of collector current while holding $v_{CE} = 2$ V.

that $v_{CE} = 2$ V). The extent of the modulated region is determined from *(18.7)*:

$$\frac{x'}{W_\nu} = 1 - \frac{J_o}{J_C} \qquad (18.41)$$

Knowing x', we can now use *(18.11)* to find the excess electron concentration at the ν-region edge of the $B-\nu$ SCL. As $n(x') \ll n_\nu(0)$, *(18.11)* becomes

$$n_\nu(0) = \frac{x'}{q2D_{e\nu}} J_C = \frac{W_\nu}{q2D_{e\nu}}(J_C - J_o) \qquad (18.42)$$

We can relate this concentration to the exponential of $v_{B\nu}$, so we can also find the concentration of electrons at the base edge of the $B-\nu$ SCL, recalling that the ν edge is in high-level injection by a factor of $\gamma = n_\nu(0)/N_{D\nu}$:

$$n_B(0) = \frac{N_{D\nu}}{N_{AB}}\gamma n_\nu(0) = \frac{1}{N_{AB}}n_\nu^2(0) \qquad (18.43)$$

Finally, we can determine the concentration of electrons at the emitter edge of the base region $n_B(-W_B)$ by relating the concentration gradient to J_C:

$$J_C = \frac{qD_{eB}[n_B(-W_B) - n_B(0)]}{W_B} \qquad (18.44)$$

and then substituting *(18.42)* and *(18.43)* into *(18.44)*:

$$n_B(-W_B) = \frac{W_B}{qD_{eB}}J_C + \frac{1}{N_{AB}}\left(\frac{W_\nu}{q2D_{e\nu}}\right)^2(J_C - J_o)^2 \qquad (18.45)$$

The base current, which consists primarily of holes injected into the emitter, is

$$J_B = \frac{qD_{hE}p_E(-W_B)}{W_E} \qquad (18.46)$$

where

$$p_E(-W_B) \approx \frac{N_{AB}}{N_{DE}}e^{\Delta E_g/kT}n_B(-W_B) \qquad (18.47)$$

We can now obtain an expression for β by combining *(18.45)*, *(18.46)*, and *(18.47)*:

$$\beta = \frac{J_C}{J_B} = \frac{\beta_o}{1 + \delta_\nu^{-1}(qV_o/4kT)[(J_C/J_o) + (J_o/J_C) - 2]} \qquad (18.48)$$

where β_o is the gain of the transistor at $J_C = J_o$, that is, at the edge of quasi-saturation:

$$\beta_o = \frac{1}{\delta_E'} = \frac{D_{eB}W_E N_{DE}}{D_{hE}W_B N_{AB}}e^{-\Delta E_g/kT} \qquad (18.49)$$

The parameter δ_ν is the injection efficiency of the $B-\nu$ junction and is equal to:

$$\delta_\nu = \frac{D_{e\nu} N_{AB} W_B}{D_{eB} N_{D\nu} W_\nu} \qquad (18.50)$$

Note that *(18.48)* is valid only for $J_C \geq J_o$. The maximum gain is therefore β_o, as shown in Fig. 18.10. For $J_C \gg J_o$, β can be approximated as:

$$\beta \approx \beta_o \left(\frac{4kT J_o}{q \delta_\nu V_o J_C} \right) \qquad (18.51)$$

which shows that the current gain is proportional to $1/J_C$ at high currents.

18.6 TEMPERATURE DEPENDENCE OF β

The current gain in the forward-active region, β_o, varies with temperature according to *(18.49)* owing to bandgap narrowing. The change in bandgap ΔE_g is itself proportional to $T^{-1/2}$, so the exponent in *(18.49)* is proportional to $T^{-3/2}$. A change in temperature from 25°C to 125°C increases β_o by a factor of 1.4.

At a constant but high value of J_C and a fixed v_{CE}, *(18.51)* shows that β is proportional to $\beta_o J_o T$. Since $J_o \propto 1/R_\nu$ for constant V_o, β decreases with increasing temperature because μ_e decreases. A good approximation of this dependence is $\mu_{e\nu} \propto T^{-2.2}$. The result is $\beta \propto \beta_o/T^{1.2}$. A change in temperature from 25°C to 125°C decreases β by a factor of 0.98, even though β_o increased by a factor of 1.4.

18.7 SWITCH TRANSITIONS

Figure 18.11 shows typical switching waveforms for a bipolar transistor. We assume that the load is a clamped inductor, as shown in Fig. 18.11(a). Note from Fig. 18.11(c) that when the transistor turns on, v_{CE} drops very quickly at first (after the full-load current has been picked up), but finishes with a long "tail." Similarly, when the transistor turns off, v_{CE} rises slowly at first, but finishes quickly, as shown in Fig. 18.11(b). In this section we describe the reasons for this behavior.

Consider the turn-off transition first. If at $t = 0$ the base current makes a step change from $+I_B$ to $-I_B$, the charge in the base and ν regions begins to decrease. We assume that this charge removal is slow enough to make the excess carrier profiles at any time have the same shape they would have in the steady state.

As i_C is constant until $v_{CE} = V_s$, the gradients of the charge distributions in the base and ν regions are also constant. During the initial phase of the turn-off transition, as charge is removed from the ν region while the gradient remains constant, the fraction of the ν region that is conductivity modulated shrinks. The voltage across the unmodulated region v'_ν therefore rises. It is this process of

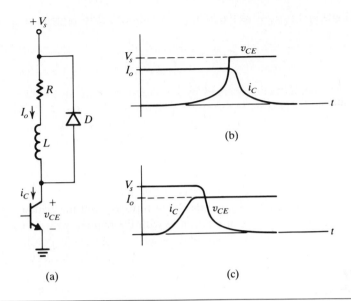

Figure 18.11   Typical switching waveforms exhibited by a power BJT: (a) the switching circuit, incorporating a clamped inductive load;   (b) the turn-off transient;   (c) the turn-on transient.

traversing the quasi-saturation region that causes the slow rise in v_{CE} evident in Fig. 18.11(b). Even though charge in the base region is decreasing during this phase, the amount of the decrease is small and we will ignore it.

The time required for the transistor to reach the edge of quasi-saturation is somewhere between the following two limits:

$$\frac{Q_o - Q_1}{2I_{Bo}} < t_1 < \frac{Q_o - Q_1}{I_{Bo}} \qquad (18.52)$$

where Q_o is the excess charge in the transistor at $t = 0$ and Q_1 is the charge at the edge of quasi-saturation. The factor of 2 in the denominator of the lower limit reflects the fact that both the negative base terminal current and the injection of holes into the emitter contribute to the removal of charge. For constant i_C, the hole injection into the emitter gets smaller as the transistor gets closer to the edge of quasi-saturation because β increases. Thus, in the worst case, all of the charge must be removed through the base terminal as base current.

As the transistor enters its forward-active operating region, the gradient of excess carriers within the base region remains constant because i_C is constant until $v_{CE} = V_s$. The base-terminal current and the hole-injection current at the emitter (which is constant for this phase) change the charge in (and therefore the voltage across) the collector–base junction capacitance. At first, this capacitance is proportional to $1/\sqrt{v_{CE}}$ as the SCL grows into the drift region:

$$C_{BC} \approx \frac{A_e \sqrt{q \epsilon_{Si} N_{D\nu}/2}}{\sqrt{V_{CE}}} \qquad (18.53)$$

This capacitance is being charged by a constant current, so the rate at which its voltage changes increases as the voltage increases. Once the SCL reaches through to the substrate, however, the value of C_{BC} becomes fixed at:

$$C_{BC} = \frac{\epsilon_{Si} A_e}{W_\nu} \qquad (18.54)$$

From this point on, the collector voltage rises linearly.

Finally, when $v_{CE} = V_s$, the diode D turns on and the remaining charge in the base region (including that associated with the emitter–base SCL) is removed. Again, both the negative base-terminal current and hole injection into the emitter contribute to this process. This phase is over when the full load current is flowing in the diode and $i_C = 0$.

At turn on, when the transistor must first pick up the load current before v_{CE} can fall, the process is simply the reverse of the one we just described. The only difference is that the injection of holes into the emitter (which is still the equivalent of *negative* base current) hinders, rather than helps, the turn-on process.

Throughout this discussion we have implicitly assumed that, when on, the transistor is not in the hard-saturation region. If it is, we must add to our description of the switch transitions a phase during which the additional charge is removed or supplied by the base current. During this time, known as the *storage delay time*, the collector–emitter voltage remains approximately constant.

EXAMPLE 18.5

Estimating the Turn-Off Time of a Transistor

Assume that the transistor of our earlier examples is used in the circuit of Fig. 18.11(a) in which $I_o = 40$ A. How long does it take the transistor to turn off? When the transistor is on, $v_{BE} = 0.69$ V, $i_C = 40$ A, and—because the area of the device is 1 cm²— $J_C = 40$ A/cm². From Example 18.2 we know that these values put the transistor at the edge of hard-saturation, so $i_C = i_{CS}$. At this operating point, $\beta_S \approx 5$, as we calculated in Example 18.3. Therefore:

$$i_B = \frac{i_C}{\beta_S} = 8 \text{ A}$$

The excess charge distribution at this time is shown in Fig. 18.12(a). We turn the transistor off by applying a negative base current $i_B = -8$ A at $t = 0$. The total stored charge at $t = 0$ is

$$Q_o = Q_B + Q_\nu \approx \frac{q n_B(-W_B) W_B}{2} + \frac{q n_\nu(0) W_\nu}{2} \qquad (18.55)$$

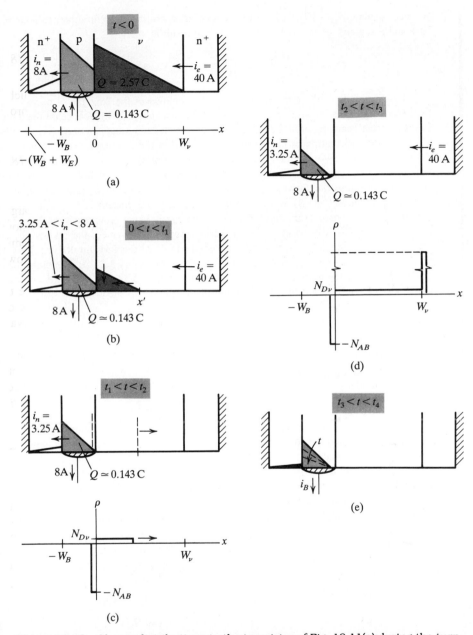

Figure 18.12 Charge distributions in the transistor of Fig. 18.11(a) during the turn-off process: (a) prior to the application of negative base current at $t = 0$; (b) during removal of excess charge in the ν region; (c) during the time the ν region is being depleted as the SCL grows. (d) after the ν region is depleted but before $v_{CE} = 400$ V; (e) after the transistor is off and $i_C = 0$.

We can calculate $n_B(-W_B)$ and $n_\nu(0)$ by relating J_C to the concentration gradients in the base and ν regions, as in *(18.20)* for $n_\nu(0)$. The result is

$$Q_o = \frac{J_{CS}W_B^2}{2D_{eB}} + \frac{J_{CS}W_\nu^2}{4D_{e\nu}} = 0.143 + 2.57 = 2.71 \ \mu C \qquad (18.56)$$

The first stage of the turn-off process is the removal of the 2.57 μC of excess charge from the ν region, as shown in Fig. 18.12(b). This occurs at $i_C = 40$ A, because the inductor keeps i_C from changing until the diode turns on, which cannot happen until $v_{CE} = 400$ V. Therefore the distribution of charge in the *base* region remains approximately unchanged during this phase. Two mechanisms remove charge (holes) from the base and ν regions: the negative base-terminal current i_B and the reverse injection of holes into the emitter $i_h(-W_B)$. Because the forward base current consisted primarily of holes to supply this reverse injection, $i_h(-W_B) = i_B = 8$ A before the base current is reversed. So at the beginning of this phase of the turn-off process, charge is being removed from the ν region at a rate of 16 C/s. But as the quasi-saturation region is traversed, the reverse injection decreases, increasing β as we showed in Example 18.4. At the forward-active region edge of quasi-saturation, $\beta = \beta_Q = 4.7$ and $i_h(-W_B) = 40/\beta_Q = 3.25$ A. Therefore the rate at which charge is being removed at the end of this phase is 11.25 C/s. If we assume that the change of rate is linear over the duration of this phase, t_1, then:

$$\frac{16 + 11.25}{2} t_1 = 2.57 \ \mu C \qquad (18.57)$$

$$t_1 = 189 \ ns$$

At t_1 the ν region has a resistance R_ν and v_{CE} has a value of:

$$v_{CE} = v_{CE(Q)} \approx R_\nu i_C = 2.68 \ V \qquad (18.58)$$

The next phase of the turn-off process is depletion of the ν region as the collector SCL grows to support the increasing value of v_{CE}. The excess charge distribution and the collector SCL charge during this phase are shown in Fig. 18.12(c). As i_C is still constrained to be 40 A, the base-charge distribution and hence $i_h(-W_B)$ remain unchanged during this phase. The charge to be removed is the equilibrium majority charge in the ν region, which for this transistor is $Q_{\nu o} = qW_\nu N_{D\nu} = 0.096 \ \mu C$. The rate at which this charge is removed is still 11.25 C/s. If this phase is complete at t_2, the duration of the phase $t_2 - t_1$ is

$$11.25(t_2 - t_1) = 0.096$$

$$t_2 - t_1 = 8.53 \ ns \qquad (18.59)$$

The transistor voltage at t_2 is

$$v_{CE}(t_2) = \frac{qN_{D\nu}W_\nu^2}{2\epsilon_{Si}} = 149 \ V \qquad (18.60)$$

Now the ν region is depleted, and as the SCL continues to grow, it depletes the p-base and n^+ collector regions in the vicinity of their junctions with the ν region, as shown in Fig. 18.12(d). These depletion regions have narrow widths because their doping concentrations are high relative to the ν region. Therefore we can consider this process to be similar to that of charging a capacitor of value C_{BC} given by *(18.54)*. The required change in voltage is $400 - 149 = 251$ V, from which we can calculate the charge to be 0.08 μC. Again, this charge is removed at a rate of 11.25 C/s, because i_C is still equal to 40 A, and we assume that the change in base width caused by the growing SCL is small. The duration of this phase is $t_3 - t_2$, and it has a value of:

$$t_3 - t_2 = \frac{0.08}{11.25} = 7.1 \text{ ns} \qquad (18.61)$$

From t_3 on, i_C can fall as Q_B is removed. The base-terminal current is still 8 A during this phase, but $i_h(-W_B) = 3.25$ A at the beginning and zero at the end. Again, approximating this variation as linear, we can calculate the duration of this phase, $t_4 - t_3$:

$$Q_{Bo} = 0.143 \ \mu\text{C} = \left(\frac{8 + 11.25}{2}\right)(t_4 - t_3) \qquad (18.62)$$

$$t_4 - t_3 = 14.9 \text{ ns}$$

The turn-off transient that we have just calculated is sketched in Fig. 18.13. It shows that the total turn-off time is approximately 212.5 ns.

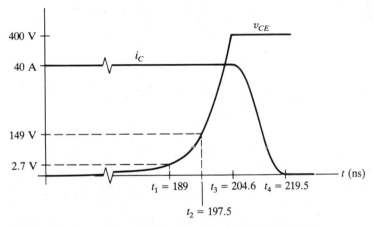

Figure 18.13 The turn-off transient in the circuit of Fig. 18.11(a).

18.8 EMITTER-CURRENT FOCUSING DURING TURN-OFF

So far we have presented the bipolar power transistor using a one-dimensional model, ignoring all two-dimensional (or three-dimensional) effects. For the turn-off transition, however, we must look beyond the one-dimensional model. Specifically, we now turn to a phenomenon called *emitter-current focusing*.

Figure 18.14 shows a more practical two-dimensional geometry for a power transistor. The base-terminal current flows laterally through the base region, and must be supported by a lateral electric field, E_y. The emitter–base *junction* voltage

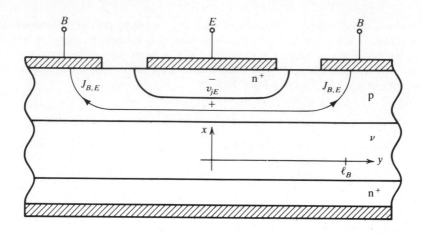

Figure 18.14 A two-dimensional transistor structure, showing emitter-current focusing caused by lateral base drops during turn-off.

v_{jE} is therefore a function of y:

$$v_{jE}(y) = v_{BE} + \int_{y}^{\ell_B} E_y \, dy \qquad (18.63)$$

The result is that the junction near the periphery of the emitter diffusion is less forward biased than the junction in the center of the emitter. Therefore the current density is higher in the center than at the edge of the emitter, and the current is said to be *focused* in the center.

Conceivably, we might avoid this problem by pulling the base terminal far enough to the negative to reverse bias the emitter–base junction along its entire length. Unfortunately, the emitter–base junction avalanches at a relatively low voltage (10 V is typical). If we try to pull the base terminal further to the negative than this voltage, the junction will avalanche near the outer edge of the emitter. The voltage drop across the lateral base resistance can then be large enough to keep the center region of the transistor forward biased.

One reason for doping the base of a power transistor so high (which is the reason it has such a low β_o) is to keep the base resistance low in order to reduce emitter focusing. In addition, modern power transistors also make use of a finely interdigitated emitter–base pattern to keep the lateral distances short.

18.9 SECOND BREAKDOWN

During turn-off when a bipolar transistor is simultaneously carrying a high current density and supporting a high collector voltage, a type of failure called *second breakdown* can occur. This breakdown is characterized by a sudden drop in the collector voltage to a lower but sustained (uncontrolled by base-terminal current) value that can lead to the thermal destruction of the transistor. For instance, as a transistor having a clamped inductive load, such as the one shown in Fig. 18.15(a), is turned off, v_{CE} must rise to its maximum value, V_s, before i_C begins to fall. The waveforms during this time are shown in Fig. 18.15(b). If V_s and I_o are high enough to cause second breakdown to occur once $v_{CE} = V_{SB}$, v_{CE} collapses to $V_{SB(sus)}$, where it will stay until the transistor is destroyed because of the high dissipation at this operating point.

Second breakdown occurs because the velocity at which mobile carriers can move eventually saturates (at about $v_{sat} = 10^7$ cm/s for electrons), rather than remaining proportional to the electric field. At values of the electric field high enough to cause velocity saturation, the drift component of (electron) current becomes

$$J_e^{drift} = qnv_{sat} \qquad (18.64)$$

From (18.64) we can calculate the concentration n necessary to carry a given current density in a situation in which the carrier velocity is saturated.

Consider the situation where the ν region of the transistor is neither depleted nor conductivity modulated (the active-region boundary of quasi-saturation), so that $n(x) = N_{D\nu} = 2 \times 10^{14}/cm^3$. At a current density high enough that $J_C > qN_{D\nu}v_{sat} = 320$ A/cm^2, the electron concentration must be larger than $N_{D\nu}$. In

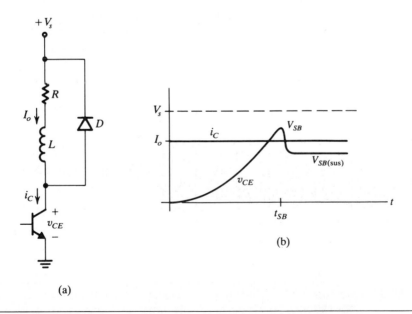

(a)

(b)

Figure 18.15   Second breakdown during turn-off: (a) switching circuit incorporating a clamped inductive load;   (b) waveforms during turn-off showing that v_{CE} fails to reach V_s, but instead collapses to $V_{SB(sus)}$.

this case, the drift region will have a net negative space charge that is uniform, and the resulting electric field will have the profile shown in Fig. 18.16(a).

Note in Fig. 18.16(a) that the electric field has a triangular profile, rather than the nearly flat profile shown in Fig. 18.1(b). Hence the peak value of the electric field will reach E_c, the level at which avalanche starts, at approximately half the voltage predicted by the flat profile. We call this voltage the *second breakdown voltage*, V_{SB}. For the conditions of Fig. 18.16(a),

$$V_{SB} = \frac{E_c W_\nu}{2} \tag{18.65}$$

The second breakdown voltage may be lower than $V_{CEO(sus)}$.

For a peak electric field of E_c in a transistor operating under the condition of velocity saturation in the ν region, the concentration of electrons in the ν region is

$$n_\nu = \frac{\epsilon_{Si} E_c}{q W_\nu} \tag{18.66}$$

From *(18.66)* we can find the current density J_{SB} at which second breakdown occurs:

$$J_{SB} = \frac{\epsilon_{Si} E_c v_{sat}}{W_\nu} \tag{18.67}$$

If the current density is higher than this value, the concentration of electrons in the drift region must be correspondingly higher than that given by *(18.66)*. The electric field profile will then have a greater slope, as shown in Fig. 18.16(b). For the same peak value E_c, the field must therefore reach zero before the end of the ν region (the remainder of this region becomes conductivity modulated). In this case, the

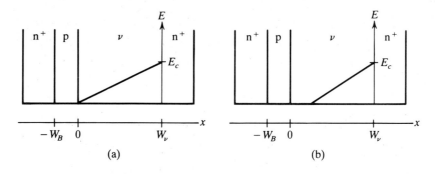

Figure 18.16   Electric field profile in the ν region under conditions of velocity saturation: (a) when the B–ν junction is reverse biased;   (b) when the B–ν junction is forward biased and the transistor is in second breakdown.

voltage drop across the ν region is less than V_{SB}, although second breakdown is still occurring.

We can now see what happens when the transistor carries a current greater than J_{SB} and v_{CE} is not kept below V_{SB}. As the excess electrons in the ν region increase, the condition of Fig. 18.16(a) is reached and avalanche begins ($v_{CE} \approx V_{SB}$). The hole current that results provides base current, which pushes the emitter–base and base–ν-region junctions further into forward bias. The modulation of part of the ν region then results in the profile of Fig. 18.16(b), which means that the collector voltage has dropped below V_{SB}. Finally, a self-sustaining voltage $V_{SB(sus)}$ is reached, where:

$$V_{SB(sus)} = \frac{\epsilon_{Si} E_c^2 v_{sat}}{2 J_C} \tag{18.68}$$

The current density required for second breakdown J_{SB} is very high compared to the density at which a transistor is typically rated. However, we have seen that during turn-off the current is focused, so it is possible to exceed the critical density during this transition.

EXAMPLE 18.6

A Second Breakdown Calculation

In Example 18.5 we considered the transistor of Fig. 18.11(a) as it turned off into a clamped inductive load with a supply voltage of $V_s = 400$ V. We now want to find the current density at which second breakdown just occurs during this transition. We assume that avalanche begins when the peak electric field in the transistor is $E_c = 1.8 \times 10^5$ V/cm.

From (18.65) and (18.67) we find that $V_{SB} = 270$ V and $J_{SB} = 1350$ A/cm^2. With $V_{SB} < V_s$, second breakdown will occur at $J_C = 1350$ A/cm^2.

If, either because of current focusing or a high load current, the current density is actually $J_C = 1500$ A/cm^2, the condition shown in Fig. 18.16(b) will exist. In that case, the slope of the electric field is

$$\frac{d|E|}{dx} = \frac{qn}{\epsilon_{Si}} = \frac{J_C}{\epsilon_{Si} v_{sat}} \tag{18.69}$$

which means that if the field has a peak value of E_c, it extends from the ν–n^+ junction into the ν region a distance:

$$x = \frac{\epsilon_{Si} E_c v_{sat}}{J_C} \tag{18.70}$$

Integrating the area under the field profile gives a sustained second breakdown voltage of:

$$V_{SB(sus)} = \frac{\epsilon_{Si} E_c^2 v_{sat}}{2 J_C} = 108 \text{ V} \tag{18.71}$$

18.10 THERMAL INSTABILITY

We have found that β_o, the gain of a power transistor in the forward-active region, increases with temperature because of the reduced effect of bandgap narrowing in the emitter. Even in the quasi-saturation region, where at high currents β is proportional to $\mu_e\beta_o$, the gain of the transistor increases with temperature. We also know that the base current is proportional to the exponential of v_{BE}. Thus we can conclude that, for a fixed set of emitter–base and collector–emitter voltages, the higher the temperature of the transistor, the more current it will carry.

This increase of current with temperature makes it difficult to parallel power transistors because the hotter one will carry a larger share of the current, thus becoming hotter still. This process is known as *thermal runaway*. Thermal runaway can also cause a problem within a single device. If one region of the transistor's die becomes hotter than another, perhaps because the thermal contact resistance of the die to its package is not uniform, the total collector current will, because of thermal runaway, focus in that area. This thermal instability of the transistor can even cause the current to focus in thin filaments. If the current density gets high enough, second breakdown will occur, destroying the device.

18.11 STRUCTURE OF THE POWER MOSFET

Figure 18.17 shows the structure of a typical n-channel power MOSFET. Note that the drain contact is on the bottom of the die, rather than on the top as it is in a signal-processing MOSFET. This structure gives maximum area to both drain

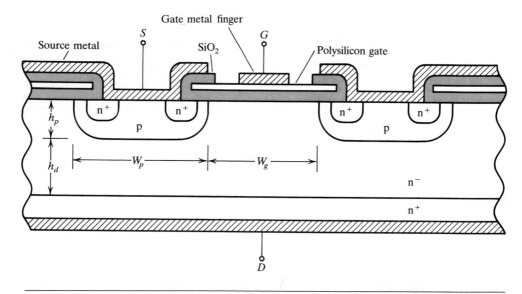

Figure 18.17 Structure of a typical vertical n-channel power MOSFET.

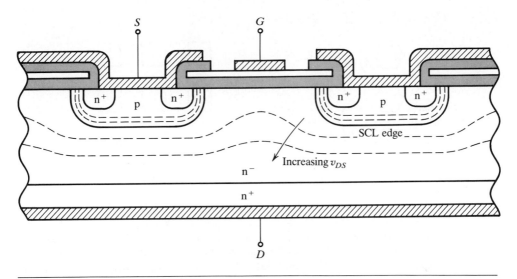

Figure 18.18 An illustration of how the SCL grows with increasing v_{DS}, pinching off the n$^-$ region between the p wells.

and source contacts in order to produce a low-resistance connection to the package terminals. The polysilicon electrode gate is insulated from the source metal that covers it by a layer of SiO_2. The polysilicon gate electrode is connected by metal fingers that make contact through windows etched in the SiO_2.

Between the source and the drain regions are *wells* of p-type material. These wells are known as the *body* region of the device. The channel of the MOSFET is formed on the surface of these p-wells just beneath the gate oxide. Note that the p wells are shorted to the source by the source electrode. We described the reason for this arrangement in Section 16.4.2.

The lightly doped n-type drain region is unique to power MOSFETs, and is provided to allow growth of a long SCL, permitting the device to block a high voltage when it is off. In this respect, the region functions like the ν region in a power BJT. This lightly doped drain region is frequently referred to as the *extended drain* or *drift region*. As the SCL grows, it *pinches off* (depletes) the region between the p wells, as shown in Fig. 18.18 (the gate electrode acts as a field plate to promote this pinch-off). This feature of a power MOSFET's structure is important because it keeps the gate oxide from being subjected to the full drain voltage. In fact, the voltage just underneath the gate oxide typically reaches only 5–10 V with respect to the gate electrode, even though the drain voltage may be 200–400 V. As a result, we can make the gate oxide relatively thin, which keeps both the gate threshold voltage V_T and the gate energy low.

Viewed from the top, the gate and the source contacts of a practical MOSFET are interleaved in a pattern of "cells." The hexagonal cell pattern of Fig. 18.19(a)

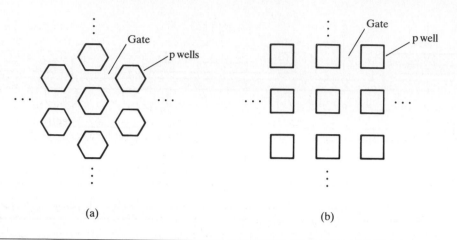

(a)   (b)

Figure 18.19   Cell pattern showing the gate regions surrounded by source contact
regions: (a) a hexagonal cell pattern;   (b) a square cell pattern.

(the source of the International Rectifier trademark HEXFET) has the greatest gate
perimeter for a given die area, but the square cell pattern of Fig. 18.19(b) does
nearly as well. The width of each cell is typically 20 to 30 μm.

18.12 ON-STATE RESISTANCE OF THE MOSFET

Figure 18.20(a) shows how the drain current flows through the MOSFET of
Fig. 18.17 when it is on. As the current flows through the extended drain, it focuses
in the area between the p wells, called the *neck* region. From there it must further

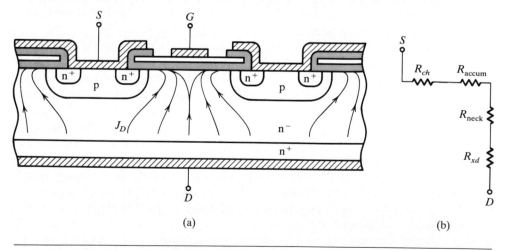

(a)   (b)

Figure 18.20   (a) Current flow in a power MOSFET.   (b) A model of R_{DS}, showing
its components.

focus in the thin entrance to the p-well channels on either side of the neck region. This latter focusing would result in a large resistance if the current had to flow through a progressively smaller cross-sectional area of the n^- region to get to the channel entrance. Fortunately, the region just below the gate oxide (and between the p wells) is *accumulated*, meaning that it has a very high concentration of mobile electrons, which makes it much more conductive than the bulk n^- region. The drain current between the p wells therefore tends to first flow directly upward to this layer (taking advantage of the full width between the p wells) and then flows horizontally (with little resistance) to the channel entrances.

Figure 18.20(b) shows how the MOSFET's total drain-source resistance, R_{DS}, can be broken down into four components: R_{xd}, R_{neck}, R_{accum}, and R_{ch}. As both the extended drain and the neck region are very lightly doped in a MOSFET designed for 400 V or above, R_{xd} and R_{neck} are typically much larger than R_{accum} and R_{ch}. For a MOSFET designed for 100 V, however, R_{ch} is typically one third to one half of R_{DS}.

18.12.1 Resistance of the Extended Drain Region

If the drain current flowed uniformly through the entire cross-sectional area of the drift region A_d, its resistance would be

$$R_{drift(min)} = \frac{h_d}{q\mu_e N_{dD} A_d} \qquad (18.72)$$

where N_{dD} is the doping concentration of the drift region and h_d is its thickness. Because the current density is not uniform across this region, the *effective* cross-sectional area of the drift region is smaller and its resistance is higher than (18.72) predicts. How much higher depends on the width of the p well W_p, compared to h_d.

Figure 18.21(a) and (b) illustrate the current distributions at the extremes of the ratio W_p/h_d. In both cases, we assume W_p to be as small as lithographic and processing techniques allow. The dashed lines represent an effective boundary between regions where the current flows and where it does not. Figure 18.21(a) corresponds to a high voltage device in which the drift region is thick compared to the width of the p wells, that is, a device with a very low W_p/h_d. Near the bottom of the drain region the full cross-sectional area of the die is carrying current, rather than just the region between the p wells. In the low-voltage device of Fig. 18.21(b), the spreading of the current in the drift region beneath the p wells is much more limited because $h_d \ll W_p$. In the first case, the effective cross-sectional area of the drift region is much closer to the actual die area than it is in the second case.

If we assume that the drain current is uniformly distributed between the dashed boundary lines of Fig. 18.21, the current density increases as the width of the current carrying region decreases. Using this assumption, we can calculate the effective resistance of the drift region. First, consider the region between the dashed lines to be a series of resistors, each having an incremental thickness dy and each carrying

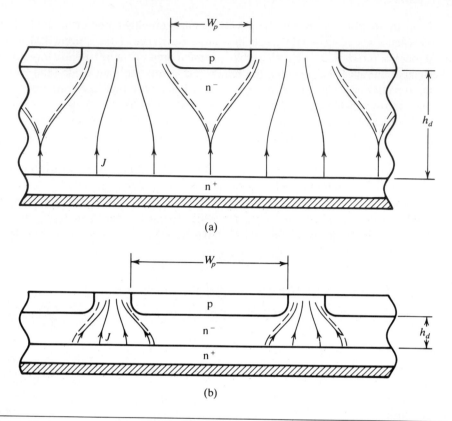

W_p

p

n^-

h_d

J

n^+

(a)

W_p

p

n^-

h_d

J

n^+

(b)

Figure 18.21   Current distributions in the drain region of a vertical MOSFET for different ratios of h_d/W_p: (a) a device in which h_d is large relative to W_p;   (b) a device in which h_d is small relative to W_p.

the same current I in the y-direction, as shown in Fig. 18.22. Each resistor has the same depth into the paper, Z, but a different width, W_r, because of the angle of the boundaries. If we define the vertical position of each resistor to be y, with $y = 0$ at the interface between the drift and the neck regions, then W_r as a function of y is

$$W_r = W_g + 2y \tan \alpha \qquad (18.73)$$

Here we have assumed that the spreading beneath the p wells is insufficient to cause current flowing through adjacent cells to meet under the p wells.

The total drift resistance is the sum of the resistances of the thin slices defined in Fig. 18.22. Taking dy to zero in the limit, this sum gives

$$R_{xd} = \int_0^{h_d} \frac{dy}{q\mu_e Z N_D W_r} = \frac{\ln\left[1 + 2(h_d/W_g)\tan\alpha\right]}{q\mu_e Z N_d \tan\alpha} \qquad (18.74)$$

The angle α of the dashed boundary lines has been shown in [10] to be well approximated for the case $h_d \geq W_g$ by:

$$\alpha = 28° - \frac{h_d}{W_g} \qquad (18.75)$$

For $h_d < W_g$, the approximation is

$$\alpha = 28° - \frac{W_g}{h_d} \qquad (18.76)$$

If the spreading were sufficient to make the dashed boundaries from adjacent cells meet under the p wells, we would simply have to calculate the drift resistance as the sum of two resistances. One we would calculate using *(18.74)* and replacing W_d with the value of y at which the boundaries meet. The other we would calculate by assuming uniform current over the remaining length of the drift region.

18.12.2 Resistance of the Neck Region

We can easily calculate the resistance of the neck region (the region between the p wells) by assuming that the current flows uniformly (vertically) through the region.

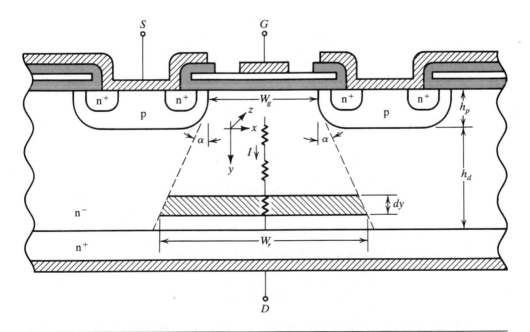

Figure 18.22   Drift-region resistance modeled as a series connection of incrementally thin resistors, each having a uniform current density.

If we ignore the curvature at the bottom of the p wells, the neck resistance is then:

$$R_{\text{neck}} = \frac{h_p}{q\mu_e Z N_D W_g} \qquad (18.77)$$

The curvature, which is approximately circular with a radius equal to the depth of the p well h_p, reduces this resistance slightly.

18.12.3 Resistance of the Accumulation Layer

We can calculate the resistance of a length of accumulation layer in the same manner as we calculated the channel conductance, (16.66), in Chapter 16:

$$R = \frac{L}{ZC_o\mu_e(v_{GS} - V_T)} \qquad (18.78)$$

where C_o is the capacitance per unit area of the gate oxide $\epsilon_{\text{ox}}/t_{\text{ox}}$. The length L is what we need to find here.

The current flows both to the left and right through the accumulation layer toward the adjacent p wells. If all the drain current entered the accumulation region at its center, we would expect the effective value of L to be $L = W_g/4$. One factor of 1/2 results from the splitting of the current to the left and right. The other factor of 1/2 accounts for the fact that the two halves of the accumulation layer are effectively in parallel. This would be the value of L if all the drain current entered the accumulation layer at the midpoint between the p wells. However, because the current enters the layer uniformly across its width, the effective length is actually a factor of 2 smaller, and the resistance of the accumulation layer is

$$R_{\text{accum}} = \frac{W_g/8}{ZC_o\mu_e(V_{GS} - V_T)} \qquad (18.79)$$

This calculation assumes that the resistivity of the accumulation layer is much lower than the neck region, which means that the gate voltage must be well above the threshold value. This condition is typical for a fully turned-on MOSFET.

18.12.4 Resistance of the Channel

We can find the resistance of the channel directly from (16.66):

$$R_{\text{ch}} = \frac{W_{\text{ch}}}{2ZC_o\mu_e(v_{GS} - V_T)} \qquad (18.80)$$

where the factor of 2 in the denominator accounts for the two channels per gate cell.

EXAMPLE 18.7

Calculating the On Resistance of a 200-V MOSFET

Typical values of the physical parameters for a MOSFET designed to withstand 200 V and having a die area of $A_d = 0.2$ cm^2 are

$$h_d = 20 \ \mu m \qquad h_p = 3 \ \mu m \qquad W_g = 15 \ \mu m \qquad W_p = 25 \ \mu m$$

$$W_{ch} = 2 \ \mu m \qquad t_{ox} = 0.1 \ \mu m \qquad Z = 100 \text{ cm} \quad \text{and} \quad N_{Dd} = 6 \times 10^{14}/\text{cm}^3$$

The gate-oxide capacitance per unit area is typically $C_o = 10$ nF/cm^2. We would typically drive the gate at 12 V, which is about 7 V above V_T.

From (18.75) we find the spreading angle in the drift region to be $\alpha = 26.7°$, and as $h_d \tan \alpha = 10 \ \mu m < W_p/2$, our assumption that the spreading was not complete applies for this device. From (18.74) we therefore obtain $R_{xd} = 117$ mΩ. This value is 1.67 times greater than $R_{xd(min)} = 70$ mΩ when $A_d = 0.2$ cm^2.

From (18.77) we find that the neck resistance is $R_{neck} = 14$ mΩ, and from (18.79) we find that the accumulation resistance is $R_{accum} = 3.65$ mΩ. The relative size of these resistances justifies our assumption that the current tends to flow vertically to the accumulation layer and then horizontally to the channels.

Finally, from (18.80) we find the channel resistance to be $R_{ch} = 3.33$ mΩ. Adding these four resistances together gives

$$R_{DS(on)} = 138 \text{ m}\Omega \qquad (18.81)$$

The resistance of the drift region dominates the total, so we cannot significantly reduce $R_{DS(on)}$ by increasing the gate voltage.

18.12.5 Temperature Dependence of $R_{DS(on)}$

Because the resistance of the high-voltage vertical MOSFET is dominated by the resistance of the extended drain, the temperature dependence of $R_{DS(on)}$ is the result of temperature dependence of μ_e. For the temperature range of interest (0–200°C), mobility decreases with temperture as $\mu_e \propto T^{-2.2}$, which means that $R_{DS(on)}$ increases with temperature. A 100°C rise produces an increase of approximately 90%.

18.13 DYNAMIC PERFORMANCE OF THE POWER MOSFET

Because the MOSFET is a majority carrier device, it does not have the buildup of excess carrier concentrations that controls the dynamic behavior of the BJT. The transient performance of the MOSFET is governed only by the oxide and SCL capacitances and by the impedances that limit our ability to charge and discharge these capacitances. (Some of these impedances are internal to the device but most are external.) Figure 18.23 shows the physical origins of the vertical MOSFET's capacitances.

Between the gate and the source are two parallel capacitances that comprise C_{GS}. One, $C_{S\text{poly}}$, is the result of source metallization actually covering the polysilicon gate, but being insulated from it by an oxide layer. Part of this capacitance is also the result of the overlap of the polysilicon and the n$^+$ source diffusion. Thus $C_{S\text{poly}}$ is an oxide capacitance, which is not a function of the voltage across it. The second capacitance making up C_{GS} is the gate to p-well capacitance of the channel $C_{p\text{poly}}$. Although there is a resistance, R_p, between the p-well side of this capacitor and the source contact, it is small enough to ignore for typical rates of change of gate voltage. As $C_{p\text{poly}}$ is the series combination of the oxide capacitance and the p-well's depletion capacitance, it is a strong function of the gate voltage when $v_{GS} < V_T$ (the capacitance gets smaller as the voltage increases). Once the gate voltage is above threshold, however, $C_{p\text{poly}}$ is dominated by the oxide capacitance, which is constant.

Between the drain and the source is the capacitance of the drift region's space-charge layer C_{DS}. Its value varies inversely with the square root of the drain-source voltage.

Finally, between the drain and the gate is C_{DG}, a series combination of two capacitors. One, C_{Gn}, is the oxide capacitance, which is independent of voltage. The other, C_{Dn}, is the SCL capacitance between the drain and the neck–oxide interface. It gets smaller as the drain voltage rises. Note that these two capacitances are "connected" at the plane in which the accumulation layer grows.

At low values of v_{DS}, the SCL does not extend into the neck region, so $C_{DG} = C_{gn}$. At higher voltages, the neck region becomes depleted, so C_{Dn} ap-

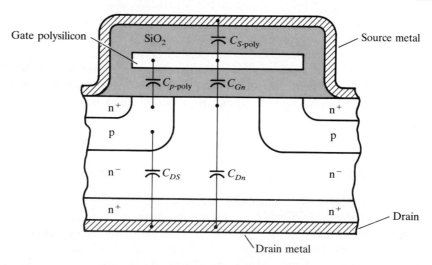

Figure 18.23   Origins of capacitances in a vertical MOSFET.

pears in series with C_{Gn} and reduces the net capacitance C_{DG}. As the drain voltage rises even higher, however, pinching off of the neck region becomes complete, and very little incremental increase in drain voltage appears at the neck–oxide interface. Therefore C_{DG} becomes very small under this condition—much smaller in comparison to C_{DS} than the relative widths of the gate W_g and the p well W_p would suggest.

Note that as the channel and the accumulation layer form, the node connecting C_{Dn} and C_{Gn} is shorted to the source by the low resistances of these regions. Under this condtion, C_{DG} becomes zero, C_{Gn} appears in parallel with the other gate-source capacitances, and C_{Dn} appears in parallel with C_{DS}.

EXAMPLE 18.8

Calculating Capacitance Values in a Vertical MOSFET

We want to determine approximate values for the capacitances of the 200-V MOSFET structure specified in Example 18.7. We need to specify the following additional parameters: $W_n = 1.0\ \mu$m, and the thickness of the oxide separating the gate from the overlaying source metal is $t_g = 1.0\ \mu$m.

Starting with the gate-source capacitances, we can make the following approximations when $v_{GS} > V_T$:

$$C_{S\text{poly}} = 2\frac{\epsilon_{\text{ox}} Z W_n}{t_{\text{ox}}} + \frac{\epsilon_{\text{ox}} Z(W_g + 2(W_n + W_{\text{ch}}))}{t_g} = 1.2 \text{ nF} \qquad (18.82)$$

$$C_{p\text{poly}} = 2\frac{\epsilon_{\text{ox}} Z W_{\text{ch}}}{t_{\text{ox}}} = 1.15 \text{ nF} \qquad (18.83)$$

We then assume that the capacitance of the drain-source SCL has the form:

$$C_{DS} = C_o\sqrt{\frac{200}{v_{DS}}} \qquad (18.84)$$

where C_o is the incremental capacitance of the junction when its voltage is 200 V. (This voltage corresponds to an SCL width of 20 μm.) At high drain voltages, where pinching off of the neck region is complete, it is reasonable to assume that the full cross-sectional area of the die contributes to C_{DS}, so that:

$$C_o = \frac{\epsilon_{\text{Si}} A_d}{20\ \mu\text{m}} = 100 \text{ pF} \qquad (18.85)$$

At lower drain voltages, where pinching off is not complete, and the cross-sectional area under the gate contributes to C_{Dn} rather than to C_{DS}, C_o might be a factor of $W_p/(W_p + W_g) \approx 2$ smaller.

Finally, we calculate C_{Gn}:

$$C_{Gn} = \frac{\epsilon_{\text{ox}} Z W_g}{t_{\text{ox}}} = 4.3 \text{ nF} \qquad (18.86)$$

These numbers correspond well to typical values exhibited by a 200-V MOSFET of this size.

18.14 THE BODY DIODE OF A VERTICAL POWER MOSFET

The connection of the p wells to the source metal gives the MOSFET an antiparallel "body" diode, as described in Section 15.3.2. You can see from Fig. 18.17 that this diode is a pin diode for a vertical power MOSFET. It therefore displays the static and dynamic characteristics of the pin diode that you studied in Section 17.2.

One important difference between the body diode and a pin power diode is that the body diode typically is not subjected to a very high current density. The cross-sectional area of the body diode is approximately the same as for that of the MOSFET, which is normally quite large for the current to be carried so that its resistance will be small. If we assume that the current density in the body diode is no larger than that in the MOSFET when on, the diode may not even be driven into high level injection. And it will not display a large forward recovery voltage because the resistance of the extended drain is so small (owing to its large area).

The body diode does display a reverse recovery phenomenon, however. Efforts to reduce the lifetime in the drift region through irradiation (which damages the crystal) have produced reverse recovery times on the order of 100 ns. But this recovery time is not as short as is possible in a separate diode, where the tradeoff of on-state resistance for reduced recovery time does not have to be made.

18.15 THE PARASITIC BIPOLAR TRANSISTOR

The presence of a parasitic bipolar transistor in a vertical power MOSFET was shown in Fig. 15.7. Even though the p-well of the MOSFET is shorted to its source, there is a lateral resistance through the p region between this short and the channel. This resistance, R_p, appears between the base and emitter terminals of the parasitic bipolar. It permits the bipolar transistor to turn on near the channel if the base current flowing through R_p is large enough to forward bias the p-well–n-source junction. Such a situation can happen when the rate of change of the drain voltage is high enough to give large capacitive charging currents. It can also happen if the device avalanches. Efforts to make W_p as small as possible to avoid wasted area also reduce R_p, so the problems associated with parasitic transistor action are becoming less of a concern for modern power MOSFETs.

18.16 THE INSULATED GATE BIPOLAR TRANSISTOR

Figure 18.24 shows the cross-sectional structure of the insulated gate bipolar transistor (IGBT). This device looks very much like the vertical MOSFET, except that the substrate is heavily doped p type rather than n type. As we showed in Fig. 15.11, the IGBT is a bipolar transistor with an integral MOSFET connected between its base and collector. Figure 18.24 shows which parts of the structure correspond to these two devices. The bipolar transistor is a pnp device, with its emitter at the substrate and its collector, the p-body region, connected to the top-layer metal, which is also the source contact for the n-channel MOSFET. The drain of the MOSFET is connected to the transistor's base (the drift region) through the neck region.

Integrating these two devices, rather than using discrete devices, has an advantage: When the IGBT is on, the BJT is also on, conductivity modulating the drift region and greatly reducing the drain resistance of the MOSFET. Had the MOSFET instead been a separate device, its unmodulated resistive drop would result in a higher collector–base, and therefore collector–emitter (on-state), voltage for the bipolar transistor.

One disadvantage of the integration is that the structure forces the transistor to have a wide, lightly doped base rather than a narrow, moderately doped base and a wide, lightly doped collector. Hence the device cannot take advantage of reach through, so the lightly doped region must be longer in the IGBT than it is in a normal bipolar transistor. This means that the IGBT can withstand its full off-state voltage rating in both directions.

Another disadvantage of the IGBT structure of Fig. 18.24 is that a pnp, rather than the superior npn, transistor structure is used. We can reverse the doping types to obtain an npn transistor, but only at the expense of producing a p-channel MOSFET—a higher resistance device than its n-channel counterpart.

A further problem is that integration of the two devices eliminates access to the base terminal of the BJT, preventing the use of negative base current to improve the turn-off transition time. Instead, when the MOSFET is turned off to initiate

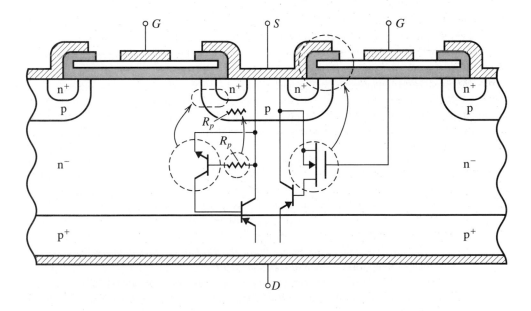

Figure 18.24   The structure of the insulated gate bipolar transistor (IGBT).

turn off of the IGBT, the IGBT current slowly tails to zero as the stored charge in the base of the BJT decreases, owing to the emitter defect. (Waveforms at turn off do show a small step in current corresponding due to cessation of the base and residual current carried by the MOSFET). Efforts to shorten this tail and to reduce its magnitude by reducing the gain of the transistor also increase the on-state voltage.

Perhaps most important, the integration of the two devices produces a parasitic thyristor. As we discuss in Chapter 19, we can model a thyristor as a regenerative connection between a pnp and an npn transistor, as shown in Fig. 18.24. Although the npn transistor's base (the p well) is shorted to its emitter, which should keep this junction from turning on, there is some resistance in this connection. If, during normal operation, the current flowing laterally through this resistor is high enough, the resulting voltage drop could turn the parasitic npn transistor on, and the parasitic thyristor structure would then latch. Once latched, there is nothing we can do by actions at the gate to turn the device off. This latching phenomenon limits the maximum current that an IGBT can carry. It is a limit that decreases as the temperature increases, in part because R_p, the resistance of the emitter–base short, increases. Also, if the rate of rise of voltage at turn off were high enough, the capacitive charging currents could trigger the thyristor.

Even with these problems, the IGBT has much to offer. It is a device well suited to high-voltage (> 400 V), moderate frequency (< 50 kHz) applications such as motor drives.

Notes and Bibliography

As in Chapter 17, Ghandi [1] and Baliga [2], give very extensive coverage of the material presented in this chapter. Ghandi provides a detailed analysis on the bipolar junction transistor, and Baliga discusses the MOSFET and IGBT. In addition, Grant and Gowar [3] is an excellent source of information on the power MOSFET and the performance tradeoffs involved in its design. Several application examples showing how the MOSFET performs in a circuit also are included.

In addition, there are several papers that provide more detail on many of the special topics of this chapter. Hower, et al [4,5,6] address the static and dynamic operation of the bipolar power transistor. Beatty, et al [7] gives a concise description of avalanche-induced second breakdown. Howard [8] and Fulop [9] give detailed descriptions of the avalanche process in a pn junction, and Sun, et al [10] show how to calculate the on-state resistance of a power MOSFET.

1. S. K. Ghandi, *Semiconductor Power Devices; Physics of Operation and Fabrication Technology* (New York: John Wiley & Sons, 1977).
2. B. J. Baliga, *Modern Power Devices* (New York: John Wiley & Sons, 1987).
3. D. A. Grant and J. Gowar, *Power MOSFETs; Theory and Applications* (New York: John Wiley & Sons, 1989).

4. P. L. Hower, "Optimum Design of Power Transistor Switches," *IEEE Trans. on Electron Devices* 20(4): 426–435 (April 1973).

5. P. L. Hower and W. G. Einthoven, "Emitter Current-Crowding in High-Voltage Power Transistors," *IEEE Trans. on Electron Devices* 25(4): 465–471 (April 1978).

6. P. L. Hower, "Application of a Charge-Control Model to High-Voltage Power Transistors," *IEEE Trans. on Electron Devices* 23(8): 863–868 (August 1976).

7. B. A. Beatty, S. Krishna, and M. S. Adler, "Second Breakdown in Power Transistors due to Avalanche Injection," *IEEE Trans. on Electron Devices* 23(8): 851–857 (August 1976).

8. N. R. Howard, "Avalanche Multiplication in Silicon Junctions," *J. Electron. Cont.* 13: 537–544 (1962).

9. W. Fulop, "Calculation of Avalanche Breakdown of Silicon pn Junctions," *J. Solid State Electron.* 10: 39–43 (1967).

10. S. C. Sun and J. D. Plummer, "Modeling of the On-Resistance of LDMOS, VDMOS, and VMOS Power Transistors," *IEEE Trans. Electron Devices* 27(2): 356–367 (February 1980).

Problems

18.1 Explain why a transistor structure that does not have a lightly doped ν-collector region, but instead has a wide base region into which the SCL grows when the collector–base junction is reverse biased, cannot take advantage of reach-through.

18.2 Assume that the base of an npn transistor is doped at $N_{AB} = 10^{17}/\text{cm}^3$ and that $W_B = 3\ \mu\text{m}$. If the distance from the centerline of the emitter to the base contact is $50\ \mu\text{m}$, calculate the net resistance (for a unit depth) for a negative base current that originates at the centerline and flows symmetrically left and right to the surrounding base contacts. What base current (per unit depth) would give a 0.6-V drop across the emitter–base junction at the centerline?

18.3 Determine the range of the ratio $V_{CEO(\text{sus})}/V_{CBO}$ of a pnp transistor having a range of gain of $5 < \beta_{\text{pnp}} < 20$.

18.4 Find R_ν for an npn transistor having $W_\nu = 20\ \mu\text{m}$, $N_{D\nu} = 2 \times 10^{14}/\text{cm}^3$, and $A_\nu = 1\ \text{cm}^2$. If this transistor is operating at $i_C = 50$ A, what is the voltage across R_ν? Find an approximate value for V_{CBO}, assuming a critical field of value $E_c = 2 \times 10^5$ V/cm.

18.5 Show that the relationship between γ and v_γ given by *(18.17)* is correct. (*Hint:* Break $v_{B\nu}$ into the sum of two parts, the first being the voltage that would make the excess carrier concentrations $n' = p' = N_{D\nu}$. The exponential of the second part therefore gives the factor γ by which the ν region is in high level injection.)

18.6 Using the equations and transistor parameters presented in Example 18.2, find J_{CS} and J_{CQ} for $v_{BE} = 0.67$ V, 0.65 V, and 0.62 V at $T = 25°\text{C}$.

18.7 Using the transistor parameters of Example 18.2 and the values for J_{CQ} and J_{CS} found in the previous problem, find v_{CE} at the forward-active and hard-saturation boundaries of quasi-saturation for $v_{BE} = 0.67$ V, 0.65 V, and 0.62 V.

540 Chapter 18 Power Transistors

18.8 Using the results of Problems 18.6 and 18.7, carefully graph and label the transistor's i_C-v_{CE} characteristics in the quasi-saturation region. *Note:* holding v_{BE} constant is the same as holding i_B constant since we are ignoring recombination in the base and ν regions. Furthermore, note that v_{CE} varies only because v_ν' varies, which itself depends on $J_C(1 - x'/W_\nu)$.

18.9 Using *(18.48)*, calculate β for $J_C = J_o$, $2J_o$, $3J_o$, and $4J_o$. Use the parameters of the transistor analyzed in Example 18.4. For each case, determine the error in the assumption $V_o = 1.8$ V (or, alternatively, $v_{BE} - v_\alpha = 0.2$ V).

18.10 Following the procedure used in Section 18.5.1, determine β for several numerical values of J_C for $v_{CE} = 1.0$ V, 1.5 V, and 3.0 V. Use the parameters of the transistor in Example 18.4. Graph these results and comment on how β at a given J_C changes with v_{CE}. *Note:* You may want to make corrections to your first assumption for V_o to get a more accurate answer, especially for the lower values of V_{CE}.

18.11 For our transistor of Examples 18.2–18.6, find how the high-current gain varies from its 300-K value as the temperature of the device is lowered to $T = -55°$C.

18.12 Qualitatively describe the bipolar transistor's turn-on transient. Derive equations that can be used to determine how long each phase of the transient takes. Using the numbers of Example 18.5, solve for these times and plot i_C and v_{CE} versus t.

18.13 Find the maximum power density $J_C V_{SB}$ above which second breakdown in a bipolar transistor occurs in terms of ϵ_{Si}, E_c, and v_{sat}.

18.14 If a MOSFET has a minimum p-well width of W_p and a minimum spacing between p wells of W_g, sketch and dimension the hexogonal cell and square cell patterns. Compare the number of cells each pattern can have per square centimeter of die. Compare the perimeter of the gate that each pattern can have per square centimeter of die. Note that the larger the gate perimeter, the lower the channel resistance.

18.15 If a MOSFET has dimensions $W_p = 25$ μm and $W_g = 15$ μm, find the length of the drift region h_d at which the boundary lines of the current flow expressed in *(18.75)* and *(18.76)* just meet under the p well. What is the approximate breakdown voltage of a MOSFET with this length of drift region?

18.16 Using the parameters provided in Examples 18.7 and 18.8, estimate the time constant $R_p C_{ppoly}$ to determine the time scale on which this internal impedance to the charging of the gate matters. Use only the oxide capacitance part for C_{ppoly}, as we did in *(18.83)*, and assume the doping concentration in the p well is 10^{17}/cm^3, and the depth of the source diffusion is 1 μm.

Chapter 19

Thyristors

IN this chapter we investigate the physics and design tradeoffs of thyristors. We begin by looking at the structure of the *silicon controlled rectifier* (SCR) and determine its reverse and forward voltage blocking capability. We then show how the regenerative turn-on process arises and determine the device's on-state characteristics. We explain the physical reasons for limiting a thyristor's dv/dt and di/dt. Next we discuss the gate turn-off (GTO) thryristor, and show how its structure differs from that of the SCR, enabling a negative drive at the gate terminal to turn the device off. Finally, we consider the structures of the reverse conducting thyristor (RCT), the triac, and the MOS controlled thyristor (MCT), a newly emerging power device.

19.1 THE SCR STRUCTURE

The SCR is a four-layer ($p^+n^-pn^+$) semiconductor device, as shown in Fig. 19.1(a). For the lower current ratings (10–100 A), the SCR is built on a small die of silicon, but for the higher current ratings (100–4000 A), the SCR is built on an entire wafer. The heavily doped p region on the bottom of the device is the *anode*, and the heavily doped n region on the top of the device is the *cathode*. The p region under the cathode is the *gate*, and it is connected to the gate metalization on the top surface as shown in Fig. 19.1(a). The interdigitation of the gate metalization with the cathode metalization is typically nonexistent or sparse in an SCR. Figure 19.1(b) shows the top view of a device whose gate is restricted to the center of the wafer. Known as a *center-gate geometry*, it is used extensively for low-frequency SCRs. The gate pattern shown in Fig. 19.1(c) is called an *involute*, which has the important property of constant spacing between two gate fingers across the entire device. A typical spacing is 3–5 mm for a wafer-sized device. Interdigitation using this or other patterns is utilized in high-frequency thyristors.

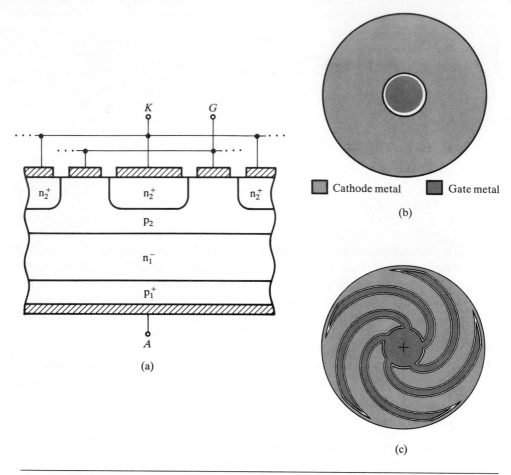

Figure 19.1   Structure of an SCR: (a) cutaway view showing the four layers and terminal connections;   (b) top view of center-gate geometry;   (c) top view of involute gate geometry.

19.2 OVERVIEW OF SCR OPERATION

The purpose of this section is to briefly describe the off-state characteristics, the regenerative turn-on process, and the on-state characteristics of the SCR. The sections that follow give greater detail on each aspect of the device's operation. Throughout these discussions we use the one-dimensional model of the SCR shown in Fig. 19.2(a). (Note that the anode is now drawn on the top.) For future reference we labeled each of the four layers (p_1, n_1, p_2, and n_2) and each of the three junctions (J_1, J_2, and J_3).

19.2.1 The Off State

When off, the SCR can block either a reverse or a forward anode–cathode voltage, v_{AK}. When v_{AK} is negative, junctions J_1 and J_3 are reverse biased (assuming that the gate terminal is left open), and J_2 is forward biased. The doping levels on each side of junction J_3 are very high, so this junction breaks down at a relatively low voltage. However, the n_1 region is both lightly doped and long. Therefore J_1 is able to block a large reverse anode–cathode voltage. As the n_1 region is much more lightly doped than the p_1 region, the SCL at J_1 grows mostly into this lightly doped n_1 region rather than into the more heavily doped p_1 region, as shown in Fig. 19.2(b).

When v_{AK} is positive, junctions J_1 and J_3 are forward biased and junction J_2, the only reverse biased junction, withstands the voltage. As the n_1 region is also more lightly doped than the p_2 region, the SCL again grows mostly into the n_1 region, as shown in Fig. 19.2(c).

The n_1 region is therefore used to block both polarities of voltage when the SCR is off. Its doping level and length must be chosen to give the desired breakdown voltage. Note that breakdown can occur in two ways: *punch through*, where the SCL extends far enough to reach the opposite edge of the n_1 region; and *avalanche*, where the peak electric field becomes high enough to cause impact ionization. Most

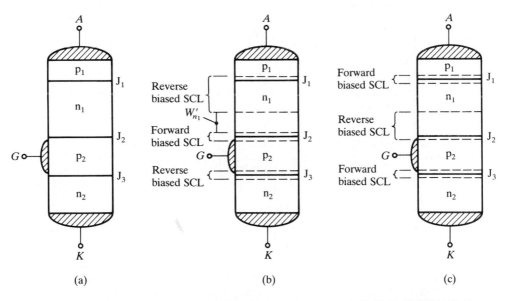

(a)   (b)   (c)

Figure 19.2   A one-dimensional model of the SCR: (a) the four layers and their junctions;   (b) space-charge layers when the SCR is blocking a reverse voltage;   (c) space-charge layers when the SCR is blocking a forward voltage.

SCRs are designed with the n_1 region long enough to cause avalanche to be the breakdown mechanism.

19.2.2 The Regenerative Turn-On Process

To understand how the SCR can go from a forward baised off state to its conducting on state, you should view the device as a cross connection of two transistors. Figure 19.3(a) shows how the four-layer device can be considered as the interconnection of two three-layer devices—one a pnp and the other an npn transistor. The base of each transistor is connected to the collector of the other. Figure 19.3(b) illustrates this two-transistor SCR model.

Note that the base of each transistor is driven with a current that is β times its collector current. As long as the product $\beta_1\beta_2 > 1$, we can show that—once some current flows—the two transistors will drive each other harder and harder until they saturate. A short pulse of current into the gate terminal is one way to start this regenerative process.

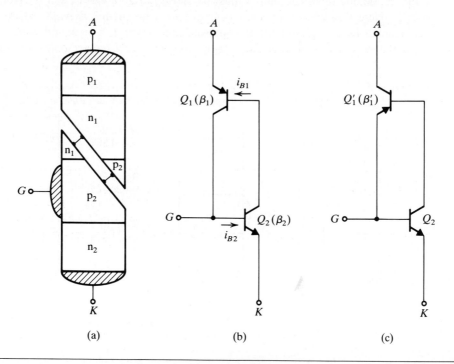

(a) (b) (c)

Figure 19.3 A two-transistor model of the SCR: (a) the four-layer device represented as the interconnection of two three-layer devices; (b) the two-transistor model of the interconnection of (a); (c) the terminals of Q_1, renamed to simplify the discussion of the role of J_1 in blocking $v_{AK} < 0$. The inverted transistor Q_1' has gain β_1'.

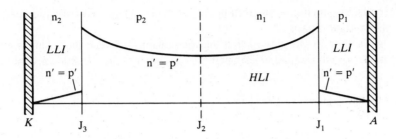

Figure 19.4 Excess carrier concentrations in an SCR in its on state. The two middle regions are in high-level injection, causing the device to behave
like a pin diode.

19.2.3 The On State

As the two transistors drive each other into saturation, the excess carrier concentrations in their base regions reach high-level injection. At this point, the actual
doping concentrations in the base regions are no longer relevant, and the SCR now
behaves as a three-layer pin diode, with the two middle layers of the thyristor,
n_1 and p_2, corresponding to the intrinsic region of the diode. Figure 19.4 shows
typical excess carrier concentrations in the SCR under this condition. Everything
you learned in Chapter 17 about the pin diode's forward voltage drop as a function
of current and design parameters applies as well to the SCR in its on state.

19.3 THE OFF STATE

We now discuss the off state of the SCR in more detail, starting first with the
reverse breakdown voltage V_{BR} and then considering the forward blocking case.
The SCR does not actually break down in the forward direction; instead, it turns
on. This process is known as *breaking over*, and the voltage at which it occurs is
the *break-over voltage* V_{BO}. Finally, we consider the temperature dependence of
these voltage limits.

19.3.1 The Reverse Blocking Voltage

Junction J_1, which supports most of the SCR's reverse voltage, is effectively the
base-emitter junction of the pnp transistor Q_1 in Fig. 19.3(b). Note that the p_2
region, which forms the collector of Q_1, is incrementally connected to the cathode
of the SCR because J_3 quickly avalanches (at about 10–30 V) when the SCR is
reverse biased.

To illustrate the effect of the pnp transistor structure on the reverse breakdown
voltage of the SCR, we now interchange the roles of the p regions of Q_1 and call
the p_1 region the collector and the p_2 region the emitter, as shown in Fig. 19.3(c).
This renaming brings us back to the familiar situation of a pnp transistor whose

base-collector junction is supporting the off-state voltage. The forward current gain of this configuration is β_1'.

In Chapter 18 we showed that the relationship between the sustaining voltage of a transistor whose base is left open ($V_{CEO(\text{sus})}$) and the breakdown voltage of the base-collector junction (V_{CBO}) is given by (18.5), repeated here for convenience:

$$V_{CEO(\text{sus})} = \frac{V_{CBO}}{(1 + \beta_F)^{1/m}} \qquad (19.1)$$

In this equation, $m = 4$ for an npn transistor and $m = 6$ for a pnp transistor. Because the base of the pnp transistor in the SCR is open (there is no external connection to the n_1 region), we can expect (19.1) to apply to the breakdown of J_1.

The structure of Q_1 is different from that of a standard pnp BJT. It does not have a lightly doped drift region in the collector. Instead, the SCL grows into the base (n_1) region. As it does, the effective length of the base decreases, and the gain of the transistor increases. This dependency keeps us from determining, in closed form, by how much the reverse breakdown voltage is reduced from V_{CBO}.

Nevertheless, we can note several points regarding the dependency of β_1', the gain of Q_1', on the reverse anode–cathode voltage and the leakage current. Recall from Chapter 16 that:

$$(\beta + 1) = \frac{1}{1 - \alpha} = \frac{1}{1 - \gamma_E \alpha_T} \qquad (19.2)$$

Because the base region is so long, particularly in very high-voltage SCRs, we can no longer assume that the base transport factor α_T is unity. Instead, it is given by:

$$\alpha_T = \frac{1}{\cosh\left(W_{n_1}' / L_h\right)} \qquad (19.3)$$

where W_{n_1}' is the width of the base region that has not been depleted (see Fig. 19.2b) and L_h is the characteristic diffusion length of holes in the base. Thus (19.3) shows that α_T approaches 1 (causing β to increase) as the reverse voltage increases and W_{n_1}' decreases. This phenomenon makes the breakdown more abrupt than (19.1) suggests.

In Chapter 16 we found that the emitter injection efficiency was

$$\gamma_E = \frac{1}{1 + \delta_E} \qquad (19.4)$$

where

$$\delta_E = \frac{D_e N_{DB} W_{n_1}'}{D_h N_{AE} W_{p_2}} \qquad (19.5)$$

Again, β_1' increases as W_{n_1}' becomes shorter with increasing reverse voltage.

Actually, *(19.5)* is correct only if we ignore recombination in the base-emitter (J_1) SCL. Additional base current is needed to supply the electrons for this recombination, and the injection efficiency is therefore reduced. This current component cannot be neglected at the very low current level at which we want to keep the leakage current of the SCR. As a result, at very small currents, γ_E is lower than *(19.5)* predicts, and therefore β_1' is smaller than we would have expected. But as breakdown occurs and the leakage current increases, this SCL recombination current becomes less important and β_1' increases. This phenomenon also makes the breakdown more abrupt than *(19.1)* suggests.

EXAMPLE 19.1

Calculating the J_2 Recombination Current

In this example we show the importance of the recombination current in the base-emitter SCL of Q_1' by comparing it to the electron current injected back into the emitter from the base. Let's assume that the doping concentration of region p_2 is $N_A = 10^{16}/cm^3$, that its length is $W_{p_2} = 50~\mu m$, that the doping concentration of region n_1 is $N_D = 5 \times 10^{13}/cm^3$, that its length is $W_{n_1} = 300~\mu m$, and that the carrier lifetime in the SCL is $\tau_{SCL} = 30~\mu s$.

Consider this SCR to be blocking a reverse voltage ($v_{AK} < 0$ at $T = 300$ K), which causes the J_1 SCL to extend 200 μm into the n_1 region ($W_{n_1}' = 100~\mu m$). If the voltage across the base-emitter junction of Q_1' (J_2) is $v_{BE} = 0.2$ V, the hole current injected into the n_1 base region is

$$J_h = \frac{qD_h n_i^2}{N_D W_{n_1}'}\left(e^{q(0.2)/kT} - 1\right) = 1.93~\mu A/cm^2 \qquad (19.6)$$

As the base transport factor is close to unity, this current is also the leakage current flowing through the SCR.

At $v_{BE} = 0.2$ V, the width of the depletion layer is

$$x_n + x_p \approx x_n \approx \sqrt{\frac{2\epsilon_{Si}(\psi_o - 0.2)}{qN_D}} = 3.16~\mu m \qquad (19.7)$$

where ψ_o, the junction's built-in potential, is 0.6 V. The recombination current within this forward biased SCL is the (negative) excess charge divided by the lifetime in the SCL:

$$J_{SCL(rec)} = \frac{qn_i x_n}{2\tau_{SCL}}\left(e^{q(0.2)/2kT} - 1\right) = 0.529~\mu A/cm^2 \qquad (19.8)$$

The electron current injected into the emitter, which diffuses across the emitter, is

$$J_e = \frac{qD_e n_i^2}{N_A W_{p_2}}\left(e^{q(0.2)/kT} - 1\right) = 0.045~\mu A/cm^2 \qquad (19.9)$$

For this example, *(19.8)* and *(19.9)* show that the recombination current dominates the electron-diffusion current at room temperature.

The emitter injection efficiency is

$$\gamma'_E = \frac{J_h}{J_h + J_e + J_{SCL(rec)}} = 0.77 \qquad (19.10)$$

If $\alpha'_T = 1$, the gain of Q'_1 is $\beta'_1 = 3.36$. With this value of β'_1, the reverse breakdown voltage of J_1 is

$$V_{BR} = \frac{V_1}{(\beta'_1 + 1)^{1/6}} = 0.78V_1 \qquad (19.11)$$

where V_1 is the voltage at which J_1 would break down if no transistor action were present.

As breakdown commences and the reverse leakage current increases, the effect of the recombination current on the emitter injection efficiency is reduced, and the breakdown characteristic becomes more abrupt.

19.3.2 The Forward Blocking Voltage

When the SCR is blocking a forward voltage ($v_{AK} > 0$), J_2 is reverse biased, while J_1 and J_3 are forward biased. Again, we can expect that transistor action reduces the breakdown voltage to below the inherent capability of junction J_2, but now we have *both* the pnp and the npn transistors operating in their forward active regions. In the discussion that follows we use the terminal definitions of Fig. 19.3(b) for the transistors Q_1 and Q_2.

Figure 19.5 shows the two-transistor model of the thyristor with a full Ebers–Moll model for each transistor. Because both transistors are well into their forward active regions, their collector currents are

$$i_{C_1} = -\alpha_{F_1} i_{E_1} - I_{CO_1} \qquad (19.12)$$
$$i_{C_2} = -\alpha_{F_2} i_{E_2} + I_{CO_2} \qquad (19.13)$$

The SCR anode current is i_{E_1} and the cathode current is $-i_{E_2}$. If $i_G = 0$, which it should be when the SCR is blocking, $i_A = -i_K$ and *(19.12)* and *(19.13)* can be solved for i_A as a function of I_{CO_1}, I_{CO_2}, α_{F_1}, and α_{F_2}:

$$i_A = \frac{I_{CO_1} + I_{CO_2}}{1 - \left(\alpha_{F_1} + \alpha_{F_2}\right)} \qquad (19.14)$$

This is the forward leakage current of the SCR and, to keep it small, $\alpha_{F_1} + \alpha_{F_2} \ll 1$. If $\alpha_{F_1} + \alpha_{F_2} = 1$, *(19.14)* shows that the SCR will enter sustained breakdown—the anode current becomes unconstrained for any values of leakage currents.

In writing *(19.2)* we assumed that $\alpha = \gamma_E \alpha_T$. Actually, if the transistor has an impact ionization multiplication rate of M in the collector SCL, then $\alpha = M\gamma_E \alpha_T$ (see Section 18.2). As Q_1 and Q_2 have a common collector SCL, they both have

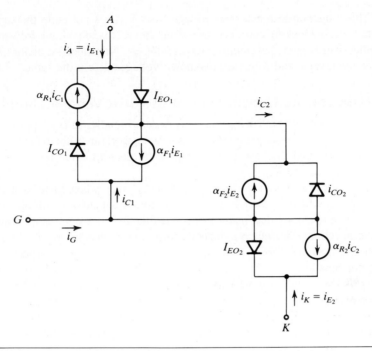

Figure 19.5 The two-transistor model of the SCR, with Q_1 and Q_2 replaced by their Ebers–Moll models.

the same M. Therefore we can rewrite the breakdown condition as:

$$M\gamma_{E_1}\alpha_{T_1} + M\gamma_{E_2}\alpha_{T_2} = 1 \qquad (19.15)$$

$$M = \frac{1}{\gamma_{E_1}\alpha_{T_1} + \gamma_{E_2}\alpha_{T_2}} \qquad (19.16)$$

Both transistors are amplifying the ionization current, so the voltage corresponding to (19.16) is lower than it would be if only one transistor were active. When the SCR is blocking a reverse voltage ($v_{AK} < 0$), the transistors do not share collector SCLs. As a result, the forward breakdown voltage of the SCR would be lower than its reverse breakdown voltage if the device were symmetrical. However, as we show later in this chapter, most SCRs have special design features that make α_{F_2} very small compared to α_{F_1} at low currents. The forward break-over voltage V_{BO} is therefore reduced by the transistor action of only Q_1 and is given by:

$$V_{BO} = \frac{V_{BR_2}}{(\beta_1 + 1)^{1/6}} \qquad (19.17)$$

where V_{BR_2} is the inherent breakdown voltage of junction J_2.

This reduction in break-over voltage from V_{BR_2} is not quite the same as it is for the reverse blocking case, because when the SCR is blocking a forward voltage, J_1 rather than J_2 is the base-emitter junction of Q_1. Nevertheless, making α_{F_2} small makes the reverse and forward breakdown voltages nearly the same.

19.3.3 Temperature Dependence of the SCR Voltage Limits

We have shown that both V_{BR} and V_{BO} become smaller as α_{F_1} approaches unity. Because $\alpha_{F_1} = \gamma_{E_1}\alpha_{T_1}$, we now look at how temperature affects the emitter injection efficiency and the base transport factor to determine the temperature dependencies of V_{BR} and V_{BO}.

The base transport factor is very close to unity when there is a large voltage across the SCR and most of the n_1 region is depleted, resulting in a narrow base. The temperature dependence of α_T is therefore not significant. However, the emitter injection efficiency γ_E depends strongly on temperature. As Example 19.1 showed, at low currents, γ_E is much smaller than we would expect because of the recombination that takes place in the SCL of the base-emitter junction. As the SCR gets hotter, this recombination current, $J_{SCL(rec)}$, increases, but not as fast as does the electron-diffusion current injected back into the emitter, or J_e given by (19.9). In the vicinity of 125°C, J_e begins to dominate $J_{SCL(rec)}$, and γ_E therefore approaches 1. The resulting increase in β drastically reduces the voltage limits of the SCR. For this reason, most SCRs have a maximum rated junction temperature of only 125°C.

EXAMPLE 19.2

Determining the Effect of Temperature on V_{BR}

For the SCR parameters and the operating conditions specified in Example 19.1, what is the reverse breakdown voltage, V_{BR} at 125°C?

We found in Example 19.1 that at $T = 25°C$, $J_{SCL(rec)} = 0.529$ $\mu A/cm^2$, $J_e = 0.045$ $\mu A/cm^2$, and $\gamma_E = 0.77$. At 125°C, $n_i = 4 \times 10^{12}/cm^3$, and these currents are

$$J_{SCL(rec)} = \frac{qn_i x_n}{2\tau_{SCL}}\left(e^{q(0.2)/2kT} - 1\right) = 58.4 \ \mu A/cm^2 \tag{19.18}$$

$$J_e = \frac{qD_e n_i^2}{N_A W_{P_2}}\left(e^{q(0.2)/kT} - 1\right) = 382 \ \mu A/cm^2 \tag{19.19}$$

In (19.19) we assumed that $D_e \propto T^{-1.2}$. Note that the electron diffusion current J_e is much larger than the recombination current.

The hole-diffusion current in the base (the n_1 region) is

$$J_h = \frac{qD_h n_i^2}{N_D W'_{n_1}}\left(e^{q(0.2)/kT} - 1\right) = 16.4 \ mA/cm^2 \tag{19.20}$$

which gives an emitter injection efficiency of:

$$\gamma_E = \frac{J_h}{J_h + J_e + J_{\text{SCL(rec)}}} = 0.974 \qquad (19.21)$$

Assuming that $\alpha_T = 1$, as we did in Example 19.1, and using the value of γ_E given by (19.21) in (19.2), we find that $\beta_1 = 37$. Substituting this value of β into (19.1) with $m = 6$ gives a reverse breakdown voltage at $T = 125°C$ of:

$$V_{BR} = 0.55V_1 \qquad (19.22)$$

where V_1 is the the breakdown voltage of J_1 if it were an isolated junction. This magnitude is substantially lower than $V_{BR} = 0.78V_1$ at 25°C, which we calculated in Example 19.1.

19.4 THE TURN-ON PROCESS

In Section 19.2.2 we described the SCR's regenerative turn on process. All that is required to turn on the SCR is that $\beta_1\beta_2 > 1$ and that a small cathode current be present to initiate regeneration. The gain requirement can be restated in terms of α as:

$$1 - (\alpha_{F_1} + \alpha_{F_2}) \leq 0 \qquad (19.23)$$

From the analysis of the off state, however, we found that we want the α's to be small to maximize V_{BR} and V_{BO}. Furthermore, when the SCR is in its forward blocking state, leakage current would be sufficient to initiate turn on if (19.23) were satisfied. Clearly, the regenerative turn-on conditions cannot exist when we want the SCR to be off. That is, when the SCR is in its forward blocking state, the condition that must be satisfied is

$$1 - (\alpha_{F_1} + \alpha_{F_2}) > 0 \qquad (19.24)$$

We can resolve this apparent conflict by looking at the dependence of the α's on emitter current. As Examples 19.1 and 19.2 showed, the α's are small at the very low current levels of the off state. However, at higher current levels, where the recombination in the base-emitter SCL is less important, the α's increase. Therefore, to obtain regeneration in the SCR, we must raise the anode current to the point where the sum of the α's exceeds 1.

One way to trigger the regenerative process is to supply some gate current to the SCR. Using the two-transistor model shown in Fig. 19.5 and following the analysis that resulted in (19.14), we can obtain the following expression relating gate current to anode current:

$$i_A = \frac{i_G + I_{CO_1} + I_{CO_2}}{1 - \left(\alpha_{F_1} + \alpha_{F_2}\right)} \qquad (19.25)$$

If i_G is sufficient to increase i_A to the value where $(\alpha_{F_1} + \alpha_{F_2}) = 1$, regeneration is initiated and the gate current no longer is needed to sustain the turn-on process. At this point the SCR is said to be *latched* in its on state.

19.4.1 The Gate-Cathode Short

One method of controlling the gain of Q_2 to be very small at low currents is to utilize a *gate-cathode short*. This short can be accomplished by placing a resistor across the base-emitter junction of Q_2, or, equivalently, across the gate-cathode junction of the SCR. This gate-cathode short, or simply *cathode short*, as it is called, is actually accomplished on the silicon die by bringing the p_2 gate region up to the surface and in contact with the cathode metalization, as shown in Fig. 19.6(a). Because this contact is removed from the active gate region by some distance, the cathode "short" actually has some resistance, R_{gk}, as the schematic representation of Fig. 19.6(b) shows.

The cathode short functions in the following manner. In the forward blocking state, the J_2 leakage current will forward bias the base-emitter junction of Q_2. As this junction's voltage rises, the cathode short diverts some of the leakage current

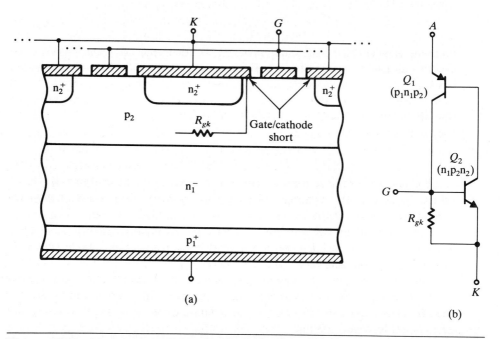

(a)

(b)

Figure 19.6   (a) A gate-cathode short accomplished on the die by making contact between the gate region and the cathode metalization.   (b) The two-transistor SCR model showing the location of the gate-cathode short R_{gk}.

out of the p_2 base, thereby reducing the current that is mulitplied by transistor action. In effect, the gain of the transistor is reduced. Let's assume that Q_2 abruptly turns on when its base-emitter voltage is 0.7 V. When the transistor is on, the current flowing through the cathode short is $0.7/R_{gk}$. If the total leakage current flowing through the SCR is smaller than this value, the base emitter voltage must be less than 0.7 V, and Q_2 will in effect be shorted by R_{gk}. Therefore the goal of the designer is to make $0.7/R_{gk}$ larger than the maximum leakage current expected when the SCR is in the forward blocking state.

To turn the SCR on, however, we need Q_2 to contribute to the regenerative process. This contribution will not occur until the current flowing through the SCR is approximately $0.7/R_{gk}$. Because this value of current is usually exceeded by i_G, the SCR will be turned on by the gate drive. But if the gate drive is removed before the anode current is large enough to keep Q_2 in its forward active region, the regenerative process stops and the SCR returns to its off state. The level of the anode current required for the SCR to remain in the on state when the gate drive is removed is called the *latching current*, I_L. Similarly, if the SCR is on and gate drive has been removed—and its anode current should fall below some critical level—it would turn off because of failure of the regenerative process. The anode current at which this occurs is called the *holding current*, I_H.

Although our simple description here suggests that $I_L = I_H \approx 0.7/R_{gk}$, the value of R_{gk} is slightly different for the turn-on and turn-off processes, owing to the differences in excess charge concentrations in the p_2 region. For a 100-A device, I_L and I_H are typically 100–300 mA, with $I_H < I_L$.

19.4.2 Current Spreading During Turn On

The SCR is normally turned on—or triggered—by gate current, which supplies base charge to Q_2, the collector of which then supplies base charge to Q_1, and anode current begins to flow. When this anode current gets large enough, $\alpha_{F_1} + \alpha_{F_2}$ approaches unity, and the regenerative process takes over. At this point the anode current quickly increases, being limited only by the external circuit, and the anode–cathode voltage collapses.

As shown in Fig. 19.7, the turn-on process first occurs near the gate fingers, because time is needed for the excess carrier concentration in the two middle base regions to spread over the regions between adjacent gate fingers. A spreading time of 0.5–5 μs is typical, depending on the finger separation. Because the anode current is initially focused into a very small cross-sectional area of the device near the gate fingers, the current density in that region can be very high, and damage can result from localized heating. Therefore we have to keep the anode current from reaching its full rated value until the SCR has had time to turn on across its full cross-sectional area. For this reason, there is a limit to di_A/dt, the rate at which the anode current of an SCR can be allowed to increase during turn on. Rates of 1000–2000 A/μs are typical for medium size (100 A) devices.

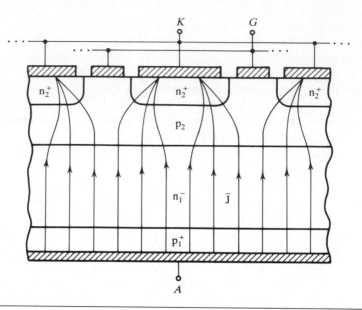

Figure 19.7   An illustration of current, focusing in the vicinity of the gate fingers during the first stages of the SCR turn-on process.

To improve the di_A/dt rating of an SCR, we minimize the distance between gate fingers. Lithographic capability and process yields limit how far this approach can be taken, but patterns such as the involute, shown in Fig. 19.1(c), help by keeping the finger spacing constant across the entire device. Of course, the closer the gate finger spacing is, the less is the area available for the cathode contact.

The amplitude of the gate current—and the rate at which it rises to this value— both influence the spreading rate. A di_A/dt rating is usually accompanied by a specified gate-current amplitude and rise time. If these specifications are not met, the device may fail if subjected to its rated di_A/dt. Many SCRs can be operated beyond their di_A/dt ratings if their gate drives provide higher currents and faster rise times than those specified with the di_A/dt rating.

19.5 THE TURN-OFF PROCESS

Once the SCR is on, it behaves as a pin diode, because the two center regions are well into high-level injection for typical anode currents. When the power circuit tries to reverse the anode current, the SCR displays a reverse recovery characteristic similar to that of the pin diode. Current will flow in the negative direction until the excess charge stored in the center regions of the SCR is either removed or lost to recombination.

The same tradeoff made between the speed of recovery and the on-state voltage of a pin diode can be made for an SCR. *Rectifier grade* SCRs trade speed for a reduced forward drop. They generally have a simple center-gate electrode geometry, and are intended for line commutation applications of up to about 400 Hz. *Inverter grade* SCRs are intended for high-frequency forced-commutation applications. These devices generally are constructed with a highly interdigitated gate geometry. The lifetime for excess carriers in the base regions is made relatively short to give a fast reverse recovery at the expense of a higher forward voltage drop. Inverter-grade SCRs generally have a forward drop that is about 50% higher than a rectifier grade device with comparable voltage and current ratings.

19.5.1 Turn-Off Time and Rate Effects

When the SCR is off and blocking a forward voltage, it can be pushed into its regenerative turn-on mode if the current flowing through the device exceeds a critical value. One way to generate this current (besides driving the gate terminal) is to apply a positive change in voltage across the SCL of junction J_2. As Fig. 19.8(a) shows, the junction can be modeled as having a parallel capacitance (because of the SCL), and the current flowing through this capacitance when dv_{AK}/dt is positive acts just like a gate current. Therefore, if dv_{AK}/dt is large enough, the regenerative turn-on mechanism in the SCR can be triggered. For this reason, there is a maximum specified off-state dv_{AK}/dt above which the device is not guaranteed to remain off. Rates of 1000–2000 V/μs are typical for inverter grade devices.

Note that the cathode short, which reduces the gain of Q_2, also helps to make the SCR less sensitive to dv_{AK}/dt. In addition, applying a negative voltage across the gate-cathode terminals when the SCR is off also helps reduce the chance of false triggering.

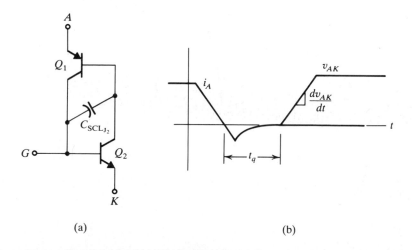

(a)   (b)

Figure 19.8   (a) Two-transistor model of an SCR showing the capacitance of the SCL at junction J_2.   (b) Reapplied forward voltage at the end of the minimum turn-off time t_q.

Because the J_2 SCL capacitance is nonlinear, the maximum allowable dv_{AK}/dt is specified at the largest value of this capacitance. This condition occurs when $v_{AK} = 0$. Moreover, the SCR is most sensitive to triggering immediately after it has been turned off, as excess charge still exists in the middle base regions. For these reasons, the maximum allowable dv_{AK}/dt is specified at a fixed time after the anode current reverses and the SCR begins its reverse recovery phase. This time is called the *circuit commutated turn-off time* and is designated by t_q. A forward voltage can safely be reapplied (meaning that retriggering will not take place) at the specified rate so long as it is delayed for a time t_q past the time that the anode current reverses. The relationships among i_A, v_{AK}, and t_q during the turn-off process are shown in Fig. 19.8(b). Typical values for t_q are 100–200 μs for rectifier-grade SCRs and 10–20 μs for inverter grade SCRs.

The circuit commutated turn-off time t_q depends on whether the circuit in which the SCR is connected permits the flow of significant reverse recovery current. For example, Fig. 19.9(a) shows a situation in which reverse recovery current is permitted to flow, aiding the SCR forward blocking recovery process by sweeping excess carriers from the base regions. Figure 19.9(b) illustrates a situation in which an antiparallel diode is connected across the SCR (a common configuration). The

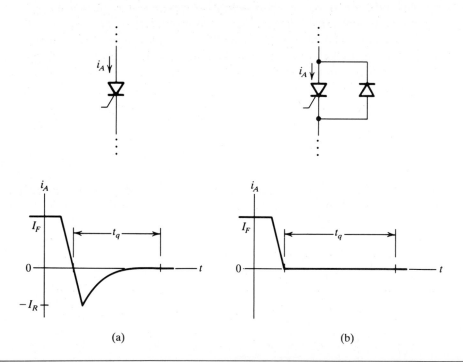

(a)   (b)

Figure 19.9 (a) A case in which reverse recovery current flows in the SCR, minimizing t_q. (b) A case in which an antiparallel diode prevents the flow of reverse recovery current, increasing t_q.

switch current will reverse through the diode, but little reverse current will be carried by the SCR. The result is that elimination of the charge stored in the base regions of the SCR is not aided by reverse current. Instead, the charge will decay through recombination at a rate determined by the excess carrier lifetimes in the base regions. In this case, t_q will be longer than in the case of Fig. 19.9(a). Data sheets generally give t_q under both conditions for inverter-grade SCRs, and their values differ by 50–100%.

19.6 THE GATE TURN-OFF (GTO) THYRISTOR

So far we have assumed that a thyristor, once it is on, will not turn off until the power circuit drives its anode current to zero. But the two-transistor thyristor model of Fig. 19.3(b) implies that if we draw enough current out of the gate terminal of the device, we can reduce i_K to a value insufficient to maintain the regenerative process and the thyristor will turn off. We cannot do this in a conventional SCR, because the gate terminal affects only the region of the device in the immediate vicinity of the gate contact. The lateral voltage drop that would accompany the flow of current from regions remote from the gate contact effectively prevents the behavior of these regions to be influenced by anything happening at the gate terminal. Therefore, while a negative gate current might stop the regenerative process in the immediate vicinity of the gate, the rest of the SCR would remain on. However, by using special techniques, including dense interdigitation, we can design a thyristor to be turned off by a large negative gate current. Such a device is known as a *gate turn-off* (GTO) *thyristor*.

We now determine the condition necessary to turn off a thyristor from its gate. If i_{GR} is the current drawn from the gate terminal—and if we ignore leakage currents—the base current for Q_2, the npn transistor, is

$$i_{B_2} = \alpha_{F_1} i_A - i_{GR} \tag{19.26}$$

If this base current is less than that needed to support the collector current of Q_2, or $\alpha_{F_2} i_K$, then Q_2 turns off and interrupts regeneration in the thyristor. In other words, turn off occurs when:

$$\alpha_{F_1} i_A - i_{GR} \leq (1 - \alpha_{F_2}) i_K \tag{19.27}$$

As $i_K = i_A - i_{GR}$, the condition in *(19.27)* occurs when:

$$i_{GR} \geq \frac{\alpha_{F_2} + \alpha_{F_1} - 1}{\alpha_{F_2}} i_A \tag{19.28}$$

A useful figure for specifying how easy it is to turn off a GTO thyristor is the *turn-off gain*, β_{off}:

$$\beta_{\text{off}} = \frac{i_A}{i_{GR}} = \frac{\alpha_{F_2}}{\alpha_{F_2} + \alpha_{F_1} - 1} \tag{19.29}$$

You should not forget that the α's are functions of current density and anode voltage, so *(19.29)* does not give a simple answer to how much reverse gate current is needed to turn off a given anode current. It does suggest, however, that to achieve large values of β_{off}, α_{F_2} should be made as close to unity as possible, and α_{F_1} as small as possible, given the competing desires to maintain a high off-state voltage rating and a low on-state voltage drop.

The preceding discription of the turn-off process is valid only for a one-dimensional model of a thyristor. In an actual device, when the negative gate current initially interrupts the regenerative process near the gate contact, the cathode current is simply focused into the regions between gate fingers, as Fig. 19.10 shows. The anode voltage rises slightly because of the higher current density that results, but it is not until the active region of the GTO thyristor is reduced to a small filament that the GTO thyristor begins to turn off. The very high current density in the filament can lead to a destructive thermal runaway. For this reason, a snubber is usually placed across a GTO thyristor to divert the anode current when the anode voltage starts to rise.

Another consequence of cathode current focusing is that the gate current must flow laterally through the p_2 base region. As it does, a lateral voltage drop, v_L, occurs from the region of focused cathode current to the gate finger, as shown in Fig. 19.10. If large enough (typically 10 V), this voltage drop will avalanche the

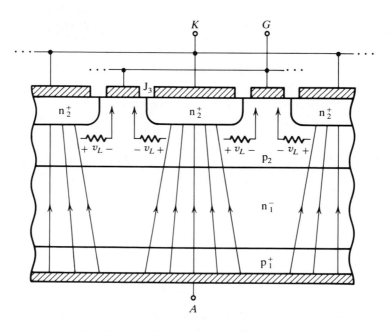

Figure 19.10 Current distribution in a GTO thyristor during turn off. The lateral voltage drop v_L caused by negative gate current focuses the anode current into a restricted region of the cathode.

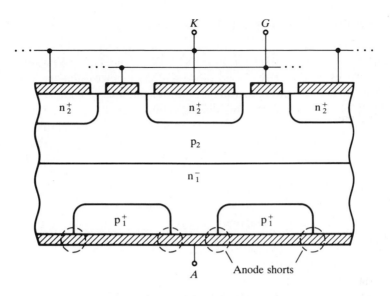

Figure 19.11   A GTO thyristor structure with anode shorts to kill the regenerative ability of the device in regions in which the anode current is focused during turn off.

gate-cathode junction, J_3, near the gate finger. Once this occurs, further increases in gate current are diverted through the avalanching region of the junction and do not contribute to the turn-off process. This limit to the maximum effective reverse gate current limits the maximum anode current that can be turned off.

To improve the turn off process as the cathode current is focused into a filament, the anode structure of a GTO thyristor is modified as shown Fig. 19.11. The n_1 base region is shorted to the anode in regions opposite the cathode contacts. This *anode short*, like the cathode short, is slightly resistive, and it serves to reduce the gain of the pnp transistor Q_1. More important, however, is the fact that when the anode current is focused during turn off, it now focuses into an area of the GTO thyristor that has only three layers, rather than four. The regenerative property of the device is therefore lost, and the turn-off process is enhanced. The penalties of using an anode short are that the on-state voltage of the device is increased slightly because of the lower gain of the pnp transistor and that the GTO thyristor can no longer block a reverse voltage larger than (typically) 50 V.

19.7 OTHER THYRISTOR STRUCTURES

To conclude this chapter, we briefly consider the structure of three other members of the thyristor family: the *reverse conducting thyristor* (RCT), the TRIAC, and the *MOS controlled thyristor* (MCT).

19.7.1 The Reverse Conducting Thyristor

A reverse conducting thyristor (RCT) is an SCR containing both a cathode short and an anode short, as shown in Fig. 19.12. When blocking a forward voltage, the RCT behaves as a reverse biased diode and does not exhibit the gain enhanced leakage current of conventional SCRs with or without cathode shorts. Therefore, for the same n_1 region length and doping concentration, the RCT exhibits a higher V_{BO} rating than that of a conventional SCR, and this rating can be specified at a higher temperature (150°–175°C, instead of 125°C). In addition, the shorts across the base-emitter junctions of both the npn and pnp transistors in the RCT give it a value of t_q that is much smaller than that for a conventional SCR.

The price paid for the improved speed and forward voltage rating of the RCT is its inability to block reverse voltage. In the reverse mode, the RCT acts like a forward biased diode composed of regions p_2 and n_1. The RCT is intended to replace the SCR–diode combination of Fig. 19.9(b) found in many applications.

19.7.2 The TRIAC

We described the operating characteristics of the TRIAC in Section 15.3.3. The device is a five-layer structure, as shown in Fig. 19.13. The MT_1 and MT_2 contacts short the p_1 region to the n_3 region and the p_2 region to the n_2 region. The result is the equivalent of two antiparallel SCR structures: $p_1n_1p_2n_2$ and $n_3p_1n_1p_2$. The n_G region is the TRIAC gate, and it is shorted to the p_2 region by the gate contact.

The possible modes of operation of the TRIAC are defined in Section 15.3.3 as I+, I−, III+, and III−. In mode I+, the device is operating as a conventional

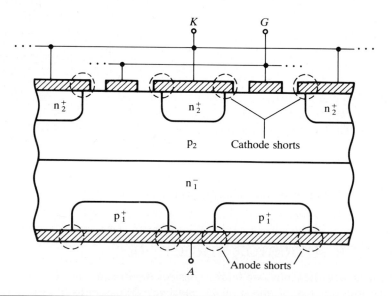

Figure 19.12   The structure of an RCT.

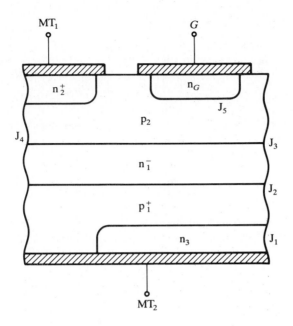

Figure 19.13 The TRIAC structure.

SCR, with the anode being the p_1 region. The gate is the p_2 region. It is connected directly to the gate terminal, because the gate metal extends beyond the edge of the n_G diffusion, shorting the terminal to the p_2 region. Therefore triggering in this mode is effected in the normal manner, and this is the mode having the most sensitive triggering characteristic.

The III− is the next most sensitive mode. In this mode the n_3 region is the cathode and the p_2 region the anode. Triggering is accomplished by pulling the gate negative with respect to MT_1. This forward biases junction J_5, which functions as the base-emitter junction of the npn transistor composed of n_1, p_2, and n_G. The injected current (electrons) is collected at J_3, which is reverse biased in quadrant III. The resulting electron flow into the n_1 region is effectively base current for the $p_1n_1p_2$ transistor, and it begins to turn on, initiating regeneration and turning the TRIAC on.

19.7.3 The MOS Controlled Thyristor

The MOS controlled thyristor (MCT) is a device that uses integral MOSFETs to either turn on or turn off the thyristor. It is a newly emerging device that combines the high current density and ruggedness of a thyristor with the ease of control of an MOS gate. Several different structures can be used to fabricate the MCT, and which of these possible structures is best is not yet clear.

Figure 19.14(a) shows one way that MOSFETs could be used to control a thyristor. The thyristor is represented by its two-transistor model, and the n-channel

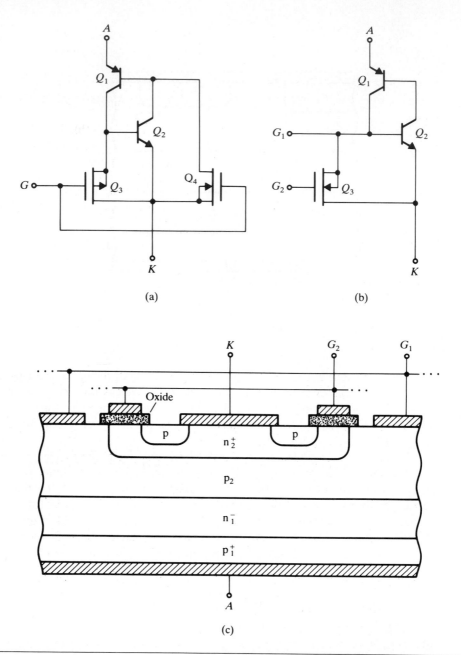

Figure 19.14 The MOS controlled thyristor: (a) a model of a structure using
two MOSFETs.   (b) a model of a structure that is turned on as
a normal SCR but is turned off by the integral MOSFET;   (c) the
structure corresponding to the model of (b).

MOSFET Q_4 is configured to connect the base of Q_1 to the cathode to turn the thyristor on. The second MOSFET, Q_3, is a p-channel device, connected between the base of Q_2 and the cathode. When Q_3 turns on, Q_2 turns off and regeneration stops. Because the MOSFETs are of complementary types, Q_4 is on when the gate terminal is positive relative to the cathode, and Q_3 is on when the gate terminal is negative.

A model of a simpler MCT design is shown in Fig. 19.14(b). This device is turned on through the normal p_2 gate connection, but turn off is achieved by turning on Q_3. A structure corresponding to this MCT design is shown in Fig. 19.14(c). In both of the designs represented in Fig. 19.14, the MOSFET Q_3, which performs the turn-off function, must be finely interdigitated throughout the device to eliminate the lateral resistive drop of the p_2 gate region. As pointed out in Section 19.6, this drop prevents a conventional SCR from being turned off by its gate.

Notes and Bibliography

A good and detailed description of the physical behavior of the thyristor can be found in Ghandi [1]. The "bible" of thyristors and their applications is the *SCR Manual*, Sixth Edition, 1979, published by General Electric Company. However, this most recent edition is not current enough to contain information about high-power GTOs or the MCT. Baliga [2] does give a short description of the MCT.

The involute gate structure shown in Fig. 19.1(c) was invented by Dr. Herbert F. Storm of the General Electric Company [3].

1. S. K. Ghandi, *Semiconductor Power Devices; Physics of Operation and Fabrication Technology* (New York: John Wiley & Sons, 1977).
2. B. J. Baliga, *Modern Power Devices* (New York: John Wiley & Sons, 1987).
3. H. F. Storm and J. G. St. Clair, "An Involute Gate-Emitter Configuration for Thyristors," *IEEE Trans. Electron Devices*, 21(8): 520–522 (1974).

Problems

19.1 Find the doping concentration and the length of the n_1 region of an SCR if it is to block a forward voltage of 3000 V and the breakdown mechanism is punch through. Ignore the multiplying effect of transistor action.

19.2 Repeat Problem 19.1 for an avalanche breakdown mechanism, with $E_c = 2 \times 10^5$ V/cm.

19.3 Show that $\beta_1 \beta_2 > 1$ if the two-transistor model of the SCR shown in Fig. 19.3(b) is to exhibit regeneration.

19.4 In Example 19.1, find the base transport factor α_T under the conditions given. Assume that the minority carrier lifetime in the n_1 region is $\tau_h = 20$ μs and that $D_h = 15$ cm^2/s. Find β and V_{BR} using this value of α_T and $\gamma_E = 0.95$.

19.5   Consider the SCR analyzed in Example 19.2. Find the emitter injection efficiency (γ_E), β, and the reduction in reverse breakdown voltage of the SCR when $T = 425K$.

19.6   Consider an npn transistor in which $\beta = 30$ and:

$$i_B = 10^{-12}(e^{qv_{BE}/kt} - 1) \text{ A} \qquad (19.30)$$

If a resistor $R_{gk} = 1\ \Omega$ is connected across its base-emitter junction, and the base terminal is driven with a current source, i_S, find as a function of i_S the effective gain of the transistor, i_C/i_S, where i_C is the collector current.

Chapter 20

Magnetic Components

INDUCTORS or transformers are present in almost every power electronic circuit. Inductors filter switched waveforms (as part of input and output filters, for example), create sinusoidal variations of voltage or current (with capacitors, as in resonant converters), limit the rate of change of current (as in snubber circuits), and limit transient current. Transformers provide isolation between two parts of a system, transform impedances, force current sharing (interphase transformer), provide phase shifting (3-phase delta–wye transformer), store and transfer energy (flyback transformer), and sense voltages and currents (potential and current transformers).

Unlike other electric components used in power electronic circuits, inductors and transformers are not generally available in their required forms. The large number of parameters that characterize a magnetic component (inductance, voltage, current, energy, frequency, turns ratio, leakage, dissipation) makes it impractical for a manufacturer to produce and stock the vast array of configurations required by the industry. Instead, individually designed magnetic components put copper and magnetic material together in a way that minimizes some combination of cost, volume, weight, and fabrication difficulty. Effective design thus requires an understanding of the principles governing the behavior of magnetic components— and a good deal of intuition.

In this chapter we discuss both the analysis and synthesis of magnetic components. Problems of analysis may be complex, but they are straightforward and yield unique answers. Synthesis, however, is an engineering challenge because of the number of parameters involved. A design is never uniquely appropriate. Rather, a design is judged "best" according to criteria established by the application. For instance, if an inductor must not exceed a certain volume, several different designs using different materials might be satisfactory. A further requirement that its cost be minimized narrows the range of acceptable designs. More than likely, however, the optimization problem is multidimensional, perhaps to minimize some combination of volume, cost, and efficiency. The solution then becomes somewhat subjective and relies heavily on experience.

20.1 THE INDUCTOR

From a field point of view, an inductor is a circuit element that stores energy only in the form of magnetic fields. Its physical basis of operation is described by the magnetostatic form of Maxwell's equations:

$$\nabla \times \vec{H} = \vec{J} \qquad \text{(Ampere's law)} \qquad (20.1)$$

$$\nabla \cdot \vec{B} = 0 \qquad \text{(divergence law)} \qquad (20.2)$$

The relationship between the magnetic field intensity H and the magnetic flux density B in a particular material is summarized by the *permeability*, μ. The permeability of a material is defined as the ratio $\mu = B/H$. In free space, $\mu = \mu_o = 4\pi \times 10^{-7}$ H/m. In various magnetic materials, μ may range from 10 to 100,000 times μ_o. The permeabilities of these materials are not constant but vary as a nonlinear function of the flux density level B. We discuss this behavior in Section 20.2.

From *(20.1)* and *(20.2)* we can calculate the magnetic field created by a current. However, in order to determine the voltage present at the terminals of the wire that carries this current, we need one more of Maxwell's equations, namely, Faraday's law:

$$\nabla \times \vec{E} = -\frac{\partial \vec{B}}{\partial t} \qquad \text{(Faraday's law)} \qquad (20.3)$$

Applying Stoke's theorem to *(20.3)* gives the familiar integral form of Faraday's law, or:

$$\oint_{\ell} \vec{E} \cdot d\vec{\ell} = -\frac{d}{dt} \iint_{S} \vec{B} \cdot d\vec{S} \qquad (20.4)$$

where S is the surface inside the closed contour ℓ. In an inductor, ℓ is the closed path formed by the winding, and S is the cross-sectional surface area of this winding, or generally the number of turns times the area enclosed by each turn. If the winding is made with a good conductor, then $\vec{E} \approx 0$ within the wire itself so that only the electric field between the terminals of the inductor contributes to the line integral of $\vec{E}$ in *(20.4)*. The left-hand side of *(20.4)* is then the negative of the terminal voltage v. Hence, we can rewrite *(20.4)* as:

$$v = \frac{d}{dt} \iint_{S} \vec{B} \cdot d\vec{S} = \frac{d\lambda}{dt} \qquad (20.5)$$

where λ is a parameter called the *flux linkage*, defined as:

$$\lambda = \iint_{S} \vec{B} \cdot d\vec{S} \qquad (20.6)$$

The difference between flux, Φ, and flux linkage, λ, can best be understood by considering a single layer coil of N turns. The flux is the integral of B over the area enclosed by a single turn. The flux linkage is the integral of B over the surface of an area whose perimeter is the wire, a surface that looks something like a screw. In this case, $\lambda = N\Phi$.

If the relationship between λ and the current that caused the magnetic field is linear (which it is if μ is constant, as can be seen from *(20.1)* and *(20.2)*), the constant of proportionality is the *inductance* L of the magnetic component, that is,

$$\lambda = Li \qquad (20.7)$$

20.1.1 Calculating Inductance

We do not discuss the details of how to solve the preceding equations to get an exact solution for the magnetic field. In most practical cases you may assume (accurate enough) that the magnetic field follows a defined path with a given cross-section area over which the field is uniform. Once you know the magnetic field intensity $\vec{H}$, you can calculate the inductance using the geometric parameters of the problem. The following steps summarize the method of solution.

1. Choose the closed-loop path ℓ along which the value of $\vec{H}$ is desired.

2. Use the divergence law, *(20.2)*, namely, that the normal component of $\vec{B}$ is continuous across a boundary between two materials having different permeabilities, to relate $\vec{H}$ on the two sides of any discontinuity of permeability along ℓ.

3. Use Ampere's law, *(20.1)*, to equate the line integral of $\vec{H}$ around the closed path to the surface integral of the current density. This approach results in an expression for $\vec{H}$ in terms of the current i, the number of turns N, and the length of the path ℓ.

4. Find the flux linked by the winding by integrating $\vec{B}$ over the surface formed by every turn. If all the turns are the same, the solution equals the number of turns times the integral of $\vec{B}$ over the surface defined by one turn. The result is the flux linkage of the winding, λ.

5. Divide λ by i to obtain the inductance, L.

The flux linkage λ specifies the state (that is, the energy) of the inductor. Faraday's law, *(20.3)*, relates the time derivative of this flux linkage to the voltage across the winding. Therefore the change in flux linkage from one time to another is the integral of the voltage over this time. Flux linkage has the units of volt-seconds and, in terms of winding terminal voltage v, can be expressed as:

$$\lambda(t) = \lambda(0) + \int_0^t v \, d\tau \qquad (20.8)$$

EXAMPLE 20.1

Applying the Concept of Flux Linkage

Let's consider an inductor originally designed to support a maximum ac voltage of $V_o \sin \omega t$. If we increase the number of turns by 50% and hold the amplitude of the voltage at V_o, how low can the frequency go before the peak flux density in the core exceeds that of the original design?

The number of turns on the new inductor N_2 is $1.5 N_1$, where N_1 is the original number of turns. The flux linkage increases by the same amount if we hold the peak flux density constant at B_o. Thus:

$$\lambda_2 \leq N_2 B_o = 1.5 N_1 B_o = 1.5 \lambda_1 \tag{20.9}$$

Using (20.8) with $\lambda(0) = -\lambda_{\max}$ (assuming that the flux is symmetrical about $\lambda = 0$), we obtain the maximum flux linkages in the two cases:

$$\lambda_1 = \frac{1}{2} \int_0^{\pi/\omega_1} V_o \sin \omega_1 t \, dt = \frac{V_o}{\omega_1} \tag{20.10}$$

$$\lambda_2 = \frac{1}{2} \int_0^{\pi/\omega_2} V_o \sin \omega_2 t \, dt = \frac{V_o}{\omega_2} \tag{20.11}$$

So, if $\lambda_2 \leq 1.5 \lambda_1$,

$$\omega_2 \geq \frac{\omega_1}{1.5}$$

Only in the case of constant permeability can we strictly define inductance. For constant permeability, the value of inductance as given by:

$$L = \frac{\lambda}{i} \tag{20.12}$$

will be directly proportional to the permeability of the material, the area of the winding, and the square of the number of turns and inversely proportional to the length of the magnetic path. If these physical parameters are constant, the inductance will not vary with time, and we express its terminal behavior by the familiar relationship:

$$v = L \frac{di}{dt} \tag{20.13}$$

EXAMPLE 20.2

An Inductance Calculation

Figure 20.1 shows an inductor constructed of windings on a doughnut shaped core. In geometry this core shape is called a *toroid*, and an inductor constructed in this way is known as a *toroidal inductor*. Let's calculate its inductance.

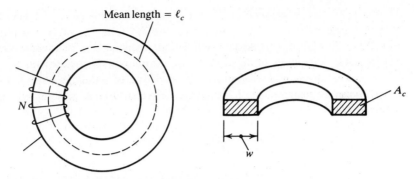

Mean length = ℓ_c

N

A_c

w

Figure 20.1 Toroidal inductor and cross section of toroid core.

We immediately confront a problem when we apply Ampere's law to determine H. The problem is that the value of H we obtain depends on whether we choose a contour at the inner or outer radius of the toroid. We can obtain an answer that is often accurate enough by assuming that, with respect to radius, H in the core is constant at the value that results from using the mean core length ℓ_c in Ampere's law. This approximation is good enough if the inner radius is large compared with the core's radial dimension w. Applying Ampere's law with this approximation and substituting $B = \mu_c H$ yields

$$B = \frac{\mu_c N i}{\ell_c} \qquad (20.14)$$

where μ_c is the permeability of the core. We are assuming B to be uniform within the core, so the flux is

$$\Phi = B A_c = \frac{\mu_c A_c N i}{\ell_c} \qquad (20.15)$$

Because the area of each turn is the same (A_c), λ is simply:

$$\lambda = N B A_c = \frac{\mu_c A_c N^2 i}{\ell_c} \qquad (20.16)$$

Dividing λ by the current yields the inductance:

$$L = \frac{\mu_c A_c N^2}{\ell_c} \qquad (20.17)$$

If we plug in some numbers, we begin to develop a feeling for the physical size of inductors. If $\mu_c = 200\mu_o$, $\ell_c = 15$ cm, $A_c = 1$ cm^2, and $N = 25$,

$$L = \frac{(200 \times 4\pi \times 10^{-7})(10^{-4})(625)}{0.15} = 105 \ \mu H$$

20.1.2 Electric Circuit Analogs

The result of any inductance calculation is an expression of the general form:

$$L = N^2 \mathcal{P} \qquad (20.18)$$

The term $\mathcal{P}$, the *permeance* of the inductor, is determined by the geometry and magnetic properties of the core material. The N^2 term is simply a scaling factor. In *(20.17)*, $\mathcal{P} = \mu_c A_c / \ell_c$. As you will see, permeance (and its inverse, *reluctance, $\mathcal{R}$*) are useful quantities in the analysis of complex magnetic structures.

If we assume a magnetic structure consisting of windings on a core made of linear magnetic material, the relationship between flux, Φ, and ampere-turns, Ni, as given by *(20.15)*, is always of the form:

$$\Phi = \frac{Ni}{\mathcal{R}} = \mathcal{P}Ni \qquad (20.19)$$

where $\mathcal{R}$ and $\mathcal{P}$ are the reluctance and permeance, respectively, of the magnetic field path used to calculate B from Ampere's law. From *(20.15)*, the reluctance of the toroidal core is

$$\mathcal{R} = \frac{\ell_c}{\mu_c A_c} \qquad (20.20)$$

The divergence law, *(20.2)*, states that magnetic field lines must close on themselves. But having the choice of a path to follow, the flux would follow the low reluctance path of a highly permeable material (such as that represented by *(20.20)*) rather than the high-reluctance path of air. The mechanism is analogous to the way in which current flows through the wires comprising a circuit rather than through the surrounding air. This observation and the form of *(20.19)* suggest an analogy with Ohm's law, where $Ni \to v$, $\Phi \to i$, and $\mathcal{R} \to R$. Such an analog is a valuable tool for analyzing or designing complex magnetic structures, for it allows exploitation of a strong intuitive understanding of electric circuits. Because of this analog, we refer to paths of high permeability as *magnetic circuits*.

The expressions for the reluctance of a magnetic path and the resistance of a conductive path have similar forms. They depend directly on the length of the path and inversely on the cross-sectional area and permeability (or conductivity) of the material. The ampere-turns (Ni) is known as the *magnetomotive force* (MMF) and is analogous to the electromotive force (EMF) in an electric circuit. The magnetic flux is analogous to the electric current.

There is, however, a practical difference between the behaviors of electric and magnetic circuits. An electric circuit is defined by copper wires and other components whose conductivities are generally more than 12 orders of magnitude greater than that of the material surrounding the circuit (air or epoxy-glass board, for example). In contrast, a magnetic circuit is composed of materials whose permeances are only a few orders of magnitude greater than that of its surrounding medium (air, for example). In fact, air frequently forms part of the magnetic circuit, as you will see. Therefore a measureable amount of flux "leaks out" of the magnetic path defined by the materials and closes on itself through alternative paths in air. This flux is known as *leakage flux*, and we frequently assume it to be negligible, at least in a first approximation to a solution of the magnetic circuit.

EXAMPLE 20.3

Using a Circuit Analog to Calculate Inductance

Figure 20.2 shows an inductor made of a three-legged core with a winding of N turns on the center leg. The two outer legs have cross-sectional areas different from that of the center leg. What is the inductance of the winding?

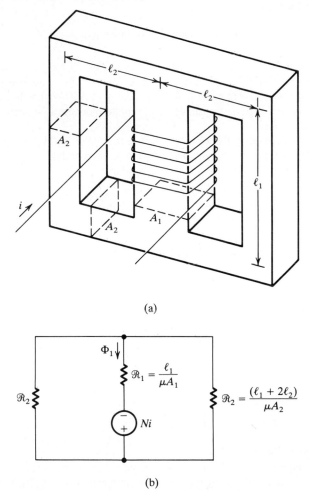

(a)

(b)

Figure 20.2 (a) Magnetic structure with one winding of N turns and three legs. The two outer legs have cross-sectional area A_2, and the inner leg has area A_1. The core material has permeability μ_c. (b) Circuit analog for the structure of (a).

We first assume that the permeability of the core is much larger than that of free space, so all the flux resides within the core. The core is the magnetic circuit, which may be broken into three branches: the center leg; the branch to the right, consisting of both horizontal and vertical core segments, and the branch to the left, which is symmetrical with that to the right. The length of each branch is approximately its mean path length. These lengths are shown in Fig. 20.2(a) as ℓ_1 and ℓ_2.

The permeance of each branch depends only on geometry and material properties of the core, so we may calculate it independent of the winding details. The result is

$$\mathcal{P}_1 = \text{permeance of center leg} = \frac{\mu_c A_1}{\ell_1} \qquad (20.21)$$

$$\mathcal{P}_2 = \text{permeance of outer legs} = \frac{\mu_c A_2}{\ell_1 + 2\ell_2} \qquad (20.22)$$

The resulting analog electric circuit is shown in Fig. 20.2(b). The winding and its current are represented by a voltage source of value Ni. Polarity is determined by the right-hand rule and the definition of positive Φ in the analog circuit. We use branch reluctances rather than permeances because we can solve this circuit more easily in terms of resistances rather than conductances.

The flux linked by the winding is that in the center leg, Φ_1. We find the inductance by calculating this flux, multiplying by N to get λ, and dividing by i:

$$\Phi_1 = Ni\frac{1}{\mathcal{R}_1 + \mathcal{R}_2/2} = \frac{2A_1 A_2 \mu_c Ni}{2A_2\ell_1 + A_1(\ell_1 + 2\ell_2)} \qquad (20.23)$$

so

$$L = \frac{N\Phi_1}{i} = \frac{2A_1 A_2 \mu_c N^2}{2A_2\ell_1 + A_1(\ell_1 + 2\ell_2)} \qquad (20.24)$$

The inductances we have calculated so far, (20.17) and (20.24), are linear functions of core permeability. This approach may be undesirable because—unless the core is made of powdered material—we cannot accurately predict its permeability, and its permeability is a function of flux level. To desensitize the value of inductance to variations in μ_c, we often place an air gap in the magnetic circuit.

EXAMPLE 20.4

Inductor with an Air Gap in the Core

An air gap in a magnetic circuit makes the inductance independent of flux level (current) and increases the energy storage density of the structure. A core with an air gap in one leg is shown in Fig. 20.3(a). What is the inductance of this structure?

We can calculate the reluctances of the core and air gap by again using the mean path length approximation:

$$\mathcal{R}_c = \frac{\ell_c}{\mu_c A} \qquad (20.25)$$

$$\mathcal{R}_g = \frac{g}{\mu_o A} \qquad (20.26)$$

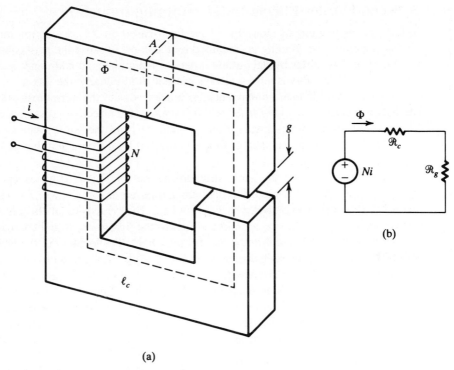

(a)

(b)

Figure 20.3   (a) An inductor made from a core with an air gap in one leg. The cross-sectional area of the core is uniform and equal to A.   (b) Circuit analog for the inductor of (a).

We now calculate the inductance, using reluctances as in Example 20.3:

$$\Phi = \frac{Ni}{\mathcal{R}_c + \mathcal{R}_g} = \frac{Ni}{(\ell_c/\mu_c A) + (g/\mu_o A)} \qquad (20.27)$$

so

$$L = \frac{N\Phi}{i} = N^2 \left(\frac{1}{\mathcal{R}_c + \mathcal{R}_g} \right) = \frac{\mu_o A N^2}{(\mu_o/\mu_c)\ell_c + g} \qquad (20.28)$$

Note that the value of the inductance becomes independent of the core's magnetic properties if:

$$g \gg \frac{\ell_c \mu_o}{\mu_c} \qquad (20.29)$$

Because the magnetic properties of most core materials vary with temperature, flux level, sample, and manufacturer, a gap is frequently required to make the inductance value predictable and stable.

20.1.3 Second-Order Effects in Determining Inductance

So far, our calculations of inductance have been based on the assumption that all the flux linked by the winding is contained by the core. Because the permeabilities of the core and air differ by only a few orders of magnitude, an additional *leakage* component of linked flux is almost always present. Additionally, the flux across the gap contains an additional component that bulges outside the core cross section. The bulge is called a *fringing field*, and it increases the effective cross section of the gap. Both these effects increase the inductance from the value calculated on the basis of the assumptions that all flux is contained by the core and that gap flux is normal to the gap faces.

Figure 20.4(a) shows the same structure as that of Fig. 20.3(a), except that the leakage flux Φ_ℓ and the gap flux including fringing Φ_g are shown. The circuit analog for this inductor is shown in Fig. 20.4(b). The reluctance of the gap, $\mathcal{R}'_g$, is now smaller than before because in addition to the flux going straight across the gap, additional flux is present in the gap fringing field. The location of the leakage branch $\mathcal{R}_\ell$, and its representation as a single branch in shunt with the source, are approximations. The reason is that some of the leakage flux passes through some of the core (the leakage flux at the inside corners, for example) and is more accurately represented by a series of L networks, each containing part of the core (series branch) and part of the leakage flux (shunt branch). Seldom is this detailed representation of leakage flux necessary.

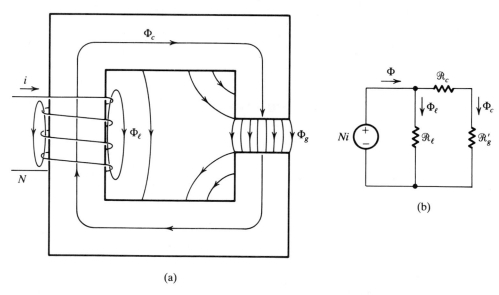

(a)

(b)

Figure 20.4   (a) Inductor showing leakage flux, Φ_ℓ, and fringing flux, Φ_g, at the edges of the gap.   (b) Circuit analog for the magnetic structure of (a).

The inductance of the inductor of Fig. 20.3(a), calculated on the basis of the model of Fig. 20.4(b), is

$$L = \frac{N\Phi}{i} = N^2 \left(\frac{1}{\mathcal{R}_\ell} + \frac{1}{\mathcal{R}_c + \mathcal{R}_g'} \right) \qquad (20.30)$$

As $\mathcal{R}_\ell > 0$ and $\mathcal{R}_g' < \mathcal{R}_g$, this calculation results in an inductance value greater than that given by (20.28).

The importance of these second-order effects depends on the aspect ratios, or relative dimensions, of the core and the gap. For instance, a gap whose length (g in Fig. 20.3a) is small compared to the length of the sides of the core cross section produces a much smaller second order increase in λ than one whose length is long compared to these dimensions. The fringing flux is a much smaller fraction of the total gap flux in the former case than in the latter case. Tabulations of inductance correction factors as a function of gap aspect ratio are available. Also, a core with a relatively short mean length exhibits a larger second-order increase from corner leakage than does a long core.

This qualitative understanding of second-order effects usually permits adequate inductor design. Performing a finite-element numerical analysis yields a more accurate design.

20.1.4 Energy Storage

In circuit-variable terms, the magnetic stored energy in an inductor W_m is

$$W_m = \frac{1}{2} Li^2 = \frac{1}{2L} \lambda^2 \qquad (20.31)$$

We can also express this energy in terms of fields as:

$$W_m = \frac{1}{2} \iiint\limits_{\text{volume}} \vec{B} \cdot \vec{H} \, d\mathcal{V} \qquad (20.32)$$

which, when we evaluate it over the volume of an ungapped core (assuming constant and uniform permeability) yields

$$W_m = \frac{B^2 \mathcal{V}_c}{2\mu} \qquad (20.33)$$

where $\mathcal{V}_c$ is the core volume.

If the core contains a gap, as in Fig. 20.4(a), the volume over which we integrate (20.32) includes both the core and the gap. Because normal $\vec{B}$ is continuous across discontinuities in μ, (that is, it has the same value in the gap as in the core) we can express W_m in terms of B as the sum of expressions for the energy stored in the core and the gap, or:

$$W_m = \frac{B^2 \mathcal{V}_c}{2\mu_c} + \frac{B^2 \mathcal{V}_g}{2\mu_o} \tag{20.34}$$

Comparing the stored energy *densities* w ($w = B^2/2\mu$) in the core and the gap, we see that:

$$\frac{w_g}{w_c} = \frac{\mu_c \mathcal{V}_g}{\mu_o \mathcal{V}_c} \tag{20.35}$$

If the cross-sectional areas of the gap and the core are equal (which may not always be true), we can rewrite *(20.35)* in terms of permeabilities and mean path lengths as:

$$\frac{w_g}{w_c} = \frac{\mu_c g}{\mu_o \ell_c} \tag{20.36}$$

For an iron core, $\mu_c \approx 10^4 \mu_o$, and the ratio *(20.36)* is typically much greater than 1. The significance of this fact is that most of the energy stored in the inductor is stored in the fields in the gap rather than in the core. That is,

$$W_m \approx \frac{B^2 \mathcal{V}_g}{2\mu_o} \tag{20.37}$$

Note that this condition is the same as *(20.29)* for the inductance to be independent of core properties. The volume of the gap times the energy density in the gap then gives a good approximation to the total stored energy in the inductor. If the core is iron with a saturation flux density of $B_{max} = 1.4$ T, the maximum energy density in the gap is

$$w_g = \frac{(1.4)^2}{2\mu_o} = 1 \text{ J/cm}^3 \tag{20.38}$$

However, if the core is ferrite, in which the saturation flux density is typically 0.3 T, the stored energy density is only 0.05 J/cm^3. These numbers are very valuable when you are designing an inductor, because specifying the values of inductance and current determines the total stored energy and thus the volume of the gap. A typical design might proceed as follows:

1. Calculate the gap volume required to store the necessary energy without causing saturation.

2. Assume that the length of the gap is small (perhaps by a factor of 10) compared to the length of the sides of the core cross section. This assumption reduces the second-order effects of fringing.

3. Determine A_c and g so that steps 1 and 2 are satisfied.

4. Determine the number of turns needed to obtain the desired value of inductance.

5. Determine the window area (the area enclosed by the core in Fig. 20.3) required to hold this number of turns for the wire size necessary to keep the resistance below an acceptable level.

6. Check to make sure that *(20.29)* is satisfied for the now known length of the magnetic path.

20.2 SATURATION, HYSTERESIS, AND RESIDUAL FLUX

In Example 20.4 we showed that one benefit of a gap in a magnetic core is that it makes the value of inductance less sensitive to variations of the magnetic properties of the core material. Another advantage of a gap is that it limits the detrimental effects of *remnant*, or *residual*, magnetization. Remnant magnetization is the flux remaining in the core when the magnetic field intensity H reaches zero. Figure 20.5 shows the relationship between B and H for a typical magnetic material. As you can see, H is proportional to the current flowing in the winding, and B, through *(20.16)* and *(20.8)*, is proportional to the integral of the voltage across the winding. This integral is the flux linkage λ.

Three important characteristics of the relationship are shown in Fig. 20.5. First, the path followed when the flux is increasing is not the same as that followed when the flux is decreasing. This behavior is called *hysteresis*. Second, note that neither path passes through the origin. When $H = 0$ the flux density is not zero but has a value $\pm B_r$ called the *remnant magnetization*, or *residual flux density*. When $B = 0$, the magnetic field is not zero but is equal to $\pm H_c$, a parameter called the *coercive force* of the material. Third, the slope of the path, which is the incremental

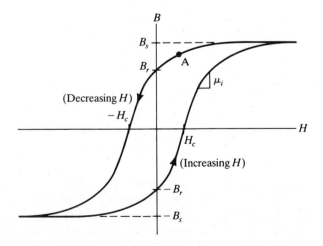

Figure 20.5   Typical B–H curve for a magnetic material.

permeability μ_i, decreases rapidly with increasing B (ultimately reaching μ_o) after some maximum $B = B_s$ is reached. This state of low incremental permeability is called *saturation* and B_s is the *saturation flux density*.

20.2.1 Saturation

Although some applications exploit the saturation characteristics of magnetic cores (for example, magnetic amplifiers, fluorescent lamp ballasts, constant voltage transformers, and self-oscillating transistor inverters), in most applications saturation is avoided. One reason for avoiding it is that in saturation the condition, *(20.29)*, necessary to make the value of L insensitive to material properties, is often no longer satisfied because μ_c has become very small. Another reason is that in saturation, H increases very rapidly with B, resulting in large changes in current for relatively small increases in the volt–time product, λ. The final reason is that saturation reduces the coupling between windings in a transformer, because the permeance of the core is no longer much larger than that of the leakage paths.

Avoiding saturation is sometimes particularly difficult during transient circuit operation. A power circuit operating in the cyclic steady-state sequentially imposes positive and negative volt-seconds of equal value across any inductors in the circuit. Stated another way, the integral of the inductor voltage over one cycle is zero, as it must be if λ, and hence the current, is not to increase without limit. When we know the value of volt-seconds to be applied, we can design the inductor for cyclic steady-state operation so that its flux density varies from just under $+B_s$ to just over $-B_s$. However, during the first cycle when the circuit is first turned on, saturation may occur because the flux may not be starting from $-B_s$ (assuming that the first half cycle is to be positive) and therefore reaches $+B_s$ after fewer volt-seconds than anticipated. For instance, if $B(0) = 0$, only half the normal change in λ is actually available for the first cycle. This is known as the *start-up problem*.

We could solve the start-up problem by designing the inductor to support twice the normal λ. This solution, however, would require four times the maximum energy storage capability and therefore an increase of approximately four times in the physical volume of the inductor. A better solution is to control the power circuit, limiting the volt-seconds to a lower than normal value during the first cycle.

20.2.2 Residual Flux Density

The start-up problem is aggravated when the core material is characterized by a B–H curve exhibiting a high value of residual flux density. Let's consider the curve of Fig. 20.5 and assume that the last time the power circuit was operated, B and H were at point A, at which time the circuit was turned off. When the current (and therefore H) in the inductor decays to zero, B decays to B_r. The result is that if the circuit were turned on again in a positive half cycle, the volt-seconds available before saturation is reached would be much smaller than in the case of starting from $B = 0$.

We can reduce the start-up problem created by residual magnetism by introducing a gap in the core. Considering the structure of Fig. 20.3, we can write

$$H_g g + H_m \ell_c = Ni \quad \text{and} \quad B = \mu_o H_g = \mu_c H_m$$

where H_m and H_g are the magnetic field intensities in the magnetic material (core) and gap, respectively. Solving these equations for H_m in terms of B yields 5

$$H_m = \frac{Ni}{\ell_c} - \frac{Bg}{\mu_o \ell_c} \tag{20.39}$$

We must satisfy both (20.39) and the B–H_m relationship shown in Fig. 20.5. We can obtain the resulting relationship between ampere–turns (Ni) and B by graphically combining (20.39) and Fig. 20.5. We do so by putting the two functions on the same axes, and plotting their intersections as we vary Ni. Figure 20.6 shows this procedure and its result. Note that the incremental inductance is smaller, B_s is unchanged because it is a property of the core material, and the residual flux at zero current is much less than before, reducing the start-up problem as we desired.

The volt-seconds required to saturate a magnetic circuit (starting from $B = 0$) is $\lambda_s = NB_s A_c = Li_s$, which is independent of the gap length g. Changing the gap length reduces the inductance, but it also increases the current i_s at which saturation occurs. The start-up problem is most frequently associated with transformers, which are often constructed on cores with no gaps.

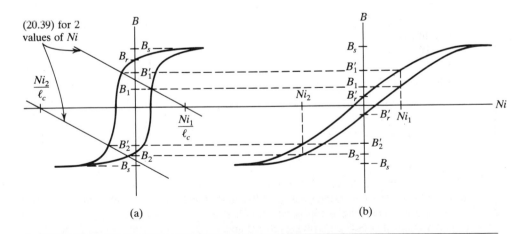

Figure 20.6   (a) Core hysteresis loop and air-gap load line.   (b) Resulting locus of operation illustrating how the value of residual flux density is reduced from B_r to B'_r by the introduction of a gap in the core.

EXAMPLE 20.5

Designing a Filter Inductor

Figure 20.7 shows the output filter of a 5-V converter. We assume that the capacitor is large enough that we can consider its voltage to be constant. The average output current of the converter is 10 A, and the filter input voltage v_d is a square wave, switching between 0 and 10 V at a frequency of 5 kHz ($T = 0.2$ ms), as shown. The 0.1-mH inductor is to be designed using a material with $B_s = 0.8$ T and $\mu_c > 10^4 \mu_o$.

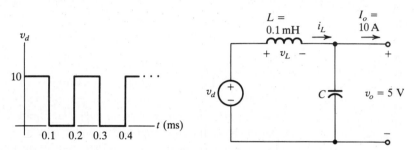

Figure 20.7   An LC filter at the output of a 5-V dc/dc converter. The effect of the converter switches is represented by the equivalent source v_d.

Under steady-state operation, the voltage across the inductor v_L is a square wave symmetrical about zero, that is, with equal positive and negative volt-seconds, having an amplitude of 5 V. The peak inductor current is

$$I_p = i(0) + \frac{1}{L} \int_0^{T/2} v_L \, dt = 7.5 + \frac{1}{10^{-4}} \int_0^{10^{-4}} 5 \, dt = 12.5 \text{ A}$$

The three physical parameters we need to specify for the inductor are N, A_c, and g. The relationships these parameters must satisfy are

$$\lambda_s = N A_c B_s = L I_p = 1.25 \times 10^{-3} \text{ V-s} \tag{20.40}$$

$$L = \frac{\mu_o N^2 A_c}{g} = 0.1 \text{ mH} \tag{20.41}$$

and

$$\frac{\ell_c}{\mu_c} \ll \frac{g}{\mu_o} \tag{20.42}$$

The easiest way to proceed is to make an educated guess at a value for g, determine A_c using the stored energy relationship (20.37), and then use (20.40) to determine N. Finally we must make sure that g is consistent with (20.42).

If we pick $g = 0.05$ cm, solving (20.37) gives us $A_c = 0.61$ cm^2, which from (20.40) results in $N = 26$ turns. These values are reasonable and give a core whose mean length is less than 10 cm, satisfying (20.42). A last consideration is wire size, which must carry the inductor current without getting too hot and which must also fit in the core window. In this case, size AWG 14 is probably a good choice, but for reasons we have not yet discussed.

This filter has a serious start-up problem, for if C is very large, its voltage at start-up rises slowly relative to the switching period. Then $v_L \approx v_d$ for many cycles, and i_L will increase well beyond the 12.5 A for which we designed the inductor. The duty ratio of the converter switches must be controlled to start from a very small duty ratio and to slowly bring the output voltage up to 5 V. This is called a *soft start*.

20.3 LOSSES IN MAGNETIC COMPONENTS

We have been assuming that magnetic components are lossless, except when we implied losses in our discussion of the relationship between core window area and wire size. However, practical magnetic components exhibit losses generated by three mechanisms: i^2R, hysteresis, and eddy currents. The first of these losses occurs in the windings, and the others occur in the core material. The size of a magnetic component is directly related to the magnitude of acceptable losses. In many applications, especially at high frequencies, magnetic component losses are the largest of any circuit component's losses.

20.3.1 Copper Loss

Determining copper, or winding, loss is straightforward in many applications. If the frequency is low enough that skin effect is not a problem, the winding resistance is simply the length of wire times its resistance per unit length. We calculate the rms current from the current waveform, and the loss is then $P_{Cu} = R_{Cu}I_{rms}^2$. Table 20.1 is a tabulation of wire data that you will find useful when specifying wire sizes for magnetic components. The criteria for selecting a wire size are functions of the application and most often relate to acceptable losses or maximum temperature rise. Typical design current densities range from 100 to 500 A/cm².

Skin depth is a measure of the lateral depth of penetration of current in a conductor. Skin depth is a result of fields created by the conductor's own current, not fields imposed by external sources, such as nearby currents caused by neighboring turns in a coil. We express skin depth δ as a function of frequency and the conductivity and permeability of the conductor, in this case copper:

$$\delta_{Cu} = \sqrt{\frac{2}{\omega \mu_{Cu} \sigma_{Cu}}} \tag{20.43}$$

The skin depth in copper is about 0.75 cm at 60 Hz. Thus not much 60 Hz current flows in the center of conductors having radii larger than a few centimeters. For this reason, aluminum conductors of power transmission lines can be reinforced with a stranded steel core without compromising current capacity ($\delta_{Al} \approx \sqrt{2}\, \delta_{Cu}$). In power electronics, skin depth is a concern primarily at frequencies above about 10 kHz, where $\delta_{Cu} < 0.5$ mm.

Where necessary, the limitations of skin depth can be avoided by using stranded wire in which each strand is insulated. Commercially available wire of this kind

Table 20.1   Copper Wire Data

AWG SIZE	DIAMETER (MM)	Ω/KM (75°C)	KG/KM	TURNS/CM^2
0	8.25	0.392	475	
1	7.35	0.494	377	
2	6.54	0.624	299	
3	5.83	0.786	237	
4	5.19	0.991	188	
5	4.62	1.25	149	
6	4.12	1.58	118	
7	3.67	1.99	93.8	
8	3.26	2.51	74.4	
9	2.91	3.16	59.0	
10	2.59	3.99	46.8	14
11	2.31	5.03	37.1	17
12	2.05	6.34	29.4	22
13	1.83	7.99	23.3	27
14	1.63	10.1	18.5	34
15	1.45	12.7	14.7	40
16	1.29	16.0	11.6	51
17	1.15	20.2	9.23	63
18	1.02	25.5	7.32	79
19	0.912	32.1	5.80	98
20	0.812	40.5	4.60	123
21	0.723	51.1	3.65	153
22	0.644	64.4	2.89	192
23	0.573	81.2	2.30	237
24	0.511	102	1.82	293
25	0.455	129	1.44	364
26	0.405	163	1.15	454
27	0.361	205	1.10	575
28	0.321	259	1.39	710
29	0.286	327	1.75	871
30	0.255	412	2.21	1090

is known as *litz* wire.[†] Using foil instead of wire can also reduce the skin-effect problem, but it introduces manufacturing complications.

20.3.2 Core Loss

Two mechanisms produce core loss: hysteresis and eddy currents. The net effect of these two mechanisms is a loss that varies approximately linearly with frequency, but that has a nonlinear flux magnitude dependency that is a function of the particular material. A useful, but approximate, dependency is

$$P_{core} \propto f B^{(1.6-2.0)} \qquad (20.44)$$

Hysteresis loss is the result of unrecoverable energy expended to rotate magnetic domains within a magnetic material. The energy lost per cycle per unit volume of material is

$$E = \oint dw_m = \oint \vec{H} \cdot d\vec{B} \qquad (20.45)$$

which is the area enclosed by the material's hysteresis loop. The total hysteresis loss is then the product of this area, the frequency, and the core volume. Manufacturers of magnetic cores provide loss data for their materials in the form of W/lb or W/cm^3 plotted as a function of B for different frequencies. In the case of materials having nonzero conductivity (laminated iron for instance), these data include eddy current losses.

Eddy current loss is simply $i^2 R$ loss caused by currents induced in the magnetic material. If this material were nonconducting, no currents would be induced in it, and the eddy current losses would be zero. But all magnetic material is more or less a conductor, except for the Ni–Zn ferrite materials, which can be considered nonconducting for most applications.

Because eddy currents are driven by a voltage induced in the material through Faraday's law, *(20.3)*, they flow in a plane perpendicular to $\vec{B}$. Nonpowdered cores are built up of laminations parallel to $\vec{B}$ to reduce the driving voltage. These laminations are insulated from one another, usually by the oxide formed when the material is annealed. Figure 20.8 shows three laminations and the paths of the eddy currents flowing in them. The loss produced by these currents is proportional to v^2/R_c, where v is the induced driving voltage and R_c is the resistance of the path through which the current flows. As R_c is proportional to the length of the path but v is proportional to the area enclosed by the path, loss will increase rapidly with increasing lamination thickness.

[†]From the German *Litzendraht*, meaning many stranded. True litz wire is also *transposed*, a construction technique in which each strand occupies each possible position in the wire bundle for a length equal to the *transposition length* divided by the number of conductors in the wire. This ensures that each strand has the same λ and that there are no circulating currents between strands.

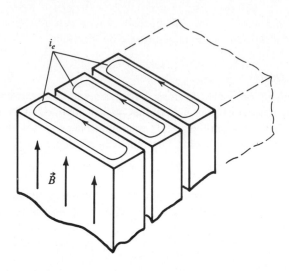

Figure 20.8   A section of a laminated core, showing the location of eddy currents i_e in the laminations.

The resistance R_c is linearly proportional to the material's resistivity. Various high-resistivity ferrous alloys have been developed to reduce the eddy current contribution to total core loss. The simplest way of increasing resistivity is to add a small amount (1–3%) of silicon to the steel. More exotic low-loss ferrous alloys contain various amounts of molybdenum, chromium, and nickel.

Another reason to laminate a core, especially for high-frequency use, is that the magnitude of B in the core falls off exponentially (from its value at the core surface) with a characteristic length equal to the skin depth of the material. The skin depth in 2.5% silicon steel at 10 kHz is approximately 0.05 mm. A solid core of this material would contain flux only in a shell about 0.05 mm thick. That is, if the core cross-sectional dimensions were much larger than 0.05 mm, the center of the core would not contain any flux.

The lower limit of practical lamination thickness is about 0.025 mm. Laminating more thinly not only makes manufacturing the core expensive, but the insulation between the laminations occupies an excessive fraction of the total core volume (approximately 17% for 0.025-mm laminations). Laminated cores can be used effectively to frequencies of about 20 kHz, but beyond this frequency, powdered or ceramic (ferrite) cores are used.

20.3.3 A Simple Loss Model

A simple model for losses in an inductor is illustrated in Fig. 20.9. The resistor R_{Cu} represents the resistance of the copper winding, and the energy dissipated in this resistor is the copper loss in the inductor. The value of R_{Cu} is determined by wire

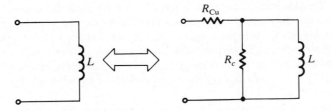

Figure 20.9 A simple circuit model for an inductor. The resistors R_{Cu} and R_c represent winding and core losses, respectively.

gauge, winding length, and any frequency dependencies, such as skin effect. Core loss is modeled by the resistor R_c, whose value is a function of both frequency and flux level. Because the inductor voltage is also a function of these variables, the most convenient location for R_c is in parallel with the inductor, as shown.

EXAMPLE 20.6

Determining Parameters in a Loss Model

Here we determine the values of R_{Cu} and R_c in the model of Fig. 20.9 for the inductor designed in Example 20.5. We assume that the core is made of a high-silicon steel having the core-loss characteristics shown in Fig. 20.10.

The inductor current, as calculated in Example 20.5, is a triangle wave having a peak–peak amplitude of 5 A added to a dc component of 10 A. The rms value of the net current

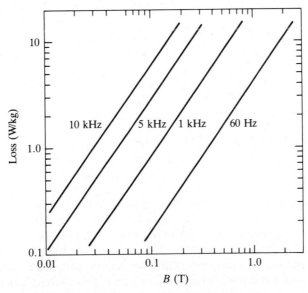

Figure 20.10 Core-loss data for 0.025 mm high-silicon steel.

is 10.1 A. The skin depth in copper at 5 kHz ($\approx$ 1 mm) is about the radius of AWG 14 wire ($\approx$ 0.8 mm) and the majority of the current is dc, so we can use solid rather than litz wire.

Assuming that the core has a square cross section and allowing for space between the core and the winding, we determine that the total length of wire comprising the 26 turns is approximately 1.25 m. From Table 20.1, we find that the resistance of AWG 14 wire is 10 Ω/km at 75°C (a reasonable operating temperature for the winding). The resistance R_{Cu} is therefore 12.5 mΩ.

Determining the resistance of R_c is more complicated. Although the peak flux in the core is 0.8 T, the ac component of this flux B_{ac} is much smaller, or:

$$B_{ac} = \frac{\lambda_{ac}}{NA_c} = \frac{2.5 \times 10^{-4}}{(26)(0.61 \times 10^{-4})} = 0.158 \text{ T} \tag{20.46}$$

Furthermore, $B(t)$ is not sinusoidal as assumed in Fig. 20.10: It is triangular at a fundamental frequency of 5 kHz. If we assume that the core loss is dominated by the fundamental, we can approximate $B(t)$ as a sinusoid with an amplitude of 0.158 T. From Fig. 20.10 we determine that the core loss of this material at 5 kHz and 0.158 T is 5 W/kg.

We can determine the weight of the core from the manufacturer's data sheet or by calulating it based on the density of iron. We have no data sheet, so we make the calculation. The specific gravities of magnetic steel alloys are all approximately 8. The volume of this core is 6.1 cm^3, so its weight is 0.049 kg. The total core loss is therefore 0.24 W. Ignoring the drop across R_{Cu}, the voltage across R_c is a square wave having a 5-V amplitude. Thus the value of R_c that gives a loss of 0.24 W is

$$R_c = \frac{V_{rms}^2}{P} = \frac{5^2}{0.24} = 104 \ \Omega \tag{20.47}$$

The resulting loss model for this inductor is shown in Fig. 20.11.

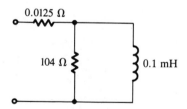

Figure 20.11   Loss model for the inductor.

20.3.4 Why Use a Core?

With the problem of saturation, the losses created by hysteresis and eddy currents, and the expense and weight of the magnetic material, you might well ask, "Why construct an inductor with a magnetic core?" After all, an air-core inductor would not present any of these problems. Although air-core inductors sometimes are used, there are several reasons for utilizing a core of magnetic material instead. One is that the magnetic fields are contained by the core. They do not extend great distances from the winding, as they would in an air-core inductor, reducing the problem of adverse effects on nearby equipment or circuits. Containing the magnetic field is also very important for a transformer if the coupling between two windings is to be good. Another reason is that, for a given value of inductance and a given overall volume, the resistance of the winding is much smaller if a core is used.

20.4 TRANSFORMERS

Transformers are inductors that are coupled through a shared magnetic circuit, that is, two or more windings that link some common flux. Figure 20.12 shows a two-winding transformer. The core provides the low-reluctance magnetic circuit through which flows most of the flux generated by the windings. Transformers come in various forms and are used in power electronic circuits to perform a number of different functions. Low-frequency (50-, 60-, or 400-Hz) power transformers are usually made of laminated steel cores and function to step the line voltage up or down, to provide electric isolation, and to provide phase shifting in systems using multiphase line supplies (12-pulse rectifiers, for instance). In high-frequency converter applications, the transformer provides isolation and voltage transformation—and also stores energy when used in circuits such as the flyback converter. Manufacturers often use powdered magnetic alloys or ferrites in high-frequency transformers to control core losses. Additionally, transformers frequently provide the coupling between gate-drive or base-drive circuits and the high-power devices being driven, and are sometimes used as voltage or current sensors in feedback control systems. Transformers are frequently the most costly components in power electronic equipment.

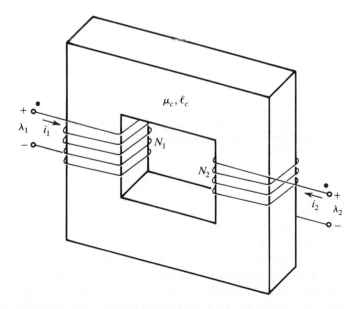

Figure 20.12   Structure of a two-winding transformer.

20.4.1 The Ideal Transformer

If the same flux is linked by both windings of a two-winding transformer—and this is the only flux they link—we say that the windings are perfectly coupled. As the flux linked by each turn is the same for both windings, the voltages induced in each turn are also identical. Moreover, the total voltage across a winding is directly proportional to its number of turns. The ratio of the two terminal voltages is therefore equal to the turns ratio, that is,

$$\frac{v_1}{v_2} = \frac{N_1}{N_2} \qquad (20.48)$$

The source of the magnetic field in a transformer is the algebraic sum of the Ni products of the windings. We use a *dot convention* to indicate winding polarity. Current into the dotted ends of windings produces aiding flux. An application of the right-hand rule to Fig. 20.12 shows that current into the dotted end of either winding produces clockwise flux. We can now use Ampere's law to calculate the magnetic field intensity:

$$H = \frac{N_1 i_1 + N_2 i_2}{\ell_c} \qquad (20.49)$$

If the reluctance of the magnetic path were zero, which would be the case if $\mu_c = \infty$ and there were no air gap, H would have to be zero to avoid infinite B. But H can be zero only if the Ni products for the two windings sum to zero. In this case, the ratio of the two currents is equal to the negative inverse of the ratio of the number of turns on the windings, or: 5

$$\frac{i_1}{i_2} = -\frac{N_2}{N_1} \qquad (20.50)$$

The directions of the currents are opposite to each other—one flows into the dot while the other flows out.

If the terminal variables are sinusoidal, we can divide *(20.48)* by *(20.50)* to illustrate the impedance transforming property of a transformer. If $\hat{V}_1$, $\hat{V}_2$,..., are the complex amplitudes of the sinusoidal terminal variables, then:

$$\frac{\hat{V}_2}{\hat{I}_2} = Z_2 = \left(\frac{N_2}{N_1}\right)^2 \frac{\hat{V}_1}{\hat{I}_1} = N^2 Z_1 \qquad (20.51)$$

where N is the turns ratio N_2/N_1.

Equations *(20.48)* and *(20.50)* describe the behavior of an *ideal transformer*, which has the circuit symbol shown in Fig. 20.13. A real transformer differs from the ideal in three ways. First, the voltages are not exactly related by *(20.48)*, because, owing to leakage, not all the flux linked by one winding is linked by the other. Second, finite permeability prevents the currents from being related by *(20.50)*. A nonzero difference between them is necessary to create flux in the core. This difference is called the *magnetizing current*. Third, the relationships *(20.48)*

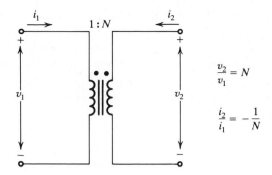

Figure 20.13 Circuit symbol that we will use and descriptive equations for the ideal transformer.

and *(20.50)* for the ideal transformer are not functions of frequency; that is, an ideal transformer works at dc, but a real transformer does not. In spite of these differences, the ideal transformer is extremely useful in modeling real transformers, in the same way that the ideal current source is useful in modeling transistors.

20.4.2 Magnetizing Inductance

In order for two windings to be magnetically coupled, a B must be created by one winding and linked by the other. As previously mentioned, only in the hypothetical case of infinite permeability can B exist without H and therefore without a net Ni. The closest we can come to achieving this ideal condition is to use a highly permeable core with no gap. In this case what we get from one winding, if the other is open circuited, is simply a very large—but finite—inductance called the *magnetizing inductance*. The magnetizing inductance will have one of two values, depending on which winding the inductance is being measured from, and these values are related by the turns ratio squared.

Figure 20.14(a) is a two-winding transformer with perfect coupling but a finite magnetizing inductance, and Fig. 20.14(b) is a model for this transformer. In this model we made use of the ideal transformer. The magnetizing inductance L_μ is in shunt with the ideal transformer. We can place L_μ on either side of the ideal transformer, so long as its value is correct for that side. As we said in Chapter 1, we have compromised the conventional ideal transformer symbol to obtain schematic clarity. The model in Fig. 20.14(b) for the real transformer is enclosed in a box, and the ideal transformer windings are identified by adjacent double lines.

The magnetizing inductor current I_μ is known as the *magnetizing current*. It is the current that prevents the Ni products of the real transformer's windings from summing to zero. The presence of the magnetizing inductance is the reason that a real transformer will not work with dc. (The magnetizing current would be infinite.)

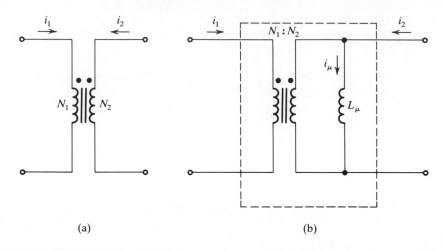

(a)   (b)

Figure 20.14   (a) A two-winding transformer with a magnetic core.   (b) Model of
the transformer of (a), assuming perfect coupling but finite magnetizing inductance L_μ.

EXAMPLE 20.7

Simple Use of the Transformer Model

Figure 20.15(a) shows a sinusoidal voltage source of amplitude V_1 connected to a resistor
of value R through a transformer with N_1 primary and N_2 secondary turns, giving a turns
ratio of $N = N_2/N_1$. As redrawn in Fig. 20.15(b) the transformer in the circuit has been

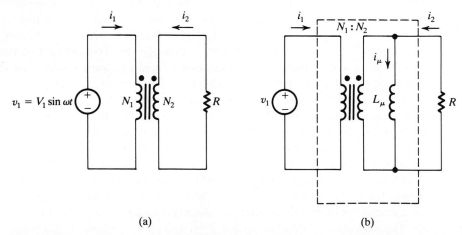

(a)   (b)

Figure 20.15   (a) Sinusoidal voltage source connected to a resistor by a transformer
with N_1 primary and N_2 secondary turns.   (b) Transformer in (a) replaced by a model consisting of an ideal transformer and a magnetizing
inductor L_μ.

replaced by a model consisting of an ideal transformer and the magnetizing inductance L_μ on the secondary side. The core has a cross-sectional area A_c and a saturation flux density B_s. We want to know the primary current i_1 and how low ω can become before the core saturates.

The current i_1 is simply the source voltage divided by the secondary impedance transformed to the primary. That is,

$$
i_1 = \text{Im}\left[V_1 e^{j\omega t} N^2 \frac{R + j\omega L_\mu}{R(j\omega L_\mu)}\right]
$$

$$
= \frac{N^2 V_1 \sqrt{R^2 + (\omega L_\mu)^2}}{R\omega L_\mu} \sin\left(\omega t - \tan^{-1}\frac{R}{\omega L_\mu}\right)
$$

(20.52)

We determine the lowest frequency at which this transformer can operate without saturating the core by equating (20.16) and (20.8) and assuming that $\lambda(0) = -\lambda_s$. (This assumption ignores the start-up problem.) The result is

$$
\lambda_s = N_x A_c B_s
$$

(20.53)

$$
\lambda_s = -\lambda_s + \frac{1}{\omega_{\min}} \int_0^\pi V_x \sin\omega t \ d(\omega t) = \frac{V_x}{\omega_{\min}}
$$

(20.54)

$$
\omega_{\min} = \frac{V_1}{N_1 A_c B_s}
$$

(20.55)

Note that we do not specify N_x as the primary or secondary turns in (20.53), or V_x as the primary or secondary voltage in (20.54). Only when we determine the side on which we will measure the winding voltage V_x do we specify N_x.

20.4.3 Leakage Inductance

Not all the magnetic flux created by one winding of a two-winding transformer follows the magnetic circuit and links the other winding. As discussed in Section 20.1.3, some of the flux "leaks" from the core and returns through the air. The result is that each winding links some flux that is not linked by the other, causing imperfect coupling. This effect is modeled by series *leakage inductances*, which are shown as L_{ℓ_1} and L_{ℓ_2} in Fig. 20.16. The transformer winding voltages are no longer related simply by the turns ratio, because we must now subtract the drop across L_ℓ from the terminal voltages to get the ideal transformer winding voltages.

Leakage is more important in transformers than in inductors. The reason is that leakage simply increases the value of an inductor, whereas it interferes with the basic operation of a transformer. Understanding the origin of leakage allows you to design transformers more effectively, minimizing this parameter. So, let's consider the core and winding structure shown in Fig. 20.17. The winding is modeled as one solid conductor wrapped once around the core, whereas in reality it may be made of N turns of round wire. From the point of view of current distribution, the

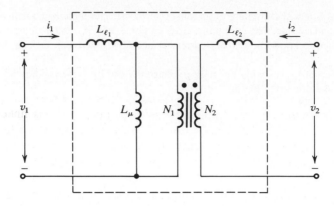

Figure 20.16 Transformer model including leakage inductances L_{ℓ_1} and L_{ℓ_2}.

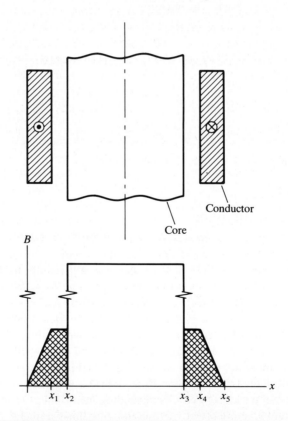

Conductor

Core

Figure 20.17 Model of a transformer winding, showing the origin of leakage flux as the double, cross-hatched area under the B versus x curve between 0 and x_2 and between x_3 and x_5.

only approximation made in the model is no space between the turns. The model more accurately represents a winding made of wire with a square rather than a round cross section.

Figure 20.17 also shows the magnetic flux density across the winding and core. The winding is of nonzero width (x_1), so a part of its flux linkage comes from flux within the conductor itself. An additional component is the result of space between the winding and the core. Neither of these flux components is linked by another winding elsewhere on the core. The linkage of these components by the winding shown thus represents one component of the leakage inductance of the winding. A second component is that caused by flux leaving the core before it reaches the other winding.

Our discussion of Fig. 20.17 suggests several ways of designing a transformer to minimize leakage inductance. The first is to place one winding over the other. This configuration ensures that all the flux *in the core* generated by one winding is linked by the other. Leakage of flux from the core then contributes only to an increase in the magnetizing inductance. Placing one winding over the other also eliminates leakage caused by the inner winding. However, some flux generated by current in the outer winding is not linked by the inner one, so the total leakage is not zero. The second approach recognizes that the leakage flux increases with x_1, the winding thickness, or *build-up*, suggesting that long windings with little buildup generally yield lower values of leakage than short, fat windings. A third technique is to use litz wire, or a conventionally twisted bundle of insulated wires, to make the winding and then connect the strands within the bundle in series or parallel to form both the primary and secondary windings. This approach is theoretically no better than overlapping the windings, but practically it yields a somewhat higher copper packing factor. The use of thin, wide foil conductors results in the lowest achievable value of leakage, because no space is wasted between conductors and the packing factor can be very high. As we mentioned in Section 20.3.1, the problem with this technique is that manufacturing the winding is more difficult than if it were made of wire.

Because the leakage flux in transformers and inductors occupies space external to the core, it can cause electromagnetic interference (EMI) in both its own circuit and external equipment. One technique used to limit the influence of these leakage fields is to place a shorted one-turn wide-foil winding around the entire magnetic circuit, including the windings, as shown in Fig. 20.18(a). Leakage flux induces currents in this winding, which in turn create bucking flux to reduce the fields outside the structure. The net magnetizing flux enclosed by this winding is approximately zero, so the winding has no effect on the magnetizing inductance or the coupling between windings. Another technique is to use what is known as a *pot core*. This is a shell-like core composed of two half core pieces that completely encloses the winding, as shown in Fig. 20.18(b). The windings in this structure are placed around the centerpost (usually on a removable plastic bobbin) and the two core pieces are held in place by a screw or clamp.

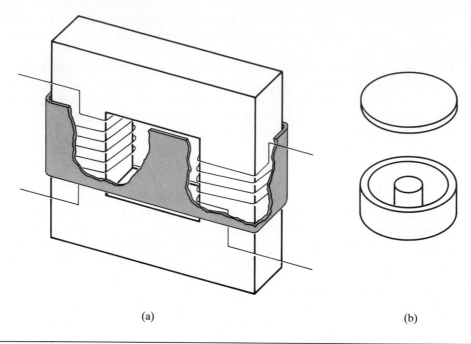

(a)

(b)

Figure 20.18 Two construction techniques for reducing EMI: (a) a transformer structure with a shorted one-turn leakage flux bucking winding (shaded);   (b) a pot-core structure, in which the windings are placed around the centerpost.

20.4.4 Equations for a Nonideal Transformer

The equations describing the relationships between the terminal voltages and currents of the two-winding transformer model of Fig. 20.16 have the same form as those for any linear, time-independent two-port circuit:

$$v_1 = L_1 \frac{di_1}{dt} + L_{12} \frac{di_2}{dt} \tag{20.56}$$

$$v_2 = L_{21} \frac{di_1}{dt} + L_2 \frac{di_2}{dt} \tag{20.57}$$

As the network is reciprocal, $L_{12} = L_{21} \equiv L_m$. By making the appropriate open- and short-circuit tests on the circuit of Fig. 20.16, we can determine the coefficients of these equations. The results are

$$L_m = L_\mu \left(\frac{N_2}{N_1} \right) \tag{20.58}$$

$$L_1 = L_{\ell_1} + L_m \left(\frac{N_1}{N_2} \right) \tag{20.59}$$

and

$$L_2 = L_{\ell_2} + L_m \left(\frac{N_2}{N_1} \right) \qquad (20.60)$$

In these equations, L_m is the *mutual inductance* between terminal pairs 1 and 2, and L_1 and L_2 are the *self-inductances* of terminal pairs 1 and 2, respectively. For a transformer constructed as in Fig. 20.12—and modeled with L_μ referred to the N_1 side, as in Fig. 20.16—the parameters L_μ and L_m are

$$L_\mu = \frac{N_1^2 A_c \mu_c}{\ell_c} \qquad (20.61)$$

$$L_m = \frac{N_1 N_2 A_c \mu_c}{\ell_c} \qquad (20.62)$$

EXAMPLE 20.8

Extracting Parameters from Transformer Measurements

The following measurements are made at the terminals of a transformer having a primary to secondary turns ratio of 5:

1. A 100-kHz voltage with an amplitude of 25 V is applied to the primary. The open-circuited secondary voltage is measured and found to be 4.6 V.

2. A 100-kHz current having an amplitude of 100 mA is applied to the primary. The short-circuit secondary current is −475 mA.

3. A 100-kHz current having an amplitude of 500 mA is applied to the secondary. The short-circuit primary current is −92 mA.

What are the parameters of this transformer if we use the model of Fig. 20.16?

To simplify our analysis, let's reflect all secondary parameters and variables to the primary of the transformer, and identify them by using a prime (for example, $L'_{\ell_2} = 25 L_{\ell_2}$). First, we measure the voltage across L_μ, which is the primary voltage reduced by the reactive divider consisting of X_{ℓ_1} and X_μ, so:

$$V'_2 = \left(\frac{X_\mu}{X_{\ell_1} + X_\mu} \right) V_1 \qquad (20.63)$$

and

$$23 = \left(\frac{X_\mu}{X_{\ell_1} + X_\mu} \right) 25$$

In this equation we have used the value of X_μ as seen from the primary.

Second, we obtain another relationship among the parameters. If we reflect our measurements to the primary side, this relationship is

$$I'_2 = -\left(\frac{X_\mu}{X_\mu + X'_{\ell_2}} \right) I_1 \qquad (20.64)$$

and

$$-0.095 = -\left(\frac{X_\mu}{X_\mu + X'_{\ell_2}}\right)0.1$$

Third, this measurement is like the second but with the terminal pairs reversed. Again, reflecting the measurement to the primary side, we get

$$I_1 = \left(\frac{X_\mu}{X_\mu + X_{\ell_1}}\right)I'_2 \qquad (20.65)$$

and

$$0.092 = \left(\frac{X_\mu}{X_\mu + X_{\ell_1}}\right)0.1$$

Solving *(20.63)*, *(20.64)*, and *(20.65)*, we obtain the 100-kHz parameters of the model:

$$L_{\ell_1} = 14.3 \ \mu H, \qquad L_{\ell_2} = 8 \ \mu H \quad \text{and} \quad L_\mu = 159 \ \mu H$$

or

$$X_{\ell_1} = 9 \ \Omega, \qquad X_{\ell_2} = \frac{X'_{\ell_2}}{N^2} = 5 \ \Omega \quad \text{and} \quad X\mu = 100 \ \Omega$$

The model, with these calculated parameters, is shown in Fig. 20.19.

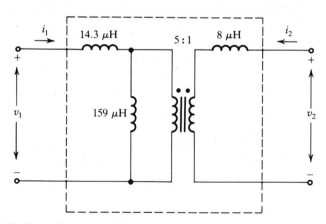

Figure 20.19   Transformer model with the calculated parameters.

20.4.5 Energy Storage Transformers

Most transformers are designed to maximize the magnetizing inductance, because this inductance seldom plays a role in the functioning of the circuit. In some types of power electronic circuits, however, the magnetizing inductance of the transformer does play an important role and is required to store appreciable energy. One type is the transformer-coupled flyback converter discussed in Chapter 7. Another is the resonant converter, in which the transformer magnetizing inductance is used as part of the resonating tank circuit.

When a transformer is used to store energy, the core must either be made of a low-permeability material (powdered iron or ferrite, for example) or contain a gap. The reason is that the stored energy density in highly permeable material is very low (as discussed in Section 20.1.4).

20.4.6 Current and Potential Transformers

The *current transformer* (CT) and *potential transformer* (PT) are generally used in instrumentation to sense a current or voltage, respectively. In some cases their secondary signals also power the sensing circuit, which usually consists of low-current gates or operational amplifiers.

Ideally, the primary and secondary currents in a CT are related by the inverse of the turns ratio. Practically, they differ by the size of the magnetizing current, which is at a minimum when the magnetizing inductance is at a maximum. Therefore the CT is usually made from an uncut core of high-permeability material, such as the Mo alloys. To minimize the size of the transformer, the secondary is usually terminated in a very low impedance to keep the volt-seconds applied to the core as small as possible. The secondary termination is known as the CT *burden*.

The PT ideally relates the primary and secondary terminal voltages by the turns ratio. In practice, the leakage inductances cause drops that prevent the terminal voltage from being related exactly by the turns ratio, so the secondary of a PT should be terminated in as high an impedance as possible. Magnetizing current is generally of little consequence, unless it becomes large enough to affect circuit operation. The techniques used to reduce leakage (discussed in Section 20.4.3) can be effectively employed in the design of PTs.

EXAMPLE 20.9

Analysis of a Current Transformer

A CT constructed of a tape-wound toroid core is designed to measure a sinusoidal current having a maximum amplitude of 5 A. The primary consists of a single turn, and the secondary is 500 turns of AWG 34. The circuit in which the transformer is used is shown in Fig. 20.20. The operational amplifier (assumed to be ideal) presents an incremental short circuit to the transformer secondary. The relationship between the primary current i_1 and the output voltage v_o is

$$v_o = -2i_1$$

The parameters of the core also are shown in Fig 20.20. We are interested in determining the minimum frequency at which the transformer will function without saturating.

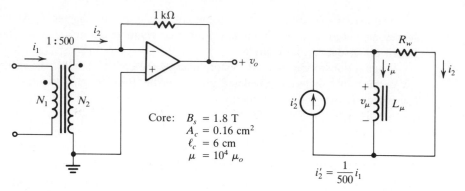

Figure 20.20   Current transformer, with an operational amplifier used as its burden, and a model of the secondary circuit.

Because the operational amplifier is the same as a short circuit to the CT secondary, the voltage at the magnetizing inductance results only from the drop across the secondary impedance (winding resistance and secondary leakage). For simplicity, we assume that the secondary impedance is dominated by winding resistance. This resistance is shown as R_w in Fig. 20.20. The resistance of AWG 34 wire is 8.56 mΩ/cm. Assuming a mean length/turn of 2 cm gives $R_w \approx 8.56\ \Omega$. We also assume that L_μ is large enough to make the magnetizing current negligible until saturation occurs.

The maximum permissible λ, calculated on the secondary side, is

$$\lambda_{max} = N_2 A_c B_s = 500(1.6 \times 10^{-3})(1.8) = 1.44 \times 10^{-2} \text{ V-s} \qquad (20.66)$$

The peak V_μ of the sinusoidal voltage v_μ across L_μ is

$$V_\mu = R_w I_2 = 8.56(0.01) = 0.086 \text{ V} \qquad (20.67)$$

Equating the integral of v_μ to λ_{max} gives

$$\int_0^{\pi/\omega} 0.086 \sin \omega t\, dt \leq 1.44 \times 10^{-2} \text{ V-s} \qquad (20.68)$$
$$\omega \geq 11.9 \text{ rad/s} = 1.9 \text{ Hz}$$

We now check our assumption that the magnetizing current is negligible. The parameters of the CT result in a magnetizing inductance of 0.84 H as measured from the 500 turn secondary. With a secondary current $i_2 = 10$ mA, the peak voltage V_μ across the magnetizing inductance is 0.086 V. This voltage gives a magnetizing current I_μ at 1.9 Hz of:

$$I_\mu = \frac{0.086}{(0.84)(11.9)} = 8.6 \text{ mA}$$

This magnetizing current produces an error of almost 100% in the measured primary current—an unacceptably high error. If we desire an error of less than 3%, the magnetizing current must be less than 0.3 mA, requiring a minimum frequency (ω_{min}) such that:

$$3 \times 10^{-4} \leq \frac{0.086}{0.84\ \omega_{min}} \qquad (20.69)$$

This constraint results in:

$$\omega_{min} \geq 341 \text{ rad/s} = 54.3 \text{ Hz} \qquad (20.70)$$

In this example, then, the error constraint is more severe than the constraint of avoiding saturation. Thus the minimum permissible frequency is approximately 54 Hz.

20.4.7 Transformers with More Than Two Windings

A transformer with more than two windings is needed for many power circuit topologies. The magnetic circuit can take two basic forms to accommodate these extra windings. The first, shown in Fig. 20.21(a), forms a magnetic circuit in which all the windings link the same flux. This is the *series approach*, as the circuit analog in Fig. 20.21(b) suggests. (The windings need not be separate and may be wound on top of one another.) If the transformer is ideal, the winding voltages are related by the corresponding turns ratios, as for a two-winding transformer. However, the currents are not necessarily related by the inverses of the turns ratios. The only constraint on the winding currents is that the Ni products of the windings sum to Φ_{μ}, which is approximately zero in a transformer not designed to store energy.

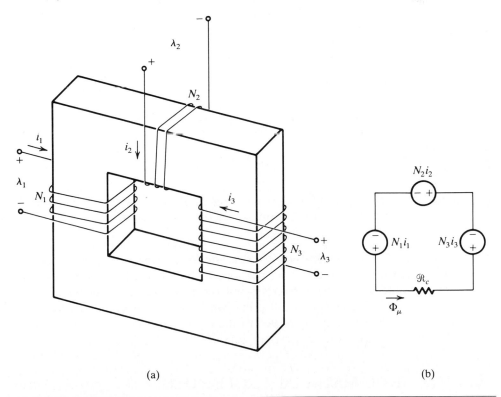

(a)   (b)

Figure 20.21   (a) Structure of a three-winding transformer with a series magnetic circuit.   (b) Circuit analog for the transformer of (a).

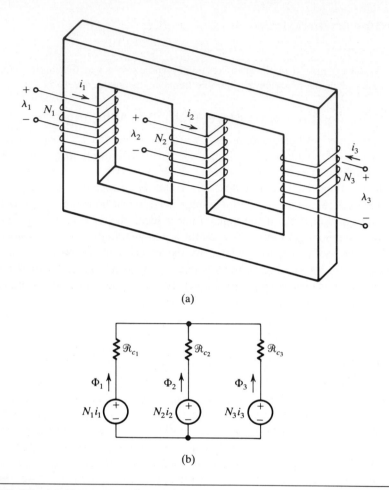

(a)

(b)

Figure 20.22 (a) Structure of a three-winding transformer with a parallel magnetic circuit. (b) Circuit analog for the transformer of (a).

Fig. 20.22 shows another possible structure for the magnetic circuit of a three-winding transformer. All the individual winding fluxes (and hence their flux linkages and voltages) can be different for this core and winding configuration, but they must sum to zero. This is called the *parallel approach*. If the transformer is ideal, the winding currents are related by the inverses of their turns ratios, and the v/N quotients of the windings sum to zero.

20.5 MAGNETIC MATERIAL PROPERTIES

The Curie temperature T_C, saturation flux density B_s, relative permeability μ, and resistivity ρ of some common magnetic materials are listed in Table 20.2. The

Table 20.2   Magnetic Properties of Materials

MATERIAL	T_C (°C)	B_s (T)	μ (μ_o)	ρ ($\mu\Omega$-cm)
Low-Si iron (0.25%)	760	2.2	2.7×10^3	10
Core iron (1%)	810	2.1	4.5×10^3	25
Si steel (2.5%)	780	2.0	5×10^3	40
48% Ni alloy	450	1.5	4×10^4	48
80% Ni, 4% Mo alloy	460	0.8	5×10^4	58
50% Co alloy	950	2.3	10^4	35
Ferrite (Mn–Zn)	150–225	0.4–0.8	1–4×10^3	—
Ferrite (Ni–Zn)	300	0.3	1.5×10^2	—
Metallic glass	370	1.6	10^4	125

various alloys of iron are relatively standard. Different material manufacturers or distributors will give them different names, but in general every manufacturer of such materials has products that match those in the table.

The low-silicon irons (Si $\leq$ 1%) are relatively high-loss, high-saturation flux density materials used primarily at line frequencies for power transformers, motors, and relays. The higher silicon content, generally grain-oriented, steels exhibit lower loss than low-silicon steels but are more expensive. They are used in high-performance line frequency applications, such as high-efficiency motors or toroidal power transformers. Specialized nickel alloys have a saturation flux density approximately 25% lower than that of silicon alloys. But nickel alloys have substantially higher resistivities, which makes them well suited for high-frequency applications. The permeabilities of these materials are the highest of any soft magnetic material. Cobalt alloys have the highest saturation flux density of any available material. Cobalt and nickel alloys are very expensive relative to silicon alloys. Ferrite is a ceramic made of various combinations of primarily ferrous oxide and manganese or nickel, and zinc. A large number of proprietary formulas exist, but in general the MnZn ferrites exhibit higher permeabilities and saturation flux densities than do the NiZn ferrites. The MnZn material can be used in power applications up to a frequency of approximately 1 MHz. The NiZn material can be used at frequencies of up to 10 MHz. Amorphous metal, or metallic glass, material has excellent magnetic properties but is available only as ribbon having a thickness of about 0.05 mm. It is very brittle and difficult to work but is available as a toroid or in a cut-core form (two U-shaped pieces).

Notes and Bibliography

One of the best and most comprehensive treatments of magnetic circuits is [1]. It provides insightful representations of magnetic-circuit models and behavior, and contains a very thorough treatment of the design of inductors. Unfortunately this book is out of print, but most technical libraries have a copy. Be careful of the units.

A wealth of practical information about the design of magnetic components is contained in [2], [3], and [4]. McLyman's book is especially useful for the design of magnetics in high-frequency converters. An array of extraordinarily useful tables, graphs, and charts illustrating material properties, design rules, units, conversion factors, and more is contained in [4].

The physics, processing, manufacturing, and properties of ferrites are well covered in [5]. This book and [4] contain good discussions of the *proximity effect*—the redistribution of current in a conductor due to the proximity of other current carrying conductors.

1. MIT EE staff, *Magnetic Circuits and Transformers* (Cambridge, Mass: The MIT Press, 1943).
2. W. T. McLyman, *Transformer and Inductor Design Handbook* (N. Y.: Marcel Dekker, 1978).
3. S. Smith, *Magnetic Components, Design, and Applications* (N. Y.: Van Nostrand Reinhold Co., 1985).
4. R. E. Tarter, *Principles of Solid-State Power Conversion* (Indianapolis: Howard W. Sams & Co., 1985).
5. E. C. Snelling, *Soft Ferrites: Properties and Applications* (London: Iliffe Books Ltd., 1969).

Problems

20.1   What is the inductance of the structure in Fig. 20.2 if the winding is placed on one of the outer legs?

20.2   Repeat Example 20.5 by expressing the peak stored energy in terms of the fields in the gap.

20.3   If the filter inductor is the only cause of loss in the converter referred to in Fig 20.7, what is the efficiency of the converter, based on the model derived in Example 20.6?

20.4   Redo the calculation of minimum frequency in Example 20.7 by using the secondary-side voltage, and show that the answer is the same as that obtained using the primary voltage.

20.5   Consider the circuit of Fig. 20.7 when the converter is first turned on. If the capacitor and load have values such that $i_o(t)$ and $v_o(t)$ can be approximated as rising *linearly* to their final values in 10 switching cycles (2 ms), plot the flux in the inductor as a function of time, assuming that $B_r = 0$.

20.6   Figure 20.23(a) shows a 12-pulse rectifier made by connecting a pair of 6-pulse rectifiers in parallel through an *interphase transformer*. The 3-phase ac sources feeding the 6-pulse rectifiers are phase shifted by $30°$ with respect to each other.

   (a) Assume that the interphase transformer is ideal, and explain how it functions to maintain continuous current in the two 6-pulse bridges.

   (b) In terms of the peak ac line voltage V_s and ω, what maximum number of volt-seconds (λ) is the interphase transformer core required to support?

   (c) In practice, the interphase transformer does not have an infinite magnetizing inductance. An appropriate model for the transformer with finite magnetizing

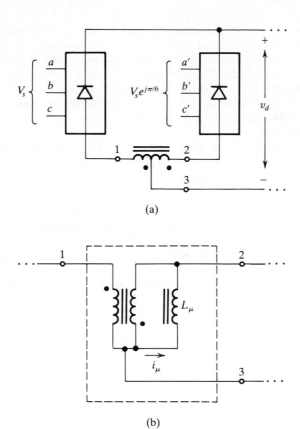

Figure 20.23 (a) The 12-pulse rectifier made by connecting two 6-pulse bridges in parallel through an interphase transformer and (b) the model of the interphase transformer with finite magnetizing inductance for Problem 20.6.

inductance, L_μ, is shown in Fig. 20.23(b). Where does the magnetizing current i_μ come from, and how does it affect circuit operation?

(d) If the peak of the line–line, 60-Hz, ac voltage is 680 V, what minimum value of L_μ allows continuous conduction of the 6-pulse bridges down to a dc load current of 10 A?

20.7 The following specifications are for an inductor needed to model the main field magnet of MIT's magnetically confined fusion research machine, Alcator-C:

$$L = 350 \text{ H} \qquad R = 133 \text{ } \Omega \quad \text{and} \quad I_{max} = 1.39 \text{ A}$$

You are to design an inductor to meet these specifications.

(a) If a maximum flux density of 1.2 T is assumed, what is the required *volume* of the air gap?

(b) What is the 60-Hz Q of this inductor?

(c) Obtain a suitable design by assuming: (1) negligible fringing at the air gap; (2) negligible leakage inductance; (3) a winding space-factor, k_w, defined as the ratio of copper area to total window area, of 0.5; and (4) a current density of no more than 2000 A/cm^2.

(d) What is the weight of your design?

   Use Table 20.1 to obtain the physical and electric characteristics of copper wire. You may specify any core size, but your design should minimize the total weight of copper and iron.

20.8   Figure 20.24 represents the basic operating principle of a self-oscillating, or regenerative, inverter. The transformer core is made of a highly permeable square-loop material. A square-loop magnetic material is one that exhibits a sharply defined saturation region. The sense winding senses when the core enters saturation. One switch is always on, and the switches change state when saturation is reached. For example, if S_1 is on, the core is being driven toward $+B_s$, and the output voltage is $(N_2/N_1)V_{dc}$. When saturation is reached, the sense winding signals the control circuit, which turns S_1 off and S_2 on. The derivative of the flux now changes sign, and the output voltage is $-(N_2/N_1)V_{dc}$.

   The core on which the transformer is wound is a toroid with the parameters shown in Figure 20.24. If $V_{dc} = 100$ V, how many primary turns, N_1 are necessary to yield an output frequency of 25 kHz?

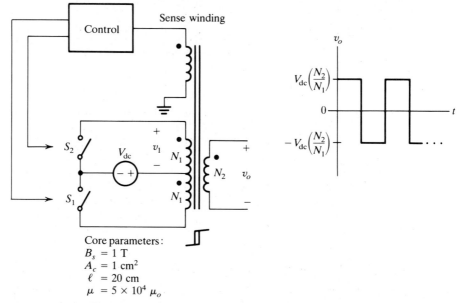

Core parameters:
$B_s = 1$ T
$A_c = 1$ cm^2
$\ell = 20$ cm
$\mu = 5 \times 10^4 \, \mu_o$

Figure 20.24   The self-oscillating, or regenerative, inverter for Problem 20.8.

20.9   Relate the reluctances, $\mathcal{R}_{c_1}$, $\mathcal{R}_{c_2}$, and $\mathcal{R}_{c_3}$ to their physical origins in the three winding transformer with parallel magnetic circuit of Fig. 20.22. That is, identify the segments of the core corresponding to each of the lumped reluctances.

20.10   The three-winding transformers of Figs. 20.21 and 20.22 are connected identically in the following manner. A sinusoidal voltage source of amplitude V_1 is connected to winding N_1, and resistors of value R_2 and R_3 are connected to windings N_2 and N_3, respectively. Determine and compare the resulting winding voltages and currents for each of these two core configurations.

20.11   A three-winding transformer is constructed on a three-legged core, as shown in Fig. 20.25(a). The cross-sectional area of the core is uniform, and the windings

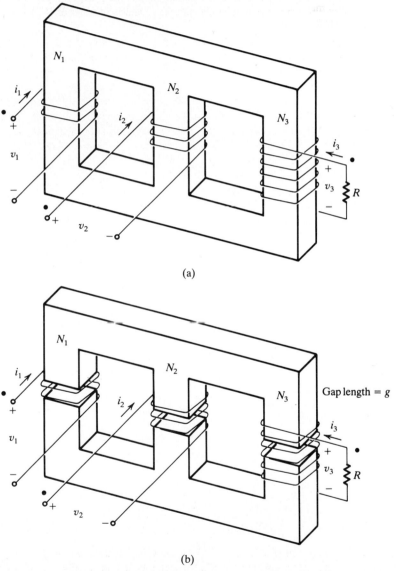

(a)

(b)

Figure 20.25   (a) The three-legged, three-winding transformer analyzed in Problem 20.11. The core has a uniform cross section.   (b) The transformer of (a) with gaps of the same dimensions placed in each of the three legs, the subject of Problem 20.12.

have N_1, N_2, and N_3 turns. Winding N_3 is terminated in a resistor of value R. If sinusoidal voltages with amplitudes V_1 and V_2 are applied to windings N_1 and N_2, respectively, what is the amplitude I_3 of the current i_3 in winding N_3? Assume that no leakage occurs.

20.12   The three legs of the transformer of Problem 20.11 are now modified to have uniform gaps as shown in Fig. 20.25(b). Assuming that the permeability of the core is infinite, determine the amplitude of the currents i_1 and i_2 in terms of V_1, V_2, N_1, N_2, and ω.

PART IV

Ancillary Issues

Chapter 21

Ancillary Issues: An Overview

THE subjects we have addressed so far—power circuits, control, and components—do not encompass all the issues encountered in designing a power electronic system. Among those we have deferred are (1) providing gate and base drives to the power semiconductor switches, (2) using forced commutation to turn off SCRs, (3) controlling the transient voltages and currents that accompany switching in any practical circuit, and (4) providing a thermo-mechanical environment that allows system components to operate within their temperature ratings. We address these four topics in Part IV.

21.1 GATE AND BASE DRIVES

The proper design and operation of the circuits that drive the gates of thyristors and the bases of transistors are crucial to the integrity of a power electronic system. They function not only to turn the switches on and off in a prescribed manner but often also to control the switches in response to sensed signals that predict impending failure—an excessive voltage or current, for instance.

21.1.1 BJT Base Drives

The purpose of a BJT drive circuit is to provide charge to the the base of the transistor. Often, the same circuit is also applicable to a MOSFET. The only real difference is that the BJT requires continuous base current, whereas the MOSFET requires only an impulsive gate-charging current.

A base-drive circuit suitable for applications requiring a controllable duty-ratio, such as dc/dc converters, is shown in Fig. 21.1. The blocks labeled Timer and Edge-triggered monostable are integrated circuit components. The timing capacitor, C_T, establishes the switching frequency, $f_s = 1/T$. The duty-ratio is determined by the monostable delay which is set by R_D and C_D. The base current i_{B_3} supplied to the power transistor Q_3 is a pulse of width DT and an amplitude of approximately $15/R_2$ A.

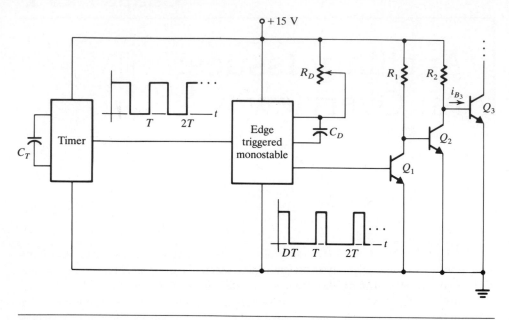

Figure 21.1   A BJT base-drive circuit suitable for applications requiring a controllable duty-ratio.

When discussing gate or base drives, we do not include the signal processing part of the circuit—the timer and monostable of Fig. 21.1. Instead we focus on the circuit that amplifies the drive signal to that level of voltage or current required by the power device. In the case of Fig. 21.1, this is the circuit containing Q_1 and Q_2.

Although the drive circuit of Fig. 21.1 is simple, it has several drawbacks. Primarily, it is inefficient, especially at small values of D. It exhibits a dissipation that is nearly constant and equal to:

$$P_{\text{diss}} = \frac{(15)^2}{R_2} \tag{21.1}$$

if we assume that $R_1 \gg R_2$. A more common drive circuit is shown in Fig. 21.2. In this circuit, Q_3 is turned on by turning on Q_1, and turned off by turning off Q_1 and turning on Q_2. The -5 V supply also forcibly removes base charge from Q_3, turning it off much faster than does the circuit of Fig. 21.1. The principal caution to be observed in driving a BJT is not to exceed the base-emitter reverse breakdown voltage when turning the device off. This breakdown voltage is typically 10 V.

21.1.2 MOSFET Gate Drives

Driving the gate of a MOSFET is a process essentially similar to charging a capacitor. Because the MOSFET will not turn on until the gate voltage reaches its threshold value V_T, typically 3–6 V, driving the gate through a low impedance to

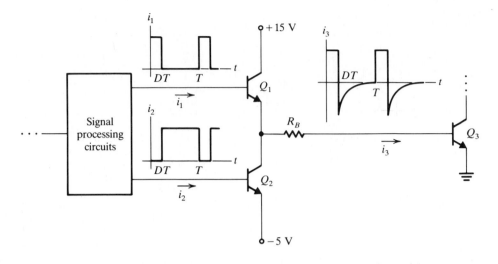

Figure 21.2 An output stage (Q_1 and Q_2) commonly employed to drive the base of a BJT.

minimize turn-on delay is important. For this reason, a circuit of the type shown in Fig. 21.1 is generally unsuitable: A small value of R_2 would cause excessive dissipation when Q_2 is on. Instead a circuit similar to that of Fig. 21.2 is almost always used. The resistor R_B can be made very small, resulting in switching times on the order of 10 ns.

A device that is particularly well suited to driving power MOSFETs is the integrated circuit known as a MOS clock driver. This IC has an output stage similar to that composed of Q_1 and Q_2 in Fig. 21.2 and is capable of both sourcing and sinking current pulses having amplitudes of up to 1 A. The DS0026 is a representative member of this class of devices.

21.1.3 Thyristor Gate Drives

Thyristor and MOSFET gate drives both must provide finite charge to the device, that is, a short pulse of current. However, whereas a negative current is required to turn off the MOSFET, a negative gate current does little to enhance the turn-off process of most thyristors. The notable exceptions are the gate-turn-off thyristor (GTO) and the mos-controlled thyristor (MCT). Therefore a circuit as simple as that shown in Fig. 21.3 is often satisfactory for triggering a thyristor. The gate-cathode resistor R_{GK} desensitizes the turn-on process to noise and dv/dt triggering.

The values of V_G and R_G in Fig. 21.3 are typically 20 V and 20 Ω for devices rated for currents below approximately 100 A. Because the thyristor can exhibit a peak value of v_{GK} during turn on that often exceeds 10 V, V_G must be large enough to ensure that the gate current does not reverse while the device is turning on.

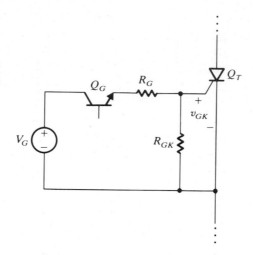

Figure 21.3   A simple thyristor gate-drive circuit. The resistor R_{GK} desensitizes the device to noise and rate-induced triggering.

21.2 THYRISTOR COMMUTATION CIRCUITS

Except for the specially constructed GTO and MCT all members of the thyristor family of swiching devices are turned off by the behavior of the power circuit in which they are connected. In phase-controlled applications, such as those discussed in Chapter 5, and cycloconverter applications discussed in Chapter 10, the SCR is turned off by the reversal of the ac line voltage. In the resonant converters of Chapter 9, turn off is accomplished by the natural behavior of the LC filter comprising part of the power circuit. However, other applications, because of their voltage and current specifications, require that thyristors be used as the switching elements. In many of these applications the thyristor must be turned off by the action of an auxiliary circuit designed to force the switch current to zero. These auxiliary circuits are known as *commutation circuits*, and the process of turning off the thyristor with a commutation circuit is called *forced commutation*.

21.2.1 A Simple Commutation Circuit

We can explore the basic issues of forced commutation by examining the behavior of the circuit of Fig. 21.4. Assume that Q is on and C is charged to $v_C = V_{dc}$ through R. The switch S is closed at $t = 0$ to initiate the turn off of Q. When we ignore the effect of R, the only purpose of which is to recharge C, the resulting waveforms of the circuit variables are shown in the figure. During the first half-cycle of the L–C ring, i_L adds to i_Q, but in the next half-cycle the ringing current reduces i_Q until $i_Q = 0$ at t_1 and Q turns off. At this time $i_d = I_{dc}$, and it remains at this value until $v_d = 0$ at t_3. Between t_1 and t_3 the voltage across L is zero (because i_L is constant) and C charges linearly. When D turns on at t_3, L and C once again ring until $i_L = 0$, at which time S opens. The commutation capacitor C now begins to recharge to V_{dc} through R in preparation for the next commutation.

21.2.2 Limitations of Forced Commutation

The waveforms of Fig. 21.4 illustrate several limiting features of forced commutation. First, Q does not turn off at $t = 0$ when the commutation process is started. A delay dependent on both the resonant frequency of the commutation circuit and the dc current I_{dc} must be considered if switch timing is critical. Some commutation circuits do not exhibit this delay, but it is present in many practical circuits. Second, the SCR current i_Q is increased from its load value to I_{Qp} during commutation. This increase affects the required rating of Q. Third, and most important, is the time t_{off} during which the SCR must recover its forward blocking capability.

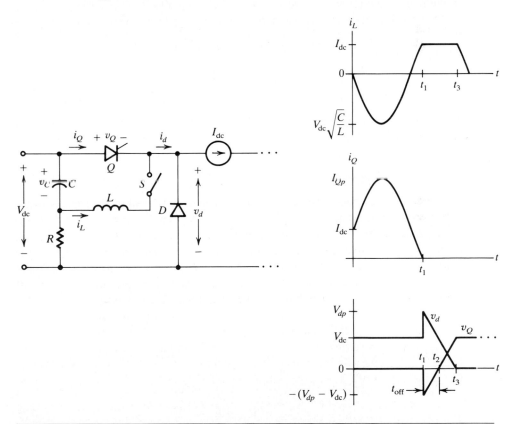

Figure 21.4 A simple commutation circuit consisting of R, L, C, and S, showing the behavior of the circuit variables during commutation.

In other words, the commutation circuit must ensure that $t_{\text{off}} > t_q$, where t_q is the manufacturer's specified circuit commutated turn-off time for Q. The time t_{off} is usually a function of L, C, and I_{dc}.

In addition to these three limitations, all of which can be inferred from the waveforms of Fig. 21.4, the commutation circuit dissipates energy, and thus represents both a cause of reduced efficiency and a source of heat. The principal cause of this energy loss in the circuit of Fig. 21.4 is the resistor R. Although we have ignored it in our description of circuit operation, this resistor has a voltage across it during the entire commutation period. It therefore dissipates energy during this entire time, not just when C is charging (or discharging) to V_{dc} to establish the proper circuit conditions for the next commutation event.

21.2.3 Types of Commutation Circuits

The commutation circuit of Fig. 21.4 performs its function by creating a current (i_L in the figure) that reduces the SCR current by being added to an appropriate node in the network. The SCR current is decreased in a defined manner, sinusoidally in this case. This process is known as *current commutation*, and the commutation circuit is known as a *current commutation circuit*.

An alternative to current commutation is *voltage commutation*. It is a commutation process characterized by a very rapid (ideally instantaneous) reversal of the anode–cathode voltage of the SCR. The reversal is usually accomplished by abruptly shunting the SCR with a capacitor charged to reverse bias the SCR. An alternative name for this process is *impulse commutation*.

21.3 SNUBBING

The term *snubbing* generally refers to the control of unwanted overcurrent or overvoltage transients, or rates of rise of current or voltage, that occur during switching. It also refers to the control of the switching trajectory locus in the switch's i–v plane. Such *switching locus control* is sometimes necessary to prevent the simultaneous presence of a high switch voltage and a high switch current.

Figure 21.5 illustrates both a condition requiring a snubber and the basic operation of a simple snubber circuit. The inductance L_p represents the parasitic inductance in series with the transistor. When the transistor is turned off, we assume that its current falls linearly to zero in time t_f, creating a voltage $L_p I_{\text{dc}}/t_f$ across L_p. The resulting transistor voltage and current waveforms are shown in Fig. 21.5(a), along with the transistor's switching locus in the i_Q–v_Q plane. When Q begins to turn off, its voltage jumps immediately to:

$$V_{p_1} = V_{\text{dc}} + L_p \frac{I_{\text{dc}}}{t_f} \qquad (21.2)$$

This transition is represented by the line from point a to point b in the i_Q–v_Q plane. The transistor current then falls from I_{dc} at point b to 0 at point c in time

t_f. During this time of high dissipation, both i_Q and v_Q have large values. When $i_Q = 0$, v_Q drops abruptly to V_{dc} at point d.

A snubber circuit designed to reduce both the large voltage transient in v_Q and the dissipation in the transistor during turn off is shown in Fig. 21.5(b). The shunt circuit consisting of R_s and C_s provides an alternative path for the inductor current when Q turns off. The resulting waveforms of v_Q and i_Q and the switching locus are also shown. The resistor R_s is necessary not only to damp the oscillation

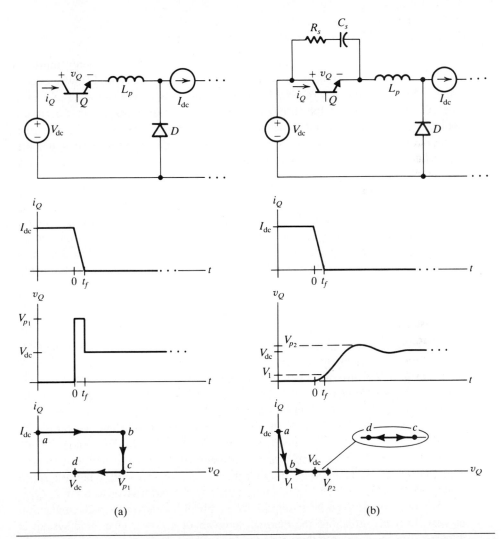

(a)   (b)

Figure 21.5 (a) A circuit containing parasitic inductance L_p that causes a transient overvoltage when Q is turned off. (b) A snubber circuit placed across Q to reduce the switching overvoltage and improve the switching locus.

between L_p and C_s but also to reduce the peak value of i_Q when Q turns on again and C_s discharges. Although this snubber has reduced the energy dissipated in the transistor, the net energy dissipated in the circuit has been increased from that in the snubberless circuit of Fig. 21.5(a). Snubbers are designed to reduce the voltage, current, and thermal stresses on switches, but they do not increase circuit efficiency unless a means is provided for recovering the energy trapped in the snubber inductors and capacitors.

21.4 THERMAL CONSIDERATIONS

Every component in a power circuit dissipates energy, and this energy must be removed from the immediate environment of the circuit. A typical power circuit might exhibit an efficiency of 90 percent when operating at a power level of 1 kW. The 100 W of loss is in the form of heat, and its removal from both the circuit components and the equipment enclosure often determines the physical size of the system. The design of the thermal aspects of a power electronic system is also the major factor determining system reliability.

21.4.1 Thermal Models

Thermal modeling is critical to the design of satisfactory heat transfer systems. These models most often take the form of electric networks, where the circuit elements are analogs for various components and processes in the thermal system. For example, steady-state heat conduction behaves according to the following relationship:

$$Q_{12} = \frac{T_1 - T_2}{R_{\theta 12}} \qquad (21.3)$$

where Q_{12} is the rate at which heat energy is being transferred from a body at temperature T_1 to a body at temperature T_2. This heat transfer rate is directly proportional to the difference between the temperatures of the two bodies, and the constant of proportionality is defined to be $1/R_{\theta 12}$. The parameter $R_{\theta 12}$ is the *thermal resistance* between bodies 1 and 2.

The relationship in *(21.3)* is of the same form as Ohm's law. Therefore we can represent it as an electric circuit, as shown in Fig. 21.6. In this analog, T_1 and T_2 are the voltages of terminals 1 and 2, respectively, measured with respect to a common reference. The current through the resistor of value $R_{\theta 12}$ is the electric analog of the rate at which heat is being transferred from body 1 to body 2, Q_{12}.

Figure 21.6 and the relationship in *(21.3)* model a steady-state heat conduction process. In many situations the thermal behavior of a power electronic system is dynamic. For instance, in a thyristor used in a pulsed application, where the anode current is high but brief and infrequent, the generation of the thermal energy is equally brief. This energy is initially stored in the thermal mass of the device,

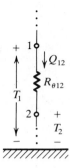

Figure 21.6 The electric circuit analog for the steady-state heat transfer relationship, *(21.3)*.

and the internal device temperature rises rapidly. The stored heat energy then is conducted through the case and its interfaces to the external environment where it is ultimately dissipated. In this situation, the thermal mass of the device plays an important role: it stores energy. The appropriate model is one that includes energy storage, such as a capacitor, and therefore exhibits dynamic behavior. This type of model is known as a *transient thermal model*.

21.4.2 Heatsinking

A ubiquitous feature of power electronic equipment is the *heatsink*, often an extruded aluminum structure with fins. Heat from components mounted on the heatsink is transferred to the sink, from which it is then tranferred to the environment. Although heatsinks are most frequently applied to cooling semiconductor devices, mounting other components, such as resistors, capacitors, and even inductors, on a heatsink to reduce their temperatures is not unusual.

Because of their complex shapes, modeling heatsinks in detail is difficult. However, for design purposes, characterizing the performance of a specific heatsink geometry by a single thermal resistance between the sink and ambient, $R_{\theta SA}$, is generally sufficient. This number, which is provided by the manufacturer, is an approximation in that the sink is assumed to be an isotherm, and the environmental conditions in which it will be used are unknown. But if the heatsink is not too large relative to the size of the heat-generating components mounted on it, the isothermal approximation is good. For example, a heatsink with a mounting surface area of 120 cm^2 would not be too far from isothermal if a device in a TO-3 package (approximate surface area of 4 cm^2) were mounted on it. Manufacturers assume that environmental conditions permit air to flow unobstructed through the fins. This flow results from the bouyancy created by the thermal gradient of the air between the (vertical) fins or is forced by a fan. The former is known as *natural convection* and the latter as *forced convection*.

Problems

21.1   The circuit of Fig. 21.7 shows a *totem pole* gate-drive circuit, consisting of Q_1 and Q_2, driving a MOSFET, Q_3, whose gate is modeled as a capacitor of constant value equal to 100 nF. The drive transistors Q_1 and Q_2 are assumed to switch in zero time. The inverter in the gate circuit of Q_2 creates a complementary drive for the totem pole: When Q_1 is on, Q_2 is off, and vice versa. If the switching frequency of Q_3 is 100 kHz, what are the powers dissipated in R_1 and R_2?

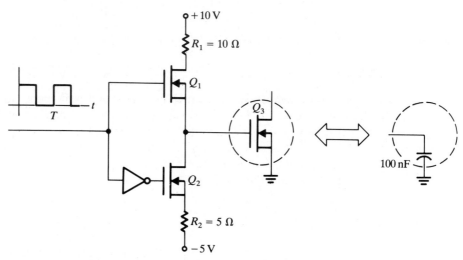

Figure 21.7   The totem-pole gate-drive circuit analyzed in Problem 21.1.

21.2   Consider the waveforms in the simple forced commutation circuit of Fig. 21.4.

(a) Determine I_{Qp}, V_{dp}, and t_{off} in terms of the circuit parameters.

(b) Determine and sketch v_C.

(c) If the switching frequency of Q were f, determine an expression for the average power dissipated in R.

21.3   Figure 21.8 illustrates the application of a *turn-off snubber*. The elements R_s and C_s act to prevent v_Q from becoming excessively high when Q turns off. If the snubber were not present, the abrupt interruption of current in the parasitic inductance L_p could cause v_Q to exceed the rating of Q.

(a) Sketch $v_Q(t)$ if the snubber, consisting of R_s and C_s, were not present and i_Q fell linearly with a fall time of 50 ns.

(b) Now assume that Q turns off instantaneously and that the snubber is in place. Draw the equivalent circuit you would use to determine $v_Q(t)$ and specify initial conditions.

(c) Determine and sketch $v_Q(t)$ for the conditions of (b).

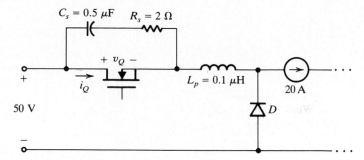

Figure 21.8 A circuit illustrating the application and performance of a turn-off snub-ber. This circuit is discussed and analyzed in Problem 21.3.

21.4 If $R_{\theta 12}$ in Fig. 21.6 represents the thermal resistance between the case of a transistor and the ambient environment, what does Q_{12} correspond to? What are the units of $R_{\theta 12}$?

Gate and Base Drives

THE three terminal power devices introduced and discussed in Part III—the bipolar junction transistor (BJT), the power MOSFET, and the thyristor—each require the application of a control terminal current or voltage to cause the device to switch. So far we have assumed the presence of the necessary gate- or base-drive waveforms. In this chapter we address the detailed drive requirements of these devices and how the necessary drive waveforms are created.

Each family of controlled power devices has unique drive requirements. The bipolar transistor needs a continuous drive current to supply the base defect. The MOSFET gate looks like a capacitor and therefore needs a gate circuit capable of driving a capacitive load. That is, it must be capable of delivering an initial pulse of current and then acting like a voltage source. The thyristor gate must be provided with enough current to initiate the regenerative process that turns the device on. Then, a lower value of continuous gate current may be required, depending on the anode circuit. The supply of the thyristor gate current must be synchronized with the correct anode–cathode voltage condition ($v_{AK} > 0$). In the case of the gate turn-off (GTO) thyristor, the turn-off drive generally requires a separate, more powerful circuit than the turn-on drive. Thyristor gate circuits also influence the susceptibility of the device to rate effects.

A gate- or base-drive circuit actually consists of two parts. The first is the signal processing circuit that creates the desired timing and wave shapes. It is usually constructed of TTL or CMOS logic devices, op-amps, and comparators, and it often receives inputs from sensors in the feedback loop. We do not address the design of these circuits because they are generally straightforward, take various forms, and can make use of new devices that will quickly outdate any material presented here. The second part of the drive circuit, the output stage, takes the low-level signals from the first part and transforms them into the levels of voltage and current required by the main power device. These gate or base variables can be tens of volts and tens of amperes, so this output stage is really a power circuit itself. It sometimes contains circuitry to give the gate or base waveforms a special shape

621

during the switching transient, and it may sense the switch voltage and current to detect and respond to overload conditions.

A frequently encountered complication in the design of gate or base drives is that power devices within a circuit do not share a common emitter, cathode, or source connection. In this case, the drives must be isolated from each other, which often requires a separate power supply for each drive circuit. A common example of this problem is the drive to the upper switches in a full-bridge circuit. Electrically isolating the control circuit from the power device it is driving also may be necessary.

In this chapter we address each of the three controlled power devices in turn. We describe considerations in designing drives and show some representative drive circuits. A number of features of drive circuits are common to all three device types. We present these features in detail, using the bipolar transistor. We do not treat base and gate drives exhaustively, for the clever designer will always be able to synthesize and adapt different circuits to specific applications.

22.1 BIPOLAR TRANSISTOR BASE DRIVES

The base-drive design for a bipolar transistor needs to achieve several goals. First, it must turn the transistor on in the shortest possible time to minimize turn-on losses. Second, it must provide sufficient steady-state base current to keep the device saturated. Third, it is generally required to turn the transistor off as quickly as possible. Sometimes switching speeds are compromised in order to prevent excessive $L\,di/dt$ voltages from appearing in the power circuit.

22.1.1 The Turn-On Problem

To turn on a bipolar transistor, an amount of majority charge sufficient to at least forward bias the base-collector junction must be supplied to the base region. Forward biasing of this junction occurs only after the base charge reaches that value necessary to support the collector current at quasi-saturation, $i_{C(\text{sat})}$. If the base current is constant at $i_B = i_{C(\text{sat})}/\beta$ (where β is the current gain at the edge of quasi-saturation), the collector current will rise asymptotically to $i_{C(\text{sat})}$ with a time constant that depends on the base defect. This rise time can be unacceptably long. We can reduce it considerably by first applying a large pulse of base current to charge quickly the base region and then reducing the drive to the value necessary to support the steady-state collector current. A base-drive waveform of this type is shown in Fig. 22.1(a) and is frequently called a *pedestal and porch* waveform.

The product $I_{B_1} t_1$ of Fig. 22.1(a) is approximately the charge Q_F needed to bring the excess base concentration up to the level necessary to support $i_{C(\text{sat})}$. The larger we make I_{B_1}, the smaller will be t_1, the transistor's turn-on time. The maximum value of I_{B_1} is constrained either by base-drive circuit limits or by the safe current limits of base leads and connections inside the transistor package.

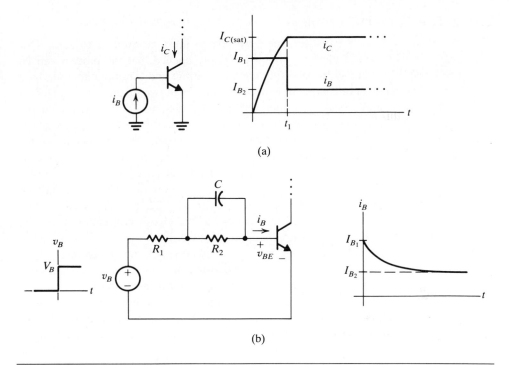

Figure 22.1 (a) Bipolar transistor base-current waveform (pedestal and porch) required for fast turn-on. (b) Standard approach to synthesizing the waveform.

We can create a practical approximation to the waveform of Fig. 22.1(a) with the circuit of Fig. 22.1(b). If we assume that $v_{BE} = 0.8$ V, the currents I_{B_1} and I_{B_2} are

$$I_{B_1} = \frac{V_B - 0.8}{R_1} \tag{22.1}$$

$$I_{B_2} = \frac{V_B - 0.8}{R_1 + R_2} \tag{22.2}$$

As β is not known precisely, and even less is usually known about Q_F, the values of R_1, R_2, and C have to be estimated or determined experimentally. Given V_B, one procedure is to use $\beta_{(min)}$ from the device data sheets and $I_{C(max)}$ for the circuit to determine the value of $R_1 + R_2$. This resistance is divided between R_1 and R_2 so that I_{B_1} is limited as discussed. The value of C is then determined experimentally. As shown in Section 22.1.2, however, these minimum values of R_1 and R_2 increase the transistor's turn-off time if its gain is larger than the minimum value $\beta_{(min)}$.

22.1.2 The Turn-Off Problem

To turn off a bipolar transistor, the base charge must be brought to zero. This condition can occur if the base circuit is simply opened, but the process can be speeded up considerably by reversing the base current and forcibly removing base charge. The removal of charge, in either case, occurs in two distinct stages. During the first stage, stored quasi-saturation charge, Q_S, is removed. This is the charge in excess of the charge Q_F needed to bring the transistor to the edge of quasi-saturation. The transistor's collector-emitter voltage remains low (but not zero) while Q_S is being removed, resulting in a delay between when the turn-off process is initiated at the base and when it becomes apparent at the collector when viewed on the scale of the full off-state voltage or on-state current. This time is known as the *storage delay time,* and is shown as t_s in Fig. 22.2(a). The second stage manifests itself as the fall in collector current as the charge concentration and its gradient at the collector junction both approach zero.

Although the collector-current fall time may be small, the storage delay time can cause control problems, such as the requested duty cycle being smaller than that realized or a phase shift occurring between control input and converter output. The problem is aggravated at higher frequencies. The obvious solution is to avoid operating deep in quasi-saturation. (Power BJTs can seldom be driven hard enough to enter the hard saturation region.) We could do so, using the circuit of

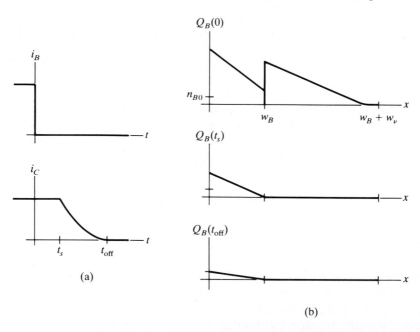

(a)

(b)

Figure 22.2   (a) Base- and collector-current profiles during turn-off from deep in quasi-saturation. (b) Base-charge distribution at three times during the turn-off process ($t = 0$, t_s, and t_{off}).

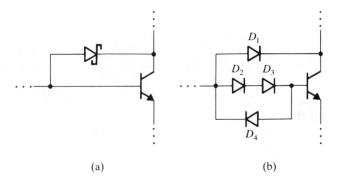

(a)   (b)

Figure 22.3 Antisaturation clamp circuits: (a) a Schottky clamp suitable for low-voltage applications; (b) bipolar diodes forming a clamp that can be used for high-voltage transistors.

Fig. 22.1(b), if we selected the value $R_1 + R_2$ with knowledge of the precise value of β. However, we cannot know β precisely, because it is a function of too many variables (temperature and current, for instance). A more practical solution is to use an *antisaturation clamp*, or a base-drive circuit that senses v_{CE} and controls the base current to keep this voltage above some minimum, that is, not too deep into quasi-saturation. The price we pay for either of these solutions is a slightly higher on-state voltage.

Figure 22.3(a) shows an antisaturation clamp circuit used for low-voltage, low-current devices. The Schottky diode prevents the collector–base junction from being forward biased by more than the forward drop of the diode, or about 0.4 V, thereby keeping the transistor at the edge of saturation. Base current beyond that necessary to maintain this collector voltage is bypassed through the Schottky diode to the collector circuit. Unfortunately, a Schottky diode cannot be used to clamp most power devices, because it must withstand the collector voltage when the transistor is off. In many applications this voltage is beyond the ratings of these diodes, so a bipolar diode must be used instead. A further problem with clamping a power device is that, because of ohmic drops, $v_{CB(\text{sat})} > 0$ (for an npn device), even though the junction itself is forward biased. The diode must clamp the collector to a voltage (measured with respect to the base) that is higher than that value to prevent a hard forward bias on the junction. Therefore the anode of the clamp is elevated by placing additional diodes in series with the base—for instance, D_2 and D_3 in Fig. 22.3(b). Diode D_4 permits the flow of negative base current during turn off.

22.1.3 The Darlington Connection

Very high-current power transistors are frequently combined with a drive transistor to form a *Darlington pair*, as shown in Fig. 22.4. The effective current gain for the pair is the product of the individual transistor gains, or $\beta_1\beta_2$. The advantage

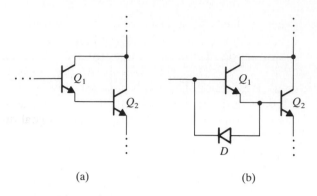

(a)   (b)

Figure 22.4   (a) A Darlington connected pair of bipolar transistors. (b) A Darlington pair with a diode connected to bypass the drive transistor during turn off.

of this connection is that the base current of the main device is drawn from its own collector supply, eliminating the need for a separate base-drive supply (V_B in Fig. 22.5) that could have a current rating on the order of the main device's collector current. For instance, a 50-A, 400-V transistor has a typical gain at rated current of approximately 5. To ensure saturation, I_{B_2} of Fig. 22.5(c) might be 15 A, which is the current that V_B would be required to supply. A Darlington driver rated at 15 A and 400 V would have a gain of about 20, giving a Darlington pair gain of 100 and a required base current on the order of 1 A.

The Darlington connection has two disadvantages. First, it exhibits a saturation voltage that is typically 1 V more than that of the single transistor driven as in Fig. 22.5. The reason is that the saturation voltage of the Darlington pair is the sum of the base-emitter drop of the main device (typically 0.8–1.0 V) and the collector-emitter voltage of the driver (0.6–1.0 V). The result is higher conduction losses. Second, the turn off transition is slower because the capacitor in Fig. 22.5 cannot be connected to the base of the main device, although this problem is reduced somewhat at turn off because the drive transistor acts like an antisaturation clamp. But in many applications the advantage of high gain is more important than these disadvantages, and manufacturers frequently market high-current devices as a Darlington pair in a single package. Many of these packaged pairs also include a diode between the two bases, as shown in Fig. 22.4(b), to provide access to the base of the main transistor during turn off.

22.1.4 Practical Implementation

Figure 22.5 shows two ways of implementing the drive waveform of Fig. 22.1(b). The single-drive transistor circuit of Fig. 22.5(a) provides drive when Q_1 is off. The capacitor C acts as a short circuit when Q_1 turns off, providing an initial peak base current of approximately $I_{B_1} = V_B/R_1$, which decays to $I_{B_2} = V_B/(R_1 + R_2)$ as

C charges to approximately $R_2 V_B/(R_1 + R_2)$. Unfortunately, in this circuit Q_1 must be on to keep the main device off. If the base drive to Q_1 fails, the main transistor will turn on, perhaps at the wrong time. Also, considerable power can be dissipated in the drive circuit, even when the main device is off. The circuit of Fig. 22.5(b) avoids this problem by placing Q_2 in series with V_B. The practical problem with this arrangement is that the base of Q_2 is elevated to the potential V_B when the device is on, requiring a level shift of the signal driving Q_2, unless Q_2 is a pnp transistor.

The base-current waveform created by either circuit is shown in Fig. 22.5(c). The new feature of this waveform is the negative pulse of current caused by the discharge of C when Q_1 turns on. This negative base current assists the main

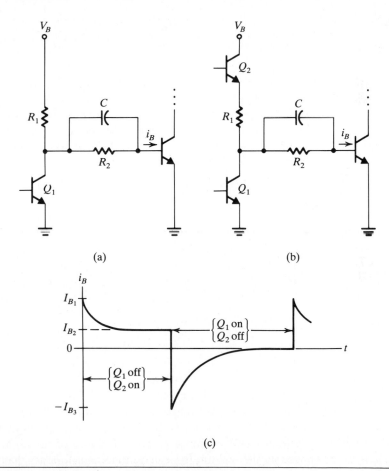

(a)

(b)

(c)

Figure 22.5 Two practical ways of implementing the drive of Fig. 22.1: (a) a single transistor base-drive circuit; (b) a more robust two transistor drive circuit; and (c) the resulting base-current waveform i_B.

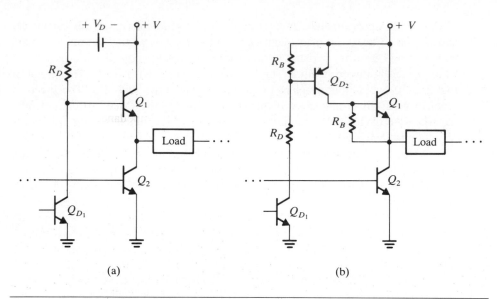

(a)   (b)

Figure 22.6   Two direct coupled circuits for driving the upper transistor in a totem
pole, or bridge leg, connection of power transistors Q_1 and Q_2: (a) a
single transistor circuit requiring a separate source V_D; and (b) a
two transistor circuit requiring a pnp device.

device in turning off quickly. Its peak value, $-I_{B3}$, is determined by the amount
of drive provided to Q_1, as well as parasitic impedances in the base circuit.

A commonly encountered connection of power transistors is the *totem pole*
shown for Q_1 and Q_2 in Fig. 22.6. It often forms one side of a full-bridge (or
H-bridge) circuit. However, this connection poses a troublesome drive problem.
The emitter of the upper transistor Q_1 is at either ground or the positive supply rail,
depending on the state of the switches, but the drive circuit usually has its ground
tied to the emitter of the lower transistor. Therefore, driving Q_2 is relatively easy,
but a level shift is required to drive Q_1.

The circuits shown in Fig. 22.6(a) and (b) for driving the upper transistor will
work, but each has drawbacks. In Fig. 22.6(a) an additional power supply, V_D, is
required to raise the base voltage above that of the collector when Q_1 is saturated.
Additionally, when Q_1 is turned off by turning on Q_{D_1}, the dissipation in R_D
is very large. The circuit of Fig. 22.6(b) avoids the need for the additional supply
V_D by having Q_{D_2} in a Darlington connection to drive Q_1. (The resistors R_B
increase the collector-emitter breakdown voltages of Q_{D_2} and Q_1 from V_{CEO}.)
In this circuit, Q_1 is turned on by turning on Q_{D_1}. Because of the high gain of
the Darlington connection, the resistor R_D can be made much larger than that in
Fig. 22.6(a), and dissipation is not a problem. But the circuit provides no means
to pull negative base current from Q_1 during turn off. Furthermore, Q_{D_2} is a pnp
transistor, which has poorer characteristics, is less available, and is more expensive
than comparable npn devices. An alternative to the direct coupled drive approaches

of Fig. 22.6 is to use a transformer to couple the upper, or both, transistor bases to the drive circuit.

22.1.5 Transformer-Coupled Drives

A transformer may be used to couple the drive circuit to the transistor, as shown in Fig. 22.7, simultaneously providing isolation and an impedance transformation. Isolation permits driving devices at potentials different from that of the drive circuit or driving several devices simultaneously (through multiple secondary windings). The impedance transformation results in reduced drive-circuit current requirements.

The use of a transformer-coupled drive presents two problems. First, transformer leakage inductance is effectively in series with the drive, limiting the achievable rates of rise and fall of base current. Second, transformer core magnetization must be reset after each pulse of base currrent to prevent the transformer from saturating. Resetting is done by driving the magnetizing current negative with $-V_2$ when Q is off (that is, when Q_1 is on).

The operating limitations imposed by core magnetization are important. Let's suppose that the core of the transformer in Fig. 22.7 exhibits saturation flux densities of $\pm B_s$, that $B = -B_s$ when Q_2 is turned on at $t = 0$, and that drive is applied to Q. The voltage across the secondary is the base-emitter drop of Q, or v_{BE}. The flux as a function of time can be found in terms of the number of secondary turns N_S, and the core area A_c:

$$B(t) = -B_s + \frac{1}{N_S A_c} \int_0^t v_{BE} \, d\tau = -B_s + \frac{v_{BE} t}{N_S A_c} \qquad (22.3)$$

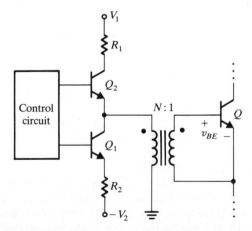

Figure 22.7 Transformer-coupled drive for Q. The drive transistors Q_1 and Q_2 are driven in a complementary fashion, with Q_1 simultaneously providing negative drive during turn off and resetting the flux in the transformer.

As v_{BE} is always approximately 1 V, this relationship shows us that the maximum base-drive duration T_D, which is the time required for the flux to change from $-B_s$ to $+B_s$, can be expressed as:

$$T_D = \frac{2B_s N_S A_c}{v_{BE}} \tag{22.4}$$

The flux in the core must be reset during the time that Q is off. Little current flows in the primary as the core is being reset, because the reverse biased base-emitter junction of Q effectively open circuits the secondary. Thus, there is a negligible drop across the resistor in the emitter of Q_1, and the voltage across the primary is approximately $-V_2$. If the number of turns on the primary is N_P, the time required to reset the core T_R is

$$T_R = \frac{2B_s N_P A_c}{V_2} \tag{22.5}$$

Dividing *(22.4)* by *(22.5)*, we see that the ratio of maximum on time to minimum off time for Q is

$$\frac{T_D}{T_R} = \frac{N_S V_2}{N_P v_{BE}} = \frac{V_2}{N v_{BE}} \tag{22.6}$$

The ratio *(22.6)* represents a serious constraint, for V_2 is limited to a maximum value of the base-emitter breakdown voltage of Q multiplied by the transformer-turns ratio N. With a typical emitter breakdown voltage of 10 V, $v_{BE} = 1$ V and $N = 5$, V_2 can have a maximum value of 50 V. In this case the off time is restricted to a minimum value of 10% of the maximum on time, and the drive circuit now requires a high-voltage supply (50 V) and high-voltage devices.

EXAMPLE 22.1

Design of a Base-Drive Transformer

Figure 22.8 shows the output stage of a transformer-coupled base-drive circuit. The driven transistor Q is to be pulse-width modulated at a frequency of 1 kHz and a maximum duty ratio of 85%. The manufacturer specified that Q has a minimum current gain of $\beta = 15$ and a minimum $v_{BE} = 0.8$ V. A toroid core made of 0.025 mm laminations of a Ni alloy is used because it exhibits high saturation ($B_s = \pm 1.2$ T), high permeability ($\mu = 2.5 \times 10^4 \mu_o$), and low enough losses at this frequency.

Voltage sources of $+5$ V and -15 V are available for the drive circuit, because they also supply logic components and operational amplifiers that precede the drive stage. These are relatively low-current supplies, so a 2:1 transformer ratio is used to reduce the primary side current requirement. This ratio results in a reverse base-emitter voltage of 7.5 V, which will not cause breakdown of the emitter junction. If we design for 50% more drive than necessary, the required base current is

$$i_B = 1.5 \times \frac{200}{10\beta} = 2 \text{ A} \tag{22.7}$$

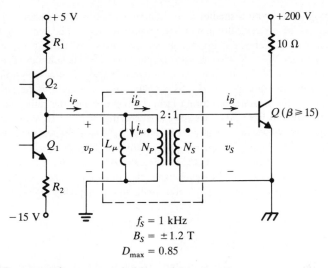

Figure 22.8   A transformer-coupled base-drive circuit.

This value of i_B corresponds to $i'_B = 1$ A on the primary side of the ideal transformer. But we must add the transformer magnetizing current i_μ to obtain i_P. For small numbers of turns i_μ can be on the order of the load current. Let's assume the peak magnetizing current $I_\mu = 0.25$ A and check it later. Thus $I_P = 1.25$ A. During turn on, the primary voltage v_P is 1.6 V, and assuming a saturation voltage of 0.4 V for Q_2, then $R_1 = 3/1.25 = 2.4$ Ω. If the core has been reset to $-B_s$ before Q_2 is turned on:

$$N_S A_c = \frac{v_{BE} T_D}{2 B_s} = \frac{0.8(0.85)(10^{-3})}{2(1.2)} = 2.83 \times 10^{-4} \qquad (22.8)$$

The choice of N_S is primarily a tradeoff between leakage and magnetizing inductance. We choose $N_S = 4$, because we know from experience that it is probably about right. The core cross-sectional area is then $A_c = 0.71$ cm^2. Referring to a manufacturer's table of available cores, we choose one with $A_c = 0.9$ cm^2 and a mean magnetic path length l of 9 cm.

We must now calculate the magnetizing current to check our assumed value for it and make sure that -15 V is sufficient to reset the core in the available off time. We check this last condition by simply plugging the circuit parameters into (22.6). The calculation yields a ratio of $15/1.6 = 9.4$, which is greater than the maximum ratio of on time to off time of $85\%/15\% = 5.67$, confirming that -15 V is ample for reset. To calculate the peak magnetizing current I_μ, we first need to determine L_μ which we can calculate for the selected core:

$$L_\mu = \frac{\mu A_c N_P^2}{l} = 2 \text{ mH} \qquad (22.9)$$

The maximum magnetizing current during turn on is

$$I_\mu = \frac{v_P T_D}{L_\mu} = \frac{2(0.8)(0.85 \times 10^{-3})}{2 \times 10^{-3}} = 0.68 \text{ A} \qquad (22.10)$$

This value is larger than we assumed. We can reduce it by increasing N_P to 12 and N_S to

6, finding a core with a smaller l, or accommodating it by reducing R_1 to 1.8 Ω. Supply-current and drive-circuit dissipation are important, but at this switching frequency, leakage inductance is probably not important. So we choose to increase the number of turns to $N_P = 12$ and $N_S = 6$. The core has ample room for this solution. The peak magnetizing current is now 0.30 A, which is close enough to our original assumption.

Finally, we must specify R_2. Its primary purpose is to limit the peak current once the transformer enters saturation during reset. Dissipation and the rating of the −15-V supply are considerations here. Also, we have assumed that $v_P = -15$ V during the time the core is being reset, which ignores any drop across R_2. The average drop across R_2 during this time is $R_2 I_\mu / 2$, which should be small relative to 15 V. For instance, if $R_2 = 10$ Ω, the reduction of T_R would be $1.5/15 = 10\%$, and we would have to determine whether this amount could be tolerated. However, when the transformer saturates, the voltage across R_2 is approximately 15 V, and if R_2 is too small, an unacceptably high drive-circuit dissipation could result.

22.1.6 Proportional Base Drives

In applications where the transistor is switched at high frequencies (PWM for instance), an effective base drive can be made by providing the base current from a transformer whose primary is in series with the emitter, as shown in Fig. 22.9. The circuit operates as follows. Transistor Q_2 is turned on momentarily, causing base current to flow in Q, which then turns on. The emitter current of Q is coupled back through the transformer to supply continuous base current to Q. (Continuous, of course, until the transformer saturates.) The turns ratio N is designed to be less than $\beta + 1$, so the gain of the feedback to the base is both positive and greater than 1. Turning on Q_1 turns Q off by shunting the secondary current from the base and also resets the flux in the transformer by applying $-V_2$ to the secondary.

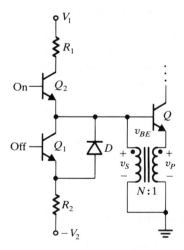

Figure 22.9 A proportional base-drive circuit for high-frequency applications.

The diode D provides a path to maintain continuity of current in the transformer leakage inductance when Q_2 turns off.

The transformer generally has a one- or two-turn primary (emitter side) and a number of secondary turns, N_S, chosen to provide the proper gain. In some cases the emitter lead may simply be slipped through a core around which the secondary turns are toroidally wound. The main consideration in designing the transformer, however, is to make sure that it is capable of supporting the volt-seconds required by the application. When Q is on, the voltage across the secondary v_S is $v_{BE}[N/(N-1)]$. For this voltage and the maximum on time T_D, we can calculate the required core cross section:

$$A_c = \frac{v_S T_D}{N_S B_s} = \frac{v_{BE} N T_D}{(N-1)N_S B_s} \qquad (22.11)$$

The proportional base-drive circuit requires only that Q_2 be pulsed to start the regenerative process, so the drive circuit power requirement is low relative to that of a circuit providing continuous gate drive. The turn-on and turn-off pulses may be coupled to the base through a third winding on the transformer, thereby isolating the drive circuit. The actual base-drive power is obtained from the emitter circuit, where the transformer primary introduces a voltage drop equal to v_{BE}/N. Although this drop increases the effective saturation voltage of Q, the increase is less than if we were to use a Darlington connection. The proportional drive is rugged, because an unexpected increase in collector current will cause a corresponding increase in base current, keeping the transistor in saturation.

EXAMPLE 22.2

Design of a Regenerative Base-Drive Transformer

Let's consider an application in which the switching frequency is 20 kHz, $I_{C(\max)} = 25$ A, $\beta_{(\min)} = 10$, and $v_{BE} = 2$ V. At this frequency, the maximum on time T_D, which occurs when $D = 1$, is 50 μs. In practice we cannot achieve a duty ratio of 1, $D = 0.9$ is more realistic, but our assumption of $D = 1$ is conservative. We want to design the transformer to overdrive the transistor by 33%, resulting in a ratio of collector to base current of 7.5. This number is often referred to as the *forced* β of the transistor when on. The required base current is then 3.3 A. Figure 22.10 shows the base-drive transformer, including the magnetizing branch. Because the magnetizing current is relatively large, it must be considered when determining the turns ratio. In this application we have previously determined that a peak magnetizing current, I_μ, as large as 20% of transformer rated current is acceptable. The result is the branch currents indicated in Fig. 22.10.

Figure 22.10 also shows the turns ratio ($N = 7$) resulting from our assumptions. Using two turns on the primary (a practical number) results in 14 on the secondary. We can now determine A_c from *(22.11)* and the mean core length l from:

$$l = \frac{\mu N_S I_\mu}{B_s} \qquad (22.12)$$

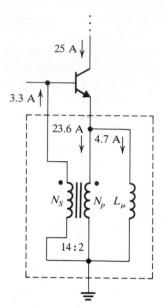

Figure 22.10   Proportional base-drive circuit, showing the transformer magnetizing branch.

For a ferrite core, we assume that $B_s = 0.5$ T and $\mu = 10^3 \mu_o$. Then the core cross-sectional area is $A_c = 0.17$ cm^2 and the core length $l = 2.3$ cm. Two turns of AWG 16 and 14 turns of AWG 24 have sufficiently low resistances and will fit on this core, so the design is practical. But for production purposes we might want to specify a core with a bit more winding room.

22.1.7 Emitter Switching

The fastest way to turn off a bipolar transistor is to draw as much current from the base as possible. Unfortunately, the maximum reverse base current is limited by the low breakdown voltage of the base-emitter junction and the lateral resistance of the base region. This limitation can be avoided by using a turn-off technique known as *emitter switching*, which is illustrated in Fig. 22.11(a). The main transistor Q_M is connected in cascode with a low-voltage device, Q_E. Transistor Q_M is turned off by turning off Q_E, thereby interrupting the emitter current of Q_M. The collector current of Q_M, or i_{CM}, must now flow out from the base, forcing $i_{BM} = -i_{CM}$. Figure 22.11(b) shows the carrier distributions in the device during this time. The slope of the concentrations at the edges of the emitter SCL must go to zero immediately when Q_E turns off. Thus there is no current of either charge type flowing into the base from the emitter, and the emitter-base junction is forward biased so breakdown is not a problem. The majority base charge is removed rapidly through the base terminal at a current equal to $-i_{CM}$. A transistor can be turned off faster using this technique than by any other method.

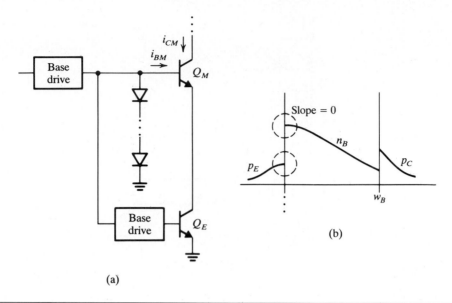

(a)

(b)

Figure 22.11   (a) Cascode connection of transistors used for emitter switching turn off of Q_M. (b) Q_M base-charge distribution and charge flows during the Q_M turn-off process after Q_E is off.

The diodes between the base of Q_M and ground provide an alternative path for the large negative base current during turn off. Thus we can avoid having to design the base-drive circuit to handle this current level. The transistor Q_E must carry the full emitter current of Q_M. However, its voltage rating needs to be only a few volts, because its maximum collector voltage is the forward drop of the diode string. Therefore we can chose Q_E to have a low $v_{CE(\text{sat})}$ relative to that of Q_M. Typically, the cascode connection increases the on-state voltage of the switch by 25%. Because the device is a low-voltage device (often a MOSFET), Q_E does not exhibit the turn-off delays resulting from charge stored in the ν region of the collector, as does Q_M. The cascode connection can increase the speed of turning off Q_M by almost a factor of 10, if the penalty of increased on-state losses can be tolerated.

A practical base-drive implementation for both Q_M and Q_E is shown in Fig. 22.12. The base circuit of Q_M should exhibit a very low series impedance during turn off, so that the voltage at the base of Q_M is kept low during the period of heavy reverse current flow. The diode D is provided to remove R_3 from the circuit during turn off, thereby speeding up the process of turning off Q_E. The diode string shown in Fig. 22.11(a) is absent from Fig. 22.12, because Q_{D_2} is present to carry the negative base current.

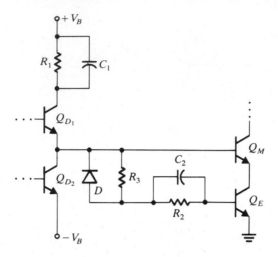

Figure 22.12   Practical implementation of the emitter switching technique.

22.2 MOSFET GATE DRIVES

One of the principal advantages of the MOSFET is that it requires no dc gate-drive power. It does, however, require the transfer of charge to and from the gate electrode in order to turn the channel on and off. This transfer must be rapid to obtain fast switching speeds, and the result is that peak gate currents can be very high.

A concern for the effects of parasitic impedances distinguishes MOSFET gate-drive circuit design from BJT base-drive circuit design. The MOSFET is capable of switching with rise and fall times of less than 10 ns, but practically achievable times are often limited to more than this by the presence of resistance and inductance in the drive circuit.

22.2.1 Gate-Drive Requirements

Switching a MOSFET requires that its gate voltage be switched between some value below the gate threshold voltage V_T, where the device is off, and some value above V_T, where the device is on. Power MOSFETs have a V_T that is typically between 3 V and 6 V. The available gate current is also a critical gate-drive parameter, because it determines how quickly the gate capacitance can be charged or discharged.

The MOSFET gate characteristics are most conveniently represented in the form of a gate-charge versus gate-source voltage graph, such as the one in Fig. 22.13. From this graph we can determine the influence of the gate-drive

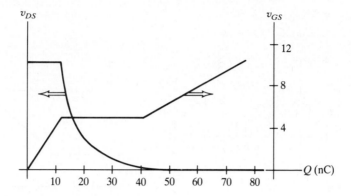

Figure 22.13 Typical gate-charge versus gate-source voltage characteristic for a power MOSFET transistor driving a resistive load.

circuit on the rise and fall times of the drain current or voltage, as well as the gate-circuit power requirements. For instance, if the gate of the device represented by Fig. 22.13 were driven by a circuit capable of sourcing 1 A, the fall time of v_{DS} is approximately:

$$t_f = \frac{40 \text{ nC}}{1 \text{ A}} = 40 \text{ ns}$$

where 40 nC is approximately the change in gate charge required to turn the device on with the specified resistive load. This calculation assumes that the driver current rise time is negligible, which often is not true.

EXAMPLE 22.3

Use of the MOSFET Gate-Charge Characteristic

Because of their ability to provide high peak currents to capacitive loads, clock drivers, such as the DS0026, are frequently used to drive power MOSFETs. Figure 22.14 shows the output stage of a typical clock driver connected to the gate of a MOSFET used in a direct converter. Also shown is the gate-charge characteristic of the MOSFET in this application. The driver is current limited to 1.5 A. Assuming that the driver provides a perfect step to 1.5 A, we want to determine the drain voltage fall time t_{vf} and the power dissipated in the drive circuit if the switching frequency is 500 kHz.

The fall time is approximately equal to the length of the plateau region in the gate-charge characteristic, because during this time the falling drain voltage causes the gate voltage to be constant due to the Miller effect. Thus:

$$t_{vf} = \frac{50 \text{ nC}}{1.5 \text{ A}} = 33 \text{ ns} \tag{22.13}$$

The required drive power is the switching frequency times the energy required to deliver

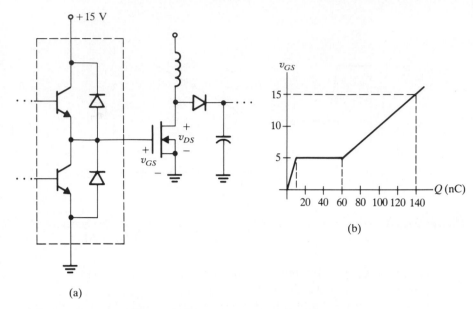

Figure 22.14 (a) A MOSFET in a direct converter being driven by an integrated MOS clock driver, whose output stage is shown in the dashed-line box. (b) The gate-charge characteristic for the MOSFET.

the charge to bring the gate voltage to 15 V:

$$P = QVf = (140 \times 10^{-9})(15)(5 \times 10^{5}) = 1 \text{ W} \qquad (22.14)$$

This power does not include all the losses in the clock driver. At high switching frequencies these integrated circuits exhibit substantial additional losses due to shoot through in the output stage.

22.2.2 Parallel Operation of MOSFETs

Because they are basically resistive devices having a positive temperature coefficient of resistance, MOSFETs do not exhibit the current-hogging behavior of paralleled bipolar transistors. However, MOSFETs do exhibit a tendency to oscillate when paralleled, and measures must be taken to suppress these oscillations to avoid breakdown of the gate dielectric.

The oscillations are generally at frequencies in excess of 100 MHz and occur only during the switching transient when the devices are in their active gain region. For this reason they are difficult to observe. The oscillations can be damped sufficiently by placing small resistors in series with the gate lead, as shown in Fig. 22.15. A value of $R_G = 5 \ \Omega$ is typically adequate and interferes little with the switching times of the devices. In some cases, the resistance of the polysilicon gate electrode is sufficient to eliminate the oscillation problem, but prudent design would include R_G anyway, because process changes made by device manufacturers cannot be anticipated.

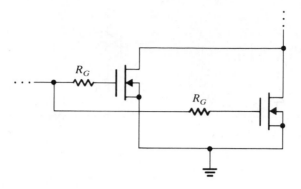

Figure 22.15   Paralleled MOSFETs illustrating the incorporation of resistors R_G to eliminate parasitic oscillations.

22.3 THYRISTOR GATE DRIVES

Thyristor gates pose a unique problem during turn on. Because the gate-cathode drop contains a component resulting from the product of anode current and the bulk resistance of the cathode region, the gate characteristic presented as a load to the drive circuit is time dependent. In particular, the gate voltage increases radically during the turn-on process, which, among other problems, can destroy a poorly designed drive circuit.

Converters using thyristors almost invariably require isolation of one or more of the thyristors' gates. The reason is that devices within a circuit seldom are arranged with common cathodes, or the cathode potential is frequently flying between ground and a high-voltage rail. Either opto-couplers or transformers are generally used to provide isolation. Opto-couplers require a power supply and amplifying circuitry on the thyristor side of the coupler. Transformers can couple enough energy to drive gates directly but require provisions for resetting the flux in cores.

22.3.1 Gate-Drive Requirements

In our discussion of the physics of thyristor operation, we showed that the turn-on process requires that sufficient charge be supplied to the p-base region to increase α_2 to the point of regeneration ($\alpha_1 + \alpha_2 = 1$). Once regeneration commences, the charge is maintained by the flow of anode current. Thus the gate current must be a pulse with an area large enough to charge the base to the device's regenerative threshold. For the SCR, the gate plays only a minor role in the turn-off process. For the GTO thyristor, however, the turn-off gate requirements are very strict. We discuss the problem of GTO turn-off drives separately.

Manufacturers generally use two parameters to characterize the gates of thyristors: V_{GT} and I_{GT}. The first is the minimum gate voltage guaranteed to trigger a thyristor. The second is the minimum gate current at which a thyristor is guaranteed

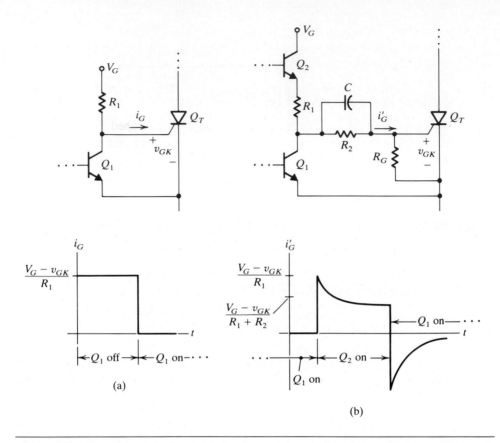

Figure 22.16 (a) Simple thyristor turn-on drive circuit; when Q_1 is on, Q_T is off. (b) Pedestal and porch gate drive with turn-off assist to maximize reapplied dv/dt.

to trigger. Note that V_{GT} is not v_G at $i_G = I_{GT}$. Instead, these parameters describe the extremes of the triggering thresholds of the gate variables, as measured for a large sample of the device. Furthermore, these values are measured *before* anode current flows.

Two straightforward gate-drive circuits are shown in Fig. 22.16. They are similar to the bipolar base drives of Fig. 22.5, but the element values and the operation of the circuits are different. The circuit of Fig. 22.16(a) provides gate current during the time when Q_1 is off. The value of this current, $i_G \approx V_G/R_1$, should be larger than I_{GT} for the particular thyristor type being used. When Q_1 is on, it holds the gate of Q_T close to the cathode potential, providing a degree of immunity to gate noise and off-state dv/dt for Q_T. But this circuit again has the drawback that, if the base drive to Q_1 fails, Q_T turns on.

The alternative circuit shown in Fig. 22.16(b) has the dual advantages that (1) it is fail safe with respect to loss of base drive; and (2) the negative pulse of gate current at turn off assists the thyristor in supporting the highest possible reapplied dv/dt. The pedestal and porch turn-on waveform generated by this circuit has the following desirable effects on the performance of the main device.

1. The pedestal gives the gate an initial charge, enabling the thyristor to support the maximum di/dt at turn on.

2. If the load is inductive and the anode current has not reached its latching value by the time the pedestal has expired, the porch provides gate current until the latching current is reached, keeping the device from turning off prematurely.

3. If the thyristor is required to remain on for low anode currents, the porch permits the anode current to fall below the holding current.

4. If Q_T were a GTO thyristor, the porch would be necessary because manufacturers generally specify a low level of continuous drive to offset the high holding current for these devices.

The element values in the circuits of Fig. 22.16 are not particularly critical. The resistor R_G in Fig. 22.16(b) desensitizes the device to noise and dv/dt turn on. Its value is usually suggested by the manufacturer and ranges from 10 Ω to 1 kΩ, depending on device size. A capacitor is sometimes placed in parallel with R_G to further enhance noise and dv/dt performance. Except for considerations of gate dissipation and efficiency of the drive circuit, there is little to discourage overdriving the thyristor. This situation contrasts with a bipolar drive, where excess base charge results in increased switching times. The only critical requirement is to ensure that the gate current remains positive during the turn-on process.

As mentioned previously, when the device is turning on with a high rate of rise of anode current di/dt, the initially restricted turned-on region presents a high resistance to anode current, resulting in a large voltage at the gate terminal. If V_G is smaller than this voltage, the gate current will actually reverse for a short time while the anode current is rising. This condition reduces the speed of the turn-on process and increases dissipation. To ensure positive gate current at all times, the manufacturer generally provides a gate-drive specification of the form "20 V behind 20 Ω" for a gate current of 1 A. Figure 22.17 is a photograph of the gate current of a thyristor subjected to a high di/dt at turn on with an improperly designed gate drive, showing a substantial negative excursion. As we have suggested, high turn-on di/dt operation of thyristors requires special care in the design of the gate drive. The design rule is to get as much charge into the gate as quickly as possible. The di/dt specification in data sheets is generally predicated on a gate-current rise time shorter than some specified maximum.

22.3.2 Optically Isolated Drives

Opto-couplers are available in a single package (the "mini DIP" being the most common) in a variety of configurations, with a light-emitting diode driving a single transistor, a Darlington connected pair, an SCR, or a TRIAC. Figure 22.18

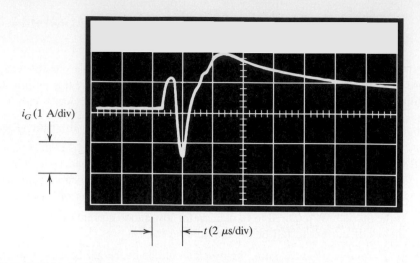

i_G (1 A/div)

$\longleftarrow t$ (2 μs/div)

Figure 22.17 Gate current of an improperly driven thyristor with a high di/dt at turn on.

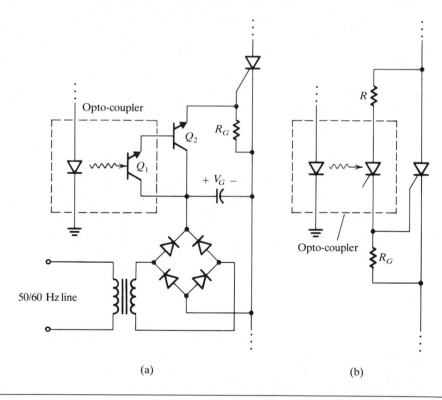

(a)

(b)

Figure 22.18 Two ways of providing an optically isolated gate drive: (a) an opto-transistor with a power supply on the gate side of the coupler; (b) an opto-SCR circuit that does not require an additional power supply.

illustrates two ways of providing an optically isolated drive. The optically coupled diode–transistor pair shown in Fig. 22.18(a) requires additional gain. Here, it is provided by the second transistor Q_2, externally connected as a Darlington, and a power supply V_G, referenced to the SCR cathode. A practical means of creating this supply is also shown. No current-limiting resistor is shown in series with the gate, because the drive transistor usually provides such limiting and there is seldom power to spare on the gate side of the circuit.

Figure 22.18(b) shows how an optically coupled SCR may be used in place of the transistor. The advantage of this arrangement is that the anode voltage of the main thyristor is the source of its gate current, eliminating the need for a separate supply. However, the opto-SCR must have the same voltage rating as the main device. Moreover, the dv/dt rating of the switch is now the lower of the opto-SCR and the main thyristor, and the dv/dt rating of the opto-SCR is often lower than that of the main thyristor.

A problem to be aware of when you use optically coupled gate drives is that the capacitance between the input side and the output side of the coupler can cause unexpected thyristor turn on. You can understand this phenomenon by referring to the half-bridge circuit of Fig. 22.19, which shows the opto-coupler modeled as a capacitance C_c between input and output. The output (transistor) side of C_c is approximately fixed at the potential of the thyristor cathode, because it is separated from the cathode by only three junction drops. Because the input of C_c moves about the signal ground by only a few volts, it is essentially fixed at the signal-ground potential. If the cathode voltage of Q_{T1} changes from positive to zero, as when Q_{T_2} turns on, then C_c would experience a large negative dv_C/dt. The resulting current in C_c is positive base current for Q_1, and because of the high gain of the Darlington, Q_{T_1} is often provided with enough gate current to turn on, causing a direct short across V_D.

To avoid this problem, some opto-couplers are manufactured with very low values of C_c. Others are provided with a *Faraday shield* between input and output. This shield should be connected to the point to which the output side of the gate drive circuit is incrementally referenced—for instance, the cathode of Q_{T_1} in Fig. 22.19(a). An appropriate shield connection and the resulting base and shield currents are shown in Fig. 22.19(b). Note that the collector of the drive transistor is incrementally tied to the cathode of Q_{T_1} because V_G is constant.

22.3.3 Transformer-Isolated Drives

An effective way to provide isolation for gate drives, particularly when two drives shifted by 180° are required, is to use a transformer. Transformers are especially suited for thyristor drives because they are most frequently required to transfer a brief pulse, unlike the continuous drive requirement of a transistor. Thus these *pulse transformers* can be small. They do not suffer from the same problem with parasitic capacitance as do opto-couplers, but interwinding capacitance can be the source of noise on the signal side of the transformer. A Faraday shield can minimize this problem.

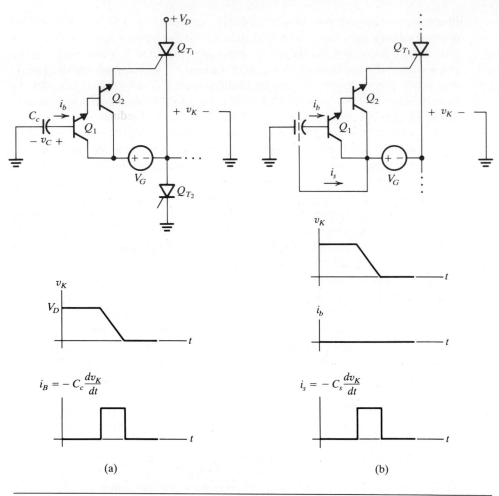

(a)   (b)

Figure 22.19   (a) Illustration of the possibility of turning on Q_{T_1} by charging the parasitic capacitance between the input and output of an opto-coupler.   (b) Proper connection of the shield in a Faraday shielded opto-coupler. The capacitance between the shield and the output terminal of the coupler is C_s.

A pulse-transformer drive for a single thyristor is shown in Fig. 22.20. The transistor Q_2 is on during the time $0 < t < t_1$, providing the SCR with gate drive. The pulse duration t_1 need only be long enough to trigger the SCR, typically 5–50 μs, depending on device size and how long it takes the anode current to reach its latching value. The transformer is usually designed to begin to saturate at t_1 to minimize its size. Between t_1 and t_2, the next time the SCR is triggered, the transformer core is reset by applying a negative voltage across the primary. The diode D opens the secondary at t_1, so $i_a(t_1) = i_\mu(t_1) = I_\mu$. During the reset period the transformer saturates at $-B_s$, and i_μ is limited by R_1. The diode also prevents negative gate current from flowing during the reset period. If the anode current were low, negative gate current could cause the SCR to drop out of conduction.

A frequently encountered power circuit configuration is two thyristors, driven alternately. Adjacent thyristors in the phase-controlled bridge is one example. A single-pulse transformer with two secondaries can be used effectively in this case, as illustrated in Fig. 22.21. The polarities of the secondary windings are arranged so that when the gate of one device is receiving a positive pulse, the other is receiving a negative pulse. Secondary side diodes can be used as in Fig. 22.20 if negative gate current is to be avoided. The bipolar primary voltage v_P eliminates the need for further means of resetting the core.

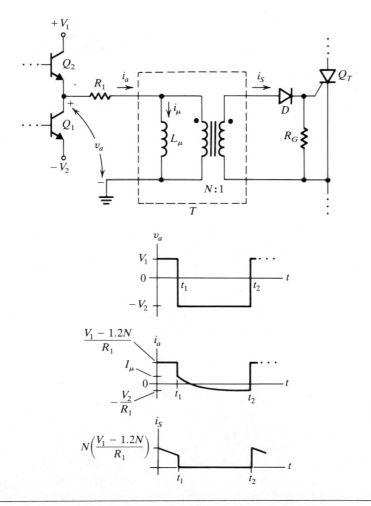

Figure 22.20 Isolated gate drive employing a pulse transformer.

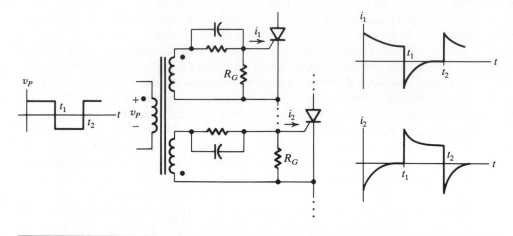

Figure 22.21   Two thyristors fired from a common pulse transformer.

Like other magnetic components in power electronic circuits, pulse transformers are usually custom designed for the application, although some manufacturers make available only a limited product line. Inexpensive transformers are usually wound on a bar core, that is, as a short solenoid, and therefore exhibit a very low remnant flux. These cores can be reset to approximately zero flux with a simple diode-resistor clamp, as shown in Fig. 22.22. The compromise is that only half the available flux swing is used, that is, 0 to B_s instead of $-B_s$ to $+B_s$.

22.3.4 GTO Gate Drives

The gate drive requirements of the GTO differ from those of the SCR in two ways. First, because the gate of the GTO is purposely made insensitive compared to the gate of an SCR, a continuous (but small) gate current is usually supplied during the entire time the GTO is on. This gate current prevents the device from

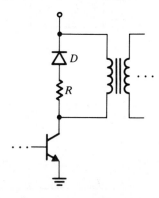

Figure 22.22   Simple reset circuit consisting of R and D suitable for pulse transformers with large air-gap cores.

dropping out of conduction at low anode currents. Second, a relatively high-power (but low-energy) negative gate drive is required to turn the GTO off.

Besides the peak value of the negative gate current, which is determined by the device's turn-off gain and the value of anode current being switched, the other important parameter of the turn-off drive is the rate of rise of the negative gate current. It must be held below the value specified by the manufacturer to avoid potentially damaging current focusing during turn off. This phenomenon is analogous to the di/dt problem at turn on. A snubber is always used with a GTO, and the anode current (which is usually constrained by lead inductance to change slowly) must commutate to this snubber as the GTO is turning off. Therefore the negative gate current must not increase at a rate that exceeds the rate at which the anode current can commutate to the snubber. An inductor is generally used in series with the turn-off drive to limit di_G/dt. Device manufacturers often specify the turn-off di_G/dt rating indirectly by specifying a minimum gate-inductance value.

A practical GTO gate-drive circuit suitable for devices with current ratings up to approximately 50 A is shown in Fig. 22.23(a). A control signal is connected

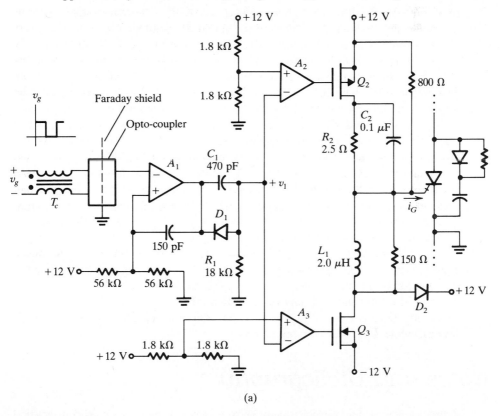

(a)

Figure 22.23 (a) A GTO gate drive suitable for use with devices having current ratings up to about 50 A.

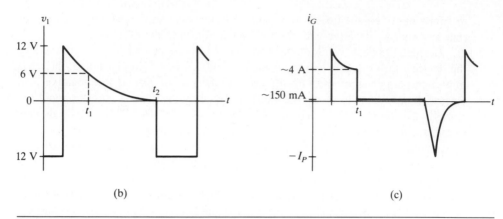

(b)   (c)

Figure 22.23 (cont.)   (b) The shaped turn-on and turn-off pulses at the inputs to the comparators A_2 and A_3.   (c) The GTO gate current.

to an opto-coupler through a *common-mode choke* T_c. The opto-coupler contains a Faraday shield. Both the choke and the shield are employed to reduce coupling of the power circuit to the circuit generating the control signal. These precautions are necessary because the reference on the power circuit side is the cathode of the GTO, a point that can undergo large and rapid excursions relative to signal ground. The output of the opto-coupler drives a comparator, A_1, which is connected with hysteresis to provide additional noise immunity and drives a pulse-forming network consisting of C_1 and R_1. The output of the pulse-forming network v_1, shown in Fig. 22.23(b), drives a pair of comparators, A_2 and A_3, which in turn drive Q_2 and Q_3, respectively. Note that Q_2 is a p-channel power MOSFET.

When v_1 goes high, the output of A_2 goes low and Q_2 turns on, driving the gate of the GTO through the R_2C_2 network, which causes a high-current pulse to turn the GTO on quickly. When v_1 decays to 6 V at t_1, A_2 changes state and Q_2 turns off. The gate is now connected to the 12-V supply through an 800-Ω resistor, which provides a small amount of continuous gate current.

To turn the GTO off, v_1 goes negative, causing the output of A_3 to switch from low to high and turn on Q_3. The rate of rise of the negative gate current is now limited by L_1 and the -12-V supply. The negative peak value of i_G, $-I_p$, is determined by the anode current being turned off, or the current limitation of Q_3. The diode D_2 and the 150 Ω resistor clamp the drain of Q_3 at 12 V should Q_3 turn off while L_1 is still carrying current.

Notes and Bibliography

A detailed presentation of bipolar transistor drive issues, along with a number of practical circuits, can be found in [1]. The most comprehensive reference on SCR gate drive requirements and circuits is [2]. A number of circuits suitable for driving the upper switch in a totem-pole connection are given in [3]. This reference and [4] provide some detailed discussion of GTO gate drive requirements and characteristics, and also present representative drive circuits. A good overview of high-speed, gate-drive circuits for power MOSFETs, and the use of MOSFETs to drive large bipolar Darlingtons are discussed in Sections 5.1 and 6.8, respectively, of [5]. An analysis of oscillations in paralleled MOSFETs and a discussion of corrective measures are presented in [6].

1. *The Power Transistor in Its Environment* (Aix-en-Provence, France: Thompson-CSF, Semiconductor Division, 1978).
2. *SCR Manual*, 6th ed. (New York: General Electric Co., 1979).
3. B. W. Williams and P. R. Palmer, "Drive and Snubber Techniques for GTO's and Power Transistors—Particularly for Inverter Bridges," in *Proc. Int. Conf. on Power Electronics and Variable Speed Drives*, 234: 42–45 (London: 1984).
4. E. Ho and P. C. Sen, "Effects of Gate Drive Circuits on GTO Thyristor Characteristic," in *IEEE IAS Annual Meeting Record* (1984), 706–714.
5. R. Severns, ed., *MOSPOWER Applications Handbook* (Santa Clara, CA: Siliconix Inc. 1984).
6. J. G. Kassakian and D. M. Lau, "An Analysis and Experimental Verification of Oscillations in Paralleled Power MOSFET's," *IEEE Trans. Electron Devices*, 31 (7): 959–963 (1984).

Problems

22.1 The circuit shown in Fig. 22.24 has the following parameters:

$$L_C = 0.2 \text{ mH} \qquad 5 \; \Omega < R_C < 25 \; \Omega \quad \text{and} \quad 12 < \beta < 20$$

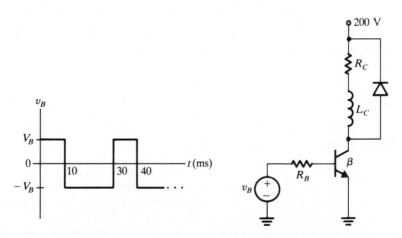

Figure 22.24 The circuit whose base drive parameters are determined in Problem 22.1.

(a) Specify values for V_B and R_B that guarantee saturation for the specified ranges of R_C and β.

(b) In practice, the base-drive voltage v_B does not have zero rise and fall times. Comment on the minimum values you would try to achieve if you were designing the complete drive circuit.

22.2 The circuit of Fig. 22.25(a) exhibits the base- and collector-current waveforms shown. In order to avoid the storage delay time of 0.1 μs, the base circuit is changed to that shown in Fig. 22.25(b). The presence of C_B still allows the base region to be charged swiftly, and R_B can be sized to avoid driving the transistor deep into quasi-saturation. This problem is concerned with determining values for R_B and C_B.

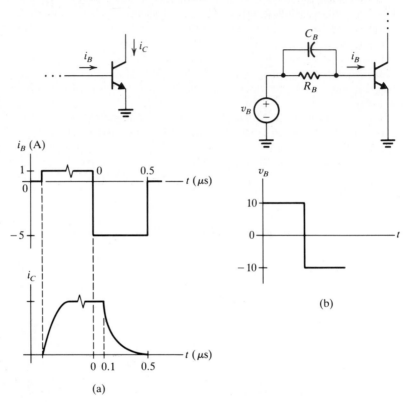

(a)

Figure 22.25 The base circuits discussed in Problem 22.2.

(a) Base current is principally composed of carriers supplying recombination in the base and ν regions and reverse injection into the emitter. Because both processes are proportional to the total stored charge in the base, the base current is also proportional to the total stored charge. This stored charge is composed of two parts: The first, Q_F, is that necessary to bring the transistor to the edge of quasi-saturation; the second, Q_S, is the charge in excess of that value. Removal of this excess stored charge causes storage delay time. Assuming that the lifetime in the base is long compared to the total switching time (0.5 μs), determine the values of Q_F and Q_S for the transistor in Fig. 22.25(a).

(b) To avoid storage delay time, we do not want to supply more base charge than necessary to keep the transistor at the edge of quasi-saturation. However, we would like to supply this charge as quickly as possible. Determine values for R_B and C_B that achieve these two goals. Sketch and dimension the resulting base current at both turn on and turn off. (*Hint:* Assume some small resistance in series with the base so that an RC time constant can be defined and then let this resistor approach zero. Make suitable approximations.)

22.3   The paralleled MOSFETs shown in Fig. 22.15 each have the gate-charge characteristic shown in Fig. 22.14(b). If they are driven at a common point by the current-limited output stage of the clock driver shown in Fig. 22.14(a), what will be the fall time, t_{vf}?

22.4   Calculate the average power dissipated in R_1 and R_2 in Fig. 22.8 of Example 22.1. Ignoring the problem of reset, but considering the effect of magnetizing current, what is the maximum possible on time T_D that this transformer will support?

22.5   Refer to the proportional base-drive circuit shown in Fig. 22.9. Comment on the advantages and disadvantages of placing the transformer primary in the collector circuit instead of the emitter circuit.

22.6   Figure 22.26 shows a proportional base-drive transformer connection that is an alternative to the one shown in Fig. 22.9. Redo Example 22.2 for this connection.

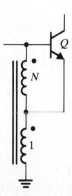

Figure 22.26   An alternative to the proportional base-drive transformer connection shown in Fig. 22.9. This connection is the subject of Problem 22.6.

22.7   Show that when Q is on in the proportional base-drive circuit of Fig. 22.9, the secondary voltage is

$$v_S = v_{BE}\left(\frac{N}{N-1}\right)$$

22.8   At the end of Section 22.1.6 we stated that "the proportional drive is rugged, because an unexpected increase in collector current will cause a corresponding increase in base current, keeping the transistor in saturation." Is this statement true for unbounded increases in i_C? Why or why not? Consider only the need for keeping Q saturated and ignore thermal problems such as bond-wire fusing, etc.

22.9   Sketch the magnetizing current i_μ in the pulse transformer T of Fig. 22.20.

22.10   Figure 22.27(a) is a simple incandescent light dimmer circuit utilizing a TRIAC, T_1, and a *silicon bilateral switch* (SBS), Q_1. The SBS is one of a class of *trigger*, or *breakover*, devices developed for simplifying TRIAC gate-drive circuits. The i–v characteristic of the SBS is shown in Fig. 22.27(b). The device is an open circuit until its voltage reaches ± 10 V, at which point the voltage *breaks back* to ± 1 V.

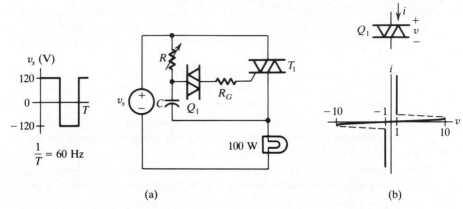

(a)   (b)

Figure 22.27   The incandescent light dimmer discussed in Problem 22.10: (a) the circuit, utilizing an SBS (Q_1) trigger device; (b) the i–v characteristic of Q_1.

To simplify the analysis, assume that the light dimmer is operating from a square-wave source instead of the sinusoidal line voltage. What is the necessary range of the variable resistor R if the *power* to the 100-W lamp is to be varied between 10% and 95%? Assume that the lamp, which is rated at 120 V_{rms}, can be modeled as a simple resistor.

22.11   When the transistor Q in the circuit of Fig. 22.28 is on, we want it to exhibit a voltage of $v_{CE(sat)} = 2.4$ V with $v_{BE} = 1.2$ V. An antisaturation clamp (ASC) is used in the circuit to prevent Q from going too deeply into quasi-saturation.

(a) Specify the network inside the box labeled ASC in Fig. 22.28.

(b) What is the minimum value that i_B can have and still maintain Q at the edge of saturation over the specified range of β?

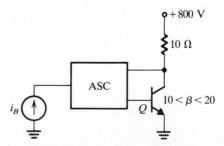

Figure 22.28   A transistor being driven through an antisaturation clamp (ASC). Design of the clamp is the subject of Problem 22.11.

Chapter 23

Thyristor Commutation Circuits

$\mathbf{IN}$ Chapters 5 and 9, we discussed the use of SCRs in circuits that were commutated either by the changing polarity of the line voltage (*line commutation*) or the natural behavior of the circuit (*natural commutation*). In this chapter we discuss the application of auxiliary circuits designed solely to turn off a conducting thyristor. This process is called *forced commutation*, and the auxiliary circuit is known as a *commutation circuit*.

In principle, the combination of an SCR and an appropriate commutation circuit could be substituted for a transistor switch in any power circuit. This substitution is made in many applications that require switch ratings exceeding those of available transistors. For example, a circuit commonly referred to as a *chopper* is often no more than a dc/dc down converter that utilizes an SCR and a commutation circuit instead of a transistor. The maximum switching frequencies of force-commutated circuits are limited to between 1 and 30 kHz because of the time required for an auxiliary commutation circuit to operate and the reverse recovery time of the thyristors. Higher power circuits are limited to lower frequencies.

A commutation circuit has two functions: (1) to force the current in a conducting thyristor to zero; and (2) to ensure that the thyristor remains reverse biased for a time greater than the reverse recovery time of the device, or t_q. Two general classes of circuits perform this function: *voltage-commutation* circuits (sometimes called *impulse-commutation* circuits) and *current-commutation* circuits. The voltage-commutation circuit causes an abrupt application of a reverse biasing voltage to the thyristor. The current-commutation circuit forces the thyristor current to zero through the more gentle action of a resonant circuit.

23.1 VOLTAGE COMMUTATION

Figure 23.1 illustrates the basic operation of a voltage-commutation circuit. The capacitor C is charged to a negative voltage, $v_C = -V_o$, by the action of some circuit not shown. Commutation is initiatied by closing the switch and forcing

653

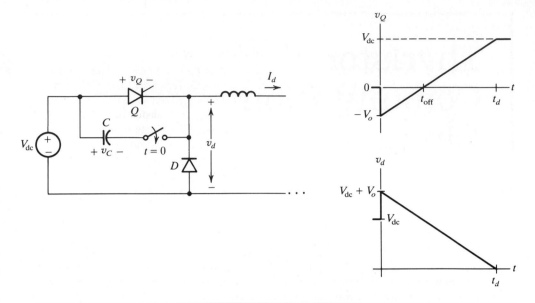

Figure 23.1   Basic voltage-commutation circuit, showing waveforms of circuit variables during the commutation interval.

a reverse bias of $-V_o$ across the thyristor. Stored charge in the thyristor is thus removed impulsively, which is the reason the process is sometimes called impulse commutation. Ideally, the current I_d immediately commutates from the SCR to the capacitor, which charges linearly until its voltage equals the source voltage V_{dc} and D turns on, clamping the capacitor voltage at V_{dc}. The SCR is reverse biased for a time equal to t_{off}, where:

$$t_{off} = \frac{V_o C}{I_d} \qquad (23.1)$$

The values of V_o and C are chosen to ensure that $t_{off} > t_q$.

Several aspects of this simple voltage-commutation circuit's operation are important to its application. First, the circuit must be *reset* before the thyristor can be turned off again. That is, the capacitor must be recharged to $-V_o$ between commutations to obtain repetitive operation. Second, the reverse voltage across the diode, v_d, increases from V_{dc} to $V_{dc} + V_o$ at the moment of commutation, requiring a diode with a higher voltage rating than would be necessary if Q were a transistor. Third, the "switch" composed of the thyristor and its commutation circuit is not really "off" until a time t_d after the initiation of commutation, and t_d is a function of the load current I_d. This condition imposes a load dependent minimum duty ratio on the chopper. And last, the thyristor is subjected to a reverse voltage, making this commutation technique inappropriate for use with thyristors having no reverse blocking capability, such as the asymmetrical SCR (ASCR). Practical commutation circuits must address these issues.

EXAMPLE 23.1

A Transformer-Coupled Voltage-Commutation Circuit

The behavior of the circuit shown in Fig. 23.2 illustrates the operating principles of a frequently used transformer-coupled commutation technique. The circuit shown is essentially a down converter having switches Q_1 and D_2. The remaining components, Q_2, D_1, C, and T comprise the auxiliary commutation circuit. We design the transformer to have a turns ratio of 1, a magnetizing inductance of value L_μ (that is, each winding acts like an inductor of value L_μ when the other is open circuited), and perfect coupling between windings. We assume that the time constant of the load circuit, a dc motor armature perhaps, is much longer than the duration of the commutation process. Thus during commutation we can consider the load current to be constant and replace the load with a current source, as shown.

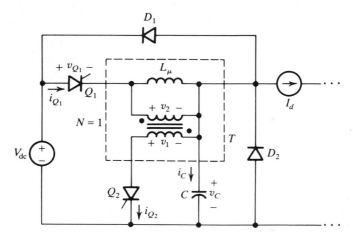

Figure 23.2   A transformer-coupled voltage-commutation circuit. The transformer turns ratio $N=1$.

The basic operation of the circuit is quite simple. Assuming that Q_1 has been on for a long time and Q_2 is off, C is charged to the supply voltage V_{dc}. When Q_2 is turned on, v_C appears across the lower winding $v_1 = -v_C$, and because of the perfect coupling v_C also appears across the upper winding $v_2 = v_C$. Therefore the cathode of Q_1 is elevated instantaneously to $2v_C = 2V_{dc}$, reverse biasing Q_1 by $-V_{dc}$ and turning it off. During both the turn-on and turn-off transients, L_μ and C exhibit ringing behavior at frequency $\omega_o = 1/\sqrt{L_\mu C}$. The diode D_1 clamps v_C to V_{dc} and stops the ringing. We now explore this behavior in detail, first "walking" through the circuit's operation, and then filling in the details.

When Q_2 is turned on, Q_1 turns off and the current I_d flowing in L_μ now flows through the ideal transformer and Q_2. If we apply KCL to the 4-terminal node comprised of T, we determine that $i_C = -2I_d$. Therefore the capacitor C is discharged by a current $2I_d$ until $v_C = 0$ and D_2 turns on. The stored energy remaining in L_μ now dissipates in the winding resistance and the forward drops of Q_2 and D_2, while the load current circulates through D_2. When the energy has been dissipated completely, $i_{Q_2} = 0$ and Q_2 turns off. The turn-off time t_{off} presented to Q_1 is the time between the firing of Q_2 and the time at which $v_C = V_{dc}/2$.

The Q_1 turn-on process is a little more complicated than the turn-off process just described. When Q_1 is turned on, V_{dc} appears across the upper half of T because D_2 is still on and carrying the load current I_d. Since the lower winding is open, the upper winding simply acts like an inductor of value L_μ. The current i_{Q_1} increases linearly with a slope V_{dc}/L_μ until it equals I_d, at which time D_2 turns off and C and L_μ begin to ring. The diode D_1 turns on when $v_C = V_{dc}$, clamping v_C at V_{dc}. At this time $i_{Q_1} > I_d$. The excess energy in L_μ dissipates in the winding resistance and the forward drops of D_1 and Q_1, at which time $i_{Q_1} = I_d$ and D_1 turns off. The circuit is now back in the state in which we started. The switching states of the circuit are summarized in Table 23.1.

Table 23.1   States of the Voltage-Commutation Circuit of Fig. 23.2
(1 indicates *on* and 0 indicates *off*)

PERIOD	Q_1	Q_2	D_1	D_2
$t < 0$	1	0	0	0
$0 < t < t_1$	0	1	0	0
$t_1 < t < t_2$	0	1	0	1
$t_2 < t < t_3$	0	0	0	1
$t_3 < t < t_4$	1	0	0	1
$t_4 < t < t_5$	1	0	0	0
$t_5 < t < t_6$	1	0	1	0
$t_6 < t$	1	0	0	0

The equivalent circuit for the time between the turn on of Q_2 and the turn on of D_2, that is, $0 < t < t_1$, is shown in Fig. 23.3. The initial conditions are $i_{Q_2}(0) = I_d$ and $v_C(0) = V_{dc}$. Solving for the circuit variables of interest during this interval ($0 < t < t_1$) yields

$$v_C = -\frac{2I_d}{\omega_o C} \sin \omega_o t + V_{dc} \cos \omega_o t$$

$$= \sqrt{\left(\frac{2I_d}{\omega_o C}\right)^2 + V_{dc}^2} \, \cos\left[\omega_o t + \tan^{-1}\left(\frac{2I_d}{\omega_o C V_{dc}}\right)\right] \tag{23.2}$$

$$v_{Q_1} = V_{dc} - 2v_C(t) \tag{23.3}$$

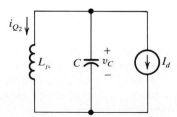

Figure 23.3   Equivalent of the transformer coupled voltage-commutation circuit of Fig. 23.2 during the time interval between the turn on of Q_2 and the turn on of D_2, that is, $0 < t < t_1$.

and

$$i_{Q_2} = -I_d - C\frac{dv_C(t)}{dt}$$

$$= -I_d + \sqrt{(2I_d)^2 + \left(\frac{V_{dc}}{\omega_o L_\mu}\right)^2}\, \sin\left[\omega_o t + \tan^{-1}\left(\frac{2I_d}{\omega_o C V_{dc}}\right)\right]$$

(23.4)

These variables are plotted in Fig. 23.4. Note that the current i_{Q_2} exceeds the load current I_d. It decays to zero during the next circuit state. The turn-off time presented to Q_1, t_{off}, is also shown in the figure, and is defined by:

$$v_C(t_{\text{off}}) = \frac{V_{dc}}{2}$$

(23.5)

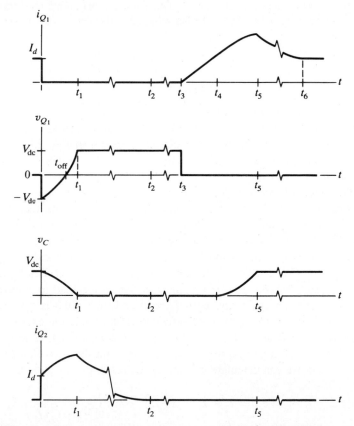

Figure 23.4 Variables in the voltage-commutation circuit of Fig. 23.2.

During the next interval $t_1 < t < t_2$, both Q_2 and D_2 are on as i_{Q_2} decays to zero. The rate of this decay is determined by the forward drops of Q_2 and D_2 and the winding resistance of the lower half of the transformer. Because the voltage driving this current to zero is small, this period is much longer than the others. Once this decay is complete at t_2, Q_2 turns off and Q_1 may be turned on again at t_3.

When Q_1 is turned on again at t_3, its current will ramp linearly up to I_d with a slope of V_{dc}/L_μ. For $t_3 < t < t_4$,

$$i_{Q_1} = \frac{V_{dc}}{L_\mu}(t - t_3) \qquad (23.6)$$

At the same time, the current in D_2 is ramping down, and when it reaches zero and D_2 turns off at t_4, C begins to ring with L_μ. For $t_4 < t < t_5$,

$$v_C = V_{dc}[1 - \cos \omega_o(t - t_4)] \qquad (23.7)$$

$$i_{Q_1} = I_d + \frac{V_{dc}}{\omega_o L_\mu} \sin \omega_o(t - t_4) \qquad (23.8)$$

At t_5, $v_C = V_{dc}$ and D_1 turns on, clamping v_C and permitting the excess energy stored in L_μ to dissipate. At t_6 the circuit has returned to the state it was in when we started our analysis.

23.2 CURRENT COMMUTATION

A current-commutation circuit is designed to bring a thyristor's current to zero with a finite slope, generally through the oscillatory action of an LC circuit. The essence of a current-commutation circuit is shown in Fig. 23.5. The capacitor is initially charged to $v_C = V_{dc}$ through R, and the switch is closed to start the commutation process. We assume that R is so large that it has no further influence on the circuit's behavior during commutation. On its first half cycle, the tank current i_L rings through Q, adding to the load current. During its second half cycle, i_Q is reduced by i_L until $i_Q = 0$ at t_1, when D_2 turns on and carries the net current $i_L - I_d$. The thyristor is reverse biased by the forward drop of D_2 while D_2 is conducting. When once again $i_L = I_d$ at t_2, D_2 turns off and i_L is clamped at I_d by the load. The capacitor then charges linearly until it reaches V_{dc} at t_3, when D_1 turns on and commutation is complete as far as the load is concerned. With D_1 on, L and C can again ring until $i_L = 0$ and the switch is opened. At this time $v_C > V_{dc}$, so some way must be provided to discharge the excess energy from C if periodic operation is desired. If the period is long enough R serves this function.

The turn-off time t_{off} presented to Q is the time during which D_2 is conducting, $t_2 - t_1$, and we can calculate it from the waveform of i_L shown in Fig. 23.5:

$$t_{\text{off}} = t_2 - t_1 = 2\sqrt{LC} \; \cos^{-1}\left(\frac{I_d}{V_{dc}\sqrt{C/L}}\right) \qquad (23.9)$$

Note that this time is dependent on the load current, as it was for voltage commutation.

As for the voltage-commutation circuit, the current-commutation circuit requires a means for resetting the capacitor voltage. The thyristor in this circuit is subjected to a large positive dv/dt at t_2, creating the potential for it to turn on again at this time. This sort of behavior is typical of current-commutation circuits, because they almost always exhibit a ringing current that is clamped—causing a step in the inductor voltage, which appears across one or more thyristors and requires the use of a snubber. Because most current-commutation circuits have a diode in antiparallel with the thyristor, they are especially suitable for commutating thyristors that have no reverse blocking capability (for example, the ASCR).

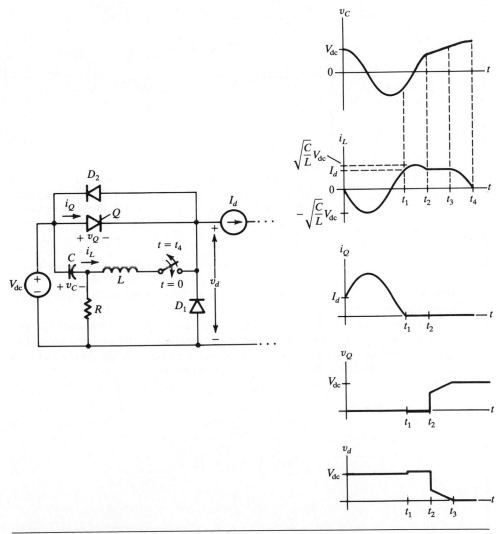

Figure 23.5 Basic current-commutation circuit, showing waveforms of circuit variables during the commutation interval.

EXAMPLE 23.2

A Practical Current-Commutation Circuit

Figure 23.6 shows a practical chopper circuit utilizing current commutation. When Q_2 is fired, the inductor and capacitor ring, first through Q_2 and then through D_2, forcing the current in Q_1 to zero and turning it off. The resistor R ensures that the capacitor voltage is reset to V_{dc} between commutations. Figure 23.7 shows the behavior of the important branch variables during a commutation period.

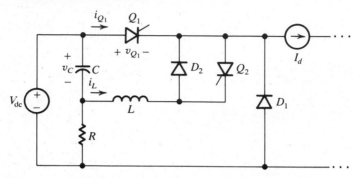

Figure 23.6 A practical current commutated chopper circuit.

The sequence of circuit states is as follows. For $t < 0$, Q_1 is conducting, all the other devices are off, and $v_C = V_{dc}$. At $t = 0$, Q_2 is turned on, and L and C ring at the frequency w_o through Q_1 and Q_2, causing i_{Q_1} to exceed I_d. The current in the tank rings for a half cycle, causing Q_2 to turn off and D_2 to turn on when the current changes sign at t_1. At t_2, $i_L = I_d$ and $i_{Q_1} = 0$, and Q_1 turns off. The inductor current is now clamped at I_d by the load, and C charges linearly until $v_C = V_{dc}$ at t_3 and D_1 turns on. The inductor and capacitor now ring through D_2 and D_1 until t_4, when i_L tries to change sign and D_2 turns off. The excess charge on C decays through R. This sequence of switching states is summarized in Table 23.2.

We now analyze the circuit in detail, assuming that R has no effect except to remove the excess charge from C at the end of the commutation interval. The circuit topology does not change during the time $0 > t > t_2$, even though the ringing current commutates from Q_2 to D_2 at t_1. Therefore the circuit variables of interest are

$$v_C = V_{dc} \cos w_o t \qquad (23.10)$$

$$i_L = -\frac{V_{dc}}{w_o L} \sin w_o t \qquad (23.11)$$

$$i_{Q_1} = I_d + \frac{V_{dc}}{w_o L} \sin w_o t \qquad (23.12)$$

$$w_o = \frac{1}{\sqrt{LC}}$$

At t_2, Q_1 turns off and i_L is clamped at I_d. Therefore C charges linearly until t_3. During this time, the voltage across L is zero because its current is constant, resulting in

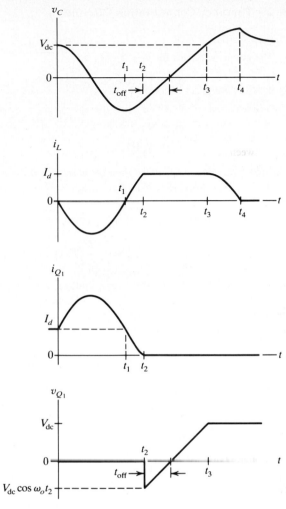

Figure 23.7 Branch variables during the commutation period for the current-commutated chopper of Fig. 23.6.

Table 23.2 States of the Current Commutation Circuit of Fig. 23.6 (1 indicates *on* and 0 indicates *off*)

PERIOD	Q_1	Q_2	D_1	D_2
$t < 0$	1	0	0	0
$0 < t < t_1$	1	1	0	0
$t_1 < t < t_2$	1	0	0	1
$t_2 < t < t_3$	0	0	0	1
$t_3 < t < t_4$	0	0	1	1
$t_4 < t$	0	0	1	0

661

$v_{Q_1} = v_C$. The circuit variables for $t_2 < t < t_3$ are

$$v_C = V_{\text{dc}} \cos \omega_o t_2 + \frac{I_d}{C}(t - t_2) \qquad (23.13)$$

$$i_L = I_d \qquad (23.14)$$

$$i_{Q_1} = 0 \qquad (23.15)$$

$$v_{Q_1} = v_C(t) \qquad (23.16)$$

The turn-off time presented to Q_1 is the time between t_2 and the next zero-crossing of v_C, as indicated in Fig. 23.7. Solving (23.13) for t_{off} at $t = t_2 + t_{\text{off}}$ yields

$$v_C(t_2 + t_{\text{off}}) = 0 = V_{\text{dc}} \cos \omega_o t_2 + \frac{I_d}{C} t_{\text{off}} \qquad (23.17)$$

$$t_{\text{off}} = -\frac{C V_{\text{dc}}}{I_d} \cos \omega_o t_2 \qquad (23.18)$$

During the interval $t_2 < t < t_3$, the reverse voltage across D_1 is equal to the difference between V_{dc} and v_C. Thus when $v_C = V_{\text{dc}}$ at t_3, D_1 turns on. The inductor current is no longer constrained to equal the load current, so L and C begin to ring again through D_2 and D_1. The circuit variables during the interval $t_3 < t < t_4$ are

$$v_C = V_{\text{dc}} + \frac{I_d}{\omega_o C} \sin \omega_o(t - t_3) \qquad (23.19)$$

$$i_L = I_d \cos \omega_o(t - t_3) \qquad (23.20)$$

$$v_{Q_1} = V_{\text{dc}} \qquad (23.21)$$

When $i_L = 0$ at t_4, D_2 turns off and v_C decays back to V_{dc} through R:

$$v_C = V_{\text{dc}} + \frac{I_d}{\omega_o C} e^{-(t - t_4)/RC} \qquad (23.22)$$

23.3 ENERGY RECOVERY

The two chopper circuits presented so far, Figs. 23.2 and 23.6, are limited in their maximum switching frequency by dissipation in their commutation circuits. In Fig. 23.2, this dissipation occurs primarily between t_{off} and t_2, during which time an amount of energy equal to $L i_{Q_2}^2(t_{\text{off}})/2$ dissipates in Q_2, D_2, R, and the resistance of the lower winding of T. In the current-commutation circuit of Fig. 23.6, a voltage across R, equal to $V_{\text{dc}} - v_C$, results in dissipation during the entire commutation interval, and the excess energy stored in C dissipates in R after t_4. In applications where these dissipations limit performance, an *energy recovery network* is utilized. We discuss energy recovery using the impulse commutated half-bridge inverter shown in Fig. 23.8(a).

The SCR-diode pairs Q_1–D_1 and Q_2–D_2 form the switches in the half-bridge circuit. They are also part of the commutation circuit. The capacitors C_1 and C_2

and the transformer T_c are present only because they form part of the commutation circuit. The SCR Q_1 is turned off by turning on Q_2, and Q_2 is turned off by turning on Q_1. This is sometimes called *complementary commutation* or *autocommutation*. We now analyze the operation of this circuit, identify the trapped energy present when commutation is complete, and then consider a network whose function it is to remove this trapped energy and return it to the dc sources.

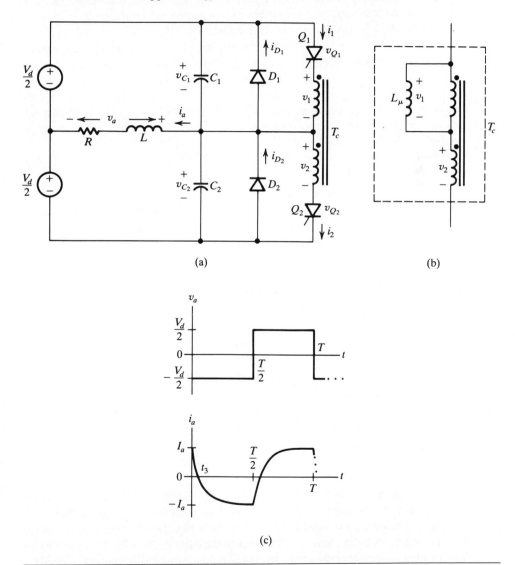

(a)

(b)

(c)

Figure 23.8 (a) An impulse commutated half-bridge inverter. (b) A model for the transformer T_c. (c) The ac load waveforms v_a and i_a.

23.3.1 Complementary Commutation

During commutation of the circuit of Fig. 23.8(a), the capacitors C_1 and C_2 resonate with the magnetizing inductance of the transformer T_c whose model is shown in Fig. 23.8(b). The transformer is constructed so that its magnetizing inductance has a specific value, L_μ, and its leakage is negligible. The value of L_μ is small enough so that, except during commutation, $L_\mu di/dt \approx 0$. That is, the voltage across either winding of the transformer is negligible except during commutation.

Initially, Q_1 is conducting the load current $i_a = I_a$, and Q_2 is off. At $t = 0$, $v_{C_1} = 0$, $v_{C_2} = V_d$, and commutation is initiated by turning on Q_2. We assume that the commutation dynamics are so much faster than the load time constant (L/R) that $i_a = I_a$ during the entire commutation period. When Q_2 is turned on, V_d is immediately impressed across the lower winding of T_c, so that $v_2 = V_d$ and therefore $v_1 = V_d$. This reverse biases Q_1 and turns it off. The current I_a that was flowing in Q_1 immediately commutates to the lower winding of T_c. The resulting equivalent circuit is shown in Fig. 23.9(a).

As far as the load is concerned, commutation is complete when $v_{C_2} = 0$ at t_1 and D_2 turns on. The voltage across the load, v_a, reverses when $v_{C_2} = V_d/2$, but because the load has a lagging power factor, i_a does not reverse until t_3. Therefore the load current flows in D_2 immediately after commutation is complete at t_1 and then flows in Q_2 when it eventually reverses. So, starting from $t = 0$ when commutation is initiated, the conduction sequence is $Q_2 - D_2 - Q_2$. We now determine t_1 by analyzing the circuit of Fig. 23.9(a) with $C_1 = C_2 = C$:

$$v_{C_2} = V_d \cos \omega_o t - \frac{I_a}{C\omega_o} \sin \omega_o t \qquad (23.23)$$

$$\omega_o = \frac{1}{\sqrt{2CL_\mu}}$$

$$v_{C_2}(t_1) = 0$$

$$t_1 = \frac{1}{\omega_o} \tan^{-1} \frac{\omega_o C V_d}{I_a} \qquad (23.24)$$

The energy trapped in L_μ at t_1, W_T, is

$$W_T = \frac{1}{2} L_\mu i_2^2(t_1) \qquad (23.25)$$

This energy dissipates in the forward drops of D_2 and Q_2 and the resistance of the winding. At t_2 all the trapped energy has been removed from L_μ, and Q_2 turns off. The equivalent circuit during this time is shown in Fig. 23.9(b). The current source has been replaced by the actual load, because the time it takes for L_μ to discharge may be long enough that the load current cannot be considered constant. After W_T has been removed from L_μ and $i_2 = 0$, D_2 is on and carries the load current until t_3 when i_a changes sign and commutates to Q_2. Gate current is usually supplied

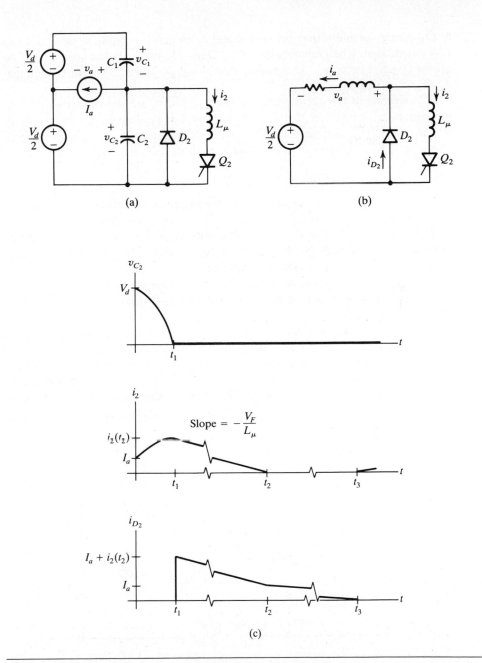

Figure 23.9   (a) The equivalent circuit of Fig. 23.8(a) during the initial commutation interval $0 < t < t_1$.   (b) Equivalent circuit for $t_1 < t < t_2$ during which time the energy trapped in L_μ dissipates.   (c) Waveforms for $0 < t < t_3$.

to Q_2 during the entire time between 0 and t_3 so that it is ready to turn on when i_a changes sign. The commutation of i_a from D_2 to Q_2 is assumed to not be affected by L_μ, because i_a is changing slowly enough and L_μ is small enough that $L_\mu di_a/dt$ is negligible.

Waveforms of the circuit variables for $0 < t < t_3$ are shown in Fig. 23.9(c). For simplicity, we assume that $i_a = I_a$ for $0 < t < t_2$. That is, L_μ discharges in a time that is short compared to L/R. We also assume that the energy dissipates primarily in the voltage V_F, which is the sum of the forward drops of Q_2 and D_2.

23.3.2 An Energy Recovery Network

Figure 23.10(a) illustrates a method of recovering some of the energy W_T trapped in L_μ at t_1. The transformer T_R is ideal and couples energy back to the dc sources through D_2 and D_{22}. We show only that portion of the network that functions immediately after commutation of the load current from Q_1 to Q_2 in Fig. 23.8(a). Although not shown, a symmetrical arrangement of components (another trans-former and diode) serves to recover energy during the alternate commutation. In its complete form, this is a variant of the *McMurray–Bedford commutation* circuit (see Problem 23.9).

We can show the operation of the energy recovery network more simply by redrawing the circuit, as in Fig. 23.10(b). Just after commutation, the initial condi-tions on v_{C_2} and i_2 are the same as those in Fig. 23.9(a). Because of the presence of T_R, D_2 now does not turn on until $v_{C_2} = -V_d/N$ at t_1'. (We use a prime to distinguish these times from those of Fig. 23.9.) At this time, v_{C_2} is clamped at $-V_d/N$, $v_2 \approx -V_d/N$, and i_2 decreases linearly until it reaches zero at t_2'. During

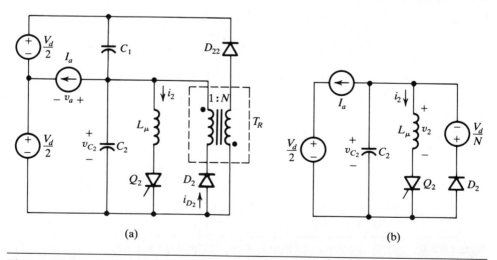

(a)   (b)

Figure 23.10   (a) An energy recovery network, consisting of T_R and D_{22}, designed to return some of the trapped energy in T_c of Fig. 23.8(a) to the source $V_d/2$.   (b) An equivalent circuit for (a) during the recovery phase.

this time $(t_2' - t_1')$, the energy $W_T = L_\mu i_2^2(t_1)/2$ is returned to the dc sources through D_2, D_{22}, and T_R. After t_2', the load current flows through D_2 and T_R, increasing the load voltage by V_d/N. When the load current changes sign at t_3', D_2 and D_{22} turn off and C_2 starts charging towards zero from $-V_d/N$. When it reaches zero at t_4', Q_2 again conducts.

If $V_d/N \gg V_F$, where V_F is the combined forward drops of D_2 and Q_2, almost no trapped energy is lost. However, to reduce the physical size of T_R, it is generally designed to saturate after t_2', so an amount of energy equal to $(C/2)(V_d/N)^2$ is lost.

Notes and Bibliography

Two excellent general references on forced commutation techniques are [1] and [2]. A detailed analysis of the McMurray–Bedford commutation circuit with energy recovery appears in Section 7.4 of [1]. McMurray's paper [3], while containing the word "impulse" in its title, addresses current commutation in a bridge circuit. This paper forms the basis for many of the commutation circuit developments that followed it. An overview of various commutation circuits also can be found in [4]. Murphy and Turnbull [5] contains an extensive bibliography at the end of Chapter 4. Many of the entries in this bibliography pertain to forced commutation circuits.

1. B. D. Bedford and R. G. Hoft *Principles of Inverter Circuits* (New York: John Wiley & Sons, 1964).
2. S. B. Dewan and A. Straughen, *Power Semiconductor Circuits* (New York: John Wiley & Sons, 1975).
3. W. McMurray, "SCR Inverter Commutated by an Auxiliary Impulse," *IEEE Trans. Communications and Electronics* 83 (75): 824–829 (1964).
4. A. J. Humphrey, "Inverter Commutation Circuits," *IEEE Trans. Industry and General Applications* 4 (1): 104–110 (1968).
5. J. M. D. Murphy and F. G. Turnbull, *Power Electronic Control of AC Motors* (Oxford: Pergamon Press, 1988).

Problems

23.1   Determine the thyristor turn-off time t_{off}, if the diode D_2 is eliminated from the circuit of Fig. 23.5.

23.2   The simple voltage-commutation circuit shown in Fig. 23.11 is useful in applications where efficiency is not important. The capacitor C is charged to V_{dc} when Q_1 is on. When Q_c is turned on, $-V_{\text{dc}}$ is immediately impressed across Q_1 and it turns off. The capacitor then charges until $v_C = -V_{\text{dc}}$, at which time D_1 turns on and Q_c turns off.

(a) Sketch v_{Q_1}, v_C, and v_d for a full cycle of operation. Assume that $R_c C \ll T$, where T is the chopping period.

(b) For the source and load resistance values given in Fig. 23.11, determine values

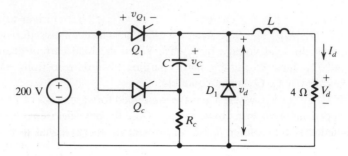

Figure 23.11 A simple, but inefficient, voltage-commutation circuit that is analyzed in Problem 23.2.

of C and R_c that permit the circuit to operate at duty ratios of between 0.2 and 0.8, if $T = 1$ ms and Q_1 has a specified circuit commutated turn-off time of $t_q = 20$ μs.

(c) Determine the efficiency of the circuit with the component values determined in (b) when it is operating at the extremes of D.

23.3 Determine t_1, the time at which D_2 turns on, and $i_2(t_1)$ for the circuit of Fig. 23.9(a).

23.4 A *load-commutated current-source inverter* is shown in Fig. 23.12. The capacitor C may be part of the load or an integral part of the inverter. (In the latter case the circuit is technically not load commutated.) If Q_1 and Q_2 have been on for time $t \gg RC$, then $v_C = I_d R$. When Q_3 and Q_4 are turned on, Q_1 and Q_2 are reverse biased by $-I_d R$ and turn off.

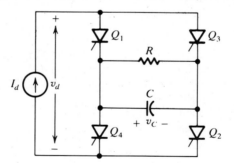

Figure 23.12 The load-commutated current-source inverter of Problem 23.4.

(a) Sketch $v_C(t)$ for several cycles of operation.
(b) Calculate the turn-off time t_{off} presented to the SCRs as a function of R, C, T, and I_d.
(c) Determine and sketch v_d, the voltage across the current source.

23.5 The current source I_d shown in Fig. 23.12 usually consists of a variable dc voltage source, V_{dc}, in series with an inductor, L_d, as shown in Fig. 23.13.

(a) In the circuit of Fig. 23.12, if $I_d = 100$ A, $R = 5$ Ω, $T = 400$ μs, and $C = 20$ μF, what is the value of V_{dc}?
(b) For the parameters used in (a), what is the minimum value of L_d if the peak–peak ripple in i_d can be no more than 20% of I_d? (You need the answer to (c) of Problem 23.4.)

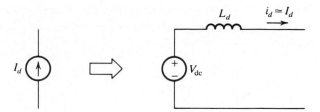

Figure 23.13 The circuit implementation of the current source in Fig. 23.12. Problem 23.5 discusses how the values of V_d and L_d are determined.

(c) What determines I_d and how is it controlled?

23.6 Describe the behavior of the circuit of Fig. 23.8(a) if the energy trapped in T_c, W_T, is not completely removed before i_a changes sign.

23.7 The complementary commutated half-bridge inverter of Fig. 23.8(a) has the following parameters:

$$V_d = 100 \text{ V} \quad I_a = 40 \text{ A} \quad C_1 = C_2 = 5 \ \mu\text{F} \quad \text{and} \quad L_\mu = 100 \ \mu\text{H}$$

What is the turn-off time that is available to the SCR being turned off? (*Hint:* Derive an expression for v_{Q_1} for $t > 0$ and solve $v_{Q_1}(t_{\text{off}}) = 0$ for t_{off} by iteration.)

23.8 In Section 23.3.2, we described the operation of the energy recovery network of Fig. 23.10(a). Determine and sketch the waveforms of v_{C_2}, i_2, and $i_{D_{22}}$ and compare them to the waveforms for the complementary commutated converter of Fig. 23.8(a).

23.9 Starting from a complete half-bridge containing two energy recovery transformers, such as the one in Fig. 23.10(a), that is, one transformer for the top and one for the bottom of the half-bridge, argue that a logical simplification results in the circuit of Fig. 23.14, which contains a single energy recovery transformer, T_R. This is known as the *McMurray–Bedford commutation circuit*.

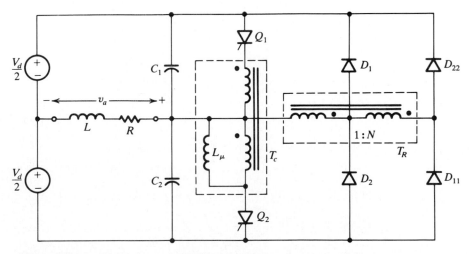

Figure 23.14 The McMurray–Bedford commutation circuit studied in Problem 23.9.

23.10   What is $i_{Q_1}(t_5)$ in the transformer coupled commutation circuit of Fig 23.2?

23.11   In *(23.25)*, W_T is expressed in terms of $i_2(t_1)$. What is W_T as a function of I_a and V_{dc}?

Snubber Circuits and Clamps

WHEN a power semiconductor switch is either on or off, its power dissipation is relatively small. However, the transition from one state to the other is not so benign, for it is a time when high voltage and high current occur simultaneously. Special efforts are often required to ensure that the device will survive this most stressful part of its operating cycle.

There are several reasons why we might need to control, or limit, the voltage and current in a switch when it makes the transition between states. One is the requirement that the voltage and current remain within the *safe operating area* (SOA) of the device. The SOA for a bipolar transistor is shown in Fig. 24.1. This area is bounded by constraints on peak current, peak voltage, peak power, and a region of voltage and current that could result in second breakdown. We discussed the bases of these limitations in Chapter 18.

A second reason is to keep the rate of change of the device voltage and current low enough to provide correct and reliable operation during the switch transition.

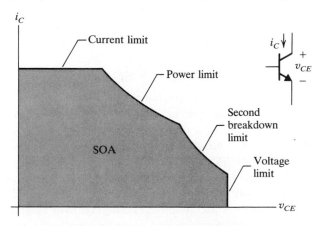

Figure 24.1 A typical safe operating area (SOA) for a bipolar junction transistor.

The false triggering of SCRs caused by an excessive dv/dt, and the destructive results of current crowding in all devices when they are exposed to an excessive di/dt are examples of why this limitation is important.

A third reason is to limit the dissipation that occurs in the device during switching. This switching loss, defined as the energy dissipated in each transition times the number of transitions that occur each second, is converted to heat that must be removed from the device to prevent the junction temperature from exceeding a specified limit. For a given thermal design, the less energy dissipated per transition, the higher the switching frequency can be.

Finally, the switching of a controllable switch is often accompanied by a change in state of a diode, such as in the PWM converter of Fig. 8.15. Thus making this transition slow enough to keep the diode's forward recovery voltage and reverse recovery current from excessively stressing the controllable switch is often important.

To satisfy all these concerns, we often add special components to the power circuit to limit the voltage and current in a switch during switching transitions. These components form what is called a *snubber circuit*. There are two basic kinds of snubbers: one to control the turn-off transition and the other to control the turn-on transition. The first controls the rate of rise of the switch voltage. The second controls the rate of rise of switch current. Both use small energy storage elements— a capacitor in the first case and an inductor in the second. In this chapter we discuss the design and operation of these snubber circuits.

24.1 THE TURN-OFF SNUBBER

The best way to understand the need, design, and operation of a snubber circuit is to investigate an application. The one we chose is the direct down converter of Fig. 24.2(a) in which $I_{dc} = 50$ A and $V_{dc} = 400$ V. It is typical of many power circuits in that an inductor current commutates between a transistor and a diode. The turn-off snubber is designed to modify the voltage and current waveforms during the turn-off transition in order to reduce the power the transistor dissipates.

24.1.1 Turn-Off Dissipation and the SOA

In the following calculations we assume that all switch waveforms make linear transitions from their starting value to their final value. Although this is not exactly correct, it is accurate enough for our present purposes.

When the transistor is turned off, its voltage and current waveforms are typically as shown in Fig. 24.2(b). Note that the current in the transistor cannot change until the diode turns on, which cannot occur until the voltage across the transistor reaches V_{dc}. The turn-off process lasts for a time t_f, which is determined by the parameters of the transistor and its drive circuit. (We will assume t_f is 0.5 μs, a typical value for a bipolar transistor that might be used in this application).

Figure 24.2(c) shows the power dissipated in the transistor, p_{diss}, as a function

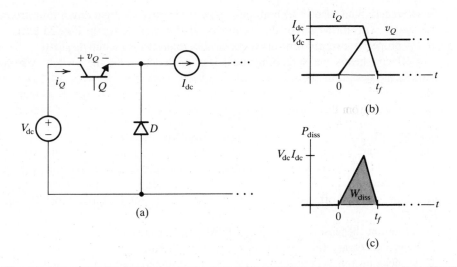

Figure 24.2   (a) The direct down converter to be used for all snubber examples. (b) Transistor voltage, v_Q, and current, i_Q, during turn off.   (c) Dissipated power versus time during turn off.

of time during the turn-off transition. Its peak value is the product of V_{dc} and I_{dc}, which is equal to 20 kW. Its time integral, the energy lost per transition, is

$$W_{diss} = \int_0^{t_f} v_Q i_Q \, dt = \frac{V_{dc} I_{dc} t_f}{2} = 5 \text{ mJ} \qquad (24.1)$$

If the switching frequency were 20 kHz, 100 W would be dissipated in the transistor because of the turn-off transition alone.

Figure 24.3 is a superposition of the v_Q–i_Q switching locus on the transistor's SOA. It shows that if the transistor is operated near its rating, the SOA can easily

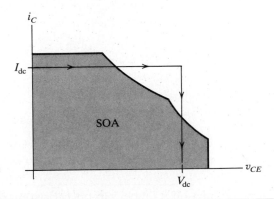

Figure 24.3   Switching locus of the turn-off transition for the waveforms of Fig. 24.2(b).

be exceeded, with respect to both peak power and second-breakdown limits, when the switch waveforms are of the form and relationship shown in Fig. 24.2(b).

A properly designed turn-off snubber both reduces the switch dissipation during turn-off and keeps the switching locus within the SOA of the switch. We now address the design of such a snubber.

24.1.2 A Basic Turn-Off Snubber

If either the peak or the average power dissipation caused by the waveforms of Fig. 24.2(b) are too high for the transistor, or if, as shown in Fig. 24.3, the switching locus exceeds the SOA, a turn-off snubber must be used to change these waveforms. The capacitor C_s in the circuit of Fig. 24.4(a) is a simple snubber. Now when the transistor is turned off, its voltage and current waveforms look like those shown in Fig. 24.4(b).

Note that because C_s supplies a third path through which I_{dc} can flow, it is no longer necessary for v_Q to rise to equal V_{dc} before i_Q can begin to fall as the diode picks up the load current. Instead, any difference between I_{dc} and i_Q flows through C_s. The charging of this capacitor, rather than the switching time of the transistor, governs the rate at which v_Q rises from zero to V_{dc}. Because the voltage across C_s increases at a rate proportional to $1/C_s$, we can choose a large enough capacitor to keep the rise in voltage to only a small fraction, γ_v, of V_{dc} by the time i_Q reaches zero. The consequence of including C_s is that the peak power and total

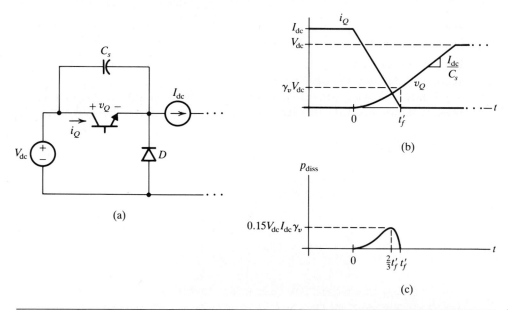

(a)

(b)

(c)

Figure 24.4   (a) A basic turn-off snubber circuit consisting of the capacitor C_s. (b) Transistor voltage and current waveforms at turn off.   (c) Power dissipated in the transistor as a function of time during turn off.

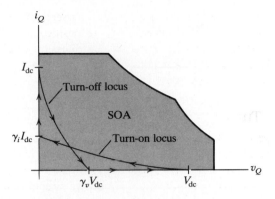

Figure 24.5   The switching loci of the transistor in the circuit of Fig. 24.4 when turn-off and turn-on snubbers are included. Both the turn-on and turn-off loci are maintained within the SOA.

dissipation in the transistor are much smaller than they were before the inclusion of C_s, as illustrated by Fig. 24.4(c).

In addition to reducing the dissipation in the transistor, the snubber capacitor also maintains the switching locus in the SOA of the transistor. Figure 24.5 shows the turn-off switching locus as modified by the presence of C_s. Also shown is the turn-on locus when a turn-on snubber is included.

EXAMPLE 24.1

Choosing a Value For C_s

When the current flowing through the transistor in the circuit of Fig. 24.4(a) reaches zero at $t = t'_f$, the voltage across the transistor is $v_Q = \gamma_v V_{dc}$. Between $t = 0$ and $t = t'_f$, v_Q is quadratic and the total energy dissipated in the transistor during the turn-off transition is

$$W_{\text{diss}} = \int_0^{t'_f} v_Q i_Q dt = \int_0^{t'_f} \gamma_v V_{dc} \left(\frac{t}{t'_f}\right)^2 I_{dc} \left(1 - \frac{t}{t'_f}\right) dt$$

$$= \left(\frac{1}{12}\right) \gamma_v V_{dc} I_{dc} t'_f \tag{24.2}$$

If we choose C_s so that $\gamma_v = 0.2$ for $V_{dc} = 400$ V, $I_{dc} = 50$ A, and if we assume $t'_f = 0.33$ μs, then $W_{\text{diss}} = 0.12$ mJ. During the time $0 < t < t'_f$, the capacitor current is

$$i_C = I_{dc} - i_Q = \frac{I_{dc}}{t'_f} t \tag{24.3}$$

The charge Q_C delivered to C_s in the time t'_f is the integral of this current between 0 and t'_f. We can now determine the size of C_s by solving the following equation:

$$C_s = \frac{Q_C(t'_f)}{v_Q(t'_f)} = \frac{I_{dc} t'_f}{2\gamma_v V_{dc}} = 0.1 \ \mu\text{F} \tag{24.4}$$

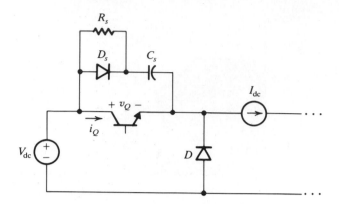

Figure 24.6   A practical turn-off snubber circuit consisting of C_s, R_s, and D_s.

24.1.3 A More Practical Snubber

A capacitor by itself is not a sufficient turn-off snubber because, when the transistor is turned on again, C_s will discharge through Q. The resulting transistor current can be very large and lead to failure. To avoid this problem, we add a resistor and a diode to C_s, as shown in Fig. 24.6. The purpose of the resistor R_s is to limit the discharge current when the transistor is turned on, and the purpose of the diode D_s is to allow the charging current to bypass the resistor during the turn-off transition.

A tradeoff is necessary in determining the value of R_s. It must be small enough to ensure that the capacitor is fully discharged during the shortest time that Q might be on but large enough to prevent the discharge current from exceeding the transistor rating.

24.2 THE TURN-ON SNUBBER

The transistor voltage and current waveforms during turn on in the converter of Fig. 24.2 are shown in Fig. 24.7(a). Initially i_Q rises, but until this current equals I_{dc}, v_Q must remain at V_{dc} because D must be on to carry the difference between I_{dc} and i_Q. When the load current has completely commutated from the diode to the transistor, D turns off and v_Q falls to zero. As during turn off, the transistor voltage and current during turn on are simultaneously nonzero, leading to switching loss and the possibility of leaving the SOA. A turn-on snubber is designed to modify the switching waveforms to reduce this loss and maintain operation within the SOA.

24.2.1 Turn-On Dissipation and the SOA

Dissipation during turn on can be calculated using (24.1), with t_f replaced by t_r. A typical value of t_r for a 400-V, 50-A transistor is 0.5 μs. This value for t_r results in a peak power of 20 kW and an energy loss of 5 mJ during each turn-on

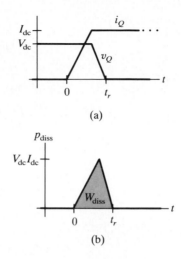

Figure 24.7   (a) Waveforms during turn on of the transistor in the converter of Fig. 24.2(a).   (b) Power dissipated in the transistor as a function of time.

transition. With a switching frequency of 20 kHz, turn on causes another 100 W of dissipation in the transistor.

The switching locus in the i_Q–v_Q plane for this transition is identical to the one shown in Fig. 24.3 for the turn–off transition, except the path is taken in the reverse direction. Once again, the transistor's peak power and second-breakdown limits are exceeded.

24.2.2 A Basic Turn-On Snubber

The solution to the turn-on problem is the dual of the solution to the turn-off problem. A snubber inductor, L_s, is placed in series with the transistor, as shown in Fig. 24.8(a), and the waveforms of Fig. 24.8(b) result. Notice that with L_s in place, it is no longer necessary at turn on for the transistor to support V_{dc} until the diode current reaches zero. Instead, v_Q can decrease to zero while the diode remains on. Whatever fraction of V_{dc} that does not appear across the transistor appears across L_s. The snubber inductor's voltage then determines the rate at which i_Q increases. When the current reaches its final value of I_{dc}, the diode turns off, and the voltage across the inductor drops to zero.

Because the rate at which the snubber inductor current rises is proportional to $1/L_s$, we can choose a value for L_s so as to limit the transistor current to a fraction γ_i of I_{dc} by the time v_Q reaches zero. Assuming that $\gamma_i = 0.33$ and $t'_r = 0.1$ μs, we can calculate the resulting dissipation as we did in (24.2):

$$W_{diss} = \left(\frac{1}{12}\right)\gamma_i V_{dc} I_{dc} t'_r = 0.06 \text{ mJ} \qquad (24.5)$$

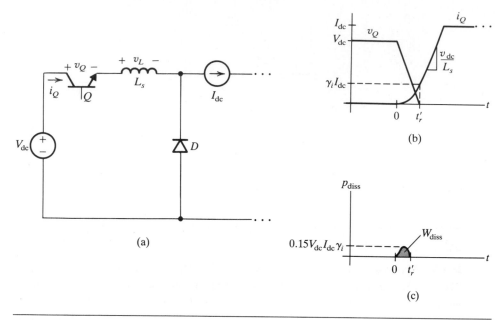

Figure 24.8   (a) A basic turn-on snubber consisting of the inductor L_s.   (b) Collector voltage and current waveforms.   (c) Power dissipated in the transistor as a function of time.

At the switching frequency of 20 kHz, the resulting average turn-on dissipation *in the transistor* is 1.2 W. This quantity is almost two orders of magnitude reduction from the transistor turn-on losses without the snubber. The peak power is also reduced, and the turn-on switching locus, shown in Fig. 24.5, now remains within the SOA.

24.2.3 A Practical Turn-On Snubber

As with the turn-off snubber, the turn-on snubber requires additional components to allow the transistor to survive the switch transition. In this case, if nothing were added to the circuit, the transistor would receive an impulse of voltage when it turned off, because the inductor current would be forced to change in a very short time. To avoid this potentially destructive transient, a resistor and a diode can be added to the snubber circuit, as shown in Fig. 24.9. The resistor R_s provides an alternative path for the inductor current when the transistor turns off. The purpose of the diode D_s is to keep the resistor from conducting during the turn-on transition.

The resistance R_s must be large enough to completely discharge L_s during the off state. However, to minimize the transistor's voltage stress, it should not be made any larger than necessary.

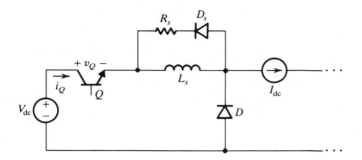

Figure 24.9 A practical turn-on snubber circuit.

EXAMPLE 24.2

Determining Element Values for the Turn-On Snubber

The minimum value of L_s can be found from the relationship $\lambda = Li$. At $t = t'_r$, $i_Q = \gamma_i I_{dc}$ and λ can be found by integrating v_L, shown in Fig. 24.8(a), between 0 and t'_r:

$$\lambda = \int_0^{t'_r} v_L \, dt = \frac{V_{dc} t'_r}{2} = L_s \gamma_i I_{dc}$$

$$L_s = \frac{V_{dc} t'_r}{2\gamma_i I_{dc}} = 1.2 \ \mu\text{H}$$

(24.6)

The resulting value of 1.2 μH is based on our previous parameter values of $V_{dc} = 400$ V, $I_{dc} = 50$ A, $\gamma_i = 0.33$, and $t'_r - 0.1 \ \mu$s.

If the shortest time that we expect the transistor to be off is 2.5 μs, or 5% of a 20 kHz cycle, we need an L_s/R_s time constant of about 0.5 μs to ensure that L_s is completely discharged. Meeting this condition requires a resistor of value $R_s \approx 2.5 \ \Omega$.

24.3 A COMBINED TURN-ON/TURN-OFF SNUBBER

When both a turn-on and a turn-off snubber are utilized in a power circuit, we do not have to include a separate diode–resistor pair for each. Instead, one snubber can be used to correct the problem that required a resistor and diode in the other snubber. Two ways of doing so are described below.

Figure 24.10(a) shows the first approach, in which a standard turn-off snubber, composed of C_s, R_s, and D_s, is used in conjunction with a turn-on snubber inductor, L_s. The behavior of the circuit at turn on is identical to that shown in Fig. 24.8, except that the discharge current of C_s must be added to i_Q. During turn off, however, the circuit behavior is quite different from that illustrated in Fig. 24.4. They are similar until t_1 when C_s is charged to V_{dc}. But at t_1 when $v_C = V_{dc}$ and

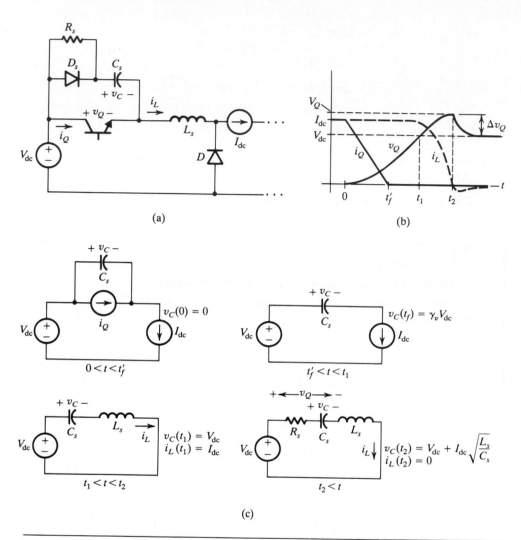

Figure 24.10 (a) A combined turn-on/turn-off snubber circuit with the resistor placed to discharge C_s.   (b) Voltage and current waveforms at turn off.   (c) Equivalent circuits and initial conditions for different time periods during the turn-off transient.

D turns on, L_s is still carrying a current $i_L = I_{dc}$, and therefore L_s and C_s will ring for a quarter of a cycle until $i_L = 0$. At this time, D_s turns off and R_s damps the transient.

Figure 24.10(b) shows the waveforms of the combined snubber at turn off, and Fig. 24.10(c) shows the equivalent circuits during different time periods. When analyzed using the specified initial conditions, these circuits yield the waveforms of Fig. 24.10(b). The incremental ringing voltage across Q is

$$\Delta v_Q = I_{dc}\sqrt{\frac{L_s}{C_s}} \qquad (24.7)$$

This incremental voltage is added to V_{dc} to give the peak transistor voltage, which is equal to the peak voltage across C_s.

EXAMPLE 24.3

Transient Voltage Produced by a Combined Snubber

In Examples 24.1 and 24.2, we found values for the snubber capacitor and inductor of $C_s = 0.066\ \mu F$ and $L_s = 1.2\ \mu H$. If we use these values for L_s and C_s in the combined snubber of Fig. 24.10(a), we can then calculate the peak transistor voltage V_Q, using (24.7):

$$V_Q = V_{dc} + \Delta v_Q = V_{dc} + I_{dc}\sqrt{\frac{L_s}{C_s}} = 400 + 50\sqrt{18} = 613\ V \qquad (24.8)$$

This value is large relative to V_{dc} and illustrates the need to carefully consider the values of elements used in the design of a combined snubber.

The second way in which a turn-on/turn-off snubber can be combined is shown in Fig. 24.11(a). The behavior of this circuit is essentially the dual of the behavior exhibited by the combined snubber of Fig. 24.10. During turn off, the waveforms of v_Q and i_Q are identical to those shown in Fig. 24.4, except that the discharge

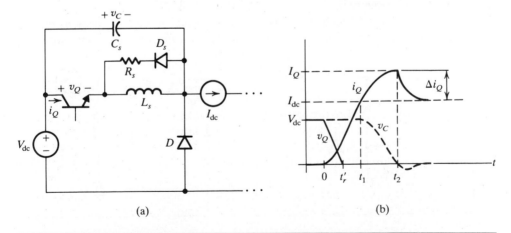

(a) (b)

Figure 24.11 (a) A combined turn-on/turn-off snubber circuit with the resistor placed to discharge L_s. (b) Voltage and current waveforms at turn on.

voltage across R_s contributes to v_Q. But at turn on the circuit exhibits a period of second-order behavior. After $v_Q = 0$ at t'_r, L_s charges linearly to I_{dc} at t_1, at which time D turns off. During this period, D_s is reverse biased and $v_C = V_{dc}$. After D turns off at t_1, L_s and C_s ring until $v_C = 0$ and D_s turns on at t_2. With R_s now in the circuit, i_Q settles to I_{dc} in an overdamped manner. The peak current in the transistor occurs at t_2 and is equal to:

$$I_Q = I_{dc} + \Delta i_Q = I_{dc} + V_{dc}\sqrt{\frac{C_s}{L_s}} \qquad (24.9)$$

The waveforms of i_Q, v_Q, and v_C during turn on are shown in Fig. 24.11(b).

24.4 ALTERNATIVE PLACEMENTS OF THE SNUBBER CIRCUITS

In our discussions so far, we have placed the turn-off snubber in parallel with the transistor and the turn-on snubber in series with the transistor. We now explore alternative placements for these snubber circuits.

The job of the turn-off snubber is to control the rise of v_Q. Because the collector of the transistor is connected to a stiff voltage source, V_{dc}, we also control v_Q if we control the voltage between the emitter of Q and ground. Therefore connecting the turn-off snubber capacitor between the emitter terminal and any other node that is an incremental ground is sufficient. For example, Fig. 24.12(a) shows an alternative placement where C_s is connected to ground. A connection to the output terminal, if there is a filter capacitor that makes this terminal an incremental ground, is another possible choice.

For the turn-off snubber to work properly when connected in positions different from directly across the switch, it is essential that the terminal of the switch to

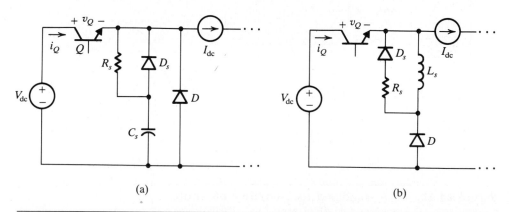

(a)   (b)

Figure 24.12   Alternative positions for the turn-on and turn-off snubber circuits: (a) turn-off snubber; and   (b) turn-on snubber.

which the snubber is not connected—the collector in the case of Fig. 24.12(a)—be at incremental ground. If there is any appreciable impedance between this switch terminal and incremental ground, such as lead inductance, the alternative placements suggested here will not protect the transistor to the same extent that a snubber connected directly across the switch would.

The turn-on snubber is meant to control the rise of the switch current. So long as it is placed in a path through which the current changes during the turn-on transition, the snubber will function properly. Figure 24.12(b) shows one example of a placement that is an alternative to putting the snubber directly in series with the switch. In this case the snubber inductor is in series with the diode D. Because I_{dc} is constant, this connection is identical to the series connection in its effect on i_Q.

It is important to recognize, especially when the switching frequency is high, that the power circuit has parasitic elements that provide some snubber action. For example, the node to which the transistor, diode, and output inductor (modeled by the current source) are connected in Fig. 24.2(a) has parasitic capacitance to other nodes. If one or more of these other nodes are at incremental ground potential, these capacitances function as a turn-off snubber. Similarly, the loop formed by the input source, transistor, and diode has a parasitic inductance that provides some turn-on snubbing action.

24.5 DISSIPATION IN SNUBBER CIRCUITS

At the beginning of a turn-on or turn-off transition, the energy storage element in the corresponding snubber circuit contains no stored energy. When the transition is over, however, the element is left with a nonzero value of stored energy. This energy is dissipated during the next switch transition. For instance, C_s in Fig. 24.10(a) is uncharged when Q begins to turn off and is charged to V_{dc} at the end of the turn-off transient. When Q turns on again, the energy stored in C_s is dissipated in R_s. Our purpose in this section is to evaluate the amount of energy lost in the process of resetting to zero the energies stored in L_s and C_s.

24.5.1 Separate Turn-On and Turn-Off Snubbers

We first consider the case of the turn-off snubber of Fig. 24.6. The energy stored in C_s at the end of the turn-off transient is

$$W_C = \frac{1}{2}C_s V_{\mathrm{dc}}^2 \qquad (24.10)$$

We choose the value of C_s according to (24.4) to give a specified value of γ_v. We use the parameters of Example 24.1 to find the stored energy:

$$W_C = \frac{t_f' I_{\mathrm{dc}} V_{\mathrm{dc}}}{4\gamma_v} = 8.25 \text{ mJ} \qquad (24.11)$$

Note that this is $t'_f/(2t_f\gamma_v)$ times larger than the energy dissipated in the transistor without a snubber capacitor, as in *(24.1)*. Increasing the snubbing action by making γ_v smaller increases the snubber loss.

The snubber loss at turn on is determined through a calculation similar to that done above for turn off. The energy stored in the snubber inductor just before turn-off is

$$W_L = \frac{1}{2}L_s I_{dc}^2 \tag{24.12}$$

Using *(24.6)* and the parameters of Example 24.2, we can calculate a numerical value for W_L:

$$W_L = \frac{t'_r I_{dc} V_{dc}}{4\gamma_i} = 1.5 \text{ mJ} \tag{24.13}$$

Again, this energy is larger than that dissipated in the transistor without a turn-on snubber.

24.5.2 The Combined Snubber

When the turn-on and turn-off snubbers are combined as shown in Fig. 24.10(a), we have shown that the capacitor is charged to a voltage that exceeds V_{dc} by Δv_Q. The peak energy stored in C_s therefore is

$$W_C = \frac{1}{2}C_s(V_{dc} + \Delta v_Q)^2 \tag{24.14}$$

Using (24.7), we can rewrite *(24.14)* as:

$$W_C = \frac{1}{2}(C_s V_{dc}^2 + L_s I_{dc}^2) + V_{dc} I_{dc}\sqrt{L_s C_s} \tag{24.15}$$

The first component of the right-hand side of *(24.15)* is exactly the amount of energy that would be lost if the two snubbers were not combined. The last term is a further, and not negligible, contribution to the energy stored in C_s. However, this additional energy is not dissipated. Instead, it is returned to the input voltage source by the current that discharges the capacitor. Therefore the actual energy lost in the combined snubber is equal to the energy that would be lost if the snubbers were not combined:

$$W_{diss} = \frac{1}{2}(C_s V_{dc}^2 + L_s I_{dc}^2) \tag{24.16}$$

24.5.3 Increased Circuit Dissipation Caused by Snubbers

If the transistor and its parallel turn-off snubber are considered as a single switching element, as shown in Fig. 24.13(a), the turn-off voltage and current waveforms for

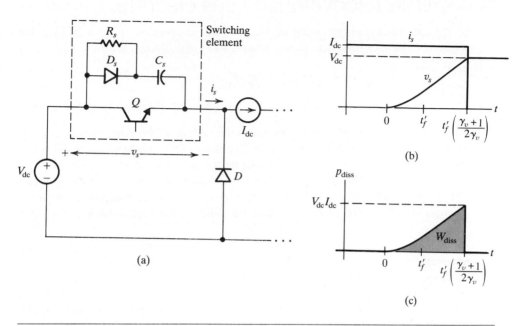

Figure 24.13   Combined switch and snubber losses: (a) circuit, showing the transistor and snubber as a single switching element;   (b) voltage and current waveforms, v_S and i_S, at the switching element; and (c) dissipated power as a function of time for the switching element.

this element are as shown in Fig. 24.13(b). The corresponding power is given in Fig. 24.13(c). This element has the characteristics of a switch whose voltage rises linearly until it reaches V_{dc}, at which point D turns on and picks up the load current I_{dc}. This switch, like the transistor itself, absorbs an energy proportional to V_{dc}, I_{dc}, and the time it takes to complete the transition. By design however, the transition takes $t'_f/(2t_f\gamma_v)$ times longer to complete with the snubber than without it, so the energy absorbed is correspondingly larger. A similar argument can be made for the reason that the energy absorbed by a transistor and a turn-on snubber is $t'_r/(2t_r\gamma_i)$ times larger than for the transistor operating without a snubber.

Even though the introduction of a dissipative snubber circuit reduces the net circuit efficiency, the loss incurred by the semiconductor switch is reduced. The additional loss occurs in other elements—notably R_s—which are designed specifically to dissipate energy. However, this dissipation must still be removed as heat, which can cause packaging problems, and the reduction in circuit efficiency can sometimes be unacceptable. In these cases, recovering some of the stored snubber energy may be practical. We next discuss snubber circuits containing energy recovery capability.

24.6 ENERGY RECOVERY SNUBBER CIRCUITS

If the combined snubber circuit of Fig. 24.10(a) is operated at $f_s = 20$ kHz, the combined snubber loss, determined from *(24.11)* and *(24.13)*, is

$$P_{\text{diss}} = f_s(\frac{1}{2}C_s V_{\text{dc}}^2 + \frac{1}{2}L_s I_{\text{dc}}^2)$$
$$= 20 \times 10^3 (5 \times 10^{-3} + 1.5 \times 10^{-3}) = 130 \text{ W}$$

(24.17)

This quantity is 2.6 percent of the converter output if the duty ratio is 25 percent. We may or may not be willing to lose this energy—depending on efficiency requirements—but even if we could tolerate the loss, the removal of the heat can still be a problem. Therefore it is sometimes desirable to provide a means by which the snubber capacitor can be reset by recovering, instead of dissipating, its stored energy. The recovered energy is transfered either to the load or back to the input source.

Figure 24.14(a) shows a turn-off snubber design that provides this energy recovery feature. We can understand how this *energy recovery snubber* operates by first assuming that the transistor Q is on and the two snubber capacitors C_1 and C_2 are each charged to V_{dc}. When Q is turned off, the two capacitors, in conjuction with their series diodes, behave exactly like two simple turn-off snubbers in parallel. The net snubber capacitance is the sum of the two individual capacitances $C_1 + C_2$. The inductor never comes into play during this transition, because D_3 is always reverse biased.

When Q is next turned on, its current rises to I_{dc} in time t_r', at which point D turns off and v_Q falls to zero in time t_r''. If t_r'' is small compared to the period of the ringing in the snubber circuit, we can assume that at this time the series circuit consisting of C_2, D_3, L, and C_1 is excited by a voltage step having an amplitude V_{dc}. An assumed state argument will show that D_3 turns on, and the response of the series circuit is a second-order ring. This ring continues for one-half cycle, at which time i_L tries to go negative, turning off D_3. At this time, the total capacitor voltage is $v_{C_1} + v_{C_2} = 2V_{\text{dc}}$. As the two capacitors are in series, they will share this total voltage in proportion to their relative values. If $C_1 = C_2$, they will both be charged to V_{dc}, and the circuit will be reset for the next turn-off transition.

Note that the snubber circuit completed a full cycle of its operation without requiring energy to be dissipated. During turn on, the source V_{dc} provided the load current as well as the current to charge the capacitors. Therefore the stored energy at this point in the cycle comes from the input voltage source. During the turn-off transition, the load current discharges the capacitors, transferring their stored energies to the load.

If the capacitors are of equal value, they are ideally charged to V_{dc} during the turn on transition. However, parasitic losses in the snubber circuit cause the final capacitor voltages to be less than this value. This means that during turn off, the rise of v_Q will not be constrained by the snubber until D_1 and D_2 turn on when v_Q reaches a value equal to the difference between V_{dc} and the capacitor voltage. To avoid this problem, we customarily use unequal capacitors in the snubber circuit. If their values are sufficiently different, the smaller capacitor always charges to V_{dc} during the turn-on transition. The smaller capacitor's voltage never exceeds V_{dc}, however, because its diode turns on at this point. Therefore the result of using unequal capacitors is that at least one of the capacitors is available to control the rise of v_Q, although the full snubber action will not be present until v_Q rises enough to let the other capacitor come into play.

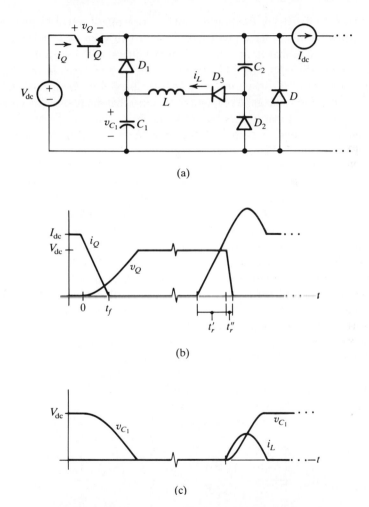

(a)

(b)

(c)

Figure 24.14 (a) An energy recovery turn-off snubber circuit. (b) Transistor voltage and current waveforms during turn on and turn off. (c) Capacitor C_1 voltage, v_{C_1}, and inductor current, i_L, during turn-on and turn off. The waveforms are based on the assumption that $C_1 = C_2$.

There are several other approaches to a lossless snubber, which we do not discuss in detail here. One is to use an auxiliary power circuit to transfer the energy stored in the snubber capacitor to either the source or the load. This method is feasible only in systems where the recovered energy is large enough to justify the added expense of the additional circuitry.

24.7 VOLTAGE CLAMPS

The turn-off snubber, by limiting the rate of rise of v_Q, protects the switch from simultaneously high voltage and current. There are situations, however, in which a protection circuit that limits the maximum voltage at the switch is needed. This special circuit, shown in Fig. 24.15, is referred to as a *voltage clamp*. In its simplest form, it consists of a capacitor, C_c, which is charged to the maximum allowable (clamp) voltage V_c, and some regulation circuitry that maintains this voltage. There is also a diode, D_c, that directs the load current I_{dc} into C_c when v_Q reaches V_c. If C_c is large enough, it absorbs this current for the rest of the switch transition without a significant change in its voltage.

The loop formed by V_{dc}, Q, and D has parasitic inductance, which is the reason for needing the voltage clamp. In the absence of this inductance, the voltage across the transistor would never exceed V_{dc}. When the loop inductance is present, however, v_Q rises above V_{dc}, as was shown in Fig. 24.10(b).

We can also limit the peak voltage across the transistor by choosing an appropriate turn-off snubber capacitance. Thus *(24.7)* shows that the larger the capacitor, the smaller is the amount by which v_Q exceeds V_{dc}. The disadvantage of making C_s large is that the dissipation, as given by *(24.16)*, also increases. Therefore if, to limit the peak value of v_Q, we must choose a value for C_s that is much larger than required just for snubbing, we then have to accept more loss than otherwise. Fortunately, for many applications, a capacitance chosen to give acceptable snubbing is close to the value required for limiting the peak voltage, and the penalty for limiting the maximum value of v_Q is not great.

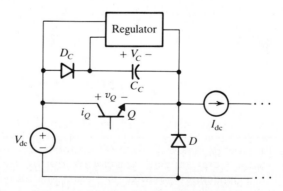

Figure 24.15 A clamp circuit designed to limit the peak transistor voltage to V_c.

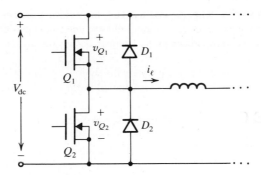

Figure 24.16 Part of a PWM bridge inverter in which the operation of D_1 and D_2 can be viewed as clamping v_{Q_2} and v_{Q_1}, respectively.

However, there are two situations in which the value of C_s determined by snubbing requirements may be too small to limit the peak voltage, thus requiring a clamp. The first occurs in circuits that use power MOSFETs. The switching speed and SOAs of these devices are such that often no turn-off snubber is needed, and a voltage clamp is therefore used to absorb the inductor energy. The second situation occurs in circuits such as switching power supplies, in which the leakage inductance of a transformer plays the role of a turn-on snubber inductor. This inductor is often much larger than normally needed for a turn-on snubber, and for this reason it is typical to find that the ideal snubber capacitor is not large enough to limit the peak voltage.

Although the clamp of Fig. 24.15 consists of a specially constructed voltage source—C_c, D_c, and the regulator—many power circuits have voltages available to which v_Q can be clamped. For instance, in the half of a PWM bridge inverter shown in Fig. 24.16, D_1 can be thought of as clamping v_{Q_2} to V_{dc} when Q_2 turns off. However, in order to function properly, the circuit construction must be such that the parasitic inductance in that part of the circuit consisting of the transistors, diodes, and V_{dc} is small enough to permit the load current i_ℓ to commutate from Q_2 to D_1 without creating a voltage transient that would manifest itself as an unacceptable increase in v_{Q_2}.

Notes and Bibliography

The snubber literature is extensive, and the references listed below represent only a cross section of papers addressing important issues. All contain additional references that may be used to pursue specific topics of interest.

McMurray's paper, [1], is the classic snubber reference. Criteria and detailed analytic techniques for designing conventional RLC snubbers are presented in [1], [2], and [3]. Only circuits resulting in second-order behavior are considered in [1] and [2], but [3] considers the effects of parasitic capacitances, such as those associated with junctions and transformer windings, and presents a technique for analyzing the resulting third-order network. Ohashi's paper, [4], is an excellent overview of the problems and techniques of snubbing GTOs. The effects of reverse recovery processes on snubber design for thyristors is treated nicely in [9].

Snubbers capable of energy recovery are the subjects of [5]–[8]. A configuration suitable for a three-phase, full-bridge converter is detailed in [5] and is accompanied by a clear and complete description of its operation.

1. W. McMurray, "Optimum Snubbers for Power Semiconductors," *IEEE Trans. Industry Applications* (5): 593–600, (September/October 1972).
2. W. McMurray, "Selection of Snubbers and Clamps to Optimize the Design of Transistor Switching Converters," *IEEE Trans. Industry Applications* 16 (4): 513–523, (July/August 1980).
3. K. Harada, T. Ninomiya, and M. Kohno, "Optimum Design of an RC Snubber for a Switching Regulator by means of the Root Locus Method," in *IEEE Power Electronics Specialists Conference Record* (Syracuse, NY: June 13–15, 1978), 158–167.
4. H. Ohashi, "Snubber Circuit for High-Power Gate Turn-Off Thyristors," *IEEE Trans. Industry Applications* 19 (4): 655–664 (July/August 1983).
5. F. C. Zach, K. H. Kaiser, J. W. Kolar, and F. J. Haselsteiner, "New Lossless Turn-On and Turn-Off (Snubber) Networks for Inverters, Including Circuits for Blocking Voltage Limitation," *IEEE Trans. Power Electronics* 1 (2): 65–75 (April 1986).
6. G. Fregien, H. G. Langer, and H.-Ch. Skudelny, "A Regenerative Snubber for a 200 kVA GTO Inverter," in *IEEE Power Electronics Specialists Conference Record* (Kyoto: April 11–14, 1988), 498–505.
7. W. McMurray, "Efficient Snubbers for Voltage-Source GTO Inverters," *IEEE Trans. Power Electronics* 2 (3): 264–272 (July 1987).
8. J. C. Bendien, H. van der Broeck, and G. Fregien, "Recovery Circuit for Snubber Energy in Power Electronic Applications with High Switching Frequencies," *IEEE Trans. Power Electronics* 3 (1): 26–30 (January 1988).
9. C. W. Lee and S. B. Park, "Design of a Thyristor Snubber Circuit by Considering the Reverse Recovery Process," *IEEE Trans. Power Electronics* 3 (4): 440–446 (October 1988).

Problems

24.1   In this chapter we assumed that, during a switch transition, the voltage and current waveforms of a transistor without a snubber changed linearly from their starting values to their final values. For this problem, assume that these waveforms change quadratically, that is, the *rate* at which the waveforms change grows linearly with time. If the total time required to complete the transition (t_f or t_r) is the same as in the linear case, what happens to the amount of energy dissipated in the transistor when these alternative shapes are assumed?

24.2   In some power circuits the load does not have the large series inductor (modeled as the current source I_{dc}) shown in Fig. 24.2. In such a case the current drawn by the load changes during the switch transitions. Draw the switch waveforms for both the turn-on and turn-off transitions when the load is purely resistive and no snubbers are used. How much energy is dissipated in the transistor during each transition?

24.3   We added the resistor R_s to the basic turn-off snubber of Fig. 24.4(a) to limit the discharge current when Q turns on. We then added D_s to bypass this resistor during turn off, giving the complete turn-off snubber of Fig. 24.6. The purpose of this problem is to illustrate the importance of D_s. Assume the circuit parameters of Example 24.1, with $R_s = 5\ \Omega$, and determine and sketch v_Q during turn off if D_s is not present. Compare this waveform to that of v_Q in Fig. 24.4(b), which is the same as v_Q at turn off in the circuit containing D_s, in Fig. 24.6.

24.4   Consider the down converter with a turn-off snubber shown in Fig. 24.6. Assume the following parameters: $V_{dc} = 400$ V, $I_{dc} = 50$ A, $C_s = 0.25\ \mu$F, and $R_s = 5\ \Omega$. If the transistor is turned on for only 2 μs, sketch and dimension v_Q during the turn-off transition. Also sketch the locus of the turn-off transition on the i_Q–v_Q plane.

24.5   Determine and sketch the voltage v_Q at turn off for the circuit of Fig. 24.9, assuming that the element values are those determined in Example 24.2.

24.6   For the combined turn-on/turn-off snubber shown in Fig. 24.10, we stated that of the total energy stored in the snubber capacitor, given by *(24.15)*, part was dissipated in the resistor and part was returned to the input voltage source. Show that this is true so long as the transistor does not turn on before the capacitor has had a chance to discharge to V_{dc}.

24.7   Consider the down converter with only a turn-on snubber inductor shown in Fig. 24.8. Assume that $V_{dc} = 400$ V, $I_{dc} = 50$ A, and $L_s = 16\ \mu$H. Without an alternative path for the snubber inductor current to flow, we stated that when the transistor turns off, an impulse of voltage appears across the inductor and, correspondingly, the transistor. Actually, the transistor voltage quickly reaches its avalanche breakdown level, at which point the snubber inductor current can continue to flow through the transistor. Assume that this transistor breakdown voltage is 500 V. Sketch as a function of time the snubber inductor's voltage and the transistor's voltage and current waveforms. When does the transistor come out of avalanche? How much energy is dissipated in the transistor while it is avalanching?

24.8   Determine and draw, as was done in Fig. 24.10(c), the equivalent circuits for the combined snubber of Fig. 24.11(a).

24.9   The combined snubber of Fig. 24.10(a) is constructed using the values for V_{dc}, I_{dc}, L_s, and C_s from Example 24.3.

  (a) What is the minimum value of R_s that gives the desired overdamped response after t_2?

  (b) Redraw the waveforms of Fig. 24.10(b) to scale. That is, calculate and identify the times t_1 and t_2 and also calculate and identify the time at which v_Q first passes through V_{dc} after t_2.

  (c) Using duality, repeat (a) and (b) for the combined snubber of Fig. 24.11(a).

24.10   In very high-frequency converters, the nonlinear junction capacitance of diodes and transistors often forms part of a turn-off snubber. Consider the down converter of Fig. 24.4(a) in which the capacitor C_s is nonlinear. Specifically, its incremental capacitance is

$$C_{si} = \frac{dQ}{dv} = C_o\sqrt{\frac{V_{dc}}{v_c}} \qquad (24.18)$$

where Q is the charge in the capacitor and C_o is the value of C_{si} when the voltage across the capacitor v_c equals V_{dc}. Assume that $V_{dc} = 400$ V, $I_{dc} = 50$ A, and $C_o = 0.1 \ \mu$F. In addition, assume that, when the transistor turns off at $t = 0$, its current drops linearly to zero in a time $t_f = 0.25 \ \mu$s.

(a) Sketch the waveform of v_Q during turn off.

(b) What is $\gamma_v = v_c(t_f)/V_{dc}$?

(c) Calculate the energy stored in C_s when $v_c = V_{dc}$.

(d) If the snubber capacitor were linear, what value of capacitance would give the same γ_v? Compare the energy stored in this linear capacitor when $v_c = V_{dc}$ to your answer for (c).

Chapter 25

Thermal Modeling and Heat Sinking

AN unfortunate consequence of our preoccupation with things electrical is that the problems of heat sinking and thermal management are frequently ignored until forced on us by sound, sight, or smell. The insatiable need to make things smaller—and the possibility of doing so by using higher frequencies and new components and materials—aggravates the problem of heat transfer, because such improvements in power densities are seldom accompanied by corresponding improvements in efficiency. Thus we are stuck with the task of getting the same heat out of a smaller volume while disallowing any increase in temperature.

The diversity of heat sources within power electronic apparatus produces a challenging cooling problem. Unlike signal processing circuits—where most heat generating components come in a common, small, and low-profile rectangular package—energy processing circuits contain components of odd shapes and orientations. Even components of the same type come in many different forms and packages. Inductors, for instance, can be small or large, round or rectangular, and with loss dominated by core or copper. Each possesses special requirements and presents a unique thermal problem. The task of integrating these parts into a reliable piece of equipment becomes as much a mechanical challenge as the circuit design was an electrical challenge.

Heat transfer occurs through three mechanisms—conduction, convection, and radiation. In conduction the heat transfer medium is stationary, and heat is transferred by the vibratory motion of atoms or molecules. Convective heat transfer occurs through mass movement—the flow of a fluid (gas or liquid) by the heat generating apparatus. The flow can be either natural or forced. In *natural convection*, the bouyancy created by temperature gradients causes the fluid to move; in a *forced convection* system, the mass flow is created by pumps or fans. Heat transfer by radiation turns the heat energy into electromagnetic radiation, which is absorbed by other elements in the environment. Except for space applications, radiation as a mechanism of heat transfer in a power electronic system is seldom of sufficient importance to justify the complexity of considering it in detail. Therefore in this chapter we focus on conduction and convection.

The material in this chapter will not give you new and novel ideas for designing heat transfer systems. The problem is too application specific to permit a discussion of that type to be of general value. Instead, we describe first the parameters governing the performance of any such system. Then we consider the modeling of both steady-state and transient thermal behavior, as applicable to power electronic systems. Some straightforward examples of specific designs will be presented to illustrate the discussions.

25.1 STATIC THERMAL MODELS

Circuit theory is the "lingua franca" of engineering for good reason. The elegance and simplicity of its canonical formulations (for example, KCL and KVL) permit complex problems to be approached in an organized way, and the insights gained through such formulations (for example, eigenvalues) are extremely valuable in predicting system behavior. Therefore many engineering problems in contexts other than electrical engineering are cast in terms of circuit models before being analyzed. One of these contexts is heat transfer.

Problems in which conduction and convection dominate radiation can be accurately transformed to electrical networks over restricted temperature ranges. Radiation is a complex mechanism that does not lend itself to linear circuit analogs. Fortunately, heat transfer from most power electronic systems is dominated by conduction and convection. If only these two mechanisms are considered, the design will be conservative, as any heat transferred through radiation will reduce the temperature of the apparatus below the design temperature. The exception is in enclosures, where radiation from hot components may be absorbed by those at lower temperatures, causing these latter components to operate at higher temperatures than anticipated. In such cases, radiation shields—shiny metal partitions—can be employed to isolate the offending or affected components. We consider here, therefore, only conduction and convection processes and the network models that represent them in the static state. In a later section we consider transient models for pulsed or short duty-cycle operation.

25.1.1 Analog Relations for the dc Case

The rate at which heat energy is transferred by conduction from a body at temperature T_1 to another at temperature T_2 is Q_{12}. It is linearly proportional to the temperature difference between the two bodies, $T_1 - T_2$, and inversely proportional to a physical parameter called the *thermal resistance* between them, R_θ:

$$Q_{12} = \frac{T_1 - T_2}{R_\theta} \qquad (25.1)$$

The analogy with Ohm's law is evident, and we can make the following assignment of variables:

$$T_{1,2} \Longleftrightarrow v_{1,2} \qquad Q_{12} \Longleftrightarrow i \quad \text{and} \quad R_\theta \Longleftrightarrow R$$

Note that the analog of thermal power is i, not vi. If heat leaves body 1 only through the interface characterized by R_θ, then i is not only analogous to Q_{12}, but, because we are considering only steady-state conditions, it also represents the rate at which energy is being converted to heat in body 1. In the context of our interests, body 1 would be a packaged electrical network, and Q_{12} would represent the rate at which electrical energy was being converted to heat (dissipated) in the package, that is,

$$p_{\text{diss}} \Longleftrightarrow i$$

The thermal management problem is to design a heat transfer system (that is, R_θ) that constrains ΔT (the temperature difference between component and ambient) to that value dictated by component ratings and ambient conditions.

Figure 25.1 illustrates the electrical analog for the simple two-body system just discussed. The bodies are at temperatures T_1 and T_2 and are connected thermally through the crosshatched interface, which can be characterized by a thermal resistance of value $R_{\theta 12}$. The units of T are °C, and as the units of Q are watts (W), thermal resistance has the units °C/W. As with electric circuits, where parallel resistances can be combined into a single equivalent resistance, parallel thermal paths, each characterizable by a thermal resistance, can be combined into an equivalent single thermal resistance.

Convection is the mechanical transport of heat by a moving fluid. The fluid (air, for instance) can move because of gravitational forces caused by density gradients, in which case the process is called natural convection. Or the fluid can be driven (perhaps by a fan), resulting in what is called forced convection. Convection is a somewhat more complex process than conduction and can be described by the relation:

$$Q_{12} = h(T, \nu)A(T_1 - T_2) \tag{25.2}$$

where ν is the fluid velocity. The parameter $h(T, \nu)$ is the *film coefficient of heat*

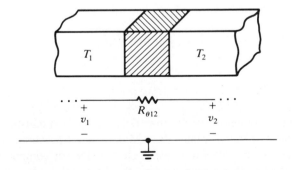

Figure 25.1 Electrical analog of simple two-body thermal system. The crosshatched region is the thermal interface characterized by the longitudinal thermal resistance $R_{\theta 12}$.

transfer and depends on temperature and fluid velocity, and A is the cross-sectional area of the interface. Over the temperature ranges of interest (approximately $-40°$ to $+100°C$), h is fairly constant. With respect to fluid velocity, significant changes in h occur when the flow regime changes from laminar to turbulent. Within each regime, however, h is relatively constant as a function of velocity. Within these limits, the product hA may be modeled as constant, giving to *(25.2)* the same form as *(25.1)*, with $R_{\theta 12} = 1/hA$. Thus the electrical analog shown in Fig. 25.1 is appropriate for representing convective as well as conductive heat transfer.

25.1.2 Thermal Resistance

A thermal resistance can be used to model both conductive and convective heat transfer, as we have just shown. The physics governing thermal conduction is much like that for electrical conduction, and the thermal resistance or conductance can be described in terms of parameters abstracted from the physics of the process (for example, conductivity, and geometry). In fact, thermal and electrical conductivity are intimately related by the *Wiedeman–Franz law*—materials of high electrical conductivity are also good thermal conductors. Convection, however, depends on parameters that are not so easily abstracted. For instance, while conductivity can be adequately described in terms of material type and temperature, h is a function not only of these parameters but also of velocity, surface characteristics, and geometry. The latter parameter often determines the *Reynolds number*, a dimensionless number that indicates whether laminar or turbulent flow will occur in a particular situation. Furthermore, the geometry of the convective part of the system is invariably complex, consisting quite often of a finned aluminum extrusion. Thus the equivalent thermal resistance model for convective transfer from a specific heat sink type in a variety of environments (for example, forced or natural convection) is tabulated by the manufacturer. For this reason, the following discussion is directed at determining the equivalent thermal resistance for those parts of the system where heat transfer is by conduction.

The analog of electrical resistivity (Ω-m, or more commonly, Ω-cm) is thermal resistivity, ρ_θ, in units of $°C$-cm/W (or $°C$-in./W in a field noted for its lack of concern about mixed units). In terms of ρ_θ and physical dimensions, the longitudinal thermal resistance of a piece of material of cross-sectional area A and length l is

$$R_\theta = \frac{\rho_\theta l}{A} \qquad (25.3)$$

The thermal resistivities of various materials used in heat transfer paths in electronic equipment are shown in Table 25.1. Mylar, and less commonly mica, is used to provide electrical isolation between electrically hot components (for example, the semiconductor device package and the heat sink). Mica has a much higher dielectric strength, is more impervious to mechanical puncture, and can be cleaved to produce thinner sheets than mylar—but is more expensive. Beryllia and alumina,

Table 25.1 Thermal Resistivities of Materials Used in Electronic Equipment

MATERIAL	RESISTIVITY (°C-cm/W)
Still air	3050
Mylar	635
Silicone grease	520
Mica	150
Filled silicone grease	130
Alumina (Al_2O_3)	6.0
Silicon	1.2
Beryllia (BeO)	1.0
Aluminum Nitride (AlN)	0.64
Aluminum	0.48
Copper	0.25

and recently aluminum nitride (AlN), are also used to provide electrical isolation, most frequently within device packages.

Silicone grease impregnated with metallic oxides, such as ZnO_2, is used to fill imperfections such as scratches on mating surfaces in a heat transfer path— between the bottom of a device package and the top of a heat sink, for instance. The need to fill these voids with something besides air is apparent from the table. Filled silicone grease is also referred to as *thermal grease* or *thermal compound* (or "goop" for reasons that become clear when you use it). Anodization is frequently used to create an attractive or black surface on aluminum components. Since the resulting oxide is very thin, its contribution to the thermal resistance of a path is negligible. Although the oxide is a good insulator, it is unwise to rely on it in lieu of mica or mylar for providing electrical isolation between surfaces. Note that the ratio of ρ_θ for copper and aluminum is the same as that for their electrical resistivities. This result relates to the similarity of the physics of both heat and charge transport in metals and is described by the Wiedemann–Franz law.

EXAMPLE 25.1

A Calculation Using an Electrical Analog

Figure 25.2(a) shows a resistor embedded in a block of aluminum. What is the temperature of the resistor if it is dissipating 50 W and the ambient temperature is $T_A = 75°C$?

As the length of the block (10 cm) is much longer than the radius of the resistor (3 mm), we can assume that the power is dissipated uniformly in a plane centered between the two heat sinks, which are at temperature T_A. This assumption says that the detailed pattern of heat flow in the vicinity of the resistor is unimportant. The resulting analog circuit model

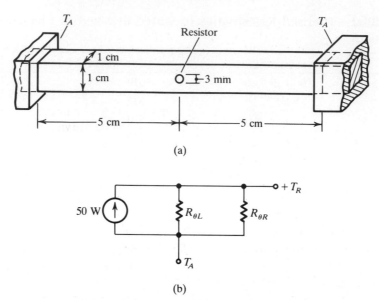

(a)

(b)

Figure 25.2   (a) A thermal system consisting of a resistor embedded in the center of an aluminum block.   (b) The electric circuit analog for the thermal system of (a).

is shown in Fig. 25.2(b), where $R_{\theta L}$ and $R_{\theta R}$ are the thermal resistances of the aluminum bar to the left and right of center. The value of these resistances are

$$R_{\theta L} = R_{\theta R} = \frac{(0.48)(5)}{1} = 2.4°C/W \qquad (25.4)$$

At a dissipated power of 50 W, the temperature of the resistor, T_R, is

$$T_R = 75 + 50\left(\frac{2.4}{2}\right) = 135°C$$

25.2 THERMAL INTERFACES

A critical part of any heat transfer system is the interface between mechanical components in the thermal path. Some issues related to these interfaces in the context of our application were raised in the previous section. The geometry of most interfaces can be modeled as two parallel planes with material of a specific thermal resistivity between. If the material is of thickness δ and of area A, the thermal resistance of the interface between the planes is

$$R_{\theta i} = \frac{\rho_\theta \delta}{A} \qquad (25.5)$$

Consider now that the interface is between a TO-3 package and a flat plate, and consists of 0.05 mm mylar. The mating surface area of a TO-3 package is approximately 5.0 cm², giving a thermal resistance between the case and the sink of:

$$R_{\theta CS} = \frac{(635)(0.005)}{5.0} = 0.64°C/W \qquad (25.6)$$

Thus the difference between the case and sink temperatures increases by 0.64°C for each watt of thermal power being transported across the interface. A dissipation of 25 W is not unusual for a device in a TO-3 package, giving a temperature rise of 16°C across just the mylar interface. This amount is a lot and illustrates the price paid for requiring isolation.

EXAMPLE 25.2

A Thermal System

The physical structure depicted in Fig. 25.3 is typical of the thermal system that results from mounting semiconductor devices. The silicon device itself is bonded to the header using solder or epoxy, the header is made part of a package that is mounted to a heat sink (perhaps with some intervening insulating material), and the heat sink is thermally connected to the ambient environment, generally through free or forced (fan) convection. An overly detailed model for this system is shown in Fig. 25.4(a), where each part of the system is explicitly represented by its equivalent thermal resistance. The model also shows the relationship between certain node voltages and the temperatures they represent. The current source represents the rate at which electrical energy is dissipated in the device, P_{diss}. The physical location within the device of this dissipation is the node to which the source is connected, the junction in this case.

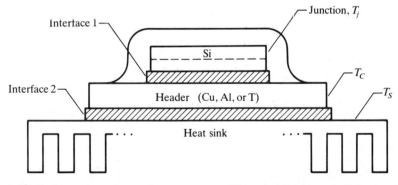

Figure 25.3 Typical mechanical structure used for mounting a semiconductor device.

Some of the thermal resistances shown in Fig. 25.4(a) are so small relative to others that they can be neglected. Because the header is made of copper or aluminum, its vertical thermal resistance is negligible ($\approx 0.03°C/W$ for an aluminum TO-3 package), as is that of the thermal grease (assuming that it is applied properly, which it often is not!). Others of the identified resistances are frequently lumped together, such as those for the silicon and bonding material, which are generally inaccessible to the circuit designer. Implicit in the element $R_{\theta SA}$ are the thermal resistance of the sink extrusion between the region on which

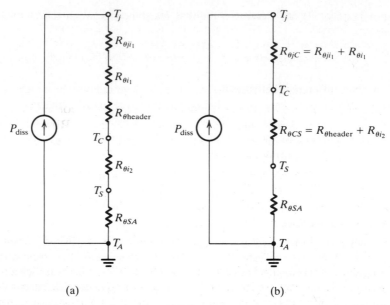

(a)   (b)

Figure 25.4 (a) A static thermal model for the structure of Fig. 25.3. (b) Simplified model of the circuit in (a).

the package is mounted and the surfaces from which heat is being convected away, and the thermal resistance representing the convection process. Figure 25.4(b) shows the simplified model.

The variable of interest in the models of Fig. 25.4 is T_j, the "junction" temperature of the device. The term "junction" is used rather loosely to represent in lumped form the source of heat in the device. In reality, this source seldom exists as a simple plane. In the MOSFET it is not a junction at all. Nevertheless the term persists, and manufacturers determine $R_{\theta jC}$ empirically, which takes into account the actual geometry of the heat-producing region. To determine T_j, we need to know P_{diss} as well as the thermal resistances between the junction and the ambient enviroment.

The dissipation in the device is a function of its electrical environment (for example, its current, voltage, and switching loci). For purposes of this example we assume that these calculations have been made for our device and that the result is $P_{\text{diss}} = 25$ W. The physical configuration is a TO-3 package mounted on an extruded, finned, free convection-cooled sink without any insulating interface but with thermal grease. Typical values of the thermal resistances are $R_{\theta jC} = 0.75$, $R_{\theta CS} = 0.12$, and $R_{\theta SA} = 1.8$, all units being °C/W. The last parameter needed is the ambient temperature, which is not "room temperature," but the temperature of the air in the vicinity of the sink. We take it to be 40°C. We determine the temperature drop between nodes in the model by using Ohm's law, obtaining

$$T_S = 40 + (25)(1.8) = 85°C$$

$$T_C = 85 + (25)(0.12) = 88°C$$

$$T_j = 88 + (25)(0.75) = 107°C$$

Is this an adequate design? The answer depends on the type of device being cooled. If it were a bipolar or MOSFET transistor, the design gives a good margin between predicted junction temperature and typical maximum limits of 150°C. For a thyristor the design is marginal. What would be the influence of an isolating interface between package and sink?

25.2.1 Practical Interfaces

Mechanical interfaces are not parallel in practice. They contain surface imperfections, such as scratches, and a characteristic called *run-out*. Run-out is the maximum deviation from flatness that a surface exhibits over a specified lateral distance. It is measured in (linear dimension)/(linear dimension), for example, cm/cm. A standard aluminum extrusion exhibits run-out that is typically 0.008 cm/cm. Both of these surface characteristics degrade the thermal performance of an interface.

The use of thermal grease has already been mentioned. It is designed to reduce the degrading effects of surface scratches and other small imperfections but is not designed to remedy the effects of run-out. Although our primary focus is on basic principles, thermal grease is so frequently misused that a brief departure from "principle" to "practice" is justified. The problem arises from a belief that if a little is good, a lot is better. However, the art of applying thermal grease is much like that of watering plants—too much and it's dead. Silicone grease is highly viscous and refuses to "squish out" when squeezed between header and sink by mounting hardware. In such cases, a thin layer of grease can remain in the interface, giving rise to a significant thermal resistance that was not anticipated in the thermal design. The grease should be applied sparingly, and then wiped off, removing almost all traces. Thermal grease oozing from under device packages is a sign of poor construction and potential thermal problems.

Run-out is usually not under the design engineer's control. Therefore we must determine the run-out characteristics of our components, and make proper allowance for them. As an example, Fig. 25.5 shows the resulting thermal resistance as a function of the sum of both header and sink run-out for several common device packages.

25.2.2 The Convective Interface

Even though we showed that both conduction and convection processes could be modeled by similar electrical analogs, our discussion so far has focussed on conduction interfaces. The fact of the matter, however, is that all conduction leads to a convective interface. Heat is removed from a conventional finned sink by air flowing over the fins. A more sophisticated system incorporating a heat exchanger probably uses a liquid to move the heat from one place to another. These are convective processes. As mentioned earlier, the physics governing these processes is beyond our scope here. But a short discussion of the application of finned sinks is helpful.

The critical issue in the application of finned sinks is to ensure that air flow through the fins is turbulent rather than laminar. Laminar flow, as the name implies,

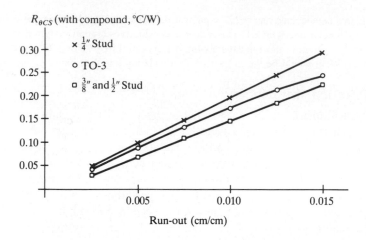

Figure 25.5   Thermal resistance of the interface between various device packages and sink surface as a function of run-out. (Data from W. E. Goldman, "An Introduction to the Art of Heat Sinking," *Electronic Packaging and Production*, July 1966.)

is the flow of a fluid in such a way that strata can be defined; that is, all flow is in one direction. The important feature of laminar flow is that no mixing of strata occurs. Turbulent flow, on the other hand, causes considerable mixing. Without such mixing, the particular stratum of fluid in contact with the fin would remain in contact with it for its entire length, resulting in a very low value for the film coefficient of heat transfer h, discussed in Section 25.1.1. Stated another way, the boundary layer next to the fin surface remains intact and virtually stationary in laminar flow, preventing efficient heat transfer from the fin to the moving stream of air (or other fluid).

The boundary between laminar and turbulent fluid flow is a function of many variables. But fin geometry and flow rate are the critical ones for our application. The essential influences of these two parameters quickly become evident to anyone who has canoed in rapids. A stream moving "slowly enough" will be smooth, a reflection of laminar flow. Increase the flow and eddies, holes, standing waves, and other hideous 3-dimensional phenomena arise to humble us. This same turbulently flowing stream, when constricted to a narrow passage (known as a "chute") becomes instantly laminar and glass smooth, even though the canoe shoots through it like a rocket. The salient feature of the chute is that it is not wide enough to permit turbulence. So it is with fins too closely spaced and turbulence won't occur.

The relationship among fin spacing, flow rate, and the onset of turbulence is given by the *Reynolds Number*. A high Reynolds Number is characteristic of turbulent flow; a low number is characteristic of laminar flow. The Reynolds Number

Re for a fluid flowing at velocity ν through a channel of width w is

$$Re = \frac{\rho \nu w}{\eta}$$

where ρ is the fluid density, and η is its coefficient of viscosity. This expression shows that fluid flowing in a wider channel will enter the turbulent flow regime at a lower velocity than that flowing through a narrower channel. The point here is that, like thermal compound, more is not necessarily better. Because of reduced turbulence and flow rate, a sink with many closely spaced fins and a large surface area could produce poorer thermal performance than one with fewer, but more widely spaced fins and a smaller surface area.

Although we have been using the context of semiconductor heat sinks for this discussion, it is equally appropriate to the cooling geometry associated with other components. Closely spaced parts impede proper convective flow for the same reasons that too closely spaced fins do.

25.3 TRANSIENT THERMAL MODELS

So far our discussion and models have been limited to systems in which both the energy being dissipated and the temperatures within the system are constant. Our models do not represent the thermal processes associated with start-up, where dissipation may be constant, but temperatures are climbing. Or pulsed operation, where temperatures may be constant, but dissipation is not. The latter situation is the more important, for under such conditions the permissible instantaneous dissipation can be much higher than predicted by static thermal models. Essentially the heat capacity of components or their constituent parts creates a low-pass filter, which in the limit of small bandwidth only responds to the dc in $p_{\mathrm{diss}}(t)$.

Heat capacity is a measure of the energy required to raise the temperature of a mass by a specific amount. In SI units it is specifically the energy in joules required to raise one kilogram of the material one centigrade degree. It has the units J/°C-kg. Water has one of the largest thermal capacities of any fluid at room temperature: 4.2×10^3 J/°C-kg.

Masses in a thermal system, then, constitute thermal energy storage devices, and thermal systems containing mass will exhibit dynamic behavior. Since thermal power is Q, the analog of thermal capacitance is electrical capacitance in the electric circuit model for heat transfer. In its simplest form, then, the dynamic model for a mass being supplied with heat energy is an RC circuit, as shown in Fig. 25.6. If the heat source is constant, the final temperature is analogous to the final voltage across the capacitor, that is, $R_\theta Q$. Previously we have dealt with the steady-state solutions to such thermal systems. Now we are concerned with the transients leading to these steady states.

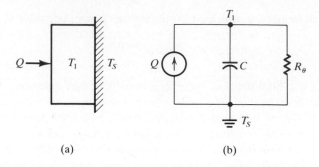

(a)   (b)

Figure 25.6   (a) A simple thermal system consisting of a mass at temperature T_1 being supplied heat Q and in contact with a sink at temperature T_S. (b) A single lump dynamic model for the system shown in (a).

25.3.1 A Lumped Model and Transient Thermal Impedance

Energy in a mass is stored in a continuum. But as is done for a transmission line, this continuum system is modeled by a series of lumped electrical elements. Consider, for instance, the mass of Fig. 25.6 with a source of heat energy applied at one end. The mass can be broken into an arbitrary number of sections, each assumed to be at a uniform temperature. These sections are each characterized by a heat capacity and are interconnected through a thermal resistance. This thermal resistance is that of the mass section between the two interfaces with other sections. This multilump model of the mass of Fig. 25.6 is shown in Fig. 25.7. The number of lumps, 5, was chosen arbitrarily.

When a continuous system is modeled by lumps, each lump displays the aggregate behavior of the physical piece of the system it represents. The model of Fig. 25.7 has been constructed so that the node voltages represent the section tem-

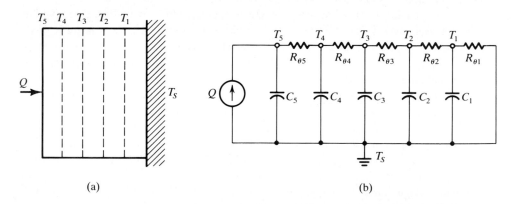

(a)   (b)

Figure 25.7   (a) The thermal system of Fig. 25.6(a) divided into five "lumps."   (b) The lumped electrical analog model for the thermal system of (a).

perature aggregated at the interface. The number of lumps chosen to represent a system depends on the bandwidth of the behavior being modeled. For instance, if Q is constant, no dynamics are excited, the bandwidth of the behavior is small, and a one-lump static model is adequate. However, if Q varies with time at a rate much greater than $(R_\theta C)^{-1}$ for the segments of Fig. 25.6, more lumps would be needed to model accurately the behavior of the system.

The step response of any of the node temperatures (voltages) in Fig. 25.7 looks like an exponential, because all eigenvalues are on the negative real axis. Figure 25.8 shows this response to a step of P_0 in thermal power, Q, for an arbitrary node, n. The initial node temperature is T_{0n}, the final node temperature is T_{fn}, and they are connected by a sum of exponential functions that looks like a single exponential. The final temperature is that which would have been predicted by the static thermal model consisting of five thermal resistances; that is,

$$T_{fn} = P_0 \sum_{m=1}^{n} R_{\theta m} \tag{25.7}$$

The temperature curve of Fig. 25.8 predicts the temperature of a node as a function of time for a given amplitude of a step in thermal power. If this curve is normalized by the step amplitude (P_0 in this case) the resulting vertical scale has the units of thermal impedance, that is, °C/W. An experimentally or theoretically determined normalized curve of this kind is useful for predicting temperatures during thermal transients. The normalized quantity is a function of time and is called the *transient thermal impedance*, denoted by $Z_\theta(t)$:

$$Z_\theta(t) = \frac{T(t)}{P_0} \tag{25.8}$$

The device and package structure of Fig. 25.3 contains several thermal masses that contribute dynamics to its thermal behavior. These dynamics are important when the device is forced to dissipate high levels of power for short periods of

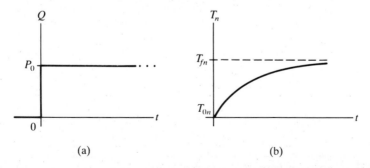

(a)   (b)

Figure 25.8   (a) A step in thermal power exciting the thermal system of Fig. 25.7(a). (b) The temperature response of node n to the excitation of (a).

time. "Short" is relative to the $R_\theta C$ time constant of the structure's electrical model. For very short pulses, the mass of the silicon is most important in determining the excursion of the junction temperature T_j. As the pulse gets longer, the mass of the header and then the heat sink become important. Manufacturers of power devices usually provide a transient thermal impedance curve in the specification sheets for their devices. One such curve for the C149 thyristor is shown in Fig. 25.9(a). Figure 25.9(b) for the IRF440 MOSFET illustrates an alternative form for displaying the transient thermal response data. Technically, only the bottom curve in this family is $Z_\theta(t)$ as defined by (25.8).

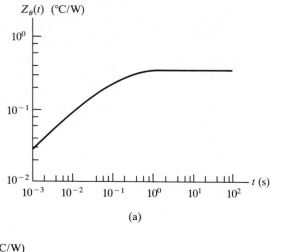

(a)

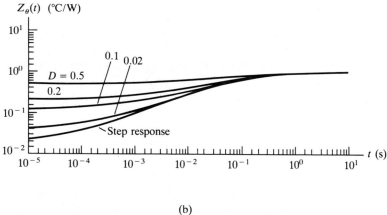

(b)

Figure 25.9 Curves of transient thermal impedance, $Z_\theta(t)$: (a) C149 thyristor; (b) IRF440 MOSFET.

EXAMPLE 25.3

Transient Thermal Design for a Thyristor

Power semiconductor devices are occasionally subjected to transient pulses of current that can considerably exceed their steady-state rating. For example, before protection circuitry can respond to a short circuit at the load, device currents may build up to values considerably above design values. The ability of a device to survive such a stress depends on the duration of the stress, as well as the device's transient thermal behavior. In this example we consider a C149 thyristor operating at a junction temperature of 80°C suddenly subjected to a 25 μs rectangular pulse of current, such as might occur between the inception of a fault and the reaction of the protective circuitry. The essential question is: What is the maximum permissible value of the current? We answer this question by determining the maximum power the device can support for 25 μs, calculating the energy dissipated during this time, and then relating this energy to a peak current through data provided by the manufacturer.

Figure 25.10 shows the current pulse and the resulting power pulse. We derive the power pulse by assuming that the device looks like a voltage source during its on state (which it does, except in the vicinity of the turn-on and turn-off edges). The maximum permissible junction temperature is 125°C, so the current pulse can cause a ΔT_j of 45°C and:

$$P_0 = \frac{\Delta T_j}{Z_\theta(25 \ \mu s)} \tag{25.9}$$

To determine $Z_\theta(25 \ \mu s)$, the transient thermal impedance curve of Fig. 25.9(a) must be extrapolated to 25 μs. A straight-line extrapolation results in the relationship:

$$Z_\theta(t_1) = \frac{Z_\theta(t_2)}{\sqrt{t_2/t_1}} \tag{25.10}$$

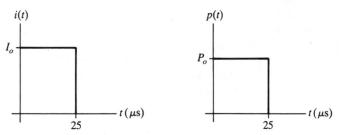

Figure 25.10 Current pulse and corresponding power pulse for the C149 thyristor.

Applying (25.10) results in $Z_\theta(25 \ \mu s) = 4.7 \times 10^{-3}$ °C/W, which gives $P_o = 9575$ W and an energy of 0.24 W-s. Figure 25.11, taken from the C149 data sheet, shows the constant energy relationship between peak current and pulse base width for an approximately rectangular pulse. From the appropriate curve, we see that I_o is approximately 600 A. Thus the transient thermal characteristics of this device permit it to carry safely a current pulse whose amplitude is ten times its steady-state current rating.

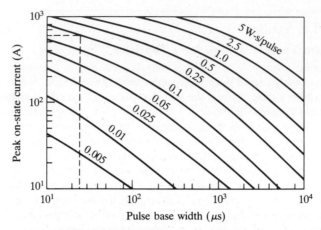

Figure 25.11　Constant dissipated energy curves for the C149 thyristor. These curves relate peak current and pulse width at a junction temperature $T_j = 125°C$. The current pulse rise time is 100 A/μs.

EXAMPLE 25.4

Transient Thermal Design for a MOSFET

The IRF440 MOSFET, whose transient thermal impedance characteristics are shown in Fig. 25.9(b), is used in an application where it is subjected to repetitive current pulses with a duty ratio of 20% at a frequency of 5 kHz, as shown in Fig. 25.12. We want to know the maximum possible device current, the peak dissipation, and the required case-to-ambient thermal resistance $R_{\theta CA}$. The maximum T_j specified by the manufacturer is 175°C, but we permit a maximum T_j of only 150°C to improve reliability. We want to design the system for a case temperature of $T_C = 65°C$ in an ambient of $T_A = 40°C$. Our procedure is first to determine the appropriate transient thermal impedance, then calculate the power dissipation, and finally relate this dissipation to the on-state drain-source resistance and drain current. We find $R_{\theta jC}$ by a straightforward steady-state calculation. For the specified duty ratio and pulse duration, the transient thermal impedance, from Fig. 25.9(b), is $Z_{\theta jC}(40\ \mu s) = 0.22°C/W$.

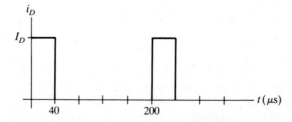

Figure 25.12　Drain current of the IRF440 MOSFET under a repetitive pulse mode of operation that might occur, for instance, in a switching power supply.

Thus the maximum power is

$$P_{max} = \frac{T_j - T_C}{Z_{\theta jC}} = 386 \text{ W} \tag{25.11}$$

The appropriate $R_{DS(on)}$ value to use for the calculation of I_D is the value at 150°C. This can be obtained from the manufacturer's data relating $R_{DS(on)}$ to T, shown in Fig. 25.13. We now calculate I_D:

$$I_D = \sqrt{\frac{P_{max}}{R_{DS(on)}}} = \sqrt{\frac{386 \text{ W}}{1.8 \text{ }\Omega}} = 14.5 \text{ A} \tag{25.12}$$

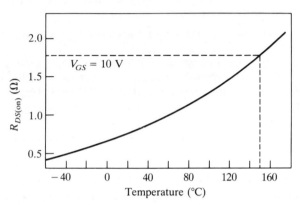

Figure 25.13 On-state drain-source resistance, $R_{DS(on)}$, as a function of temperature for the IRF440 MOSFET.

The required thermal resistance of the heat sink is calculated on the basis of steady-state heat transfer. The reason for using a steady-state model is that the device case temperature will be constant because of its mass. (This conclusion is based on intuition, not analysis. Our sense for the behavior of the physical world tells us that we will not be able to change the temperature of the TO-3 case at a 5 kHz rate.) Thus:

$$R_{\theta CA} = \frac{\Delta T_{CA}}{\langle P \rangle} = \frac{25}{386(0.2)} = 0.32°\text{C}/\text{W} \tag{25.13}$$

Notes and Bibliography

The volume of work published on the general area of heat transfer is massive. The references selected here are representative of those that are accessible to the nonspecialist. Jacob and Hawkins, [1], is an excellent introductory text that addresses conduction, convection, and radiation. It includes a large number of calculated examples and well conceived problems. Jacob, [2], is a classic text—comprehensive and highly analytical. It is recommended if you want a deeper understanding of fundamental heat transfer issues.

An undergraduate text covering most topics of interest to the designer of electronic equipment, although not in this context, is [3]. The book is liberally illustrated and contains numerous examples. Like [3], [4] is designed as an introductory undergraduate text in heat transfer. One of its unique inclusions are photographs of Ludwig Prandtl, Osborne Reynolds, and Ernst Kraft Wilhelm Nusselt, whose namesakes are the Prandtl, Reynolds, and Nusselt numbers, important parameters in heat transfer. This book is somewhat more abstract than [3].

Steinberg, [5], is short on theory but long on practical applications. The numerous examples reflect the author's own experience in the military/avionics area. A lot of practical data is presented, and there is a good discussion of fluid-based heat transfer systems, including heat pipes.

A concise discussion and mathematical statement of the Wiedman-Franz law can be found on p. 150 of Kittel, [6].

1. Max Jacob and George A. Hawkins, *Elements of Heat Transfer*, 3rd ed. (New York: John Wiley & Sons, 1958).
2. Max Jacob, *Heat Transfer*, vol. 1 (New York: John Wiley & Sons, 1949).
3. Frank M. White, *Heat Transfer* (Reading, Massachusetts: Addison-Wesley, 1984).
4. John H. Lienhard, *A Heat Transfer Textbook* (New Jersey: Prentice Hall, 1981).
5. Dave S. Steinberg, *Cooling Techniques for Electronic Equipment* (New York: John Wiley & Sons, 1980).
6. Charles Kittel, *Introduction to Solid State Physics*, 6th ed. (New York: John Wiley & Sons, 1986).

Problems

25.1   A double-insulated window is made of panes of glass 4 mm thick spaced 1 cm apart. Window glass has approximately the same thermal resistivity as SiO_2, $100°C$-cm/W. If the interior temperature of the building is $25°C$ and the outside temperature is $0°C$, what is the rate of heat lost by conduction in kW/m^2?

25.2   The CRC *Handbook of Chemistry and Physics* (35th ed.) gives thermal conductivity of materials as "the quantity of heat in calories which is transmitted per second through a plate 1 cm thick across an area of 1 cm^2 when the temperature difference is $1°C$." The value specified for dry compact snow is 0.00051. What is the thermal resistivity of dry compact snow in units of $°C$-cm/W?

25.3   An isolating interface of alumina having a thickness of 1 mm is placed between the device package and the heat sink in Fig. 25.3 (Example 25.2). What is the junction temperature T_j, if other parameters of the example remain unchanged?

25.4   Figure 25.14 shows two identical devices, Q_1 and Q_2, mounted on a common heat sink. The devices are in TO-220 packages and have a thermal resistance from junction to case of $R_{\theta jC} = 1.2°C/W$. The interface between the case and sink has a thermal resistance of $R_{\theta CS} = 0.20°C/W$, and the thermal resistance between the sink and ambient is $R_{\theta SA} = 0.8°C/W$.

(a) Draw the static thermal model for the thermal system of Fig. 25.14.

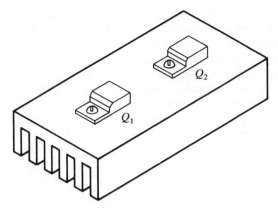

Figure 25.14   Two devices mounted on a common heat sink. This thermal system is analyzed in Problem 25.4.

(b) If the devices are dissipating the same power, and $T_A = 40°C$, what is the maximum total power that can be dissipated if $T_{j(max)} = 150°C$?

(c) What is the maximum power that can be dissipated if only one of the devices is operating?

25.5 Figure 25.15 shows the internal structure and dimensions of a power diode mounted in an axial lead package. The diode is cooled by conduction through its leads, which are soldered to terminals that are assumed to be at temperature T_A. Heat is generated at the junction of the diode, which is planar and centered between the two surfaces.

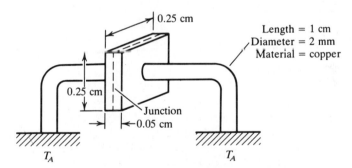

Figure 25.15   The axial lead packaged diode analyzed in Problem 25.5.

(a) Draw the analog circuit model for the thermal system of Fig. 25.15.

(b) If the maximum permissible junction temperature of the diode is $T_j = 225°C$, what is the maximum permissible dissipation for $T_A = 75°C$?

25.6 A transistor in a TO-3 case has a junction-to-case thermal resistance of $1°C/W$ and is to be used in an environment having an ambient temperature of $60°C$. The transistor is to be isolated from its heat sink by a mylar spacer having a thickness of 0.1 mm, and the available heat sink has a specified value of sink-to-ambient thermal resistance of $R_{\theta SA} = 2°C/W$.

(a) Determine and draw the static thermal model for this system.

(b) What is the maximum power that can be dissipated by the device if its junction temperature must be less than 150°C?

25.7 The IRF440 MOSFET operating as described in Example 25.4 is subjected to an overload condition that is cleared by a protection circuit in 10 μs. The MOSFET had been operating at a junction temperature of $T_j \doteq 150$°C. How much energy can the device be allowed to dissipate during the fault if $T_j \leq 200$°C?

List of Symbols

A_c core area

B_s saturation flux density

D dissipation factor

D duty ratio

$d(t)$ continuous duty ratio

D_a ambipolar diffusion constant

$D_{e,h}$ carrier diffusion constant

g gap length

$g(t)$ pulse train

H magnetic field intensity

h film coefficient

h_{FE} dc current gain

I identity matrix

I_F rated forward current

I_H SCR holding current

I_L SCR latching current

I_R rated leakage current

$I_{C(\text{max})}$ rated collector current

I_{rr} reverse recovery current

k depth of modulation

k_d distortion factor

k_p power factor

k_w winding space-factor

k_θ displacement factor

L_a ambipolar diffusion length

L_m mutual inductance

L_ℓ leakage inductance

L_μ magnetizing inductance

$L_{e,h}$ carrier diffusion length

n_i intrinsic carrier concentration

Q quality factor

Q reactive power

$q(t)$ switching function

Q_F forward charge store

Q_S quasi-saturation charge

Q_{12} rate of heat transfer

Re Reynolds Number

R^* Richardson constant

R_θ thermal resistance

$R_{DS(\text{on})}$ on-state resistance

S apparent power

T switching period, averaging interval

T_C Curie temperature

t_q circuit-commutated turn-off time

t_s storage delay time

t_{fr} forward recovery time

t_{off} reverse-biased time

t_{rr} reverse recovery time

t_{vf} voltage fall time

u commutation period

u_{max} maximum commutation period

V_T gate threshold voltage

V_{BO} forward breakover voltage

V_{BR} reverse blocking voltage

V_{CBO} collector-base breakdown voltage

$V_{CEO(\text{sus})}$ sustaining voltage

V_{CEX} collector-emitter breakdown voltage

V_{do} unregulated rectifier output voltage

V_{fp} forward recovery voltage

$V_{SB(\text{sus})}$ sustained second breakdown voltage

V_{SB} second breakdown voltage

w magnetic energy density

W_C stored electric energy

W_L stored magnetic energy

W_m stored magnetic energy

W_T trapped energy

X_μ magnetizing reactance

$Z_\theta(t)$ transient thermal impedance

α firing angle

α_T base transport factor

α_{design} design maximum firing angle

$\alpha_{\max}$ maximum firing angle

β common emitter current gain

β_{off} GTO gate turn-off gain

$\mathcal{P}$ permeance

$\mathcal{R}$ reluctance

δ skin depth

δ_E emitter defect

δ_R recombination defect

$\ell(j\omega)$ loop gain

$\ell(s)$ loop transfer function

γ margin angle

γ_E emitter efficiency

γ_i current fraction

γ_v voltage fraction

λ flux linkage

μ permeability

$\mu'(s)$ complementary sensitivity function

$\mu(s)$ sensitivity function

μ_c core permeability

μ_i incremental permeability

μ_o permeability of free space

$\mu_{e,h}$ carrier mobility

ν region

ω_c unity-gain crossover frequency

ω_d damped resonant frequency

ω_o undamped resonant frequency

Φ magnetic flux

ϕ_m work function (Schottky barrier)

ψ_o built-in potential

ρ_θ thermal resistivity

τ_a ambipolar lifetime

$\tau_{e,h}$ minority carrier lifetime

$\mathcal{R}_C$ voltage ripple ratio

$\mathcal{R}_L$ current ripple ratio

$\mathcal{V}_c$ core volume

$\mathcal{V}_g$ gap volume

Index